Lecture Notes in Computer Science 9537

Commenced Publication in 1973
Founding and Former Series Editors:
Gerhard Goos, Juris Hartmanis, and Jan van Leeuwen

More information about this series at http://www.springer.com/series/7407

Sergei Artemov · Anil Nerode (Eds.)

Logical Foundations of Computer Science

International Symposium, LFCS 2016
Deerfield Beach, FL, USA, January 4–7, 2016
Proceedings

Springer

Editors
Sergei Artemov
City University of New York
New York, NY
USA

Anil Nerode
Cornell University
Ithaca, NY
USA

ISSN 0302-9743   ISSN 1611-3349 (electronic)
Lecture Notes in Computer Science
ISBN 978-3-319-27682-3   ISBN 978-3-319-27683-0 (eBook)
DOI 10.1007/978-3-319-27683-0

Library of Congress Control Number: 2015957091

LNCS Sublibrary: SL1 – Theoretical Computer Science and General Issues

Printed on acid-free paper

This Springer imprint is published by SpringerNature
The registered company is Springer International Publishing AG Switzerland

Preface

The Symposium on Logical Foundations of Computer Science series provides a forum for the fast-growing body of work in the logical foundations of computer science, e.g., those areas of fundamental theoretical logic related to computer science. The LFCS series began with "Logic at Botik," Pereslavl-Zalessky, 1989, which was co-organized by Albert R. Meyer (MIT) and Michael Taitslin (Tver). After that, the organization passed to Anil Nerode. Currently LFCS is governed by a Steering Committee consisting of Anil Nerode (General Chair), Stephen Cook, Dirk van Dalen, Yuri Matiyasevich, J. Alan Robinson, Gerald Sacks, and Dana Scott. The 2016 Symposium on Logical Foundations of Computer Science (LFCS 2016) took place in the Wyndham Deerfield Beach Resort, Deerfield Beach, Florida, USA, during January 4–7. This volume contains the extended abstracts of talks selected by the Program Committee for presentation at LFCS 2016.

The scope of the symposium is broad and includes constructive mathematics and type theory; homotopy type theory; logic, automata, and automatic structures; computability and randomness; logical foundations of programming; logical aspects of computational complexity; parameterized complexity; logic programming and constraints; automated deduction and interactive theorem proving; logical methods in protocol and program verification; logical methods in program specification and extraction; domain theory logics; logical foundations of database theory; equational logic and term rewriting; lambda and combinatory calculi; categorical logic and topological semantics; linear logic; epistemic and temporal logics; intelligent and multiple-agent system logics; logics of proof and justification; non-monotonic reasoning; logic in game theory and social software; logic of hybrid systems; distributed system logics; mathematical fuzzy logic; system design logics; other logics in computer science.

We thank the authors and reviewers for their contributions. We acknowledge the support of the US National Science Foundation, Cornell University, the Graduate Center of the City University of New York, and Florida Atlantic University.

October 2015

Anil Nerode
Sergei Artemov

Organization

Steering Committee

Stephen Cook	University of Toronto, Canada
Yuri Matiyasevich	Steklov Mathematical Institute, St. Petersburg, Russia
Anil Nerode	Cornell University, USA - General Chair
J. Alan Robinson	Syracuse University, USA
Gerald Sacks	Harvard University, USA
Dana Scott	Carnegie-Mellon University, USA
Dirk van Dalen	Utrecht University, The Netherlands

Program Committee

Sergei Artemov	CUNY Graduate Center, New York, USA - Chair
Eugene Asarin	Université Paris Diderot – Paris 7, France
Steve Awodey	Carnegie Mellon University, USA
Matthias Baaz	The Vienna University of Technology, Austria
Alexandru Baltag	University of Amsterdam, The Netherlands
Lev Beklemishev	Steklov Mathematical Institute, Moscow, Russia
Andreas Blass	University of Michigan, Ann Arbor, USA
Samuel Buss	University of California, San Diego, USA
Robert Constable	Cornell University, USA
Thierry Coquand	University of Gothenburg, Sweden
Ruy de Queiroz	The Federal University of Pernambuco, Recife, Brazil
Nachum Dershowitz	Tel Aviv University, Israel
Melvin Fitting	CUNY Graduate Center, New York, USA
Sergey Goncharov	Sobolev Institute of Mathematics, Novosibirsk, Russia
Denis Hirschfeldt	University of Chicago, USA
Martin Hyland	University of Cambridge, UK
Rosalie Iemhoff	Utrecht University, The Netherlands
Hajime Ishihara	Japan Advanced Institute of Science and Technology, Kanazawa, Japan
Bakhadyr Khoussainov	The University of Auckland, New Zealand
Roman Kuznets	The Vienna University of Technology, Austria
Daniel Leivant	Indiana University Bloomington, USA
Robert Lubarsky	Florida Atlantic University, Boca Raton, USA
Victor Marek	University of Kentucky, Lexington, USA
Lawrence Moss	Indiana University Bloomington, USA
Anil Nerode	Cornell University, USA

Hiroakira Ono	Japan Advanced Institute of Science and Technology, Kanazawa, Japan
Ramaswamy Ramanujam	The Institute of Mathematical Sciences, Chennai, India
Michael Rathjen	University of Leeds, UK
Jeffrey Remmel	University of California, San Diego, USA
Helmut Schwichtenberg	University of Munich, Germany
Philip Scott	University of Ottawa, Canada
Alex Simpson	University of Ljubljana, Slovenia
Sonja Smets	University of Amsterdam, The Netherlands
Sebastiaan Terwijn	Radboud University Nijmegen, The Netherlands
Alasdair Urquhart	University of Toronto, Canada

Additional Reviewers

Marc Bagnol
Pablo Barenbaum
Can Baskent
Gianluigi Bellin
Marta Bilkova
Spencer Breiner
Merlin Carl
Thomas Colcombet
Pilar Dellunde
Eric Finster
Jose Gil-Ferez
Stefan Göller
Makoto Kanazawa
Ryo Kashima
Bjoern Lellmann

Peter Lefanu Lumsdaine
Bob Milnikel
Joshua Moerman
Edward Morehouse
Rafael Peñaloza
Andrew Polonsky
Damien Pous
Revantha Ramanayake
Arnaud Sangnier
Katsuhiko Sano
Mirek Truszczynski
Benno van den Berg
Michael Warren
Pascal Weil
Fan Yang

Contents

Modal Logics with Hard Diamond-Free Fragments

Antonis Achilleos[(✉)]

The Graduate Center of The City University of New York, New York, USA
aachilleos@gradcenter.cuny.edu

Abstract. We investigate the complexity of modal satisfiability for certain combinations of modal logics. In particular we examine four examples of multimodal logics with dependencies and demonstrate that even if we restrict our inputs to diamond-free formulas (in negation normal form), these logics still have a high complexity. This result illustrates that having D as one or more of the combined logics, as well as the interdependencies among logics can be important sources of complexity even in the absence of diamonds and even when at the same time in our formulas we allow only one propositional variable. We then further investigate and characterize the complexity of the diamond-free, 1-variable fragments of multimodal logics in a general setting.

Keywords: Modal logic · Satisfiability · Computational complexity · Diamond-free fragments · Multi-modal · Lower bounds

1 Introduction

The complexity of the satisfiability problem for modal logic, and thus of its dual, modal provability/validity, has been extensively studied. Whether one is interested in areas of application of Modal Logic, or in the properties of Modal Logic itself, the complexity of modal satisfiability plays an important role. Ladner has established most of what are now considered classical results on the matter [17], determining that most of the usual modal logics are PSPACE-hard, while more for the most well-known logic with negative introspection, S5, satisfiability is NP-complete; Halpern and Moses [12] then demonstrated that KD45-satisfiability is NP-complete and that the multi-modal versions of these logics are PSPACE-complete. Therefore, it makes sense to try to find fragments of these logics that have an easier satisfiability problem by restricting the modal elements of a formula – or prove that satisfiability remains hard even in fragments that seem trivial (ex. [4,11]). In this paper we present mostly hardness results for this direction and for certain cases of multimodal logics with modalities that affect each other. Relevant syntactic restrictions and their effects on the complexity of various modal logics have been examined in [13,14]. For more on Modal Logic and its complexity, see [10,12,20].

A (uni)modal formula is a formula formed by using propositional variables and Boolean connectives, much like propositional calculus, but we also use two

© Springer International Publishing Switzerland 2016
S. Artemov and A. Nerode (Eds.): LFCS 2016, LNCS 9537, pp. 1–13, 2016.
DOI: 10.1007/978-3-319-27683-0_1

additional operators, $\Box$ (box) and $\Diamond$ (diamond): if ϕ is a formula, then $\Box\phi$ and $\Diamond\phi$ are formulas. Modal formulas are given truth values with respect to a Kripke model (W, R, V),[1] which can be seen as a directed graph (W, R) (with possibly an infinite number of vertices and allowing self-loops) together with a truth value assignment for the propositional variables for each world (vertex) in W, called V. We define $\Box\phi$ to be true in a world a if ϕ is true at every world b such that (a, b) is an edge, while $\Diamond$ is the dual operator: $\Diamond\phi$ is true at a if ϕ is true at some b such that (a, b) is an edge.

We are interested in the complexity of the satisfiability problem for modal formulas (in negation normal form, to be defined later) that have no diamonds – i.e. is there a model with a world at which our formula is true? When testing a modal formula for satisfiability (for example, trying to construct a model for the formula through a tableau procedure), a clear source of complexity are the diamonds in the formula. When we try to satisfy $\Diamond\phi$, we need to assume the existence of an extra world where ϕ is satisfied. When trying to satisfy $\Diamond p_1 \wedge \Diamond p_2 \wedge \Box\phi_n$, we require two new worlds where $p_1 \wedge \phi_n$ and $p_2 \wedge \phi_n$ are respectively satisfied; for example, for $\phi_0 = \top$ and $\phi_{n+1} = \Diamond p_1 \wedge \Diamond p_2 \wedge \Box\phi_n$, this causes an exponential explosion to the size of the constructed model (if the model we construct for ϕ_n has k states, then the model for ϕ_{n+1} has $2k+1$ states). There are several modal logics, but it is usually the case that in the process of satisfiability testing, as long as there are no diamonds in the formula, we are not required to add more than one world to the constructed model. Therefore, it is natural to identify the existence of diamonds as an important source of complexity. On the other hand, when the modal logic is D, its models are required to have a serial accessibility relation (no sinks in the graph). Thus, when we test $\Box\phi$ for D-satisfiability, we require a world where ϕ is satisfied. In such a unimodal setting and in the absence of diamonds, we avoid an exponential explosion in the number of worlds and we can consider models with only a polynomial number of worlds.

Several authors have examined the complexity of combinations of modal logic (ex. [9,15,18]). Very relevant to this paper work on the complexity of combinations of modal logic is by Spaan in [20] and Demri in [6]. In particular, Demri studied $L_1 \oplus_{\subseteq} L_2$, which is $L_1 \oplus L_2$ (see [20]) with the additional axiom $\Box_2\phi \rightarrow \Box_1\phi$ and where L_1, L_2 are among K, T, B, S4, and S5 – modality 1 comes from L_1 and 2 from L_2. For when L_1 is among K, T, B and L_2 among S4, S5, he establishes EXP-hardness for $L_1 \oplus_{\subseteq} L_2$-satisfiability. We consider $L_1 \oplus_{\subseteq} L_2$, where L_1 is a unimodal or bimodal logic (usually D, or D4). When L_1 is bimodal, $L_1 \oplus_{\subseteq} L_2$ is $L_1 \oplus L_2$ with the extra axioms $\Box_3\phi \rightarrow \Box_1\phi$ and $\Box_3\phi \rightarrow \Box_2\phi$.

The family of logics we consider in this paper can be considered part of the much more general family of *regular grammar logics (with converse)*. Demri and De Nivelle have shown in [8] through a translation into a fragment of first-order logic that the satisfiability problem for the whole family is in EXP (see also [7]). Then, Nguyen and Szałas in [19] gave a tableau procedure for the general satisfiability

---

[1] There are numerous semantics for modal logic, but in this paper we only use Kripke semantics.

problem (where the logic itself is given as input in the form of a finite automaton) and determined that it is also in EXP.

In this paper, we examine the effect on the complexity of modal satisfiability testing of restricting our input to diamond-free formulas under the requirement of seriality and in a multimodal setting with connected modalities. In particular, we initially examine four examples: $D_2 \oplus_\subseteq K$, $D_2 \oplus_\subseteq K4$, $D \oplus_\subseteq K4$, and $D4_2 \oplus_\subseteq K4$.[2] For these logics we look at their diamond-free fragment and establish that they are PSPACE-hard and in the case of $D_2 \oplus_\subseteq K4$, EXP-hard. Furthermore, $D_2 \oplus_\subseteq K$, $D \oplus_\subseteq K4$, and $D4_2 \oplus_\subseteq K4$ are PSPACE-hard and $D_2 \oplus_\subseteq K4$ is EXP-hard even for their 1-variable fragments. Of course these results can be naturally extended to more modal logics, but we treat what we consider simple characteristic cases. For example, it is not hard to see that nothing changes when in the above multimodal logics we replace K by D, or K4 by D4, as the extra axiom $\Box_3\phi \to \Diamond_3\phi$ ($\Box_2\phi \to \Diamond_2\phi$ for $D \oplus_\subseteq K4$) is a derived one. It is also the case that in these logics we can replace K4 by other logics with positive introspection (ex. S4, S5) without changing much in our reasoning.

Then, we examine a general setting of a multimodal logic (we consider combinations of modal logics K, D, T, D4, S4, KD45, S5) where we include axioms $\Box_i\phi \to \Box_j\phi$ for some pairs i, j. For this setting we determine exactly the complexity of satisfiability for the diamond-free (and 1-variable) fragment of the logic and we are able to make some interesting observations. The study of this general setting is of interest, because determining exactly when the complexity drops to tractable levels for the diamond-free fragments illuminates possibly appropriate candidates for parameterization: if the complexity of the diamond-free, 1-variable fragment of a logic drops to P, then we may be able to develop algorithms for the satisfiability problem of the logic that are efficient for formulas of few diamonds and propositional variables; if the complexity of that fragment does not drop, then the development of such algorithms seems unlikely (we may be able to parameterize with respect to some other parameter, though). Another argument for the interest of these fragments results from the hardness results of this paper. The fact that the complexity of the diamond-free, 1-variable fragment of a logic remains high means that this logic is likely a very expressive one, even when deprived of a significant part of its syntax.

A very relevant approach is presented in [13,14]. In [13], Hemaspaandra determines the complexity of Modal Logic when we restrict the syntax of the formulas to use only a certain set of operators. In [14], Hemaspaandra et al. consider multimodal logics and all Boolean functions. In fact, some of the cases we consider have already been studied in [14]. Unlike [14], we focus on multimodal logics where the modalities are not completely independent – they affect each other through axioms of the form $\Box_i\phi \to \Box_j\phi$. Furthermore in this setting we only consider diamond-free formulas, while at the same time we examine the cases where we allow only one propositional variable. As far as our results are concerned, it is interesting to note that in [13,14] when we consider frames with

---

[2] In general, in $A \oplus_\subseteq B$, if A a bimodal (resp. unimodal) logic, the modalities 1 and 2 (resp. modality 1) come(s) from A and 3 (resp. 2) comes from logic B.

serial accessibility relations, the complexity of the logics under study tends to drop, while in this paper we see that serial accessibility relations (in contrast to arbitrary, and sometimes reflexive, accessibility relations) contribute substantially to the complexity of satisfiability.

Another motivation we have is the relation between the diamond-free fragments of Modal Logic with Justification Logic. Justification Logic can be considered an explicit counterpart of Modal Logic. It introduces justifications to the modal language, replacing boxes ($\Box$) by constructs called justification terms. When we examine a justification formula with respect to its satisfiability, the process is similar to examining the satisfiability of a modal formula without any diamonds (with some extra nontrivial parts to account for the justification terms). Therefore, as we are interested in the complexity of systems of Multi-modal and Multijustification Logics, we are also interested in these diamond-free fragments. For more on Justification Logic and its complexity, the reader can see [3,16]; for more on the complexity of Multi-agent Justification Logic and how this paper is connected to it, the reader can see [2].

It may seem strange that we restrict ourselves to formulas without diamonds but then we implicitly reintroduce diamonds to our formulas by considering serial modal logics – still, this is not the same situation as allowing the formula to have any number of diamonds, as seriality is only responsible for introducing at most one accessible world (for every serial modality) from any other. This is a nontrivial restriction, though, as we can see from this paper's results. Furthermore it corresponds well with the way justification formulas behave when tested for satisfiability.

For an extended version with omitted proofs the reader can see [1].

2 Modal Logics and Satisfiability

For the purposes of this paper it is convenient to consider modal formulas in negation normal form (NNF) – negations are pushed to the atomic level (to the propositional variables) and we have no implications. Note that for all logics we consider, every formula can be converted easily to its NNF form, so the NNF fragment of each logic we consider has exactly the same complexity as the full logic. We discuss modal logics with one, two, and three modalities, so we have three modal languages, $L_1 \subseteq L_2 \subseteq L_3$. They all include propositional variables, usually called $p_1, p_2, \ldots$ (but this may vary based on convenience) and $\bot$. If p is a propositional variable, then p and $\neg p$ are called literals and are also included in the language and so is $\neg\bot$, usually called $\top$. If ϕ, ψ are in one of these languages, so are $\phi \lor \psi$ and $\phi \land \psi$. Finally, if ϕ is in L_3, then so are $\Box_1\phi, \Box_2\phi, \Diamond_1\phi, \Diamond_2\phi, \Box_3\phi, \Diamond_3\phi$. L_2 includes all formulas in L_3 that have no $\Box_3, \Diamond_3$ and L_1 includes all formulas in L_2 that have no $\Box_2, \Diamond_2$. In short, L_n is defined in the following way, where $1 \leq i \leq n$: $\phi ::= p \mid \neg p \mid \bot \mid \neg\bot \mid \phi \wedge \phi \mid \phi \vee \phi \mid \Diamond_i\phi \mid \Box_i\phi$. If we consider formulas in L_1, $\Box_1$ may just be called $\Box$.[3]

A Kripke model for a trimodal logic (a logic based on language L_3) is a tuple $\mathcal{M} = (W, R_1, R_2, R_3, V)$, where $R_1, R_2, R_3 \subseteq W \times W$ and for every propositional variable p, $V(p) \subseteq W$. Then, (W, R_1, V) (resp. (W, R_1, R_2, V)) is a Kripke model for a unimodal (resp. bimodal) logic. Then, (W, R_1), (W, R_1, R_2), and (W, R_1, R_2, R_3) are called frames and R_1, R_2, R_3 are called accessibility relations. We define the truth relation $\models$ between models, worlds (elements of W, also called states) and formulas in the following recursive way:

$\mathcal{M}, a \not\models \bot$;
$\mathcal{M}, a \models p$ iff $a \in V(p)$ and $\mathcal{M}, a \models \neg p$ iff $a \notin V(p)$;
$\mathcal{M}, a \models \phi \wedge \psi$ iff both $\mathcal{M}, a \models \phi$ and $\mathcal{M}, a \models \psi$;
$\mathcal{M}, a \models \phi \vee \psi$ iff $\mathcal{M}, a \models \phi$ or $\mathcal{M}, a \models \psi$;
$\mathcal{M}, a \models \Diamond_i\phi$ iff there is some $b \in W$ such that aR_ib and $\mathcal{M}, b \models \phi$;
$\mathcal{M}, a \models \Box_i\phi$ iff for all $b \in W$ such that aR_ib it is the case that $\mathcal{M}, b \models \phi$.

In this paper we deal with five logics: K, $\mathsf{D_2} \oplus_\subseteq \mathsf{K}$, $\mathsf{D_2} \oplus_\subseteq \mathsf{K4}$, $\mathsf{D} \oplus_\subseteq \mathsf{K4}$, and $\mathsf{D4_2} \oplus_\subseteq \mathsf{K4}$. All except for K and $\mathsf{D} \oplus_\subseteq \mathsf{K4}$ are trimodal logics, based on language L_3, K is a unimodal logic (the simplest normal modal logic) based on L_1, and $\mathsf{D} \oplus_\subseteq \mathsf{K4}$ is a bimodal logic based on L_2. Each modal logic M is associated with a class of frames C. A formula ϕ is then called M-satisfiable iff there is a frame $\mathcal{F} \in C$, where C the class of frames associated to M, a model $\mathcal{M} = (\mathcal{F}, V)$, and a state a of $\mathcal{M}$ such that $\mathcal{M}, a \models \phi$. We say that $\mathcal{M}$ satisfies ϕ, or a satisfies ϕ in $\mathcal{M}$, or $\mathcal{M}$ models ϕ, or that ϕ is true at a.

K is the logic associated with the class of all frames;

$\mathsf{D_2} \oplus_\subseteq \mathsf{K}$ is the logic associated with the class of frames (W, R_1, R_2, R_3) for which R_1, R_2 are serial (for every a there are b, c such that aR_1b, aR_2c) and $R_1 \cup R_2 \subseteq R_3$;

$\mathsf{D_2} \oplus_\subseteq \mathsf{K4}$ is the logic associated with the class of frames $\mathcal{F} = (W, R_1, R_2, R_3)$ for which R_1, R_2 are serial, $R_1 \cup R_2 \subseteq R_3$, and R_3 is transitive;

$\mathsf{D} \oplus_\subseteq \mathsf{K4}$ is the logic associated with the class of frames $\mathcal{F} = (W, R_1, R_2)$ for which R_1 is serial, $R_1 \subseteq R_2$, and R_2 is transitive;

$\mathsf{D4_2} \oplus_\subseteq \mathsf{K4}$ is the logic associated with the class of frames $\mathcal{F} = (W, R_1, R_2, R_3)$ for which R_1, R_2 are serial, $R_1 \cup R_2 \subseteq R_3$ and R_1, R_2, R_3 are transitive.

Tableau. A way to test for satisfiability is by using a tableau procedure. A good source on tableaux is [5]. We present tableau rules for K and for the diamond-free fragments of $\mathsf{D_2} \oplus_\subseteq \mathsf{K}$ and then for the remaining three logics. The main reason we present these rules is because they are useful for later proofs and because they help to give intuition regarding the way we can test for satisfiability. The ones for K are classical and follow right away. Formulas used in the tableau are

---

[3] It may seem strange that we introduce languages with diamonds and then only consider their diamond-free fragments. When we discuss K, we consider the full language, so we introduce diamonds for L_1, L_2, L_3 for uniformity.

Table 1. Tableau rules for K.

$\dfrac{\sigma\ \phi\vee\psi}{\sigma\ \phi\ \mid\ \sigma\ \psi}$	$\dfrac{\sigma\ \phi\wedge\psi}{\begin{array}{c}\sigma\ \phi\\ \sigma\ \psi\end{array}}$	$\dfrac{\sigma\ \Box\phi}{\sigma.i\ \phi}$ where $\sigma.i$ has already appeared in the branch.	$\dfrac{\sigma\ \Diamond\phi}{\sigma.i\ \phi}$ where $\sigma.i$ has not yet appeared in the branch.

given a prefix, which intuitively corresponds to a state in a model we attempt to construct and is a string of natural numbers, with . representing concatenation. The tableau procedure for a formula ϕ starts from $0\ \phi$ and applies the rules it can to produce new formulas and add them to the set of formulas we construct, called a branch. A rule of the form $\frac{a}{b\mid c}$ means that the procedure nondeterministically chooses between b and c to produce, i.e. a branch is closed under that application of that rule as long as it includes b or c. If the branch has $\sigma\ \bot$, or both $\sigma\ p$ and $\sigma\ \neg p$, then it is called propositionally closed and the procedure rejects its input. Otherwise, if the branch contains $0\ \phi$, is closed under the rules, and is not propositionally closed, it is an accepting branch for ϕ; the procedure accepts ϕ exactly when there is an accepting branch for ϕ. The rules for K are in Table 1.

For the remaining logics, we are only concerned with their diamond-free fragments, so we only present rules for those to make things simpler. As we mention in the Introduction, all the logics we consider can be seen as regular grammar logics with converse ([8]), for which the satisfiability problem is in EXP. This already gives an upper bound for the satisfiability of $D_2\oplus_\subseteq K4$ (and for the general case of $(N,\subset,F)$ from Sect. 4). We present the tableau rules anyway (without proof), since it helps to visually give an intuition of each logic's behavior, while it helps us reason about how some logics reduce to others.

To give some intuition on the tableau rules, the main differences from the rules for K are that in a frame for these logics we have two or three different accessibility relations (lets assume for the moment that they are R_1, R_3, and possibly R_2), that one of them (R_3) is the (transitive closure of the) union of the others, and that we can assume that due to the lack of diamonds and seriality, R_1 and R_2 are total functions on the states. To establish this, notice that the truth of diamond-free formulas in NNF is preserved in submodels; when R_1, R_2 are not transitive, we can simply keep removing pairs from R_1, R_2 in a model as long as they remain serial. As for the tableau for $D4_2\oplus_\subseteq K4$, notice that for $i=1,2$, R_i can map each state a to some c such that for every $\Box_i\psi$, subformula of ϕ, $c\models\Box_i\psi\rightarrow\psi$. If a is such a c, we map a to a; otherwise we can find such a c in the following way. Consider a sequence $bR_ic_1R_ic_2R_i\cdots$; if some $c_j\not\models\Box_i\psi\rightarrow\psi$, then $c_j\models\Box_i\psi$, so for every $j'>j$, $c_{j'}\models\Box_i\psi\rightarrow\psi$. Since the subformulas of ϕ are finite in number, we can find some large enough $j\in\mathbb{N}$ and set $c=c_j$. Notice that using this construction on c, R_i maps c to c, is transitive and serial.

The rules for $D_2\oplus_\subseteq K$ are in Table 2. To come up with tableau rules for the other three logics, we can modify the above rules. The first two rules that cover

Table 2. The rules for $D_2 \oplus_C K$

$$\frac{\sigma \; \phi \vee \psi}{\sigma \; \phi \; | \; \sigma \; \psi} \qquad \frac{\sigma \; \phi \wedge \psi}{\begin{array}{c}\sigma \; \phi \\ \sigma \; \psi\end{array}} \qquad \frac{\sigma \; \Box_1 \phi}{\sigma.1 \; \phi} \qquad \frac{\sigma \; \Box_2 \phi}{\sigma.2 \; \phi} \qquad \frac{\sigma \; \Box_3 \phi}{\begin{array}{c}\sigma.1 \; \phi \\ \sigma.2 \; \phi\end{array}}$$

the propositional cases are always the same, so we give the remaining rules for each case. In the following, notice that the resulting branch may be infinite. However we can simulate such an infinite branch by a finite one: we can limit the size of the prefixes, as after a certain size (up to $2^{|\phi|}$, where ϕ the tested formula) it is guaranteed that there will be two prefixes that prefix the exact same set of formulas. Thus, we can either assume the procedure terminates or that it generates a full branch, depending on our needs. In that latter case, to ensure a full branch is generated, we can give lowest priority to a rule when it generates a new prefix.

The rules for the diamond-free fragment of $D_2 \oplus_C K4$ are in Table 3; the rules for the diamond-free fragment of $D \oplus_C K4$ in Table 4; and the rules for the diamond-free fragment of $D4_2 \oplus_C K4$ are in Table 5.

Proposition 1. *The satisfiability problem for the diamond-free fragments of* $D_2 \oplus_C K$, *of* $D \oplus_C K4$, *and of* $D4_2 \oplus_C K4$ *is in* PSPACE; *satisfiability for the diamond-free fragment of* $D_2 \oplus_C K4$ *is in* EXP.

The cases of $D \oplus_C K4$ and $D4_2 \oplus_C K4$ are especially interesting. In [6], Demri established that $D \oplus_C K4$-satisfiability (and because of the following section's results also $D4_2 \oplus_C K4$-satisfiability) is EXP-complete. In this paper, though, we establish that the complexity of these two logics' diamond-free (and one-variable) fragments are PSPACE-complete (in this section we establish the PSPACE upper bounds, while in the next one the lower bounds), which is a drop in complexity

Table 3. Tableau rules for the diamond-free fragment of $D_2 \oplus_C K4$

$$\frac{\sigma \; \Box_1 \phi}{\sigma.1 \; \phi} \qquad\qquad \frac{\sigma \; \Box_2 \phi}{\sigma.2 \; \phi} \qquad\qquad \frac{\sigma \; \Box_3 \phi}{\begin{array}{c}\sigma.1 \; \phi \\ \sigma.2 \; \phi \\ \sigma.1 \; \Box_3 \phi \\ \sigma.2 \; \Box_3 \phi\end{array}}$$

Table 4. Tableau rules for the diamond-free fragment of $D \oplus_C K4$

$$\frac{\sigma \; \Box_1 \phi}{\sigma.1 \; \phi} \qquad\qquad \frac{\sigma \; \Box_2 \phi}{\begin{array}{c}\sigma.1 \; \phi \\ \sigma.1 \; \Box_2 \phi\end{array}}$$

Table 5. Tableau rules for the diamond-free fragment of $D4_2 \oplus_{\subseteq} K4$

$$\frac{\sigma \ \Box_1 \phi}{n_1(\sigma) \ \phi} \qquad\qquad \frac{\sigma \ \Box_2 \phi}{n_2(\sigma) \ \phi} \qquad\qquad \frac{\sigma \ \Box_3 \phi}{\begin{array}{l} n_1(\sigma) \ \phi \\ n_2(\sigma) \ \phi \\ n_1(\sigma) \ \Box_3 \phi \\ n_2(\sigma) \ \Box_3 \phi \end{array}}$$

where $n_i(\sigma) = \sigma$ if $\sigma = \sigma'.i$ for some σ' and $n_i(\sigma) = \sigma.i$ otherwise.

(assuming PSPACE $\neq$ EXP), but not one that makes the problem tractable (assuming P $\neq$ PSPACE).

3 Lower Complexity Bounds

In this section we give hardness results for the logics of the previous section – except for K. In [4], the authors prove that the variable-free fragment of K remains PSPACE-hard. We make use of that result here and prove the same for the diamond-free, 1-variable fragment of $D_2 \oplus_{\subseteq} K$. Then we prove EXP-hardness for the diamond-free fragment of $D_2 \oplus_{\subseteq} K4$ and PSPACE-hardness for the diamond-free fragments of $D \oplus_{\subseteq} K4$ and of $D4_2 \oplus_{\subseteq} K4$, which we later improve to the same result for the diamond-free, 1-variable fragments of these logics.

Proposition 2. *The diamond-free, 1-variable fragment of* $D_2 \oplus_{\subseteq} K$ *is* PSPACE-*complete.*

For the remaining logics we first present a lower complexity bound for their diamond-free fragments and then we can use translations to their 1-variable fragments to transfer the lower bounds to these fragments. We first treat the case of $D_2 \oplus_{\subseteq} K4$.

Lemma 1. *The diamond-free fragment of* $D_2 \oplus_{\subseteq} K4$ *is* EXP-*complete, while the diamond-free fragments of* $D \oplus_{\subseteq} K4$ *and of* $D4_2 \oplus_{\subseteq} K4$ *are* PSPACE-*complete.*

From Lemma 1, with some extra work, we can prove the following.

Proposition 3. *The 1-variable, diamond-free fragment of* $D_2 \oplus_{\subseteq} K4$ *is* EXP-*complete; the 1-variable, diamond-free fragments of* $D \oplus_{\subseteq} K4$ *and of* $D4_2 \oplus_{\subseteq} K4$ *are* PSPACE-*complete.*

One may wonder whether we can say the same for the variable-free fragment of these logics. The answer however is that we cannot. The models for these logics have accessibility relations that are all serial. This means that any two models are bisimilar when we do not use any propositional variables, thus any satisfiable formula is satisfied everywhere in any model, thus we only need one prefix for our tableau and we can solve satisfiability recursively on ϕ in polynomial time.

Then what about $D4 \oplus_{\subseteq} K4$? Maybe we could attain similar hardness results for this logic as for $D4_2 \oplus_{\subseteq} K4$. Again, the answer is no. As frames for $D4$ come with a serial and transitive accessibility relation, frames for D4 $\oplus_\subseteq$ K4 are of the form (W, R_1, R_2), where $R_1 \subseteq R_2$, R_1, R_2 are serial, and R_1 is transitive. It is not hard to come up with the following tableau rule(s) for the diamond-free fragment, by adjusting the ones we gave for D4$_2$ $\oplus_\subseteq$ K4 to simply produce 0.1 ϕ from every σ $\square_i\phi$. This drops the complexity of satisfiability for the diamond-free fragment of D4 $\oplus_\subseteq$ K4 to NP (and of the diamond-free, 1-variable fragment to P), as we can only generate two prefixes during the tableau procedure. The following section explores when we can produce hardness results like the ones we gave in this section.

4   A General Characterization

In this section we examine a more general setting and we conclude by establishing tight conditions that determine the complexity of satisfiability of the diamond-free (and 1-variable) fragments of such multimodal logics.

A general framework would be to describe each logic with a triple $(N, \subset, F)$, where $N = \{1, 2, \ldots, |N|\} \neq \emptyset$, $\subset$ a binary relation on N, and for every $i \in N$, $F(i)$ is a modal logic; a frame for $(N, \subset, F)$ would be $(W, (R_i)_{i \in N})$, where for every $i \in N$, (W, R_i) a frame for $F(i)$ and for every $i \subset j$, $R_i \subset R_j$. It is reasonable to assume that $(N, \subset)$ has no cycles – otherwise we can collapse all modalities in the cycle to just one – and that $\subset$ is transitive. Furthermore, we also assume that all $F(i)$'s have frames with serial accessibility relations – otherwise there is either some $j \subseteq i$ for which $F(j)$'s frames have serial accessibility relations and $R(i)$ would inherit seriality from R_j, or when testing for satisfiability, $\square_i\psi$ can always be assumed true by default (the lack of diamonds means that we do not need to consider any accessible worlds for modality i), which allows us to simply ignore all such modalities, making the situation not very interesting from an algorithmic point of view. Thus, we assume that $F(i) \in \{D, T, D4, S5\}$.[4,5] The cases for which $\subset = \emptyset$ have already had the complexity of their diamond-free (and other) fragments determined in [14]. For the general case, we already have an EXP upper bound from [8].

The reader can verify that $(N, \subset, F)$ is, indeed, a (fragment of a) regular grammar modal logic with converse. For example, D$_2 \oplus_\subseteq$ D4 can easily be reduced to K$_2 \oplus_\subseteq$ K4 by mapping ϕ to $\lozenge_1\top \wedge \lozenge_2\top \wedge \square_3(\lozenge_1\top \wedge \lozenge_2\top) \wedge \phi$ to impose seriality, for which the corresponding regular languages would be $\square_1$, $\square_2$, and $(\square_1 + \square_2 + \square_3)^*$ (see [8] for more on regular grammar modal logics with converse and their complexity and the extended version of this paper, [1], for more details on why $(N, \subset, F)$ belongs in that category).

---

[4] We can consider more logics as well, but these ones are enough to make the points we need. Besides, it is not hard to extend the reasoning of this section to other logics (ex. B, S4, KD45 and due to the observation above, also K, K4), especially since the absence of diamonds makes the situation simpler.

[5] Frames for D have serial accessibility relations; frames for T have reflexive accessibility relations; frames for D4 have serial and transitive accessibility relations; frames for S5 have accessibility relations that are equivalence relations (reflexive, symmetric, transitive).

Table 6. Tableau rules for the diamond-free fragment of $(N, \subset, F)$

$\dfrac{\sigma\ \Box_i\phi}{\sigma\ \Box_j\phi}$	$\dfrac{\sigma\ \Box_i\phi}{n_i(\sigma)\ \phi}$	$\dfrac{\sigma\ \Box_i\phi}{\sigma\ \phi}$	$\dfrac{\sigma\ \Box_i\phi}{n_j(\sigma)\ \Box_i\phi}$
where $j \subset i$	where $i \in \min(N)$	where the frames of $F(i)$ have reflexive acc. relations	where $j \in \min(i)$ and $F(i)$'s frames have transitive acc. relations

For every $i \in N$, let

$$\min(i) = \{j \in N \mid j \subset i \text{ or } j = i, \text{ and } \not\exists j' \subset j\}$$

and $\min(N) = \bigcup_{i \in N} \min(i)$. We can now give tableau rules for $(N, \subset, F)$. Let

- $n_i(\sigma) = \sigma$, if either
  - the accessibility relations of the frames for $F(i)$ are reflexive, or
  - $\sigma = \sigma'.i$ for some σ' and the accessibility relations of the frames for $F(i)$ are transitive;
- $n_i(\sigma) = \sigma.i$, otherwise.

The tableau rules appear in Table 6.

From these tableau rules we can reestablish EXP-upper bounds for all of these cases (see the previous sections). To establish correctness, we only show how to construct a model from an accepting branch for ϕ, as the opposite direction is easier. Let W be the set of all the prefixes that have appeared in the branch. The accessibility relations are defined in the following (recursive) way: if $i \in \min(N)$, then $R_i = \{(\sigma, n_i(\sigma)) \in W^2\} \cup \{(\sigma, \sigma) \in W^2 \mid n_i(\sigma) \notin W$ or $F(i)$ has reflexive frames$\}$; if $i \notin \min(N)$ and the frames of $F(i)$ do not have transitive or reflexive accessibility relations, then $R_i = \bigcup_{j \subseteq i} R_j$; if $i \notin \min(N)$ and the frames of $F(i)$ do have transitive (resp. reflexive, resp. transitive and reflexive) accessibility relations, then R_i is the transitive (resp. reflexive, resp. transitive and reflexive) closure of $\bigcup_{j \subseteq i} R_j$. Finally, (as usual) $V(p) = \{w \in W \mid w\ p$ appears in the branch$\}$. Again, to show that the constructed model satisfies ϕ, we use a straightforward induction.

By taking a careful look at the tableau rules above, we can already make some simple observations about the complexity of the diamond-free fragments of these logics. Modalities in $\min(N)$ have an important role when determining the complexity of a diamond-free fragment. In fact, the prefixes that can be produced by the tableau depend directly on $\min(N)$.

Lemma 2. *If for every $i \in \min(N)$, $F(i)$ has frames with reflexive accessibility relations ($F(i) \in \{\mathsf{T}, \mathsf{S5}\}$), then the satisfiability problem for the diamond-free fragment of $(N, \subset, F)$ is NP-complete and the satisfiability problem for the diamond-free, 1-variable fragment of $(N, \subset, F)$ is in P.*

Corollary 1. *If* $\min(N) \subseteq \{i\} \cup A$ *and* $F(i)$ *has frames with transitive accessibility relations* $(F(i) \in \{\mathsf{D4}, \mathsf{S5}\})$ *and for every* $j \in A$, $F(j)$ *has frames with reflexive accessibility relations, then the satisfiability problem for the diamond-free fragment of* $(N, \subset, F)$ *is* NP*-complete and the satisfiability problem for the diamond-free, 1-variable fragment of* $(N, \subset, F)$ *is in* P.

In [6], Demri shows that satisfiability for $\mathsf{L}_1 \oplus_\subset \mathsf{L}_2 \oplus_\subset \cdots \oplus_\subset \mathsf{L}_n$ is EXP-complete, as long as there are $i < j \leq n$ for which $\mathsf{L}_i \oplus_\subset \mathsf{L}_j$ is EXP-hard. On the other hand, Corollary 1 shows that for all these logics, their diamond-free fragment is in NP, as long as L_1 has frames with transitive (or reflexive) accessibility relations.

Finally, we can establish general results about the complexity of the diamond-free fragments of these logics. For this, we introduce some terminology. We call a set $A \subset N$ *pure* if for every $i \in A$, $F(i)$'s frames do not have the condition that their accessibility relation is reflexive (given our assumptions, $F[A] \cap \{\mathsf{T}, \mathsf{S5}\} = \emptyset$). We call a set $A \subset N$ *simple* if for some $i \in A$, $F(i)$'s frames do not have the condition that their accessibility relation is transitive (given our assumptions, $F[A] \cap \{\mathsf{D}, \mathsf{T}\} \neq \emptyset$). An agent $i \in N$ is called pure (resp. simple) if $\{i\}$ is pure (resp. simple).

Theorem 1. *1. If there is some* $i \in N$ *and some pure* $A \subseteq \min(i)$ *for which* $F(i)$ *has frames with transitive accessibility relations* $(F(i) \in \{\mathsf{D4}, \mathsf{S5}\})$ *and either*
  - $|A| = 2$ *and* A *is simple, or*
  - $|A| = 3$,
  then the satisfiability problem for the diamond-free, 1-variable fragment of $(N, \subset, F)$ *is* EXP*-complete;*
2. *otherwise, if there is some* $i \in N$ *and some pure* $A \subseteq \min(i)$ *for which either*
  - $|A| = 2$ *and there is some pure and simple* $j \in \min(N)$, *or*
  - $|A| = 3$,
  then the satisfiability problem for the diamond-free, 1-variable fragment of $(N, \subset, F)$ *is* PSPACE*-complete;*
3. *otherwise, if there is some* $i \in N$ *and some pure* $A \subseteq \min(i)$ *for which* $F(i)$ *has frames with transitive accessibility relations* $(F(i) \in \{\mathsf{D4}, \mathsf{S5}\})$ *and either*
  - $|A| = 1$ *and* A *is simple or*
  - $|A| = 2$,
  then the satisfiability problem for the diamond-free (1-variable) fragment of $(N, \subset, F)$ *is* PSPACE*-complete;*
4. *otherwise the satisfiability problem for the diamond-free (resp. and 1-variable) fragment of* $(N, \subset, F)$ *is* NP*-complete (resp. in* P*).*

5 Final Remarks

We examined the complexity of satisfiability for the diamond-free fragments and the diamond-free, 1-variable fragments of multimodal logics equipped with an inclusion relation $\subset$ on the modalities, such that if $i \subset j$, then in every frame $(W, R_1, \ldots, R_n)$ of the logic, $R_i \subseteq R_j$ (equivalently, $\square_j \rightarrow \square_i$ is an axiom).

We gave a complete characterization of these cases (Theorem 1), determining that, depending on $\subset$, every logic falls into one of the following three complexity classes: NP (P for the 1-variable fragments), PSPACE, and EXP – Theorem 1 actually distinguishes four possibilities, depending on the way we prove each bound. We argued that to have nontrivial complexity bounds we need to consider logics based on frames with at least serial accessibility relations, which is a notable difference in flavor from the results in [13,14].

One direction to take from here is to consider further syntactic restrictions and Boolean functions in the spirit of [14]. Another would be to consider different classes of frames. Perhaps it would also make sense to consider different types of natural relations on the modalities and see how these results transfer in a different setting. From a Parameterized Complexity perspective there is a lot to be done, such as limiting the modal depth/width, which are parameters that can remain unaffected from our ban on diamonds. For the cases where the complexity of the diamond-free, 1-variable fragments becomes tractable, a natural next step would be to examine whether we can indeed use the number of diamonds as a parameter for an FPT algorithm to solve satisfiability.

Another direction which interests us is to examine what happens with more/ different kinds of relations on the modalities. An example would be to introduce the axiom $\Box_i \phi \rightarrow \Box_j \Box_i \phi$, a generalization of Positive Introspection. This would be of interest in the case of the diamond-free fragments of these systems, as it brings us back to our motivation in studying the complexity of Justification Logic, where such systems exist. Hardness results like the ones we proved in this paper are not hard to transfer in this case, but it seems nontrivial to immediately characterize the complexity of the whole family.

Acknowledgments. The author is grateful to anonymous reviewers, whose input has greatly enhanced the quality of this paper.

References

1. Achilleos, A.: Modal Logics with Hard Diamond-free Fragments. CoRR abs/1401.5846 (2014)
2. Achilleos, A.: Interactions and Complexity in Multi-Agent Justification Logic. Ph.D. thesis, The City University of New York (2015)
3. Artemov, S.: The logic of justification. Rev. Symbolic Logic **1**(4), 477–513 (2008)
4. Chagrov, A.V., Rybakov, M.N.: How many variables does one need to prove PSPACE-hardness of modal logics. In: Advances in Modal Logic, pp. 71–82 (2002)
5. Agostino, M., Gabbay, D.M., Hähnle, R., Posegga, J. (eds.): Handbook of Tableau Methods. Springer, Heidelberg (1999)
6. Demri, S.: Complexity of simple dependent bimodal logics. In: Dyckhoff, R. (ed.) TABLEAUX 2000. LNCS, vol. 1847, pp. 190–204. Springer, Heidelberg (2000)
7. Demri, S.: The complexity of regularity in grammar logics and related modal logics. J. Logic Comput. **11**(6), 933–960 (2001). http://logcom.oxfordjournals.org/content/11/6/933.abstract

8. Demri, S., De Nivelle, H.: Deciding regular grammar logics with converse through first-order logic. J. Logic Lang. Inform. **14**(3), 289–329 (2005). http://dx.doi.org/10.1007/s10849-005-5788-9
9. Gabbay, D.M., Kurucz, A., Wolter, F., Zakharyaschev, M. (eds.): Many-Dimensional Modal Logics Theory and Applications. Studies in Logic and the Foundations of Mathematics, vol. 148. Elsevier, North Holland (2003)
10. Fagin, R., Halpern, J.Y., Moses, Y., Vardi, M.Y.: Reasoning About Knowledge. The MIT Press, Cambridge (1995)
11. Halpern, J.Y.: The effect of bounding the number of primitive propositions and the depth of nesting on the complexity of modal logic. Artif. Intell. **75**, 361–372 (1995)
12. Halpern, J.Y., Moses, Y.: A guide to completeness and complexity for modal logics of knowledge and belief. Artif. Intell. **54**(3), 319–379 (1992)
13. Hemaspaandra, E.: The complexity of poor Man's logic. J. Logic Comput. **11**(4), 609–622 (2001)
14. Hemaspaandra, E., Schnoor, H., Schnoor, I.: Generalized modal satisfiability. J. Comput. Syst. Sci. **76**(7), 561–578 (2010). http://www.sciencedirect.com/science/article/pii/S0022000009001007
15. Kurucz, A.: Combining modal logics. Stud. Logic Pract. Reasoning **3**, 869–924 (2007)
16. Kuznets, R.: Complexity Issues in Justification Logic. Ph.D. thesis, CUNY Graduate Center, May 2008
17. Ladner, R.E.: The computational complexity of provability in systems of modal propositional logic. SIAM J. Comput. **6**(3), 467–480 (1977). http://link.aip.org/link/?SMJ/6/467/1
18. Marx, M., Venema, Y.: Multi-dimensional Modal Logic. Springer, Heidelberg (1997)
19. Nguyen, L.A., Szałas, A.: EXPTIME tableau decision procedures for regular grammar logics with converse. Stud. Logic. **98**(3), 387–428 (2011)
20. Spaan, E.: Complexity of Modal Logics. Ph.D. thesis, University of Amsterdam (1993)

Pairing Traditional and Generic Common Knowledge

Evangelia Antonakos$^{(\boxtimes)}$

Bronx Community College, CUNY, Bronx, NY, USA
evangelia.antonakos@bcc.cuny.edu

Abstract. Common Knowledge C is a standard tool in epistemic logics. Generic Common Knowledge J is an alternative which has desirable logical behavior such as cut-elmination and which can be used in place of C in the analysis of many games and epistemic senarios. In order to compare their deductive strengths directly we define the multi-agent logic $\mathsf{S4}_n^{CJ}$ built on a language with both C and J operators in addition to agents' K_is so that any finite prefix of modal operators is acceptable. We prove $\mathsf{S4}_n^{CJ}$ is complete, decidable, and that $J\varphi \rightarrow C\varphi$ though not $C\varphi \rightarrow J\varphi$. Additional epistemic scenarios may be investigated which take advantage of this dual layer of common knowledge agents.

Keywords: Generic common knowledge · Common knowledge · Epistemic logic · Modal logic

1 Introduction

In systems of multiple knowers, or agents, it is natural to consider what information is publicly known. The most investigated such concept is that of common knowledge. Informally, if a sentence or proposition φ is common knowledge, $C\varphi$, then everyone knows it ($E\varphi$), and everyone knows everyone knows it ($EE\varphi$), and everyone knows everyone knows everyone knows it, etc., i.e., iterated knowledge of φ, $I\varphi$. Common knowledge has overwhelmingly been formalized as an equivalence of $C\varphi$ and $I\varphi$ via a finite set of axioms. In each multi-agent system, C is unique.

However, there is a more general and eventually simpler conception of common knowledge, generic common knowledge, J. While $J\varphi$ is sufficient to yield iterated knowledge, it is not necessarily equivalent to $I\varphi$. This alternative offers a broader view of common knowledge as it allows for a choice between multiple logically non-equivalent common knowledge operators. Moreover, generic common knowledge which is not the traditional common knowledge naturally appears in some canonical epistemic scenarios. For example, a public announcement of an atomic fact A creates not common knowledge but rather universal knowledge (an instance of generic common knowledge) of A since A, a posteriori, holds at all worlds, not only at all reachable worlds. In the belief revision situations, such as the well-known Stalnaker-Halpern game, the revision function

© Springer International Publishing Switzerland 2016
S. Artemov and A. Nerode (Eds.): LFCS 2016, LNCS 9537, pp. 14–26, 2016.
DOI: 10.1007/978-3-319-27683-0_2

overspills to another reachability cluster of worlds and hence no longer should obey the common knowledge assumption [4].

The generic common knowledge was introduced by McCarthy in [11] as 'any fool knows' and independently by Artemov in [6] as 'justified common knowledge' who later termed it 'generic common knowledge.' In [6] it was the implicit 'forgetful projection' counterpart to the explicit constructive knowledge LP component of $S4_n LP$, a logic in the family of justification logics. J differs from C in logical behavior: its addition to a system does not hinder straightforward completeness proofs and as the cut-rule can be eliminated the way is paved for its Realization to an explicit justification logic counterpart e.g. the realization of $S4_n^J$ in $S4_n LP$ [5,6] or in $LP_n(LP)$ [2]. These realizations impart a rich semantics: $J\varphi$ asserts that φ is common knowledge arising from a proof of φ. In applications, J can be used in place of C whenever common knowledge is assumed as a premise, rather than being the desired outcome [3]. The cut-rule for traditional common knowledge has been investigated in [1] and syntactic elimination obtained for some systems as in [9].

This paper defines a multi-agent epistemic logic $S4_n^{CJ}$ which expands on the n-agent logic $S4_n$ to encompass two formulations common knowledge C and J. Completeness for this logic is shown, providing a basis for direct comparison of the deductive strength of J and C. We shall see that $J\varphi \to C\varphi$ though not the converse.

2   Axiomatization of $S4_n^{CJ}$

In $S4_n^{CJ}$ we can consider formulas which may contain both C and J as well as K_i modalities.

Definition 1. The *language* $\mathcal{L}_{S4_n^{CJ}}$ is an extension of the propositional language:

$$\mathcal{L}_{S4_n^{CJ}} := \{ \mathit{Var}, \wedge, \vee, \to, \neg, K_i, C, J \}$$

for $i \in \{1, 2, \ldots, n\}$ where *Var* is the set of propositional variables. *Formulas* are defined by the grammar

$$\varphi := p \mid \varphi \wedge \varphi \mid \varphi \vee \varphi \mid \varphi \to \varphi \mid \neg\varphi \mid K_i\varphi \mid C\varphi \mid J\varphi$$

where $p \in \mathit{Var}$.

The formula $K_1 K_2 \varphi$ has the intended semantics of 'agent 1 knows that agent 2 knows φ' while $C\varphi$ and $J\varphi$ have the intended semantics of 'φ is common knowledge' and 'φ is generic common knowledge' respectively.

Definition 2. The *axioms and rules of* $S4_n^{CJ}$, for $i \in \{1, 2, \ldots, n\}$ where $\square$ is K_i or J or C:

CLASSICAL PROPOSITIONAL CALCULUS:
A.   axioms of classical propositional calculus
R1.   modus ponens: $\vdash \varphi,\ \varphi \to \psi \Rightarrow \vdash \psi$
        S4 AXIOMS FOR EACH MODALITY:
K. $\Box(\varphi \to \psi) \to (\Box\varphi \to \Box\psi)$
T. $\Box\varphi \to \varphi$
4. $\Box\varphi \to \Box\Box\varphi$
        ADDITIONAL KNOWLEDGE AXIOMS:
Con. $J\varphi \to K_i\varphi$
ConC. $C\varphi \to K_i\varphi$

IA. $\varphi \wedge C(\varphi \to E\varphi) \to C\varphi$, where $E\varphi = \bigwedge\limits_{i=1}^{n} K_n\varphi$

        NECESSITATION FOR EACH MODALITY:
R2. $\vdash \varphi \Rightarrow \vdash \Box\varphi$.

Proposition 1. *Both $C\varphi$ and $J\varphi$ satisfy X in the Fixed Point Axiom,*

$$X \leftrightarrow E(\varphi \wedge X).$$

Proof. $J\varphi \leftrightarrow E(\varphi \wedge J\varphi)$:

$(\to)$

1 $JJ\varphi \to EJ\varphi$		from Con and definition of E
2 $J\varphi \to JJ\varphi$		4 for J
3 $J\varphi \to EJ\varphi$		from 2. and 1
4 $J\varphi \to E\varphi$		from Con and definition of E
5 $J\varphi \to (E\varphi \wedge EJ\varphi)$		from 3. and 4
6 $J\varphi \to E(\varphi \wedge J\varphi)$	from 5. as normal modalities commute with $\wedge$	

$(\leftarrow)$

1 $E(\varphi \wedge J\varphi) \to E\varphi \wedge EJ\varphi$	normal modalities commute with $\wedge$	
2 $E\varphi \wedge EJ\varphi \to EJ\varphi$		
3 $EJ\varphi \to K_iJ\varphi$		definition of E
4 $K_iJ\varphi \to J\varphi$		T for K_i
5 $E(\varphi \wedge J\varphi) \to J\varphi$		from 1. – 4

Normal modals are those with K axiom and subject to necessitation (R2). Each J axiom or rule has a C counterpart. Thus, as J satisfies the fixed point axiom, so does C.

Proposition 2. $S4_n^{CJ} \vdash J\varphi \to C\varphi$.

Proof. Reasons from propositional calculus are not listed.

1. $J\varphi \to EJ\varphi$	from 4 for J, Con, and definition of E
2. $C(J\varphi \to EJ\varphi)$	from 1. by R2 for C
3. $J\varphi \to C(J\varphi \to EJ\varphi)$	from 2.
4. $J\varphi \to J\varphi$	
5. $J\varphi \to J\varphi \wedge C(J\varphi \to EJ\varphi)$	from 3. and 4.
6. $J\varphi \wedge C(J\varphi \to EJ\varphi) \to CJ\varphi$	IA on $J\varphi$
7. $J\varphi \to CJ\varphi$	from 5. and 6.
8. $J\varphi \to \varphi$	T for J
9. $CJ\varphi \to C\varphi$	from 8. by R2, K, R1 for C
10. $J\varphi \to C\varphi$	from 7. and 9.

That the converse does not hold must wait till Proposition 4, after $\mathsf{S4}_n^{CJ}$ is shown to be sound and complete.

We will use the following proposition in the completeness proof (Theorem 2).

Proposition 3. $\mathsf{S4}_n^{CJ} \vdash C\varphi \rightarrow EC\varphi.$

Proof. Just as lines 1.– 3. in the forward direction of proof of Proposition 1.

Definition 3. An $\mathsf{S4}_n^{CJ}$-model is $M^{CJ} = \langle W, R_1, \ldots, R_n, R_C, R_J, \Vdash \rangle$ such that

- $W \neq \emptyset$ is a set of worlds;
- $R_i \subseteq W \times W$ is reflexive and transitive for $i \in \{1, \ldots, n\}$;
- $R_C = \left(\bigcup_{i=1}^{n} R_i \right)^{TC}$, the transitive closure of the union of R_is;
- $R_J \subseteq W \times W$ is reflexive and transitive and $R_C \subseteq R_J$;
- $\Vdash \subseteq W \times Var$ so that for $w \in W, p \in Var,\ w \Vdash p$ iff p holds at w;
- $\Vdash$ is extended to correspond with Boolean connectives at each world and so the asccessibility relations R_i, R_C, and R_J corresponds to the modalities K_i, C, and J respectively, so that in M^{CJ}

$$u \Vdash K_i\varphi \text{ iff } (\forall v \in W)(uR_iv \Rightarrow v \Vdash \varphi),$$

$$u \Vdash C\varphi \text{ iff } (\forall v \in W)(uR_Cv \Rightarrow v \Vdash \varphi),$$

$$u \Vdash J\varphi \text{ iff } (\forall v \in W)(uR_Jv \Rightarrow v \Vdash \varphi).$$

Note that the accessibility relation of C corresponds to reachability in each connected component of the model and is exactly prescribed by the agents' relations. On the other hand there is flexibility for R_J to be any reflexive transitive relation as small as R_C or a large as the total relation.

Theorem 1. $\mathsf{S4}_n^{CJ}$ *is sound with respect to* M^{CJ} *models.*

Proof (Soundness). Let M be an arbitrary $\mathsf{S4}_n^{CJ}$-model. Assume χ is provable and show it holds in each world of M. It is enough to show that all the axioms and rules are valid.

- χ is a propositional variable: $u \Vdash \chi$ for all worlds in the model M implies χ is valid by definition.
- $\chi = \neg\varphi \mid \varphi \wedge \psi \mid \varphi \vee \psi \mid \varphi \rightarrow \psi$. If χ is formed by Boolean connectives, it is valid by the definition of these connectives at each world.
- modus ponens: Suppose $u \Vdash \varphi \rightarrow \psi$. Then by the definition of the connectives, either $u \not\Vdash \varphi$ or $u \Vdash \psi$. If also $u \Vdash \varphi$, then $u \Vdash \psi$. So if $\varphi \rightarrow \psi$ and φ hold at any world, so does ψ.

- K axioms: Shown for K_i but analogous for C and J. $\chi = K_i(\varphi \to \psi) \to (K_i\varphi \to K_i\psi) = (K_i(\varphi \to \psi) \wedge K_i\varphi) \to K_i\psi$. Suppose $u \Vdash K_i(\varphi \to \psi) \wedge K_i\varphi$, then for all v such that uR_iv, $v \Vdash \varphi \to \psi$ and $v \Vdash \varphi$. So as modus ponens is valid $v \Vdash \psi$, and hence $u \Vdash K_i\psi$. Therefore $u \Vdash K_i(\varphi \to \psi) \to (K_i\varphi \to K_i\psi)$ is valid.
- T axioms: Shown for K_i but analogous for C and J. $\chi = K_i\varphi \to \varphi$. Suppose $u \Vdash K_i\varphi$, then for all v such that uR_iv, $v \Vdash \varphi$. Since R_i is reflexive, uR_iu, and so $u \Vdash \varphi$. Thus $u \Vdash K_i\varphi \to \varphi$ is valid. R_C is reflexive as it is the transitive closure of a union of reflexive relations.
- 4 axioms: Shown for K_i but analogous for C and J. Suppose $u \Vdash K_i\varphi$, then for all v such that uR_iv, $v \Vdash \varphi$. As R_i is transitive, for all w such that vR_iw, uR_iw and so $w \Vdash \varphi$ and so $v \Vdash K_i\varphi$ and hence $u \Vdash K_iK_i\varphi$. Therefore $u \Vdash K_i\varphi \to K_iK_i\varphi$ is valid.
- modal necessitation: Shown for K_i but analogous for C and J. Assume φ is valid in M, then it is true at each world so $u \Vdash \varphi$, and for all worlds v such that vR_iu, $v \Vdash \varphi$. Thus $u \Vdash K_i\varphi$. As the world u was arbitrary, $K_i\varphi$ holds at all worlds and so is valid in the model. Therefore $\vdash \varphi \Rightarrow \vdash K_i\varphi$ is valid.
- Con axiom: $\chi = J\varphi \to K_i\varphi$. Suppose $u \Vdash J\varphi$ so that for all v such that uR_Jv, $v \Vdash \varphi$. For all i, $R_i \subseteq R_J$ by definition, so for all w such that uR_iw, also uR_Jw and so $w \Vdash \varphi$, thus $u \Vdash K_i\varphi$.
- ConC: Analogous to the proof shown above for J's connection axiom Con.
- IA: $\chi = \varphi \wedge C(\varphi \to E\varphi) \to C\varphi$. Suppose $u \Vdash \varphi \wedge C(\varphi \to E\varphi)$. Then for all v such that uR_Cv, $v \Vdash \varphi \to E\varphi$ (∗∗). We want to show $u \Vdash C\varphi$, i.e. $v \Vdash \varphi$ for all v reachable from u. Proceed by induction on length of path l along R_is from u to v. It is sufficient to show this for paths of length l along the R_is as then the R_C paths are of length $\leq l$ (and in fact of length 0 or 1 along R_C).
  - If $l = 0$ then $u = v$ and by assumption, $u \Vdash \varphi$.
  - Induction Hypothesis: Assume $s \Vdash \varphi$ holds for worlds s reachable from u by a path of length l.
  - Suppose that v is reachable from u by a path of length $l + 1$. Then there is a world t reachable from u in l steps and tR_iv for some i. By the induction hypothesis, $t \Vdash \varphi$ but also by (∗∗) and modus ponens, $t \Vdash E\varphi$. But tR_iv, so $v \Vdash \varphi$. Thus $u \Vdash C\varphi$.

3   Completeness of $\mathsf{S4}_n^{CJ}$

Theorem 2. $\mathsf{S4}_n^{CJ}$ *is complete with respect to M^{CJ} models.*

To show completeness, the usual approach would be to construct the canonical model. However, here the canonical structure turns out not to be a model of $\mathsf{S4}_n^{CJ}$. So, instead of a single large model which acts as a counter-model for all non-provable φ, for each non-provable φ we construct a finite model with a world at which φ does not hold. Filtration techniques on the canonical structure yield these counter-models. The proof of Theorem 2 is delayed until the end of Sect. 3.2 after the presentation on filtrations.

Definition 4. The *canonical structure* for $\mathsf{S4}_n^{CJ}$ is $M' = \langle W, R_1, \ldots, R_n, R_C, R_J, \Vdash \rangle$ where

- $W = \{\Gamma \mid \Gamma \text{ is a maximally consistent set of } \mathsf{S4}_n^{CJ} \text{ formulas}\}$;
- $\Vdash \subseteq W \times Var$ such that $\Gamma \Vdash p$ iff $p \in \Gamma$ for $p \in Var$;
- $\Gamma R_i \Delta$ iff $\Gamma^i \subseteq \Delta$, where $\Gamma^i := \{\varphi \mid K_i\varphi \in \Gamma\}$;
- $\Gamma R_C \Delta$ iff $\Gamma^C \subseteq \Delta$, where $\Gamma^C := \{\varphi \mid C\varphi \in \Gamma\}$;
- $\Gamma R_J \Delta$ iff $\Gamma^J \subseteq \Delta$, where $\Gamma^J := \{\varphi \mid J\varphi \in \Gamma\}$.

Lemma 1 (Truth Lemma). *M' satisfies the Truth Lemma: for all Γ*

$$M', \Gamma \Vdash \varphi \Leftrightarrow \varphi \in \Gamma. \tag{1}$$

Proof. The proof by induction on φ is standard and mimics the $\mathsf{S4}_n$ case but we reproduce it here.

- base case: $\varphi = p$ for $p \in Var$. Holds by definition of $\Vdash$.
- Induction Hypothesis: Assume that the Truth Lemma holds for formulas of lower complexity.
- Boolean cases: by extension of $\Vdash$, the induction hypothesis, and maximality of Γ.
- modal case: Shown for K_i but analogous for C and J. $\varphi = K_i\varphi$ ($\Leftarrow$) Assume $K_i\varphi \in \Gamma$. Then for all Δ such that $\Gamma R_i \Delta$, $\varphi \in \Delta$ so by the induction hypothesis, $\Delta \Vdash \varphi$. Thus $\Gamma \Vdash K_i\varphi$. ($\Rightarrow$) Assume $K_i\varphi \notin \Gamma$. Then $\Gamma^i \cup \{\neg\varphi\}$ must be consistent by the maximality of Γ, for otherwise φ would be provable and hence (by necessitation) so would $K_i\varphi$, which would contradict the consistency of Γ. If Δ is any maximally consistent set containing $\Gamma^i \cup \{\neg\varphi\}$, then $\Gamma R_i \Delta$ by definition of R_i. So $\Gamma \nVdash K_i\varphi$.

Corollary 1. *As a consequence of the Truth Lemma, any maximal consistent set of formulas is satisfiable in M'.*

Thus $\mathsf{S4}_n^{CJ} \vdash \varphi \Rightarrow M', \Gamma \Vdash \varphi$, so soundness holds for the canonical structure.

Lemma 2. *The canonical structure M' is not a model of $\mathsf{S4}_n^{CJ}$ (cf. [12] p. 50).*

In M', all accessibility relations are reflexive and transitive and $R_C \subseteq R_J$. However, $R_C \neq \left(\bigcup_{i=1}^{n} R_i \right)^{TC}$ as we only have $\left(\bigcup_{i=1}^{n} R_i \right)^{TC} \subset R_C$, thus M' is not a model of $\mathsf{S4}_n^{CJ}$.

Proof. It suffices to show that $R_C \not\subseteq \left(\bigcup_{i=1}^{n} R_i \right)^{TC}$. Consider a set of formulas

$$\Phi = \{Ep, EEp, EEEp, \ldots\} \cup \{\neg Cp\} \tag{2}$$

for some $p \in Var$ and abbreviate $EEEp$ as E^3p, etc.

Claim. Φ is $\mathsf{S4}_n^{CJ}$-consistent.

Proof (of Claim). Suppose Φ is inconsistent. Then there is a finite $\Delta \subset \Phi$ which is already inconsistent so say $\Delta = \{E^{k_1}p, E^{k_2}p, \ldots, E^{k_m}p \mid k_i < k_{i+1} \text{ for } i < m\} \cup \{\neg Cp\}$. (If Δ were already inconsistent, including $\{\neg Cp\}$ would keep Δ inconsistent.) Consider the model $N = \langle W, R_1, R_2, R_C, R_J, \Vdash \rangle$t where

- $W = \mathbb{N}$;
- $R_1 = \{(n, n) \mid n \in \mathbb{N}\} \cup \{(n, n+1), (n+1, n) \mid n \in \mathbb{N} \text{ and } n \text{ even}\}$;
- $R_2 = \{(n, n) \mid n \in \mathbb{N}\} \cup \{(n, n+1), (n+1, n) \mid n \in \mathbb{N} \text{ and } n \text{ odd}\}$;
- $R_J = R_C = (R_1 \cup R_2)^{\mathrm{TC}}$;
- $x \Vdash p$ iff $x \leq k_m + 1$.

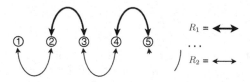

Fig. 1. This shows the frame of N with the reflexive arrows of R_1 and R_2 suppressed.

For this model R_C is an equivalence relation with one class, $m R_C n$ for all $m, n \in \mathbb{N}$. But $N, 1 \Vdash \Delta$. To see why, consider an example where $k_m = 3$ thus $1, 2, 3, 4 \Vdash p$,

$$1, 2, 3 \Vdash K_1 p \wedge K_2 p \wedge Ep \qquad \text{though } 4 \Vdash \neg K_1 p,$$
$$1, 2 \Vdash K_1 Ep \wedge K_2 Ep \wedge EEp \qquad \text{though } 3 \Vdash \neg K_2 K_1 p,$$
$$1 \Vdash K_1 EEp \wedge K_2 EEp \wedge EEEp \quad \text{though } 2 \Vdash \neg K_1 K_2 K_1 p, \text{ and}$$
$$1 \Vdash \neg C\varphi \text{ as } 5 \Vdash \neg p \text{ and } 1 R_C 5.$$

Since $1 \Vdash E^3 p \wedge \neg Cp$, this Δ is satisfied and hence is consistent. Since no finite subset of Δ is inconsistent, Φ is consistent.$_{Claim}$

We now finish the proof of Lemma 2. Since Φ is consistent, it is contained in some maximal consistent set Φ'. Let $\Theta = \{\neg p\} \cup \{\theta \mid C\theta \in \Phi'\}$. Note that Θ is consistent. As $\{\theta \mid C\theta \in \Phi'\} \subseteq \Phi'$ which is maximal consistent, Θ could only be inconsistent if $\neg p \wedge p \in \Theta$. As $\neg Cp \in \Phi$, Cp is not in Φ', so p is not in Θ, so Θ is consistent, and so contained in some maximal consistent set Θ'. Observe that $\Phi'^C \subseteq \Theta'$ so that in M', $\Phi' R_C \Theta'$. However $(\Phi', \Theta') \notin \left(\bigcup_{i=1}^{n} R_i \right)^{\mathrm{TC}}$ as for each m, $E^m p \in \Phi'$, but $\neg p \in \Theta'$. Therefore, M' is not a model of $\mathsf{S4}_n^{CJ}$.

Essentially, M' fails to be an appropriate model because $Ip \not\rightarrow Cp$, where I is iterated knowledge.

3.1   Filtrations: The General Modal Case

Filtration is an established technique for producing a finite model from an infinite one so that validity of subformulas is maintained. As in M^{CJ} there are already only a finite number of R_i, a finite model must be one in which W is finite. Each world in the finite model will be an equivalency class of worlds in the original model. We look first at a general modal case, where our modality is '$\Box$.' In the following section we apply these techniques to M' to produce finite counter-models to those formulas not provable in $\mathsf{S4}_n^{CJ}$, concluding the proof of completeness.

Definition 5. For a given finite set of formulas Φ, say two worlds in a model M are equivalent if they agree on all formulas in Φ:

$$s \equiv_\Phi t \text{ iff } (\forall \psi \in \Phi)(M, s \Vdash \psi \Leftrightarrow M, t \Vdash \psi)$$

and define an equivalence class of worlds

$$[s]_\Phi := \{t \mid s \equiv_\Phi t\},$$

or simply $[s]$ if Φ is clear.

Note that $\equiv_\Phi$ is indeed an equivalence relation.

Definition 6. A model $N = \langle S, T_1, \ldots, T_n, \Vdash_N \rangle$ is a *filtration of M through Φ* if M is a model $\langle W, R_1, \ldots, R_n, \Vdash \rangle$ and the following hold:

- Φ is a finite set of formulas closed under subformulas;
- $S = \{[w] \mid w \in W\}$, which is finite as Φ is finite;
- $w \Vdash p \Leftrightarrow [w] \Vdash_N p$ for $p \in Var \cap \Phi$ and $\Vdash_N$ is extended to all formulas;
- Each relation T_i satisfies the following two properties for all modals $\Box$:
$min(T_i/R_i) : (\forall [s], [t] \in S)(\text{if } s'R_it', \ s' \in [s], \text{ and } t' \in [t], \text{ then } [s]T_i[t])$
$max(T_i/R_i) : (\forall [s], [t] \in S)(\text{if } [s]T_i[t], \text{ then } (\forall \Box \psi \in \Phi)[M, s \Vdash \Box \psi \Rightarrow M, t \Vdash \psi])$.

The condition $min(T_i/R_i)$ ensures that T_i simulates R_i while $max(T_i/R_i)$ permits adding pairs to T_i independently of R_i if it respects $\Box$. Note that a filtration will always exist as you can define the T_i by reconsidering either condition as a bi-implication. This will give the smallest and largest (not necessarily distinct) filtrations, respectively [8].

Theorem 3. *Let N be a filtration of M through Φ, then*

$$(\forall \psi \in \Phi)(\forall s \in W)(M, s \Vdash \psi \Leftrightarrow N, [s] \Vdash_N \psi). \tag{3}$$

Proof. By induction on the complexity of $\psi \in \Phi$.

- $\psi = p$: by definition of $\Vdash_N$.
- I.H.: As Φ closed under subformulas, $M, s \Vdash \psi \Leftrightarrow N, [s] \Vdash_N \psi$ holds for ψ of lower complexity.
- $\psi = \neg \varphi$: $M, s \Vdash \neg \varphi \Leftrightarrow M, s \not\Vdash \varphi \Leftrightarrow$ (by I.H.) $N, [s] \not\Vdash_N \varphi \Leftrightarrow N, [s] \Vdash_N \neg \varphi$.
- $\psi = \varphi \wedge \varphi'$: $M, s \Vdash_M \varphi \wedge \varphi' \Leftrightarrow M, s \Vdash_M \varphi$ and $M, s \Vdash_M \varphi' \Leftrightarrow$ (by I.H.) $N, [s] \Vdash_N \varphi$ and $N, [s] \Vdash_N \varphi' \Leftrightarrow N, [s] \Vdash_N \varphi \wedge \varphi'$.

– $\psi = \Box\varphi$: ($\Rightarrow$) Suppose $M,s \Vdash \Box\varphi$. Let $[t]$ be such that $[s]T[t]$. By $max(T/R)$, $M,t \Vdash \varphi$. By I.H. $N,[t] \Vdash_N \varphi$. As $[t]$ was arbitrary, $N,[s] \Vdash_n \Box\varphi$. ($\Leftarrow$) Suppose $N,[s] \Vdash_N \Box\varphi$, so $\forall[t]$ such that $[s]T[t]$, $N,[t] \Vdash_N \varphi$. Let $u \in W$ be any state such that sRu, then by $min(T/R)$ $[s]T[u]$ so that $N,[u] \Vdash_N \varphi$. By I.H. $M,u \Vdash \varphi$ and since u was an arbitrary world accessible from s, $M,s \Vdash \Box\varphi$.

3.2 Filtrations: The Canonical Structure M' Case

We now consider filtrations in the context of $\mathsf{S4}_n^{CJ}$.

Definition 7. A formula φ has a *suitable set* of subformulas Φ if $\Phi = \Phi_1 \cup \Phi_2 \cup \Phi_3 \cup \Phi_4$ where for $i \in \{1,\ldots,n\}$:

$\Phi_1 = \{\psi, \neg\psi \mid \psi$ is a subformula of $\varphi\}$;
$\Phi_2 = \{K_iK_i\psi, \neg K_iK_i\psi \mid K_i\psi \in \Phi_1\}$;
$\Phi_3 = \{K_iJ\psi, \neg K_iJ\psi, K_i\psi, \neg K_i\psi \mid J\psi \in \Phi_1\}$;
$\Phi_4 = \{K_iC\psi, \neg K_iC\psi, K_i\psi, \neg K_i\psi \mid C\psi \in \Phi_1\}$.

Crucially, a suitable set is finite and closed under subformulas.

Corollary 2. *Let Φ be a suitable set for φ and M a model such that $M,s \Vdash \varphi$. If N is a filtration of M through Φ, then $N,[s] \Vdash_N \varphi$.*

Proof. By Theorem 3 and $\varphi \in \Phi$.

Definition 8. For $M' = \langle W, R_1, \ldots, R_n, R_C, R_J, \Vdash \rangle$, the canonical structure of $\mathsf{S4}_n^{CJ}$, and a suitable set Φ for a consistent formula φ, define a model $N = \langle S, T_1, \ldots, T_n, T_C, T_J, \Vdash_N \rangle$ such that, for $i \in \{1,2,\ldots,n\}$:

– $S = \{[w] \mid w \in W\}$, which is finite as Φ is finite;
– $w \Vdash p \Leftrightarrow [w] \Vdash_N p$ for $p \in Var \cap \Phi$ and $\Vdash_N$ is extended to all formulas;
– $T_i \subseteq S \times S$ such that $[s]T_i[t]$ iff $(s \Vdash K_i\psi \Rightarrow t \Vdash \psi)$ for those $K_i\psi \in \Phi$;
– $T_C = \left(\bigcup_{i=1}^{n} T_i \right)^{TC}$;
– $T_J \subseteq S \times S$ such that $[s]T_J[t]$ iff $(s \Vdash J\psi \Rightarrow t \Vdash \psi)$ for those $J\psi \in \Phi$.

We now drop the subscript on $\Vdash_N$ to simplify notation. As worlds in N are equivalence classes, it will be clear as to which model is in question.

Lemma 3. *N is a model of $\mathsf{S4}_n^{CJ}$ (see Definition 3).*

Proof. All accessibility relations are reflexive and transitive and $T_i \subseteq T_C \subseteq T_J$.

– T_i is reflexive: For an arbitrary $s \in [s]$, $(s \Vdash K_i\psi \Rightarrow s \Vdash \psi)$ always holds. If the antecedent is true, then by the definition of $\Vdash$ and the reflexivity of R_i, the consequence follows. If the antecedent fails, the implication is vacuously true. Thus for all $[s] \in S$, $[s]T_i[s]$, so T_i is reflexive.

- T_i is transitive: Suppose $[s]T_i[t]$ and $[t]T_i[u]$ and $s \Vdash K_i\psi$ for $K_i\psi \in \Phi$. As Φ is suitable, also $K_iK_i\psi \in \Phi$. As R_i is transitive, the 4 axiom is sound so we have $(s \Vdash K_i\psi \Rightarrow s \Vdash K_iK_i\psi)$, so as $[s]T_i[t]$ and $K_iK_i\psi \in \Phi$, $(s \Vdash K_iK_i\psi \Rightarrow t \Vdash K_i\psi)$. Since $K_i\psi \in \Phi$ and $[t]T_i[u]$, $(t \Vdash K_i\psi \Rightarrow u \Vdash \psi)$ so $u \Vdash \psi$. Thus for $K_i\psi \in \Phi$, $(s \Vdash K_i\psi \Rightarrow u \Vdash \psi)$ holds so $[s]T_i[u]$, hence T_i is transitive.
- T_C is reflexive as for every $[s] \in S$, $[s]T_i[s]$ and $T_i \subseteq T_C$. T_C is transitive by definition.
- T_J is reflexive and transitive by the same reasoning as for T_i. It must also be shown that $T_C \subseteq T_J$. Suppose $[s]T_i[t]$, then we want to show $[s]T_J[t]$, i.e. for $J\psi \in \Phi$, $(s \Vdash J\psi \Rightarrow t \Vdash \psi)$ holds. If $J\psi \in \Phi$, then as Φ is suitable, $K_iJ\psi \in \Phi$. Suppose $s \Vdash J\varphi$, then as M' is sound and $\mathbf{S4}_n^{CJ} \vdash J\varphi \rightarrow K_iJ\varphi$, $s \Vdash K_iJ\varphi$. Then since $[s]T_i[t]$, $(s \Vdash K_iJ\psi \Rightarrow t \Vdash J\psi)$ holds, so $t \Vdash J\varphi$ holds, and since R_J is reflexive, $t \Vdash \psi$. Thus for $J\psi \in \Phi$ and $[s]T_i[t]$, $(s \Vdash J\psi \Rightarrow t \Vdash \psi)$ holds, so $[s]T_J[t]$. Since $T_i \subseteq T_J$, $T_C \subseteq T_J$.

Lemma 4 (Definability Lemma). *Let $S = \{[s] \mid s \in W\}$ for some suitable set Φ. Then for each subset $D \subseteq S$ there is some characteristic formula χ_D such that for all $[s] \in S$, $s \Vdash \chi_D$ iff $[s] \in D$. Note that all D are finite as S is.*

Proof. Let the set $\bigwedge\{s\}$ be the conjunction of all $\psi \in \Phi$ that are true at s. By definition of $[s]$, $t \Vdash \bigwedge\{s\}$ iff $[s] = [t]$. Let $\chi_D = \bigvee_{[t]\in D} (\bigwedge\{s\})$.

$$s \Vdash \chi_D \Leftrightarrow s \Vdash \bigvee_{[t]\in D} \left(\bigwedge\{s\}\right) \Leftrightarrow s \Vdash \bigwedge\{t\} \text{ for some } t \in [t] \in D$$

$$\Leftrightarrow [s] = [t] \text{ for some } t \in [t] \in D \Leftrightarrow [s] \in D.$$

Theorem 4. *N of Definition 8 is a filtration of M' through Φ (cf. [12]).*

A relation T is a *filtration of R* if it satisfies $min(T/R)$ and $max(T/R)$.

Proof. It needs only to be confirmed that the accessibility relations T_i, T_C, and T_J meet the conditions $min(T/R)$ and $max(T/R)$.

- T_i : T_i satisfies $max(T_i/R_i)$ by definition so it remains to check $min(T_i/R_i)$. Suppose $[s], [t] \in S$ with $s' \in [s]$ and $t' \in [t]$ such that $s'R_it'$. For $K_i\psi \in \Phi$ we have

$$s \vdash K_i\psi \Leftrightarrow s' \vdash K_i\psi \Rightarrow t' \vdash \psi \Rightarrow t \vdash \psi.$$

Thus $[s]T_i[t]$ by definition, satisfying $min(T_i/R_i)$.
- T_J : T_J is a filtration of R_J by the same reasoning as in the T_i case, thus $min(T_J/R_J)$ and $max(T_J/R_J)$ are satisfied.
- T_C : To see that T_C satisfies $min(T_C/R_C)$, suppose that sR_Ct. Let $D = \{[w] \in S \mid [s]T_C[w]\}$, the set of worlds reachable from $[s]$ by T_C. It is sufficient to show

$$s \Vdash C\chi_D, \tag{4}$$

as then sR_Ct gives $t \Vdash \chi_D$ and so by definition of χ_D, $[t] \in D$ and so $[s]T_C[t]$.
Now we show (4).
As IA is valid in the canonical structure,

$$s \Vdash C(\chi_D \to E\chi_D) \to (\chi_D \to C\chi_D). \tag{5}$$

To see that $s \Vdash C(\chi_D \to E\chi_D)$ holds, consider the following. Suppose for some
w, sR_Cw and $w \Vdash \chi_D$. We want to show $w \Vdash E\chi_D$, i.e., for all i, $w \Vdash K_i\chi_D$,
i.e. for all u, wR_iu, $u \Vdash \chi_D$. Since $w \Vdash \chi_D$, $[w] \in D$ so $[s]T_C[w]$. This means
there is a path of length l from $[s]$ to $[w]$ along the union of T_is. As each T_i is a
filtration of R_i we also have for all those worlds u accessible from w, $[w]T_i[u]$.
Thus there is a path of length $l+1$ along the T_is from $[s]$ to $[u]$ and so $[s]T_C[u]$.
This means that $u \Vdash \chi_D$, so $w \Vdash E\chi_D$. Since the antecedent of (5) holds, we
have $s \Vdash \chi_D \to C\chi_D$ so in order to conclude (4), we must show $s \Vdash \chi_D$.
Which we have by the reflexivity of T_C. Thus T_C satisfies $min(T_C/R_C)$.
T_C must also satisfy $max(T_C/R_C)$. Suppose that $[s]T_C[t]$ and for some $s \in [s]$,
$s \Vdash C\psi$ for $C\psi \in \Phi$. We must show that $t \Vdash \psi$. Note that as Φ is suitable, for
each i, $K_iC\psi$, i.e. $EC\psi \in \Phi$ as well. Recall from Proposition 3 that $\mathsf{S4}_n^{CJ} \vdash$
$C\psi \to EC\psi$ so by soundness, $s \Vdash C\psi \Rightarrow s \Vdash EC\psi$. As $[s]T_C[t]$ and T_C is
built from filtrations of the R_is, there is a path of length l along the R_is
from s to t. As $s \Vdash EC\psi$ and $EC\psi \in \Phi$, $C\psi$ also holds at the next world on
this path towards t, for whichever R_i used. By induction on the length of the
path we get $t \Vdash C\psi$. Since T_C is reflexive we have $t \Vdash \psi$. Thus T_C satisfies
$max(T_C/R_C)$.

We can now finish the proof of Theorem 2 that $\mathsf{S4}_n^{CJ}$ is sound and complete
with respect to $\mathsf{S4}_n^{CJ}$-models. Soundness was shown in Theorem 1.

Proof (Proof of Completeness). Suppose $\mathsf{S4}_n^{CJ} \nvdash \varphi$. Then $\{\neg\varphi\}$ is contained
in some maximal consistent set Θ and for the canonical structure M' we have
$M', \Theta \Vdash \neg\varphi$. Defining a suitable set Φ of subformulas of $\neg\varphi$, we can construct an
$\mathsf{S4}_n^{CJ}$-model N (Lemma 3), which, as it happens to be a filtration of M' through
Φ (Theorem 4), agrees with M' on formulas of Φ (Theorem 3) and so $N, [\Theta] \nVdash \varphi$.

Corollary 3. $\mathsf{S4}_n^{CJ}$ *exhibits the Finite Model Property and so is decidable.*

Soundness yields the following two propositions.

Proposition 4. $\mathsf{S4}_n^{CJ} \nvdash C\varphi \to J\varphi$, *as was promised after Proposition 1.*

Proof. Consider a model of $\mathsf{S4}_2^{CJ}$ with $W = \{a, b\}$ such that $R_1 = R_2 = R_C =$
$\{(a, a), (b, b)\}$ and $R_J = \{(a, a), (b, b), (a, b)\}$. Let only $a \Vdash p$ and all other propositional variable fail at both worlds. While $a \Vdash Cp$, $a \nVdash Jp$ so $a \nVdash Cp \to Jp$ so
$a \Vdash \neg(C\varphi \to J\varphi)$, so by soundness $\mathsf{S4}_n^{CJ} \nvdash (C\varphi \to J\varphi)$.

Proposition 5. $\mathsf{S4}_n^{CJ}$ *is a conservative extension of both* $\mathsf{S4}_n^J$ *and* $\mathsf{S4}_n^C$.

The axiomatization and models of $\mathsf{S4}_n^J$ and $\mathsf{S4}_n^C$ can be obtained by removing C
or J, respectively, from the language and axiomatization of $\mathsf{S4}_n^{CJ}$ and R_C or R_J
from its models. For more on $\mathsf{S4}_n^J$ see [2,5,6]. For more on $\mathsf{S4}_n^C$ see [8,10,12].

Proof. Conservativity of $\mathsf{S4}_n^{CJ}$ over $\mathsf{S4}_n^{J}$: We need to show that for each $\mathsf{S4}_n^{J}$-formula F, if $\mathsf{S4}_n^{CJ} \vdash F$, then $\mathsf{S4}_n^{J} \vdash F$. By contraposition, suppose $\mathsf{S4}_n^{J} \not\vdash F$. Then, by the completeness theorem for $\mathsf{S4}_n^{J}$, there is a $\mathsf{S4}_n^{J}$-model M such that F does not hold in M. Now we transform M into an $\mathsf{S4}_n^{CJ}$-model M^* by adding the reachability relation R_C. This can always be done and leaves the other components of M unaltered. Since the modal C does not occur in F, the truth values of F in M and in M^* remains unchanged at each world, hence F does not hold in M^*. By soundness of $\mathsf{S4}_n^{CJ}$, $\mathsf{S4}_n^{CJ} \not\vdash F$.

Conservativity of $\mathsf{S4}_n^{CJ}$ over $\mathsf{S4}_n^{C}$: Let G be an $\mathsf{S4}_n^{C}$-formula not derivable in $\mathsf{S4}_n^{C}$. We have to show that G is not derivable in $\mathsf{S4}_n^{CJ}$ either. By completeness of $\mathsf{S4}_n^{C}$ there is an $\mathsf{S4}_n^{C}$-countermodel N for G. Make N into an $\mathsf{S4}_n^{CJ}$-model N^* by the addition of R_J as the total relation (alternatively, we could put $R_J = R_C$). As G contains no J, at each world, these models agree on the valuation of G, thus G does not hold in N^* either. By soundness, $\mathsf{S4}_n^{CJ} \not\vdash G$.

4 Conclusions

$\mathsf{S4}_n^{CJ}$ is a sound and complete system in which we can directly compare J and C. As $J\varphi \to C\varphi$, J can be used in place of C in situations in which common knowledge is used, such as in the assumption of common knowledge about game rules or public announcement statements. One advantage of using J over C is the possibility to realize these statements in a explicit justification logic, another is that it maybe a more accurate representation of these scenarios [4,6]. Another opportunity this logic provides is to examine or develop an interesting class of epistemic scenarios which exploit these nested yet distinct forms of common knowledge. Keeping in mind the connection axioms Con and ConC, C might represent an oracle-like agent while J might also be an infallible agent but one whose statements can be confirmed by evidence if needed.

There is also potential for future syntactic developments. As can be noted by the models of $\mathsf{S4}_n^{CJ}$, the logical strength of J can be chosen independently from that the other agents. For instance, while the K_i and C remain $\mathsf{S4}$ modalities, J could be $\mathsf{S5}$ or perhaps weaker such as $\mathsf{K4}$ to represent belief, while maintaining Con. The logic can also be expanded to encompass multiple distinct J operators.

$\mathsf{S4}_n^{CJ}$ is a logic which provides a context in which to investigate distinct forms of common knowledge. This, together with the conservativity results of Proposition 5, indicate that generic common knowledge is useful generalization of common knowledge with technical and semantic advantages.

References

1. Alberucci, L., Jaeger, G.: About cut elimination for logics of common knowledge. Ann. Pure Appl. Logic **133**(1–3), 73–99 (2005)
2. Antonakos, E.: Explicit generic common knowledge. In: Artemov, S., Nerode, A. (eds.) LFCS 2013. LNCS, vol. 7734, pp. 16–28. Springer, Heidelberg (2013)

3. Antonakos, E.: Justified and common knowledge: limited conservativity. In: Artemov, S., Nerode, A. (eds.) LFCS 2007. LNCS, vol. 4514, pp. 1–11. Springer, Heidelberg (2007)
4. Artemov, S.: Robust Knowledge and Rationality. Technical Report TR-2010010, CUNY Ph.D. Program in Computer Science (2010)
5. Artemov, S.: Justified common knowledge. Theor. Comput. Sci. **357**(1–3), 4–22 (2006)
6. Artemov, S.: Evidence-Based Common Knowledge. Technical Report TR-2004018, CUNY Ph.D. Program in Computer Science (2004)
7. Artemov, S., Fitting, M.: Justification Logic. In: E.N. Zalta (ed), The Stanford Encyclopedia of Philosophy (Fall 2012 Edition)
8. Blackburn, P., de Rijke, M., Venema, Y.: Modal Logic. Cambridge Tracts in Theoretical Computer Science, vol. 53. Cambridge University Press, Cambridge (2001)
9. Brünnler, K., Studer, T.: Syntactic cut-elimination for common knowledge. Ann. Pure Appl. Logic **160**(1), 82–95 (2009)
10. Fagin, R., Halpern, J., Moses, Y., & Vardi, M.: Reasoning About Knowledge. MIT Press, Cambridge, MA (1995, 1st MIT paperback ed., 2004)
11. McCarthy, J., Sato, M., Hayashi, T., Igarishi, S.: On the Model Theory of Knowledge. Technical Report STAN-CS-78-657, Stanford University (1978)
12. Meyer, J.-J., van der Hoek, W.: Epistemic Logic of AI and Computer Science. Cambridge Tracts in Theoretical Computer Science, vol. 41. Cambridge University Press, Cambridge (1995)

On Aggregating Probabilistic Evidence

Sergei Artemov$^{(\boxtimes)}$

The City University of New York, The Graduate Center, 365 Fifth Avenue,
New York City, NY 10016, USA
sartemov@gc.cuny.edu

Abstract. Imagine a database – a set of propositions $\Gamma = \{F_1, \ldots, F_n\}$ with some kind of probability estimates, and let a proposition X logically follow from Γ. What is the best justified lower bound of the probability of X? The traditional approach, e.g., within Adams' Probability Logic, computes the numeric lower bound for X corresponding to the worst-case scenario. We suggest a more flexible parameterized approach by assuming probability events $u_1, u_2, \ldots, u_n$ which support Γ, and calculating **aggregated evidence** $e(u_1, u_2, \ldots, u_n)$ for X. The probability of e provides a tight lower bound for any, not only a worst-case, situation. The problem is formalized in a version of justification logic and the conclusions are supported by corresponding completeness theorems. This approach can handle conflicting and inconsistent data and allows the gathering both positive and negative evidence for the same proposition.

Keywords: Probability · Evidence · Aggregation · Justification logic

1 Introduction

Probability aggregation is a well-known problem which appears naturally in many areas. Some classical approaches to this problem can be found, e.g., in [1,2,5,9,11]. We offer a different logic-based method of aggregating probabilistic evidence. Let proposition X logically follow from assumptions

$$\Gamma = \{F_1, F_2, \ldots, F_n\};$$

symbolically,

$$\Gamma \models X. \tag{1}$$

This means that X is true whenever all propositions from Γ are true. In standard set-theoretical semantics, this states that the truth set of X is the whole space if the truth set of each proposition from Γ is the whole space. What probabilistic conclusions can be drawn from (1)? A similar observation shows that if the probability of all propositions from Γ is 1, then the probability of X is also 1. But what happens in a general case when formulas from Γ have arbitrary probabilities? Does logical entailment (1) yield meaningful estimates of the probability of X?

© Springer International Publishing Switzerland 2016
S. Artemov and A. Nerode (Eds.): LFCS 2016, LNCS 9537, pp. 27–42, 2016.
DOI: 10.1007/978-3-319-27683-0_3

The aforementioned logical observation "X is true whenever Γ is true" yields an estimate: X holds on the intersection of all events F_i, $i = 1, 2, \ldots, n$:

$$l = P(F_1 \cap F_2 \cap \ldots \cap F_n) \tag{2}$$

hence

$$l \leq P(X). \tag{3}$$

This approach is reflected in inductive probability reasoning, cf. [1,13]. The well-known Suppes' rule

$$\frac{P(A) \geq r \quad P(B \mid A) \geq p}{P(B) \geq rp}$$

is basically (3) in the special case of a single use of the logical rule *Modus Ponens*

$$A, \ A \to B \vdash B.$$

In either case we draw a probability estimate for B given probability estimates for A and $A \cap B$[1].

Approach (2), however, is not a well-principled way to aggregate probabilistic information provided by logical entailment (1): we have to find a way of accumulating evidence for X throughout the whole database. The traditional Adams' Probability Logic (cf. [1,2]) deals with this problem by introducing weights based on *degree's of essentialness* of premises from Γ and calculating the tight lower bound for the probability of X.

However, the very concept of drawing probability estimates for X only from probability estimates for Γ is limited: this format forces us to consider worst-case scenarios and in all other cases its estimates are not optimal.

Consider a simple derivation

$$A, B, C \models A \wedge B \wedge C. \tag{4}$$

For "high end" probabilities close to 1, probability-based estimates make sense: in this case, if

$$P(A), P(B), P(C) \geq 0.99,$$

then it is easy to check that the tight low estimate is

$$P(A \wedge B \wedge C) \geq 0.97.$$

In social situations, however, "medium range" probabilities are more typical and for them, the probability-based approach fails. For the same example (4) and probabilities

$$P(A), P(B), P(C) \geq 2/3,$$

the corresponding **tight** low bound for $P(A \wedge B \wedge C)$ is 0, which is meaningless.

Instead of using a numerical lower bound p for the probability of F

$$p \leq P(F)$$

---

[1] Since $P(B \mid A) = P(A \cap B)/P(A)$.

for some known p, we suggest using the evidence format:

$$u \subseteq F$$

for some event u. Given evidence $u_1, u_2, \ldots, u_n$ for $F_1, F_2, \ldots, F_n$, we build **aggregating evidence** $e(u_1, u_2, \ldots, u_n)$ for X,

$$e(u_1, u_2, \ldots, u_n) \subseteq X,$$

which provides a parameterized lower bound of probability of X:

$$P(e(u_1, u_2, \ldots, u_n)) \leq P(X).$$

For the same example (4), we introduce evidence variables u, v, w denoting events (subsets of a probability space) on which A, B, C respectively hold. The logical derivation suggests that $A \wedge B \wedge C$ is secured on event $e(u, v, w)$ (we write st for $s \cap t$, for better readability):

$$e(u, v, w) = uvw,$$

which we offer as **aggregated evidence** for $A \wedge B \wedge C$:

$$uvw \subseteq A \wedge B \wedge C.$$

If $P(u), P(v), P(w) = 2/3$, $P(uvw)$ ranges from 0 to $2/3$.

1.1 Specified and Unspecified Events

Let $(\Omega, \mathcal{F}, P)$ be a probability space (cf. [7,10,12]) where Ω is a set of outcomes; $\mathcal{F}$ is an algebra of measurable events; and P is a probability function from $\mathcal{F}$.

As a simple example, consider a six-sided symmetric die with faces from 1 to 6. Here Ω can be identified as the set of outcomes $\{1, 2, 3, 4, 5, 6\}$, each value being equally likely. $\mathcal{F}$ here is the set of all $2^6 = 64$ subsets of Ω (events) and for each event X, its probability, $P(X)$, is $1/6$ times the cardinality of X.

Within a given scenario, some events are constructively defined and considered **specified**, e.g.,

- $E = \{6\}$;
- $E = $ *all even faces* $= \{2, 4, 6\}$;
- $E = $ *all outcomes less than 3* $= \{1, 2\}$.

In other words, we may regard as specified, events which we can reliably present as concrete sets of outcomes (with known probabilities).

Some sets of outcomes can be described in a way that does not assign a specific event to them. Suppose a number $i \in \{1, 2, 3\}$ is unknown (e.g., is kept secret by an adversary, or is some yet unknown value from tomorrow's stock market). A set

$$X = \textit{all outcomes greater than } i \tag{5}$$

is an **unspecified** event. In social scenarios, we may regard as unspecified, events about which we have only partial specification, which cannot be identified *a priori* as concrete sets of outcomes.

Probabilities of events that are constructed from specified events can be calculated using basic probability theory. The likelihood of an unspecified event X can be evaluated by its **evidence**, a specified event t about which we can establish that $t \subseteq X$, symbolically,

$$t{:}X.$$

For example, for unspecified event X from (5), the best evidence t is $\{4, 5, 6\}$.

One has to distinguish "specified" and "certain" events. A certain event X is a probabilistic notion stating that X has probability 1, $P(X) = 1$. "Specified" is an epistemic feature of a description of an event X for a given agent; this has nothing to do with the probability of X.

1.2   Motivating Example

Assume (1) and suppose that X and all F_i's are some, possibly unspecified, events in a probability space and each F_i has **evidence** – a specified event u_i such that u_i guarantees F_i, i.e., $u_i \subseteq F$, symbolically

$$\{u_1{:}F_1, u_2{:}F_2, \ldots, u_n{:}F_n\}. \tag{6}$$

What is the best aggregated evidence $e(u_1, u_2, \ldots, u_n)$ for X,

$$e(u_1, u_2, \ldots, u_n){:}X,$$

and the best probability estimate for X which is justified by (1) and (6)?

Suppose we are given events u, v, and w each with probability $1/3$, which are supportive evidence for F, $F \to X$, and X respectively, i.e.,

1. $u{:}F$;
2. $v{:}(F \to X)$;
3. $w{:}X$.

Here $\Gamma = \{F, F \to X, X\}$ and Γ logically yields X, i.e., (1) holds. Basic facts from propositional logic suggest that (1) is equivalent to

$$\Gamma \vdash X \tag{7}$$

stating that **X can be derived from Γ by reasoning in propositional logic.**

What is the best aggregated evidence for this X and what is its probability? The answer depends on the set-theoretical configuration of u, v, and w.

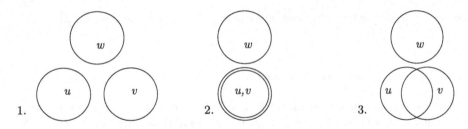

Fig. 1. Some configurations of u, v, and w.

In configuration 1, Fig. 1, events u and v are incompatible and hence do not contribute to the aggregated evidence, which is therefore equal to w and has probability $1/3$. In configuration 2, $u = v$ and hence the whole event $u \cup w$ supports X: the probability of X is at least $2/3$. In configuration 3, the contributing sections to the aggregated evidence for X are uv and w, with the probability

$$P(uv \cup w)$$

which, with an additional assumption that $P(uv) = 1/6$, is equal to $1/2$.

These and all other configurations are covered by a uniform "evidence term" $e(u, v, w)$

$$e(u, v, w) = uv \cup w \tag{8}$$

which can be obtained by **logical reasoning from** Γ: there are two ways to justify X, either from F and $F \to X$, which is valid on uv, or directly from w, hence the aggregated evidence (8).

Again, we may compare the Probability Logic answer:

"the lower bound is $1/3$"

with the aggregated evidence answer

"for a given configuration of evidence parameters u, v, w, the lower bound is $P(uv \cup w)$ which ranges from $1/3$ to $2/3$."

We put this type of reasoning on solid logical and mathematical ground. In particular, in Sect. 6, we prove that (8) is indeed the best logically justified evidence for X here.

2   General Setting

We are interested in estimating a probability $P(X)$ of an unspecified event X. As before, suppose X logically follows from a set of probabilistic assumptions which are not necessarily specified events:

$$\Gamma = \{F_1, F_2, \ldots, F_n\}.$$

Assume that for each F_i we have an evidence u_i; we denote this situation

$$\boldsymbol{u}{:}\Gamma = \{u_1{:}F_1, u_2{:}F_2, \ldots, u_n{:}F_n\}.$$

Since each u_i is considered specified, we have a lower bound of $P(F_i)$, namely $P(u_i)$.

One could wonder why we introduce different vocabularies for evidence and events, given that both are events. The answer lies in their different epistemic and functional roles: evidence $u_1, u_2, \ldots, u_n$ are events which are presumed **specified** and hence legitimate building material, or inputs, for aggregated evidence, probability estimates, etc. General events $F_1, F_2, \ldots, F_n$ are not presumed specified, and they play a role of logical types of $u_1, u_2, \ldots, u_n$ but do not serve as specified inputs of computations.

> An obvious analogy here would be with polynomial inequalities, for example, $x^2 + px + q \leq 0$, where x is unknown and coefficients p, q are known real values. The solution is presented by terms of coefficients providing known lower and upper bounds for x which itself, generally speaking, remains unknown.

A crude (correct) evidence for X is the intersection of all u_i's

$$u = u_1 u_2 \ldots u_n.$$

Indeed, since X logically follows from Γ, whenever all F_i's hold, X holds as well. Apparently, Γ holds on u, hence u is evidence for X. This is a correct, but not very useful argument, since intuitively, $u_1 u_2 \ldots u_n$ can often be $\emptyset$, as in all examples 1 – 3 from the previous section.

We offer a more refined way of aggregating evidence given in the form of probabilistic events. By tracking the logical dependences of X from Γ, we can build an aggregated witness

$$e = e(u_1, u_2, \ldots, u_n)$$

for X

$$e{:}X$$

as tight as is warranted by the data and obtain a better probability estimate for $P(X)$,

$$P(e).$$

Examples from Sect. 1.2 suggest that aggregated evidence should be produced symbolically, and only then should we proceed with calculating probabilities.

3   Logic of Probabilistic Evidence PE

The evidence terms in PE are built from variables $u_1, u_2, \ldots, u_n$, and constants **0, 1** by operations "$\cap$" and "$\cup$." As agreed, we will write st for $s \cap t$.

By $\mathcal{L}_n$ we understand a free distributive lattice (which we call **the evidence lattice**) over $u_1, u_2, \ldots, u_n, \mathbf{0}, \mathbf{1}$ with operations "∩," "∪," and lattice order $\preceq$. Note that for each n, $\mathcal{L}_n$ has only a finite number of equivalence classes w.r.t. the lattice equality "=." We fix a canonical representative in each class which is either $\mathbf{0}$ or $\mathbf{1}$, or a sum of products

$$u_{i_1} u_{i_2} \ldots u_{i_k}.$$

Free distributive lattices $\mathcal{L}_n$ for $n = 0, 1, 2$ are shown in Fig. 2.

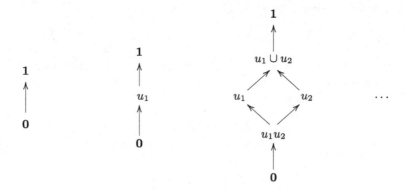

Fig. 2. Free distributive lattices.

Let

$$\mathcal{L} = \bigcup_{n \geq 0} \mathcal{L}_n.$$

Formulas are generated from propositional letters $p, q, r, \ldots$ by the usual logical connectives. We also allow formulas $t{:}F$ where F is a purely propositional formula and t an evidence term. The intended reading of evidence terms t is measurable events from $\mathcal{F}$ in a given probability space $(\Omega, \mathcal{F}, P)$, constants $\mathbf{0}$ and $\mathbf{1}$ are interpreted as $\emptyset$ and Ω respectively, and $t{:}F$ is understood as

t is an event supporting F.

The logical postulates of PE are

1. *axioms and rules of classical logic in the language of* PE;
2. $s{:}(A \to B) \to (t{:}A \to [st]{:}B)$;
3. $(s{:}A \wedge t{:}A) \to [s \cup t]{:}A$;
4. $\mathbf{1}{:}A$, *where A is a propositional tautology,*
   $\mathbf{0}{:}F$, *where F is a propositional formula*;
5. $t{:}X \to s{:}X$, *for any evidence terms s and t such that* $s \preceq t$ *in* $\mathcal{L}$.[2]

---

[2] This axiom can be replaced by an explicit list of its instances corresponding to a standard algorithm for deciding $s \preceq t$ (cf. [14]).

Definition 1. *Consider a specific evaluation consisting of*

(a) a probability space $(\Omega, \mathcal{F}, P)$;
(b) a mapping $*$ of propositions to Ω and evidence terms to $\mathcal{F}$.

Assume
$$\mathbf{0}^* = \emptyset, \quad \mathbf{1}^* = \Omega,$$
$$(st)^* = s^* \cap t^*,$$
$$(s \cup t)^* = s^* \cup t^*.$$

On propositions, interpretation $*$ is Boolean, i.e.,

$$(X \wedge Y)^* = X^* \cap Y^*, \quad (X \vee Y)^* = X^* \cup Y^*, \quad (\neg X)^* = \overline{X^*}$$

where $\overline{Y}$ is the complement of Y in Ω. The key point is interpreting an evidence assertion $t{:}X$ as the set-theoretical form of "t^* is a subset of X^*," or, equivalently, "X^* holds whenever t^* does":

$$(t{:}X)^* = \overline{t^*} \cup X^*.$$

For a set of formulas Γ,

$$\Gamma^* = \bigcap \{F^* \mid F \in \Gamma\}.$$

In particular, $\emptyset^* = \Omega$.

Proposition 1. *For each axiom A of* PE*, and each interpretation $*$,*

$$A^* = \Omega.$$

Proof. Let us check axiom 5. Since set-theoretical operations respect free distributive lattice identities, $s \preceq t$ yields $s^* \subseteq t^*$. The rest is trivial.

Definition 2. *Let Δ be a set of* PE *formulas and X a* PE *formula. We say that X probabilistically follows from Δ, notation*

$$\Delta \Vdash X,$$

if, for each interpretation $*$, $\Delta^* \subseteq X^*$. In particular, if $\Delta \Vdash X$, and $\Delta^* = \Omega$, e.g., when $\Delta = \emptyset$, then $X^* = \Omega$ as well.

We say that t **is evidence for** X **under interpretation** $*$ iff $t^* \subseteq X^*$ or, equivalently,
$$(t{:}X)^* = \Omega,$$

i.e., each outcome from t^* supports X^*. In particular, under such interpretation $*$, the probability of X^*, if defined, is at least $P(t^*)$.

The following theorem states the soundness of PE w.r.t. the aforementioned set-theoretical/probabilistic interpretation: all theorems of PE hold at each outcome from Ω.

Theorem 1 (Soundness with Respect to Probabilistic Semantics).
Logical entailment yields probabilistic entailment:

$$\Gamma \vdash F \quad \Rightarrow \quad \Gamma \Vdash F.$$

Proof. Induction on derivations in Γ. If F is an axiom of PE, then, by Proposition 1, $F^* = \Omega$ and hence $\Gamma^* \subseteq F^*$. The case $F \in \Gamma$ is trivial. The only rule of inference *Modus Ponens* $A, A \to B \vdash B$ preserves the property $\Gamma^* \subseteq F^*$.

The following internalization property describes evidence tracking by PE: whenever F logically follows from a set of assumptions in propositional logic, there is a non-zero evidence term that witnesses this fact in PE. Let CPC stand for classical propositional calculus.

Theorem 2. (Internalization). *If*

$$F_1, F_2, \ldots, F_n \vdash F$$

in CPC, *then*

$$u_1{:}F_1, u_2{:}F_2, \ldots, u_n{:}F_n \vdash (u_1 u_2 \ldots u_n){:}F$$

in PE.

Proof. ($\Rightarrow$). By induction on the derivation of F from $F_1, F_2, \ldots, F_n$, we build a non-zero evidence term t such that

$$u_1{:}F_1, u_2{:}F_2, \ldots, u_n{:}F_n \vdash t{:}F.$$

The case when F is an axiom of CPC is treated by axiom $\mathbf{1}{:}F$; it suffices to put $t = \mathbf{1}$. The case when F is one of F_i is trivial; we just put $t = u_i$. Finally, if F is obtained by *Modus Ponens* from $X \to F$ and X, then, by the induction hypothesis,

$$u_1{:}F_1, u_2{:}F_2, \ldots, u_n{:}F_n \vdash p{:}(X \to F)$$

and

$$u_1{:}F_1, u_2{:}F_2, \ldots, u_n{:}F_n \vdash q{:}X$$

for some p and q. By axiom 2,

$$u_1{:}F_1, u_2{:}F_2, \ldots, u_n{:}F_n \vdash (pq){:}F,$$

and it suffices to put $t = pq$.

So, we have found a non-zero evidence term $t \in \mathcal{L}_n$ (actually, t is either $\mathbf{1}$ or a product of some u_i's) such that

$$u_1{:}F_1, u_2{:}F_2, \ldots, u_n{:}F_n \vdash t{:}F.$$

Since $u_1 u_2 \ldots u_n \preceq t$ for each such t, we also have

$$u_1{:}F_1, u_2{:}F_2, \ldots, u_n{:}F_n \vdash (u_1 u_2 \ldots u_n){:}F.$$

If F is a theorem of CPC (i.e., a classical tautology), then

$$u_1{:}F_1, u_2{:}F_2, \ldots, u_n{:}F_n \vdash \mathbf{1}{:}F.$$

4   Aggregated Evidence

Definition 3. *Let X be a propositional formula and Γ a set of propositional formulas:*

$$\Gamma = \{F_1, F_2, \ldots, F_n\}.$$

Evidence for a proposition X given Γ *is a term $t \in \mathcal{L}_n$ such that in* PE,

$$u_1{:}F_1, u_2{:}F_2, \ldots, u_n{:}F_n \vdash t{:}X.$$

By the adopted notation,

$$\boldsymbol{u}{:}\Gamma \vdash t{:}X.$$

The **aggregated evidence $AE^\Gamma X$ for a proposition X given Γ** *is the evidence term*

$$AE^\Gamma(X) = \bigcup \{t \mid t \text{ is an evidence for } X \text{ given } \Gamma\}. \tag{9}$$

Given X logically follows from Γ, aggregated evidence $AE^\Gamma(X)$ is the collection of all evidence terms supporting X which logically follow from $\boldsymbol{u}{:}\Gamma$. Since the evidence lattice $\mathcal{L}_n$ is finite, the union in (9) is finite.

The finiteness of Γ cannot be dismissed: for the following infinite Γ

$$\{q_1, q_1 \to p, q_2, q_2 \to p, \ldots, q_n, q_n \to p, \ldots\},$$

where $p, q_1, q_2, \ldots, q_n, \ldots$ are propositional variables, the aggregated evidence for p, $AE^\Gamma(p)$ cannot be exhausted by a single evidence term.

Proposition 2. *The aggregated evidence $e = AE^\Gamma(X)$ is evidence for X given Γ, i.e., $e \in \mathcal{L}_n$ and*

$$\boldsymbol{u}{:}\Gamma \vdash e{:}X.$$

Proof. By Axioms 3 and 5 of PE,

$$(s{:}A \wedge t{:}A) \leftrightarrow [s \cup t]{:}A.$$

Therefore, the union of terms in $\mathcal{L}_n$ is evidence for X iff each of these terms is evidence for X.

Corollary 1. *A lattice term $t \in \mathcal{L}_n$ is evidence for X given Γ iff*

$$t \preceq AE^\Gamma(X).$$

This Corollary shows that aggregated evidence term $AE^\Gamma(X)$ is the largest term in the evidence lattice $\mathcal{L}_n$, which is evidence for X in Γ.

Theorem 3 (Completeness for Evidence Aggregation). *Let X be a propositional formula, Γ a set of propositional formulas $\{F_1, F_2, \ldots, F_n\}$ and $t \in \mathcal{L}_n$. Then*

$$\boldsymbol{u}{:}\Gamma \Vdash t{:}X \quad \Rightarrow \quad \boldsymbol{u}{:}\Gamma \vdash t{:}X.$$

Proof. By contrapositive: assume $\boldsymbol{u}{:}\Gamma \not\vdash t{:}X$. By Corollary 1, $t \not\preceq AE^\Gamma(X)$ in $\mathcal{L}_n$. As we have already discussed, each evidence term $s \in \mathcal{L}_n$ is equal to the union of some products

$$s = \bigcup_{i_1, i_2, \ldots, i_k} (u_{i_1} u_{i_2} \ldots u_{i_k}).$$

Since $t \not\preceq AE^\Gamma(X)$, t contains such a product which is not in $AE^\Gamma(X)$; without loss of generality we assume that $t = u_1 u_2 \ldots u_k$ for some k less than or equal to n. Since $\boldsymbol{u}{:}\Gamma \not\vdash t{:}X$, we also have

$$u_1{:}F_1, u_2{:}F_2, \ldots, u_k{:}F_k \not\vdash u_1 u_2 \ldots u_k{:}X.$$

By Internalization Theorem 2, in CPC,

$$F_1, F_2, \ldots, F_k \not\vdash X.$$

By completeness of CPC, there is a Boolean assignment $\sharp$ of truth values 0 and 1 which makes all F_i true and X false. Take an arbitrary probability space $(\Omega, \mathcal{F}, P)$ and evaluation $*$ of propositional letters such that if $p^\sharp = 0$, then $p^* = \emptyset$ and if $p^\sharp = 1$, then $p^* = \Omega$. Apparently, all $F_i^* = \Omega$, $i = 1, 2, \ldots, k$, and $X^* = \emptyset$.

   Extend $*$ to evidence variables by setting $u_i^* = \Omega$ for $i = 1, 2, \ldots, k$ and $u_i^* = \emptyset$ for $i = k+1, k+2, \ldots, n$, which makes $(u_i{:}F_i)^* = \Omega$ for all $i = 1, 2, \ldots, n$. Furthermore, $t^* = \Omega$ as well, and $(t{:}X)^* = \emptyset$, which means that $\boldsymbol{u}{:}\Gamma \Vdash t{:}X$ fails.

   The following Corollary 2 shows that for each Γ and X, the approximation provided by aggregated evidence $AE^\Gamma(X)$ cannot be improved uniformly for all probability spaces.

Corollary 2. $\boldsymbol{u}{:}\Gamma \Vdash t{:}X \quad \Leftrightarrow \quad t \preceq AE^\Gamma(X).$

Proof. By Corollary 1, if $t \preceq AE^\Gamma(X)$, then t is evidence for X in Γ, hence, by Theorem 1, $\boldsymbol{u}{:}\Gamma \Vdash t{:}X$, i.e., t is evidence for X in any probabilistic model of Γ. Now let $\boldsymbol{u}{:}\Gamma \Vdash t{:}X$. By Theorem 3, $\boldsymbol{u}{:}\Gamma \vdash t{:}X$, hence, by Corollary 1, $t \preceq AE^\Gamma(X)$.

5   General Picture

5.1   Model-Theoretical View

To build the aggregated evidence term $AE^\Gamma(X)$, find all (set-theoretically minimal) subsets Γ' of Γ such that

$$\Gamma' \models X$$

and form **lattice products** $u_{i_1} u_{i_2} \ldots u_{i_k}$ of evidence variables corresponding to all such Γ's. The aggregated evidence term $AE^\Gamma(X)$ is the **union** of these products.

5.2   Proof-Theoretical View

Alternatively, consider all possible connected derivations of X from Γ in tree-like form with axioms and assumptions at the leaf nodes and instances of *Modus Ponens* at all other nodes. It is easy to see that in each of these derivations, the aggregated evidence $s(v)$ of the root formula X is the **product** of all variables v and evidence constant $\mathbf{1}$, if any, that are evidence terms for the leaf nodes. The desired aggregated evidence term $e(u)$ is the **union** of all these $s(v)$'s. Finiteness of the evidence lattice $\mathcal{L}_n$ guarantees that $AE^\Gamma(X)$ is a specific term in $\mathcal{L}_n$.

6   Example of Aggregated Evidence

Let us return to the example from Sect. 1.2 with

$$\Gamma = \{F,\ F\to X,\ X\}$$

and the evidence variables assignment

$$\Delta = \{u{:}F,\ v{:}(F\to X),\ w{:}X\}.$$

We claim that $uv \cup w$ is the aggregated evidence term for X in Γ,

$$uv \cup w = AE^\Gamma(X).$$

As we have already seen, $uv \cup w$ is evidence for X in Γ.

Suppose there is an evidence term t such that

$$t \ \not\leq_\Gamma (uv \cup w)$$

but $\Delta \vdash t{:}X$. Of all possible products of generators u, v, w in the evidence lattice $\mathcal{L}_3$, all but u and v are less then or equal to $uv \cup w$, hence t is either $\mathbf{1}$ or a sum containing at least one of u or v. Therefore, at least one of $\mathbf{1}$, u, or v should be evidence for X in Γ, which is not the case.

Term u is not evidence for X in Γ since with $F^* = u^* = \Omega$ and $X^* = v^* = w^* = \emptyset$, all formulas from Δ are evaluated as Ω and $u{:}X$ is evaluated as $\emptyset$. This evaluation also rules out the possibility that $\mathbf{1}$ is evidence for X in Γ.

Term v is not evidence for X in Γ since with $v^* = \Omega$ and $F^* = X^* = u^* = w^* = \emptyset$, all formulas from Δ are evaluated as Ω and $v{:}X$ is evaluated as $\emptyset$.

7   Computational Summary

We consider the problem of finding the best justified probability estimate p for a proposition X in a given probability space $\mathcal{P} = (\Omega, \mathcal{F}, P)$ given

1. a set $\Gamma = \{F_1, F_2, \ldots, F_n\}$ of propositions not necessarily specified;
2. specified "evidence" events $u_1^*, u_2^*, \ldots, u_n^*$ such that $u_i^* \subseteq F_i^*$ for $i = 1, 2, \ldots, n$ in $\mathcal{P}^3$.

Our findings suggest the following procedure for calculating this p.

1. By logical tools, find term $e = AE^\Gamma(X)$, cf. Sect. 5.
2. Given evaluations $u_1^*, u_2^*, \ldots, u_n^*$ and term $e = e(u_1, u_2, \ldots, u_n)$, calculate set-theoretically the event e^* in $\mathcal{P}$.
3. Calculate p as the probability of e^*:

$$p = P(e^*).$$

8 Computational Example

This will be a specific instance of the "motivating example" from Sect. 1.2.

1. A probability space $\mathcal{P} = (\Omega, \mathcal{F}, P)$: $\Omega = \{1, 2, 3, 4, 5, 6\}$, all outcomes are equally probable.
2. $\Gamma = \{F, F \to X, X\}$, evidence variables u, v, w (for $u{:}F, v{:}(F \to X), w{:}X$).
3. $F^* = \{1, 2, 3\}$, $X^* = \{3, 4, 5\}$, hence $(F \to X)^* = \{3, 4, 5, 6\}$. These events are not assumed known to the reasoning agent; we provide them for a complete picture.
4. Specified evidence $u^* = \{1, 3\}$, $v^* = \{3, 4\}$, $w^* = \{4, 5\}$. It is easy to check consistency:
   - $u^* \subseteq F^*$,
   - $v^* \subseteq X^*$,
   - $w^* \subseteq (F \to X)^*$.

The first computational step is to calculate $AE^\Gamma(X)$, the aggregated evidence for X in Γ, which, as we know from Sect. 6, in this case is

$$AE^\Gamma(X) = uv \cup w.$$

The second step is to evaluate $AE^\Gamma(X)$ for a given $*$:

$$[AE^\Gamma(X)]^* = (uv \cup w)^* = (u^* \cap v^*) \cup w^* = \{3, 4, 5\}.$$

As predicted by the theory,

$$[AE^\Gamma(X)]^* \subseteq X^*.$$

In this case, we were able to accidentally recover X entirely by its aggregated evidence, e.g.,

$$[AE^\Gamma(X)]^* = \{3, 4, 5\} = X^*.$$

Finally, we calculate the justified lower bound p of probability of X:

$$p = P([AE^\Gamma(X)]^*) = P(\{3, 4, 5\}) = 1/2.$$

---

[3] Γ is not necessarily compatible with set $u_1^*, u_2^*, \ldots, u_n^*$ but we ignore this question for now by assuming that the given evidence is consistent with Γ.

9   Further Suggestions: Handling Inconsistent Data

In a general setting, one should not expect propositional data Γ to be logically consistent: in realistic situations, we have to deal with sets of assumptions which may contradict each other. Furthermore, we may want to gather evidence for X and for $\neg X$ from the same data. The framework of Probabilistic Evidence logic PE naturally accommodates these needs: we can track both positive and negative evidence for X from the same Γ:

$$AE_+^\Gamma(X) = AE^\Gamma(X) = \bigcup\{t \mid t \text{ is evidence for } X \text{ in } \Gamma\},$$

$$AE_-^\Gamma(X) = AE^\Gamma(\neg X) = \bigcup\{t \mid t \text{ is evidence for } \neg X \text{ in } \Gamma\},$$

with positive and negative justified ratings of X in Γ for a given interpretation $*$ being the probabilities

$$P([AE_+^\Gamma(X)]^*)$$

and

$$P([AE_-^\Gamma(X)]^*),$$

respectively.

Once $*$ makes all formulas from $\boldsymbol{u}{:}\Gamma$ true, i.e., $[\boldsymbol{u}{:}\Gamma]^* = \Omega$, $[AE_+^\Gamma(X)]^* \subseteq X^*$ and $[AE_-^\Gamma(X)]^* \subseteq \overline{X^*}$, hence $[AE_+^\Gamma(X)]^*$ and $[AE_-^\Gamma(X)]^*$ are disjoint and

$$P([AE_+^\Gamma(X)]^*) + P([AE_-^\Gamma(X)]^*) \leq 1.$$

Obviously, positive and negative ratings do not necessarily sum to 1.

As a computational example, consider the same probability space as in Sect. 8, with extended database

$$\Gamma = \{F,\ F{\to}X,\ X,\ \neg X\}, \tag{10}$$

evidence variables $\{u, v, w, y\}$ (meaning that y is evidence for $\neg X$), evaluation $*$ as before on F, X, u, v, w, and

$$y^* = \{1, 2\}.$$

Note that Γ is logically inconsistent but still yields a meaningful evidence aggregation picture.

Obviously,

$$y^* \subset \{1, 2, 3\} = (\neg X)^*.$$

The previous calculations suggest that

$$AE_+^\Gamma(X) = uv \cup w.$$

It is also easy to check that the aggregated negative evidence term for X in Γ is

$$AE_-^\Gamma(X) = y.$$

Corresponding positive evidence for X is, as before,

$$[AE_+^\Gamma(X)]^* = \{3, 4, 5\}$$

with probability $1/2$ (the positive rating of X in Γ). Negative evidence for X is

$$[AE_-^\Gamma(X)]^* = \{1, 2\}$$

with probability $1/3$ (the negative rating of X in Γ).

This possibility of measuring both positive and negative ratings is an important feature of a language with justification assertions $t{:}F$. Logic PE, as well as its predecessor, basic Justification Logic J (cf. [4]), naturally handles logical inconsistency. In the usual propositional logic, the combination of data $A, \neg A$ is inconsistent whereas in Justification Logic systems J and PE, the corresponding combination

$$u{:}A, v{:}\neg A$$

is consistent both intuitively and formally since it states that "u is evidence for A whereas v is evidence for $\neg A$."

Further issues, such as the model theory and proof theory of PE, feasible algorithms of computing aggregated evidence and recognizing consistency, etc., are left for future studies.

10   On Logical Properties of PE

The logic of probabilistic evidence PE may be regarded as a modification of the Logic of Proofs/Justification Logic JL (cf. [3,4,8]) in which we are interested in formulas of evidence depth not more than one, and in which justification terms, in addition to operations typical for JL, have a meaningful lattice-order relation, capturing the idea of the relative strength of justifications. This feature brings PE closer to the formal theory of argumentation [6]. Such a connection is certainly worthy of further exploration.

As formulated, PE deals with formulas of evidence depth 1, but it can be easily extended to formulas of arbitrary nested depth with straightforward extension of set-theoretical semantics. Such "nested" PE might be an interesting logic system to explore.

If Γ is formally inconsistent, then $\Gamma \vdash X$ for any X. It appears that such Γ can justify any proposition X: term $e = u_1 u_2 \ldots u_n$ is evidence for any X. However, under any probabilistic interpretation $*$ respecting $\boldsymbol{u}{:}\Gamma$, $e^* = \emptyset$ and hence yields no probabilistic evidence.

Acknowledgements. The author is very grateful to Melvin Fitting, Vladimir Krupski, Elena Nogina, Tudor Protopopescu, Çağıl Taşdemir, and participants of the Trends in Logic XV conference in Delft for inspiring discussions and helpful suggestions. Special thanks to Karen Kletter for editing and proofreading this text.

References

1. Adams, E.W.: A Primer of Probability Logic. CSLI Publications, Stanford (1998)
2. Adams, E.W., Levine, H.P.: On the uncertainties transmitted from premisses to conclusions in deductive inferences. Synthese **30**, 429–460 (1975)
3. Artemov, S.: Explicit provability and constructive semantics. Bull. Symbolic Log. **7**(1), 1–36 (2001)
4. Artemov, S.: The logic of justification. Rev. Symbolic Log. **1**(4), 477–513 (2008)
5. Clemen, R., Winkler, R.: Aggregating probability distributions. In: Advances in Decision Analysis: From Foundations to Applications, pp. 154–176 (2007)
6. Dung, P.M.: On the acceptability of arguments and its fundamental role in non-monotonic reasoning, logic programming and n-person games. Artif. Intell. **77**(2), 321–357 (1995)
7. Feller, W.: An Introduction to Probability Theory and its Applications. Wiley, New York (1968)
8. Fitting, M.: The logic of proofs, semantically. Ann. Pure Appl. Logic **132**(1), 1–25 (2005)
9. Halpern, J.: Reasoning About Uncertainty. MIT Press, Cambridge (2003)
10. Kolmogorov, A.: Grundbegriffe der Wahrscheinlichkeitrechnung. Ergebnisse Der Mathematik. Springer, Heidelberg (1993). Translated as: Foundations of Probability. Chelsea Publishing Company, New York (1950)
11. List, C.: The theory of judgment aggregation: an introductory review. Synthese **187**(1), 179–207 (2012)
12. Shiryaev, A.: Probability. Graduate Texts in Mathematics, vol. 95. Springer, New York (1996)
13. Suppes, P.: Probabilistic inference and the concept of total evidence. In: Hintikka, J., Suppes, P. (eds.) Aspects of inductive logic, pp. 49–65. Elsevier, Amsterdam (1966)
14. Whitman, P.: Free lattices. Ann. Math. **42**(1), 325–330 (1941)

Classical Logic with Mendler Induction
A Dual Calculus and Its Strong Normalization

Marco Devesas Campos$^{(\boxtimes)}$ and Marcelo Fiore

Computer Laboratory, University of Cambridge, Cambridge, UK
maf58@cam.ac.uk, marcelo.fiore@cl.cam.ac.uk

Abstract. We investigate (co-)induction in Classical Logic under the propositions-as-types paradigm, considering propositional, second-order, and (co-)inductive types. Specifically, we introduce an extension of the Dual Calculus with a Mendler-style (co-)iterator that remains strongly normalizing under head reduction. We prove this using a non-constructive realizability argument.

Keywords: Mendler induction · Classical logic · Curry-howard isomorphism · Dual Calculus · Realizability

1 Introduction

The Curry-Howard Isomorphism. The interplay between Logic and Computer Science has a long and rich history. In particular, the Curry-Howard isomorphism, the correspondence between types and theorems, and between typings and proofs, is a long established bridge through which results in one field can fruitfully migrate to the other. One such example, motivating of the research presented herein, is the use of typing systems based on Gentzen's sequent calculus LK [10]. At its core, LK is a calculus of the dual concepts of necessary assumptions and possible conclusions—which map neatly, on the Computer Science side, to required inputs (or computations) and possible outputs (or continuations).

Classical Calculi. The unconventional form of LK belies an extreme symmetry and regularity that make it more amenable to analysis than other systems that can be encoded in it. Indeed, Gentzen introduced LK as an intermediate step in his proof that Hilbert-style derivation systems and his own system of Natural Deduction, NK, were consistent. Curry-Howard descendants of LK are Curien and Herbelin's $\lambda\mu\tilde{\mu}$ [6] and Wadler's Dual Calculus [19]. As an example of the kind of analysis that can be done using sequents, these works focused on establishing syntactically the duality of the two most common evaluations strategies for

M.D. Campos—This work was supported by the United Kingdom's Engineering and Physical Sciences Research Council [grant number EP/J500380/1].
M. Fiore—Partially supported by ERC ECSYM.

© Springer International Publishing Switzerland 2016
S. Artemov and A. Nerode (Eds.): LFCS 2016, LNCS 9537, pp. 43–59, 2016.
DOI: 10.1007/978-3-319-27683-0_4

the lambda-calculus: call-by-name and call-by-value. While originally Classical calculi included only propositional types—i.e. conjunction, disjunction, negation, implication and subtraction (the dual connective of implication)—they were later extended with second-order types [13, 17], and also with *positive* (co-)inductive types [13]; the latter fundamentally depended on the map operation of the underlying type-schemes.

Mendler Induction. In continuing with this theme, we turn our attention here to a more general induction scheme due to Mendler [15]. Originally, this induction scheme was merely seen as an ingenious use of polymorphism that allowed induction to occur without direct use of mapping operations. However, it was later shown that with Mendler's iterator one could in fact induct on data-types of *arbitrary variance*—i.e. data-types whose induction variable may also appear negatively [14, 18]. Due to its generality, Mendler Induction has been applied in a number of different contexts, amongst which we find higher-order recursive types [1, 2] and automated theorem proving [12].

Classical Logic and *Mendler Induction.* Can one export Mendler Induction to non-functional settings without introducing unexpected side-effects? Specifically, can one extend Classical Logic with Mendler Induction without losing consistency? Note that Classical Logic has been shown to be quite misbehaved if not handled properly [11]; and certain forms of Mendler Induction have been shown to break strong normalization at higher-ranked types [2].

This paper answers both questions affirmatively. In summary, we:

– extend the second-order Dual Calculus with functional types—viz., with arrow and subtractive types (Sect. 2);
– prove its strong normalization (Sect. 3) via a realizability argument (a lattice-theoretic distillation of Parigot's proof for the Symmetric Lambda-calculus [3, 16]);
– recall the idea underlying Mendler Induction in the functional setting (Sect. 4);
– present our extension of the Dual Calculus with Mendler (co-)inductive types and argue why functional types are indispensable to its definition (Sect. 5); and
– extend the aforementioned realizability argument to give a non-constructive proof that the extension is also strongly normalizing (Sect. 6).

2   Second-Order Dual Calculus

The Base Calculus Our base formalism is Wadler's Dual Calculus [19]—often abbreviated DC. We begin by reviewing the original propositional version extended with second-order types [13] and subtractive types [5, 6]. Tables 1[1], 2, and 3 respectively summarize the syntax, the types and typing rules, and the reduction rules of the calculus.

---

[1] Unlike Wadler's presentation, we keep the standard practice of avoiding suffix operators; whilst lexical duality is lost, we think it improves readability.

Syntax. The sequent calculus LK is a calculus of multiple assumptions and conclusions, as witnessed by the action of the right and left derivation rules. Similarly, the two main components of DC are split into two kinds: *terms* (or *computations*) which, intuitively, produce values; and *co-terms* (or *continuations*), which consume them. However, whereas in the sequent calculus one can mix the different kinds of rules in any order, to keep the computational connection, the term and co-term formation rules are restricted in what phrases they expect—e.g. pairs should combine values, while projections pass the components of a pair to some other continuation. This distinction also forces the existence of two kinds of variables: variables for terms and co-variables for co-terms. We assume that they belong to some disjoint and countably infinite sets Var and $Covar$, respectively.

Table 1. Syntax of the second-order Dual Calculus.

Terms

$$t := x, y, \ldots \in Var \mid \underbrace{\langle t, t' \rangle \mid i_1\langle t \rangle \mid i_2\langle t \rangle \mid not\langle k \rangle \mid \lambda x.(t) \mid (t \,\#\, k) \mid a\langle t \rangle \mid e\langle t \rangle}_{\text{Introductions}} \mid \alpha.(c)$$

Co-terms

$$k := \alpha, \beta, \ldots \in Covar \mid \underbrace{[k, k'] \mid fst[k] \mid snd[k] \mid not[t] \mid (t \,@\, k) \mid \mu\alpha.(k) \mid a[k] \mid e[k]}_{\text{Eliminations}} \mid x.(c)$$

Cuts

$$c := t \bullet k$$

Cuts and Abstractions. The third and final kind of phrase in the Dual Calculus are *cuts*. Recall the famous dictum of Computer Science:

$$\text{Data-structures} + \text{Algorithms} = \text{Programs}.$$

In DC, where terms represent the creation of information and co-terms consume it, we find that cuts, the combination of a term with a continuation, are analogous to programs:

$$\text{Terms} + \text{Co-terms} = \text{Cuts};$$

they are the entities that are capable of being executed. Given a cut, one can consider the computation that would ensue if given data for a variable or co-variable. The calculus provides a mechanism to express such situations by means of *abstractions* $x.(c)$ and of *co-abstractions* $\alpha.(c)$ on any cut c. Abstractions are continuations—they expect values in order to proceed with some execution—and, dually, co-abstractions are computations.

Subtraction. One novelty of this paper is the central role given to subtractive types, $A - B$ [5]. Subtraction is the dual connective to implication; it is to continuations what implication is to terms: it allows one to abstract co-variables in co-terms—and thereby *compose continuations*. Given a continuation k where

Table 2. Typing for the second-order propositional Dual Calculus (with the structural rules omitted).

Types

$$T, A, B := X \mid A \wedge B \mid A \vee B \mid \neg A \mid A \to B \mid A - B \mid \forall X.T \mid \exists X.T$$

Identity

$$x : A \vdash x : A \mid \qquad \mid \alpha : A \dashv \alpha : A$$

Abstractions

$$\frac{\Gamma \vdash c \dashv \Delta, \alpha : A}{\Gamma \vdash \alpha.(c) : A \mid \Delta} \qquad \frac{x : A, \Gamma \vdash c \dashv \Delta}{\Gamma \mid x.(c) : A \dashv \Delta}$$

Cut

$$\frac{\Gamma \vdash t : A \mid \Delta \qquad \Gamma \mid k : A \dashv \Delta}{\Gamma \vdash t \bullet k \dashv \Delta}$$

Conjunction

$$\frac{\Gamma \vdash t : A \mid \Delta \qquad \Gamma \vdash t' : B \mid \Delta}{\Gamma \vdash \langle t, t' \rangle : A \wedge B \mid \Delta}$$

$$\frac{\Gamma \mid k : A \dashv \Delta}{\Gamma \mid fst[k] : A \wedge B \dashv \Delta} \qquad \frac{\Gamma \mid k : B \dashv \Delta}{\Gamma \mid snd[k] : A \wedge B \dashv \Delta}$$

Disjunction

$$\frac{\Gamma \vdash t : A \mid \Delta}{\Gamma \vdash i_1\langle t \rangle : A \vee B \mid \Delta} \qquad \frac{\Gamma \vdash t : B \mid \Delta}{\Gamma \vdash i_2\langle t \rangle : A \vee B \mid \Delta}$$

$$\frac{\Gamma \mid k : A \dashv \Delta \quad \Gamma \mid k' : B \dashv \Delta}{\Gamma \mid [k, k'] : A \vee B \dashv \Delta}$$

Negation

$$\frac{\Gamma \mid k : A \dashv \Delta}{\Gamma \vdash not\langle k \rangle : \neg A \mid \Delta} \qquad \frac{\Gamma \vdash t : A \mid \Delta}{\Gamma \mid not[t] : \neg A \dashv \Delta}$$

Implication

$$\frac{x : A, \Gamma \vdash t : B \mid \Delta}{\Gamma \vdash \lambda x.(t) : A \to B \mid \Delta} \qquad \frac{\Gamma \vdash t : A \mid \Delta \quad \Gamma \mid k : B \dashv \Delta}{\Gamma \mid (t @ k) : A \to B \dashv \Delta}$$

Subtraction

$$\frac{\Gamma \vdash t : A \mid \Delta \quad \Gamma \mid k : B \dashv \Delta}{\Gamma \vdash (t \# k) : A - B \mid \Delta} \qquad \frac{\Gamma \mid k : A \dashv \Delta, \alpha : B}{\Gamma \mid \mu\alpha.(k) : A - B \dashv \Delta}$$

Universal Quantification

$$\frac{\Gamma \vdash t : F(X) \mid \Delta}{\Gamma \vdash a\langle t \rangle : \forall X.F(X) \mid \Delta} \, (X \text{ not free in } \Gamma, \Delta) \qquad \frac{\Gamma \mid k : F(A) \dashv \Delta}{\Gamma \mid a[k] : \forall X.F(X) \dashv \Delta}$$

Existential Quantification

$$\frac{\Gamma \vdash t : F(A) \mid \Delta}{\Gamma \vdash e\langle t \rangle : \exists X.F(X) \mid \Delta} \qquad \frac{\Gamma \mid k : F(X) \dashv \Delta}{\Gamma \mid e[k] : \exists X.F(X) \dashv \Delta} \, (X \text{ not free in } \Gamma, \Delta)$$

a co-variable α might appear free, the subtractive abstraction (or catch, due to its connection with exception handling) is defined as $\mu\alpha.\,(k)$, the idea being that applying (read, cutting) a continuation k' and value t to it, *packed* together as $(t\,\#\,k')$, yields a cut of the form $t \bullet k\,[k'/\alpha]$.

Typing Judgments. We present the types and the typing rules in Table 2; we omit the structural rules here but they can be found in the aforementioned paper by Wadler [19]. We have three forms of typing judgments that go hand-in-hand with the three different types of phrases: $\Gamma \vdash t : A \mid \Delta$ for terms, $\Gamma \mid k : A \dashv \Delta$ for co-terms, and $\Gamma \vdash c \dashv \Delta$ for cuts. In all cases, the entailment symbols point to the phrase under judgment, and they appear in the same position as they would appear in the corresponding sequent of LK. *Typing contexts* Γ assign variables to their assumed types; dually, *typing co-contexts* Δ assign co-variables to their types. Tacitly, we assume that they always include the free (co-)variables in the phrase under consideration. Type-schemes $F(X)$ are types in which a distinguished type variable X may appear free; the instantiation of such a type-scheme to a particular type T is simply the substitution of the distinguished X by T and is denoted $F(T)$.

Example: Witness the Lack of Witness. We can apply the rules in Table 2 to bear proof of valid formulas in second-order Classical Logic. One such example at the second-order level is $\neg\forall X.T \to \exists X.\neg T$:

$$\mid not\,[a\,\langle\alpha.\,(e\,\langle not\,\langle\alpha\rangle\rangle \bullet \beta)\rangle)] : \neg\forall X.T \dashv \beta : \exists X.\neg T.$$

Note how the existential does not construct witnesses but simply diverts the flow of execution (by use of a co-abstraction).

Head Reduction. The final ingredient of the calculus is the set of (head) reduction rules (Table 3). They are non-deterministic—as a cut made of abstractions and co-abstractions can reduce by either one of the abstraction rules—and non-confluent. Confluence can be reestablished by prioritizing the reduction of one type of abstraction over the other; this gives rise to two confluent reduction disciplines that we term *abstraction prioritizing* and *co-abstraction prioritizing*. In any case, reduction of well-typed cuts yields well-typed cuts.[2]

Table 3. Head reduction for the second-order Dual Calculus.

$$\langle t,t'\rangle \bullet fst[k] \rightsquigarrow t \bullet k \qquad\qquad \langle t,t'\rangle \bullet snd[k] \rightsquigarrow t' \bullet k$$

$$i_1\langle t\rangle \bullet [k,k'] \rightsquigarrow t \bullet k \qquad\qquad i_2\langle t\rangle \bullet [k,k'] \rightsquigarrow t \bullet k'$$

$$not\langle k\rangle \bullet not[t] \rightsquigarrow t \bullet k$$

$$\lambda x.(t) \bullet (t'\,@\,k) \rightsquigarrow t[t'/x] \bullet k \qquad\qquad (t\,\#\,k) \bullet \mu\alpha.(k') \rightsquigarrow t \bullet k'[k/\alpha]$$

$$a\langle t\rangle \bullet a[k] \rightsquigarrow t \bullet k \qquad\qquad e\langle t\rangle \bullet e[k] \rightsquigarrow t \bullet k$$

$$\alpha.(c) \bullet k \rightsquigarrow c[k/\alpha] \qquad\qquad t \bullet x.(c) \rightsquigarrow c[t/x]$$

---

[2] As we are not looking at call-by-name and call-by-value we do not use the same reduction rule for implication as Wadler [19]; the rule here is due to Curien and Herbelin [6].

3 Strong Normalization of the Second-Order Dual Calculus

The Proof of Strong Normalization. Having surveyed the syntax, types and reduction rules of DC, we will now give a proof of its strong normalization—i.e., that all reduction sequences of well-typed cuts terminate in a finite number of steps—for the given *non-deterministic reduction* rules. It will follow, then, that the deterministic sub-calculi, where one prioritizes the reduction of one kind abstraction over the other, are also strongly normalizing.

The proof rests on a realizability interpretation for terms. Similar approaches for the propositional fragment can be found in the literature [9,17]; however, the biggest influence on our proof was the one by Parigot for the second-order extension of the Symmetric Lambda-Calculus [16]. Our main innovation is the identification of a complete lattice structure with fix-points suitable for the interpretation of (co-)inductive types. We will, in fact, need to consider two lattices: $\mathcal{OP}$ and $\mathcal{ONP}$. In $\mathcal{OP}$, we find, intuitively, all the terms/co-terms of types. In the lattice $\mathcal{ONP}$ we find *only* terms/co-terms that are introductions/eliminations; these correspond, again intuitively, to values/co-values of types. Between these two classes we have type-directed *actions* from $\mathcal{OP}$ to $\mathcal{ONP}$, and a completion operator $\amalg$ from $\mathcal{ONP}$ to $\mathcal{OP}$ that generates all terms/co-terms compatible with the given values/co-values.

$$\mathcal{OP} \underset{\amalg}{\overset{\wedge,\vee,\neg,\ldots}{\rightleftarrows}} \mathcal{ONP} \tag{1}$$

In this setting, we give (two) mutually induced interpretations for types (one in $\mathcal{ONP}$ and the other in $\mathcal{OP}$, Table 4) and establish an adequacy result (Theorem 4) from which strong normalization follows as a corollary. The development is outlined next.

Sets of Syntax. The set of all terms formed using the rules in Table 1 will be denoted by $\mathcal{T}$; similarly, co-terms will be $\mathcal{K}$ and cuts $\mathcal{C}$. We will also need three special subsets of those sets: $\mathcal{IT}$ for those terms whose outer syntactic form is an introduction; $\mathcal{EK}$, dually, for the co-terms whose outer syntactic form is an eliminator; and $\mathcal{SN}$ for the set of strongly-normalizing cuts.[3]

Syntactic Actions on Sets. The syntactic constructors give rise to obvious *actions* on sets of terms, co-terms, and cuts; e.g.

$$- \bullet - : \mathcal{P}(\mathcal{T}) \times \mathcal{P}(\mathcal{K}) \to \mathcal{P}(\mathcal{C}), \quad T \bullet K = \{t \bullet k \mid t \in T, k \in K\}\,.$$

By abuse of notation these operators shall be denoted as their syntactic counterparts; they are basic to our realizability interpretation.

[3] A non-terminating, non-well-typed cut: $\alpha.\,(\mathit{not}\,\langle\alpha\rangle \bullet \alpha) \bullet \mathit{not}\,[\alpha.\,(\mathit{not}\,\langle\alpha\rangle \bullet \alpha)]$.

Restriction under Substitution. The substitution operation lifts point-wise to the level of sets as a monotone function $(-)\,[(=)/\phi] : \mathcal{P}(U) \times \mathcal{P}(V) \to \mathcal{P}(U)$ for V the set of terms (resp. co-terms), ϕ a variable (resp. co-variable), and U either the set of terms, co-terms, or cuts. We will make extensive use of the right adjoint $(-)\big|_\phi^Q$ to $(-)\,[Q/\phi]$ characterized by

$$R\,[Q/\phi] \subseteq P \quad \textit{iff} \quad R \subseteq P\big|_\phi^Q\,,$$

and that we term the *restriction under substitution*. With it we can, e.g., express the set of cuts that are strongly normalizing when free occurrences of a co-variable α are substituted by co-terms from a set K:

$$\mathcal{SN}\big|_\alpha^K = \{\, c \in \mathcal{C} \mid \text{for all } k \in K\,.\,c\,[k/\alpha] \in \mathcal{SN}\,\}\,.$$

Orthogonal Pairs. Whenever a term t and a co-term k form a strongly normalizing cut $t \bullet k$, we say that they are *orthogonal*. Similarly, for sets T of terms and K of co-terms, we say that they are orthogonal if $T \bullet K \subseteq \mathcal{SN}$. We call pairs of such sets *orthogonal pairs*, and the set of all such pairs $\mathcal{OP}$. For any orthogonal pair $P \in \mathcal{OP}$, its set of terms is denoted $(P)^\mathsf{T}$ and its set of co-terms by $(P)^\mathsf{K}$. Note that no type restriction is in play in the definition of orthogonal pairs; e.g. a cut of an injection with a projection is by definition orthogonal as no reduction rule applies.

Lattices. Recall that a lattice S is a partially ordered set such that any *non-empty* finite subset $S' \subseteq S$ has a least upper bound (or join, or lub) and a greatest lower-bound (or meet, or glb), respectively denoted by $\bigvee S'$ and $\bigwedge S'$. If the bounds exist for any subset of S one says that the lattice is *complete*. In particular, this entails the existence of a bottom and a top element for the partial order. The powerset $\mathcal{P}(S)$ of a set S is a complete lattice under inclusion; the dual L^{op} of a (complete) lattice L (where we take the opposite order and invert the bounds) is a (complete) lattice, as is the point-wise product of any two (complete) lattices.

Proposition 1 (Lattice Structure of $\mathcal{OP}$). *The set of orthogonal pairs is a sub-lattice of $\mathcal{P}(\mathcal{T}) \times \mathcal{P}(\mathcal{K})^{op}$. Explicitly, for $P, Q \in \mathcal{OP}$,*

$$P \leq Q \quad \textit{iff} \quad (P)^\mathsf{T} \subseteq (Q)^\mathsf{T} \text{ and } (P)^\mathsf{K} \supseteq (Q)^\mathsf{K};$$

the join and meet of arbitrary non-empty sets $S \subseteq \mathcal{OP}$ are

$$\bigvee S \equiv \left(\bigcup_{P \in S} (P)^\mathsf{T},\, \bigcap_{P \in S} (P)^\mathsf{K} \right) \qquad \bigwedge S \equiv \left(\bigcap_{P \in S} (P)^\mathsf{T},\, \bigcup_{P \in S} (P)^\mathsf{K} \right).$$

Moreover, it is complete with empty join and meet given by $\bot \equiv (\emptyset, \mathcal{K})$ and $\top \equiv (\mathcal{T}, \emptyset)$.

Orthogonal Normal Pairs. The other lattice we are interested in is the lattice $\mathcal{ONP}$ of what we call *orthogonal normal pairs*. These are orthogonal pairs which are made out at the outermost level by introductions and eliminators. Logically speaking, they correspond to those proofs whose last derivation is a left or right operational rule. Computationally, they correspond to the narrowest possible interpretations of values and co-values. Orthogonal normal pairs inherit the lattice structure of $\mathcal{OP}$ but for the empty lub and glb which become $\bot \equiv (\emptyset, \mathcal{EK})$ and $\top \equiv (\mathcal{IT}, \emptyset)$.

Type Actions. Pairing together the actions of the introductions and eliminations of a given type allows us to construct elements of $\mathcal{ONP}$ whenever we apply them to orthogonal sets—in particular, then, when these sets are the components of elements of $\mathcal{OP}$—as witnessed by the following proposition.

Proposition 2. *For $P, Q \in \mathcal{OP}$ and $S \subseteq \mathcal{OP}$, the following definitions determine elements of $\mathcal{ONP}$:*

$$P \wedge Q = \left(\left\langle (P)^\top, (Q)^\top \right\rangle, \mathit{fst} \left[(P)^\mathcal{K} \right] \cup \mathit{snd} \left[(Q)^\mathcal{K} \right] \right)$$

$$P \vee Q = \left(i_1 \left\langle (P)^\top \right\rangle \cup i_2 \left\langle (Q)^\top \right\rangle, \left[(P)^\mathcal{K}, (Q)^\mathcal{K} \right] \right)$$

$$\neg P = \left(\mathit{not} \left\langle (P)^\mathcal{K} \right\rangle, \mathit{not} \left[(P)^\top \right] \right)$$

$$P \rightarrow Q = \bigvee_{x \in \mathit{Var}} \left(\lambda x. ((Q)^\top |_x^{(P)^\top}), \left((P)^\top @ (Q)^\mathcal{K} \right) \right)$$

$$P - Q = \bigwedge_{\alpha \in \mathit{Covar}} \left(\left((P)^\top \# (Q)^\mathcal{K} \right), \mu\alpha. \left((P)^\mathcal{K} |_\alpha^{(Q)^\mathcal{K}} \right) \right)$$

$$\forall S = \bigwedge_{P \in S} \left(a \left\langle (P)^\top \right\rangle, a \left[(P)^\mathcal{K} \right] \right) \qquad \exists S = \bigvee_{P \in S} \left(e \left\langle (P)^\top \right\rangle, e \left[(P)^\mathcal{K} \right] \right)$$

Orthogonal Completion. Now that we have interpretations for the actions that construct values/co-values of a type in $\mathcal{ONP}$, we need to go the other way (cf. Diagram 1, above) to $\mathcal{OP}$, so that we also include (co-)variables and (co-)abstractions in our interpretations. So, for orthogonal sets of values T and of co-values K, the term and co-term completions of T and K are respectively defined as:

$$[T](L) = \mathit{Var} \cup T \cup \bigcup_{\alpha \in \mathit{Covar}} \alpha. \left(\mathcal{SN} |_\alpha^L \right), \quad [K](U) = \mathit{Covar} \cup K \cup \bigcup_{x \in \mathit{Var}} x. \left(\mathcal{SN} |_x^U \right).$$

Due to the non-determinism associated with the reduction of (co-)abstractions, we need guarantee that all added (co-)abstractions are compatible not only with the starting set of values, but also with any (co-)abstractions that have been added in the process—and vice-versa. In other words, we need to iterate this process by taking the least fix-point:

$$(\underline{T \| K}) = \left(\mathit{lfp} \left([T] \circ [K] \right), [K] \left(\mathit{lfp} \left([T] \circ [K] \right) \right) \right).$$

(In fact, as has been remarked elsewhere [3, 16], all one needs is *a* fix-point.)

Theorem 3. *Let $N \in \mathcal{ONP}$ be an orthogonal normal pair; its structural completion $\amalg N$ is an orthogonal pair:*

$$\amalg N = (\underline{(N)^T \parallel (N)^K}) \in \mathcal{OP}.$$

Interpretations. Given a type T and a (suitable) mapping γ from its free type variables, ftv (T), to $\mathcal{ONP}$—called the interpretation context—we define (Table 4) two interpretations, as orthogonal pairs and as orthogonal normal pairs, by mutual induction on the structure of T. They both satisfy the weakening and substitution properties. The extension of an interpretation context γ where a type-variable X is mapped to $N \in \mathcal{ONP}$ is denoted by $\gamma[X \mapsto N]$.

Theorem 4 (Adequacy). *Let t, k and c stand for terms, co-terms and cuts of the Dual Calculus. For any typing context Γ and co-context Δ, and type T such that*

$$\Gamma \vdash t : T \mid \Delta, \qquad \Gamma \mid k : T \dashv \Delta, \qquad \Gamma \vdash c \dashv \Delta,$$

and for any suitable interpretation context γ for Γ, Δ and T, and any substitution σ satisfying

$$(x : A) \in \Gamma \implies \sigma(x) \in ((A)(\gamma))^T \quad and \quad (\alpha : A) \in \Delta \implies \sigma(\alpha) \in ((A)(\gamma))^K,$$

we have that

$$t[\sigma] \in ((T)(\gamma))^T, \quad k[\sigma] \in ((T)(\gamma))^K, \quad c[\sigma] \in \mathcal{SN}.$$

Corollary 5 (Strong Normalization). *Every well-typed cut of DC is strongly normalizing.*

Table 4. Interpretations of the second-order Dual Calculus in $\mathcal{ONP}$ and $\mathcal{OP}$.

$[\![T]\!](\gamma) : \mathcal{ONP}$	$(T)(\gamma) : \mathcal{OP}$
$[\![X]\!](\gamma) = \gamma(X)$	$(T)(\gamma) = \amalg([\![T]\!](\gamma))$
$[\![A \wedge B]\!](\gamma) = (A)(\gamma) \wedge (B)(\gamma)$	
$[\![A \vee B]\!](\gamma) = (A)(\gamma) \vee (B)(\gamma)$	
$[\![\neg A]\!](\gamma) = \neg(A)(\gamma)$	
$[\![A \to B]\!](\gamma) = (A)(\gamma) \to (B)(\gamma)$	
$[\![A - B]\!](\gamma) = (A)(\gamma) - (B)(\gamma)$	
$[\![\forall X . A]\!](\gamma) = \forall\{(A)(\gamma[X \mapsto N]) \mid N \in \mathcal{ONP}\}$	
$[\![\exists X . A]\!](\gamma) = \exists\{(A)(\gamma[X \mapsto N]) \mid N \in \mathcal{ONP}\}$	

4 Mendler Induction

Having covered the first theme of the paper, Classical Logical in its Dual Calculus guise, let us focus in this section on the second theme we are exploring: Mendler Induction. As the concept may be rather foreign, it is best to review it informally in the familiar functional setting.

Inductive Definitions. Roughly speaking, an inductive definition of a function is one in which the function being defined can be used in its own definition provided that it is applied only to values of strictly smaller character than the input. The fix-point operator

$$fix : \big((\mu X.F(X) \to A) \to \mu X.F(X) \to A\big) \to \mu X.F(X) \to A$$
$$fix\, f\, x = f\, (fix\, f)\, x$$

associated to the inductive type $\mu X.F(X)$ arising from a type scheme $F(X)$, clearly violates induction, and indeed breaks strong normalization: one can feed it the identity function to yield a looping term. One may naively attempt to tame this behavior by considering the following modified fix-point operator

$$fix' : \big((\mu X.F(X) \to A) \to F\big(\mu X.F(X)\big) \to A\big) \to \mu X.F(X) \to A$$
$$fix'\, f\, (in\, x') = f\, (fix'\, f)\, x'$$

in which, for the introduction $in : F\big(\mu X.F(X)\big) \to \mu X.F(X)$, one may regard x' as being of strictly smaller character than $in(x')$. Of course, this is still unsatisfactory as, for instance, we have the looping term $fix'\, (\lambda f.\ f \circ in)$. The problem here is that the functional $\lambda f.\ f \circ in : (\mu X.F(X) \to A) \to F\big(\mu X.F(X)\big) \to A$ of which we are taking the fix-point takes advantage of the concrete type $F\big(\mu X.F(X)\big)$ of x' used in the recursive call.

Mendler Induction. The ingenuity of Mendler Induction is to ban such perversities by restricting the type of the functionals that the iterator can be applied to: these should not rely on the inductive type but rather be abstract; in other words, be represented by a fresh type variable X as in the typing below[4]:

$$mitr : \big((X \to A) \to F(X) \to A\big) \to \mu X.F(X) \to A$$
$$mitr\, f\, (min\, x) = f\, (mitr\, f)\, x$$

for min the introduction $F\big(\mu X.F(X)\big) \to \mu X.F(X)$.

    Note that if the type scheme $F(X)$ is endowed with a polymorphic mapping operation $map_F : (A \to B) \to F(A) \to F(B)$, every term $a : F(A) \to A$ has as associated catamorphism $cata(a) \equiv mitr\,\big(\lambda f.\ a \circ (map_F f)\big) : \mu X.F(X) \to A$. In particular, one has $cata(map_F\, min) : \mu X.F(X) \to F\big(\mu X.F(X)\big)$.

5   Dual Calculus with Mendler Induction

Mendler Induction. We shall now formalize Mendler Induction in the Classical Calculus of Sect. 2. Additionally, we shall also introduce its dual, Mendler co-Induction. This requires: type constructors; syntactic operations corresponding

---

[4] We note that the original presentation of this inductive operator [15] was in System F and, accordingly, the operator considered instead functionals of type $\forall X.(X \to A) \to F(X) \to A$. Cognoscenti will recognize that this type is the type-theoretic Yoneda reformulation $\forall X.(X \to A) \to T(X)$ of $T(A) = F(A) \to A$ for $T(X) = F(X) \to A$.

to the introductions and eliminations, and their typing rules; and reduction rules. These are summarized in Table 5. First, we take a type scheme $F(X)$ and represent its inductive type by $\mu X.F(X)$—dually, we represent the associated co-inductive type by $\nu X.F(X)$.

Syntax. As usual, the inductive introduction, $min\langle-\rangle$, witnesses that the values of the unfolding of the inductive type $F(\mu X.F(X))$ are injected in the inductive type $\mu X.F(X)$. It is in performing induction that we consume values of inductive type and, hence, the *induction operator* (or *iterator*, or *inductor*), $mitr_{\rho,\alpha}[k,l]$ corresponds to an elimination. It is comprised of an iteration step k, an output continuation l, and two distinct induction co-variables, ρ and α. We postpone

Table 5. Extension of the second-order Dual Calculus with Mendler Induction

Types

$$T := \ldots \mid \mu X.F(X) \mid \nu X.F(X)$$

Syntax

$$t := \ldots \mid \underbrace{\ldots \mid min\langle t\rangle \mid mcoitr_{r,x}\langle t,t'\rangle}_{\text{Introductions}} \mid \ldots$$

$$k := \ldots \mid \underbrace{\ldots \mid mitr_{\rho,\alpha}[k,k'] \mid mout[k]}_{\text{Eliminations}} \mid \ldots$$

Reduction

$$min\langle t\rangle \bullet mitr_{\rho,\alpha}[k,l] \rightsquigarrow t \bullet k[\mu\alpha.(mitr_{\rho,\alpha}[k,\alpha])/\rho][l/\alpha]$$
$$mcoitr_{r,x}\langle t,u\rangle \bullet mout[k] \rightsquigarrow t[\lambda x.(mcoitr_{r,x}\langle t,x\rangle)/r][u/x] \bullet k$$

Typing rules

$$\frac{\Gamma \vdash t : F(\mu X.F(X)) \mid \Delta}{\Gamma \vdash min\langle t\rangle : \mu X.F(X) \mid \Delta}$$

$$\frac{\Gamma \mid k : F(X) \dashv \Delta, \rho : X - A, \alpha : A \qquad \Gamma \mid l : A \dashv \Delta}{\Gamma \mid mitr_{\rho,\alpha}[k,l] : \mu X.F(X) \dashv \Delta}$$

$$(X \text{ not free in } \Gamma, \Delta, A)$$

$$\frac{x : A, r : A \to X, \Gamma \vdash t : F(X) \mid \Delta \qquad \Gamma \vdash u : A \mid \Delta}{\Gamma \vdash mcoitr_{r,x}\langle t,u\rangle : \nu X.F(X) \mid \Delta}$$

$$(X \text{ not free in } \Gamma, \Delta, A)$$

$$\frac{\Gamma \mid k : F(\nu X.F(X)) \dashv \Delta}{\Gamma \mid mout[k] : \nu X.F(X) \dashv \Delta}$$

the explanation of their significance for the section on reduction below, but note now that the iterator binds ρ and α in the iteration continuation but not in the output continuation; thus, e.g.,

$$\left(mitr_{\rho,\alpha}\,[k,l]\right)[k'/\rho]\,[l'/\alpha] = mitr_{\rho,\alpha}\,[k,l\,[k'/\rho]\,[l'/\alpha]].$$

The co-inductive operators, $mcoitr_{r,x}\,\langle t,u\rangle$ and $mout\,[k]$, are obtained via dualization. In particular, the co-inductive eliminator, $mout\,[k]$, witnesses that the co-values k of type $F(\nu X.F(X))$ translate into co-values of $\nu X.F(X)$.

Reduction. To reduce an inductive cut $min\,\langle t\rangle \bullet mitr_{\rho,\alpha}\,[k,l]$, we start by passing the unwrapped inductive value t to the induction step k. However, in the spirit of Mendler Induction, the induction step must be instantiated with the induction itself *and*, because we are in a Classical calculus, the output continuation— this is where the parameter co-variables come into play. The first co-variable, ρ, receives the induction; the induction step may call this co-variable (using a cut) arbitrarily and it must also be able to capture the output of those calls— in other words, it needs to *compose this continuation* with other continuations; therefore one needs to pass $\mu\alpha.\,(mitr_{\rho,\alpha}\,[k,\alpha])$, the induction with the output continuation (subtractively) abstracted. The other co-variable, α, represents in k the output of the induction—which for a call $mitr_{\rho,\alpha}\,[k,l]$ is l[5]. For co-induction, we dualize—in particular, the co-inductive call expects the lambda-abstraction of the co-inductive step.

Typing. Lastly, we have the typing rules that force induction to be well-founded. Recall that this was achieved in the functional setting by forcing the inductive step to take an argument of arbitrary instances of the type scheme $F(X)$. Here we do the same. In typing $mitr_{\rho,\alpha}\,[k,l]$ for $\mu X.F(X)$ we require k to have type $F(X)$ where X is a variable that appears nowhere in the derivation except in the (input) type of the co-variable ρ.

Example: Naturals. Let us look at a concrete example: natural numbers under the abstraction prioritizing strategy. We posit a distinguished type variable $\mathcal{B}$, and from it construct the type $1 \equiv \mathcal{B} \vee \neg\mathcal{B}$, which is inhabited by the witness of the law of the excluded middle, $* \equiv \alpha.\,(i_2\,\langle not\,\langle x.\,(i_1\,\langle x\rangle \bullet \alpha)\rangle\rangle \bullet \alpha)$. The base type scheme for the naturals is $F(X) \equiv 1 \vee X$, and the naturals are then defined as $\mathcal{N} \equiv \mu X.F(X)$. Examples of this type are:

$$\text{zero} \equiv min\,\langle i_1\,\langle *\rangle\rangle, \quad \text{one} \equiv min\,\langle i_2\,\langle\text{zero}\rangle\rangle, \text{ and} \quad \text{two} \equiv min\,\langle i_2\,\langle\text{one}\rangle\rangle.$$

For any continuation k on $\mathcal{N}$, the successor "function" is defined as the following continuation for $\mathcal{N}$

$$succk^k \equiv x.\,(min\,\langle i_2\,\langle x\rangle\rangle \bullet k) \qquad (x \notin \text{fv}\,(k)).$$

---

[5] One may wonder if the output continuation is strictly necessary. As outputs appear on the right of sequents, and the induction is already a left-rule, the only possible alternative would be to add a co-variable to represent it. However, under this rule the system would no longer be closed under substitution [13].

Example: Addition The above primitives are all we need to define addition of these naturals. The inductive step "add m to" is

$$Step^m_{\rho,\alpha} \equiv \left[x.\,(m \bullet \alpha)\,,x.\big(\,(x \# succk^\alpha) \bullet \rho\big)\right].$$

Theorem 6. *Let n and m stand for the encoding of two natural numbers and the encoding of their sum be (by abuse of notation) $n+m$. Under the abstraction prioritizing reduction rule,*

$$n \bullet mitr_{\rho,\alpha}\left[Step^m_{\rho,\alpha}, l\right] \leadsto^* (n+m) \bullet l.$$

6   Strong Normalization for Mendler Induction

We now come to the main contribution of the paper: the extension of the Orthogonal Pairs realizability interpretation of the second-order Dual Calculus (Sect. 3) to Mendler Induction, which establishes that the extension is strongly normalizing.

Lattice Structure. The extension begins with the reformulation of the sets $\mathcal{SN}$, $\mathcal{T}$, $\mathcal{K}$, $\mathcal{C}$, $\mathcal{IT}$, and $\mathcal{EK}$ so that they accommodate the (co-)inductive operators. Modulo these changes, the definitions of $\mathcal{OP}$ and $\mathcal{ONP}$ remain the same; so do the actions for propositional and second order types, and the orthogonal completion, $\underline{\underline{\text{ll}}}$. All that remains, then, is to give suitable definitions for the (co-)inductive actions and the interpretations of (co-)inductive types.

Inductive Restrictions. The reduction rule for Mendler Induction is unlike any other of the calculus. When performing an inductive step for $mitr_{\rho,\alpha}[k,l]$, the bound variable ρ will be only substituted by one specific term: $\mu\alpha.\,(mitr_{\rho,\alpha}[k,\alpha])$. One needs a different kind of restriction to encode this invariant: take K and L to be sets of co-terms (intuitively, where the inductive step and output continuation live) and define the inductive restriction by

$$K/\!\!/^\rho_\alpha L \equiv \{\,k \in \mathcal{K} \mid \text{for all } l \in L,\ k\left[\mu\alpha.\,(mitr_{\rho,\alpha}[k,\alpha])\,/\rho\right][l/\alpha] \in K\,\};$$

and also for co-induction, for sets of terms T and U:

$$T/\!\!/^r_x U \equiv \{\,t \in \mathcal{T} \mid \text{for all } u \in U,\ t\left[\lambda x.(mcoitr_{r,x}\langle t,x\rangle)/r\right][u/x] \in T\,\}.$$

Mendler Pairing. Combining the inductive restriction with the inductive introduction/elimination set operations, we can easily create orthogonal normal pairs—much as we did for the propositional actions—from two given orthogonal pairs: one intuitively standing for the interpretation of $F(\mu F.F(X))$ and the other for the output type. However, the interpretation of the inductive type should not depend on a specific choice of output type but should accept *all* instantiations of output, as well as *all* possible induction co-variables; model-wise this corresponds to taking a meet over all possible choices for the parameters:

$$\text{MuP}(P) = \bigwedge_{\substack{Q \in \mathcal{OP} \\ \rho \neq \alpha \in Covar}} \left(min\left\langle (P)^\mathsf{T}\right\rangle, mitr_{\rho,\alpha}\left[(P)^\mathsf{K}/\!\!/^\rho_\alpha (Q)^\mathsf{K}, (Q)^\mathsf{K}\right]\right) \in \mathcal{ONP};$$

and similarly for its dual, NuP:

$$\mathrm{NuP}(P) = \bigvee_{\substack{Q \in \mathcal{OP} \\ r \neq x \in Var}} \left(mcoitr_{r,x} \left\langle (P)^\mathsf{T} \big/_x^r (Q)^\mathsf{T}, (Q)^\mathsf{T} \right\rangle mout \left[(P)^\mathsf{K} \right] \right) \in \mathcal{ONP}.$$

Monotonization. The typing constraints on Mendler Induction correspond—model-wise—to a monotonization step. This turns out to be what we need to guarantee that an inductive type can be modeled by a least fix-point; without this step, the interpretation of a type scheme would be a function on lattices that would not necessarily be monotone. There are two possible universal ways to induce monotone endofunctions from a given endofunction f on a lattice: the first one, $\lceil f \rceil$, we call the monotone extension and use it for inductive types, the other one, the monotone restriction $\lfloor f \rfloor$, will be useful for co-inductive types. Their definitions[6] are:

$$\lceil f \rceil\, x \equiv \bigvee_{y \leq x} fy \qquad \text{and} \qquad \lfloor f \rfloor\, x \equiv \bigwedge_{x \leq y} fy.$$

They are, respectively, the least monotone function above and the greatest monotone function below f. Necessarily, by Tarski's fix-point theorem, they both have least and greatest fix-points; in particular we have $lfp\,(\lceil f \rceil)$ and $gfp\,(\lfloor f \rfloor)$.

Inductive Actions. Combining the above ingredients, one can define the actions corresponding to inductive and to co-inductive types. They are parametrized by functions $f : \mathcal{ONP} \to \mathcal{OP}$,

$$\mu f \equiv lfp\,(\lceil \mathrm{MuP} \circ f \rceil) \in \mathcal{ONP} \quad \text{and} \quad \nu f \equiv gfp\,(\lfloor \mathrm{NuP} \circ f \rfloor) \in \mathcal{ONP}.$$

Interpretations For (co-)inductive types associated to a type-scheme $F(X)$ and mappings $\rho : \mathrm{ftv}\,(\mu X.F(X)) \to \mathcal{ONP}$ (the context) we set

$$[\![\mu X.F(X)]\!]\,(\gamma) = \mu(\![F(X)]\!)(\gamma\,[X \mapsto -]), \quad [\![\nu X.F(X)]\!]\,(\gamma) = \nu(\![F(X)]\!)(\gamma\,[X \mapsto -]);$$

while their *orthogonal interpretation* is as before. These interpretations also satisfy the weakening and substitution properties.

Classically Reasoning about Mendler Induction. Mendler's original proof of strong normalization for his induction principle in a functional setting was already classical [15]. For us, this issue centers around the co-term component of the interpretation of inductive types (and, dually, the term component of co-inductive types). Roughly, the induction hypothesis of the adequacy theorem states that for any $N \in \mathcal{ONP}$, $m \in ((\![X - A]\!)(\gamma\,[X \mapsto N]))^\mathsf{K}$, $l \in ((\![A]\!)(\gamma))^\mathsf{K}$, and realizability substitution σ we have

$$k\,[\sigma]\,[m/\rho]\,[l/\alpha] \in ((\![F(X)]\!)(\gamma\,[X \mapsto N]))^\mathsf{K}, \tag{2}$$

---

[6] Cognoscenti will recognize that they are point-wise Kan extensions.

and if we were to prove that $mitr_{\rho,\alpha}[k[\sigma], l[\sigma]] \in ([\![\mu X.F(X)]\!](\gamma))^K$ just by the fix-point property of the interpretation, we would need to have

$$(k[\sigma])[\mu\alpha.(mitr_{\rho,\alpha}[k[\sigma],\alpha])/\rho][l/\alpha] \in (\langle\!| F(X)|\!\rangle(\gamma[X \mapsto [\![\mu X.F(X)]\!](\gamma)]))^K$$

for arbitrary $l \in (\langle\!| A|\!\rangle(\gamma))^K$. Instantiating Formula 2 to the case when N is the interpretation of our fix-point, $[\![\mu X.F(X)]\!](\gamma)$, we see that in order to prove that $mitr_{\rho,\alpha}[k[\sigma], l[\sigma]] \in ([\![\mu X.F(X)]\!](\gamma))^K$ we would need to prove that for any $l' \in (\langle\!| A|\!\rangle(\gamma))^K$ we have that $mitr_{\rho,\alpha}[k[\sigma], l'] \in (\langle\!|\mu X.F(X)|\!\rangle(\gamma))^K$—a circularity!

For ω-complete posets there is an alternative characterization of the least fix-point of a *continuous* function as the least upper bound of a countable chain. The completion operation $\bigsqcup$ used in the definition of the $\mathcal{OP}$ interpretation is not continuous. However, classically, the least fix-point of any monotone function f on a complete lattice lies in the transfinite chain [7]

$$d_{\alpha+1} = f(d_\alpha) \qquad \text{and} \qquad d_\lambda = \bigvee_{\alpha<\lambda} d_\alpha \text{ (for limit } \lambda\text{)}$$

(and dually for co-induction).

A set (or property) $\mathbf{P} \subseteq \mathcal{ONP}$ is said to be *admissible iff* (*i*) preserves lubs: $S \subseteq \mathbf{P} \implies \mathbf{P}(\bigvee S)$; and (*ii*) is downward closed: $a \leq b$ and $\mathbf{P}(b) \implies \mathbf{P}(a)$.

Theorem 7 (Scott Induction for Monotone Extensions of Endofunctions). *Let $f : L \to L$ be an endofunction (not necessarily a homomorphism) and $\mathbf{P}$ be an admissible property on a complete lattice L. If f preserves property $\mathbf{P}$, i.e. $\mathbf{P}(a) \implies \mathbf{P}(fa)$, then $\mathbf{P}$ holds for the least fix-point of its monotone extension, i.e. $\mathbf{P}(lfp(\lceil f \rceil))$.*

With this proof principle and its dual one shows that the interpretation of DC with Mendler (co-)induction via realizability as orthogonal pairs satisfies the adequacy theorem (Theorem 4), and obtains the following result as a corollary.

Theorem 8 (Strong Normalization). *Every well-typed cut of the Dual Calculus with Mendler Induction is strongly normalizing.*

7 Concluding Remarks

We have investigated Classical Logic with Mendler Induction, presenting a Classical calculus with very general (co-)inductive types. Our work borrows from and generalizes systems based on Gentzen's *LK* under the Curry-Howard correspondence. Despite its generality, and as outlined by means of a realizability interpretation, our Dual Calculus with Mendler Induction is well-behaved in that its well-typed cuts are guaranteed to terminate. We expect—but have yet to fully confirm—that other models fit within our framework for interpreting Mendler Induction; our prime example is based on inflationary fix-points like those used in complexity theory [8] and which also apply to non-monotone functionals.

It is known that LK-based calculi can encode various other calculi [6,19]. Our calculus supports map operations for all positive (co-)inductive types. In an extended version of the paper, we expect to use these to encode Kimura and Tatsuta's extension of the Dual Calculus with positive (co-)inductive types [13].

One avenue of research that remains unexplored is how one may extract proofs from within our system—in previous work, Berardi, et al. [4] showed how, embracing the non-determinism of reduction inherent in the Symmetric Lambda-calculus (and also present in DC), one could express proof witnesses that behave like processes for a logic based on Peano arithmetic. A further direction would be to direct these investigations into the realm of linear logic, where the connection with processes may be more salient.

Acknowledgments. Thanks to Anuj Dawar, Tim Griffin, Ohad Kammar, Andy Pitts, and the anonymous referees for their comments and suggestions.

References

1. Abel, A., Matthes, R., Uustalu, T.: Iteration and coiteration schemes for higher-order and nested datatypes. Theor. Comput. Sci. **333**(1), 3–66 (2005)
2. Ahn, K.Y., Sheard, T.: A hierarchy of Mendler style recursion combinators: Taming inductive datatypes with negative occurrences. In: Proceedings of the 16th ACM SIGPLAN International Conference on Functional Programming, ICFP 2011, pp. 234–246. ACM, New York (2011)
3. Barbanera, F., Berardi, S.: A symmetric lambda calculus for classical program extraction. Inf. Comput. **125**(2), 103–117 (1996)
4. Barbanera, F., Berardi, S., Schivalocchi, M.: "Classical" programming-with-proofs in λ_{PA}^{Sym} : An analysis of non-confluence. In: Abadi, M., Ito, T. (eds.) Theoretical Aspects of Computer Software. Lecture Notes in Computer Science, vol. 1281, pp. 365–390. Springer, Berlin Heidelberg (1997)
5. Crolard, T.: A formulae-as-types interpretation of subtractive logic. J. Log. Comput. **14**(4), 529–570 (2004)
6. Curien, P.L., Herbelin, H.: The duality of computation. In: Proceedings of the Fifth ACM SIGPLAN International Conference on Functional Programming ICFP 2000, pp. 233–243. ACM, New York (2000)
7. Davey, B.A., Priestley, H.A.: Introduction to Lattices and Order. Cambridge University Press, Cambridge (2002)
8. Dawar, A., Gurevich, Y.: Fixed point logics. Bull. Symb. Log. **8**(01), 65–88 (2002)
9. Dougherty, D.J., Ghilezan, S., Lescanne, P., Likavec, S.: Strong normalization of the dual classical sequent calculus. In: Sutcliffe, G., Voronkov, A. (eds.) LPAR 2005. LNCS (LNAI), vol. 3835, pp. 169–183. Springer, Heidelberg (2005)
10. Gentzen, G.: Investigations into logical deduction. Am. Philos. Q. **1**(4), 288–306 (1964)
11. Harper, B., Lillibridge, M.: ML with callcc is unsound. Post to TYPES mailing list (1991)
12. Hur, C.K., Neis, G., Dreyer, D., Vafeiadis, V.: The power of parameterization in coinductive proof. In: Proceedings of the 40th Annual ACM SIGPLAN-SIGACT Symposium on Principles of Programming Languages POPL 2013, pp. 193–206. ACM, New York (2013)

13. Kimura, D., Tatsuta, M.: Dual Calculus with inductive and coinductive types. In: Treinen, R. (ed.) RTA 2009. LNCS, vol. 5595, pp. 224–238. Springer, Heidelberg (2009)
14. Matthes, R.: Extensions of System F by Iteration and Primitive Recursion on Monotone Inductive Types. Ph.D. thesis, Ludwig-Maximilians Universität, May 1998
15. Mendler, N.: Inductive types and type constraints in the second-order lambda calculus. Ann. Pure Appl. Log. **51**(1), 159–172 (1991)
16. Parigot, M.: Strong normalization of second order symmetric λ-calculus. In: Kapoor, S., Prasad, S. (eds.) FST TCS 2000. LNCS, vol. 1974, pp. 442–453. Springer, Heidelberg (2000)
17. Tzevelekos, N.: Investigations on the Dual Calculus. Theor. Comput. Sci. **360**(1), 289–326 (2006)
18. Uustalu, T., Vene, V.: Mendler-style inductive types, categorically. Nord. J. Comput. **6**(3), 343 (1999)
19. Wadler, P.: Call-by-value is dual to call-by-name. In: Proceedings of the Eighth ACM SIGPLAN International Conference on Functional Programming ICFP 2003, pp. 189–201. ACM, New York (2003)

Index Sets for Finite Normal Predicate Logic Programs with Function Symbols

Douglas Cenzer[1]([✉]), Victor W. Marek[2], and Jeffrey B. Remmel[3]

[1] Department of Mathematics, University of Florida, Gainesville, FL 32611, USA
cenzer@ufl.edu
[2] Department of Computer Science, University of Kentucky,
Lexington, KY 40506, USA
marek@cs.uky.edu
[3] Department of Mathematics, University of California at San Diego,
La Jolla, CA 92903, USA
jremmel@ucsd.edu

Abstract. We study the *recognition problem* in the metaprogramming of finite normal predicate logic programs. That is, let $\mathcal{L}$ be a computable first order predicate language with infinitely many constant symbols and infinitely many n-ary predicate symbols and n-ary function symbols for all $n \geq 1$. Then we can effectively list all the finite normal predicate logic programs $Q_0, Q_1, \ldots$ over $\mathcal{L}$. Given some property $\mathcal{P}$ of finite normal predicate logic programs over $\mathcal{L}$, we define the index set $I_{\mathcal{P}}$ to be the set of indices e such that Q_e has property $\mathcal{P}$. Then we shall classify the complexity of the index set $I_{\mathcal{P}}$ within the arithmetic hierarchy for various natural properties of finite predicate logic programs.

Keywords: Logic programming · Index sets · Recursive trees

1 Introduction

Past research has demonstrated that logic programming with the stable model semantics and, more generally, answer-set semantics, is an expressive knowledge representation formalism. It can be safely stated that there is a consensus in the Knowledge Representation community that stable models are the correct generalization of the least model of Horn program for the class of normal programs. Although stable model semantics is considered the correct one, past research has shown that the use of arbitrary normal logic programs admitting function symbols is not a reasonable choice for real-life programming. For example, Apt and Blair [2] proved that all arithmetic sets can be defined by using stratified programs. The import is that in general, it is impossible to query the unique stable model of such programs. Marek, Nerode, and Remmel [16,17] constructed finite predicate logic programs whose stable models could code up the paths through any infinitely branching recursive tree so that the problem of deciding whether a finite predicate logic program has a stable model is Σ_1^1-complete.

© Springer International Publishing Switzerland 2016
S. Artemov and A. Nerode (Eds.): LFCS 2016, LNCS 9537, pp. 60–75, 2016.
DOI: 10.1007/978-3-319-27683-0_5

For such reasons, researchers have focused on finite predicate logic programs without function symbols. There are a number of highly effective implementations of search engines to find stable models of finite normal predicate logic programs [11,13,16,18].

Nevertheless, researchers have searched for *some* natural classes $\mathcal{K}$ of finite normal predicate logic programs with function symbols where programming is both useful and possible. Actually, it should be clear that finding such a class $\mathcal{K}$ involves two tasks. (1) $\mathcal{K}$ needs to be *processable*. That is, given a program $P \in \mathcal{K}$, we need to have an algorithm that identifies one or more stable models of P which can be effectively queried. That is, one can effectively answer questions such as whether a given atom is in a given stable model of P or whether a given atom is in all stable models of P. (2) $\mathcal{K}$ needs to be *recognizable*. That is, we need to be able to answer the query whether a given program P belongs to $\mathcal{K}$. For instance, the class of stratified programs is recognizable (one of the fundamental results of Apt, Blair and Walker [3]), but not processable. A number of classes $\mathcal{K}$ of such programs which are both processable and recognizable have been found, see [4–6,14,21]. In particular [5] provides an extensive discussion of the reasons why researchers try to find classes of normal predicate logic programs admitting function symbols which are both recognizable and processable.

The goal of this paper is develop a systematic approach the recognition problem for the class of finite normal predicate logic programs over a computable first order predicate language $\mathcal{L}$ with infinitely many constant symbols and infinitely many n-ary predicate symbols and n-ary function symbols for all $n \geq 1$. Let $Q_0, Q_1, \ldots$ be an effective list of all the finite normal predicate logic programs over $\mathcal{L}$. Given some property $\mathcal{P}$ of finite normal predicate logic programs over $\mathcal{L}$, we define the *index set* $I_{\mathcal{P}}$ to be the set of indices e such that Q_e has property $\mathcal{P}$. For example, suppose that $\mathcal{P}$ is the property that a finite normal predicate logic program has a recursive stable model. Then the tools of this paper will allow one to classify the complexity of $I_{\mathcal{P}}$ within the arithmetic hierarchy. We will show in [8] that $I_{\mathcal{P}}$ is Σ_3^0-complete so that one can not effectively recognize the set of finite predicate logic programs which have recursive stable models.

Our approach is to extend the work of Marek, Nerode, and Remmel in [16,17], who showed that the problem of finding a stable of model of a recursive normal propositional logic program is essentially equivalent to finding an infinite path through an infinite recursive tree. That is, they showed that given any recursive normal propositional logic program P, one could construct a recursive tree such T_P such that there is an effective one-to-one degree preserving correspondence between the set of stable models of P and the set of infinite paths through T_P. Vice versa, given any recursive tree T, they constructed a recursive normal propositional logic program P_T such that there is an effective one-to-one degree preserving correspondence between the set of stable models of P_T and the set of infinite paths through T. Such correspondences also helped to motivate the definition of various natural properties of normal logic programs such as having the finite support property or the recursive finite support property (described below) since these properties correspond to natural properties of recursive trees such as being finitely branching or being highly recursive. The main goal of this paper is to provide similar constructions when we replace recursive normal propositional logic programs by finite normal predicate logic programs. This requires us to significantly modify the original constructions in [17].

To define index sets for primitive recursive trees, we need some notation. Let $\Sigma \subseteq \omega$ where $\omega = \{0, 1, 2, \ldots, \}$. Then $\Sigma^{<\omega}$ denotes the set of finite strings of letters from Σ and Σ^{ω} denotes the set of infinite sequences of letters from Σ. If $\sigma = (\sigma_1, \ldots, \sigma_n) \in \Sigma^{<\omega}$ and $a \in \Sigma$, then we let $\sigma^\frown a = (\sigma_1, \ldots, \sigma_n, a)$. A *tree* T over Σ is a set of finite strings from $\Sigma^{<\omega}$ which contains the empty string $\emptyset$ and is closed under initial segments. We say that $\tau \in T$ is an *immediate successor* of a string $\sigma \in T$ if $\tau = \sigma^\frown a$ for some $a \in \Sigma$. One can easily assign Gödel numbers to the elements of $\omega^{<\omega}$. That is, we can effectively assign a unique code $c(\sigma) \in \omega$ to each $\sigma \in \omega^{<\omega}$ such that we can effectively recover σ from $c(\sigma)$. We will identify T with the set of codes $c(\sigma)$ for $\sigma \in T$. Thus we say that T is primitive recursive, recursive, r.e., etc. if $\{c(\sigma) : \sigma \in T\}$ is primitive recursive, recursive, r.e., etc. If each node of T has finitely many immediate successors, then T is said to be *finitely branching*. We say a tree T is *highly recursive* if it is recursive and there is a recursive function f such that for any $\sigma \in T$, there are $f(\sigma)$ immediate successors of σ. An *infinite path* through a tree T is a sequence $(x(0), x(1), \ldots)$ such that $(x(0), \ldots x(n)) \in T$ for all n. Let $[T]$ be the set of infinite paths through T and $[T]_r$ denote the set of infinite recursive paths through T. We let $Ext(T)$ denote the set of all $\sigma \in T$ such that σ is an initial segment of x for some $x \in [T]$. We say that T is *decidable* if T is recursive and $Ext(T)$ is recursive. We let $T_0, T_1, \ldots$ be an effective list of all primitive recursive trees contained in $\omega^{<\omega}$. It follows that $[T_0], [T_1], \ldots$ is an effective list of all Π_1^0 classes, see [9]. Then for any property $\mathcal{P}$ of trees, we let $T_{\mathcal{P}}$ denote the set of all i such that T_i has property $\mathcal{P}$.

Our main result is to show that we can modify the constructions of Marek, Nerode, and Remmel [17] to construct recursive functions f and g such that for all e, (i) there is a one-to-one degree preserving correspondence between the set of stable models of Q_e and the set of infinite paths through $T_{f(e)}$ and (ii) there is a one-to-one degree preserving correspondence between the set of infinite paths through T_e and the set of stable models $Q_{g(e)}$. One can often use these two recursive functions to reduce the complexity of the index set $I_{\mathcal{P}}$ for various properties $\mathcal{P}$ of finite normal predicate logic programs to the complexity of the index set $T_{\mathcal{P}'}$ for an appropriate property of $\mathcal{P}'$ of primitive recursive trees. Actually, in practice, we take the reverse point of view. That is, we shall start with $T_{\mathcal{P}'}$ and try to find an appropriate property $\mathcal{P}$ of finite normal predicate logic programs such that $T_{\mathcal{P}'}$ and $I_{\mathcal{P}}$ are one-to-one equivalent.

We shall consider the following natural properties of trees contained in $\omega^{<\omega}$. Suppose that $g : \omega^{<\omega} \to \omega$. Then we say that

(I) T is **g-bounded** if for all σ and all integers i, $\sigma^\frown i \in T$ implies $i < g(\sigma)$,

(II) T is **almost always g-bounded** if there is a finite set $F \subseteq T$ of strings such that for all strings $\sigma \in T \setminus F$ and all integers i, $\sigma^\frown i \in T$ implies $i < g(\sigma)$,

(III) T is **nearly g-bounded** if there is an $n \geq 0$ such that for all strings $\sigma \in T$ with $|\sigma| \geq n$ and all integers i, $\sigma^\frown i \in T$ implies $i < g(\sigma)$,

(IV) T is **bounded** if it is g-bounded for some $g : \omega^{<\omega} \to \omega$,

(V) T is **almost always bounded** ($a.a.b.$) if it is almost always g-bounded for some $g : \omega^{<\omega} \to \omega$,

(VI) T is **nearly bounded** if it is nearly g-bounded for some $g : \omega^{<\omega} \to \omega$,

(VII) T is **recursively bounded** ($r.b.$) if T is g-bounded for some recursive $g : \omega^{<\omega} \to \omega$,

(VIII) T **almost always recursively bounded** ($a.a.r.b.$) if it is almost always g-bounded for some recursive $g : \omega^{<\omega} \to \omega$, and

(IX) T **nearly recursively bounded** (nearly $r.b.$) if it is nearly g-bounded for some recursive $g : \omega^{<\omega} \to \omega$.

For each of the properties $\mathcal{P}$ above, one can classify the index sets of the set of primitive recursive trees T satisfying property $\mathcal{P}$ and one of the following properties: $[T]$ ($[T]_r$) is empty, $[T]$ ($[T]_r$) is non-empty, $[T]$ ($[T]_r$) has cardinality c ($< c, \geq c$) for some natural number c, $[T]$ ($[T]_r$) is finite, or $[T]$ ($[T]_r$) is infinite.

To be able to precisely state our results, we must briefly review the basic concepts of recursion theory, normal logic programs and recursive trees.

We shall assume the reader is familiar with the basics of recursive and recursively enumerable sets, Turing degrees, and the arithmetic hierarchy of Σ_n^0 and Π_n^0 subsets of ω as well as Σ_1^1 and Π_1^1 sets; see Soare's book [20]. We shall generally use the terminology *recursive* rather than the equivalent term *computable* and likewise use *recursively enumerable* rather than *computably enumerable*. The former terms are standard in the logic programming community, which is an important audience for our paper. A subset A of ω is said to be D_n^m if it is the set-difference of two Σ_n^m sets. A set $A \subseteq \omega$ is said to be an *index set* if for any a, b, $a \in A$ and $\phi_a = \phi_b$ imply that $b \in A$ where $\phi_0, \phi_1, \ldots$ is an effective list of all partial recursive functions. For example, $Fin = \{a : W_a \text{ is finite}\}$ is an index set. We are particularly interested in the complexity of such index sets. Recall that a subset A of ω is said to be Σ_n^m-complete (respectively, Π_n^m-complete, D_n^m-complete) if A is Σ_n^m (respectively, Π_n^m, D_n^m) and any Σ_n^m (respectively, Π_n^m, D_n^m) set B is many-one reducible to A. For example, the set $Fin = \{e : W_e \text{ is finite}\}$ is Σ_2^0-complete.

Then, for example, Cenzer and Remmel [9] proved the following results:

(1) $\{e : T_e \text{ is r.b. and} [T_e] \text{is empty}\}$ is Σ_2^0-complete.

(2) $\{e : T_e \text{ is r.b. and} [T_e] \text{is nonempty}\}$ is Σ_3^0-complete.

(3) $\{e : T_e \text{ is bounded and} [T_e] \text{is empty}\}$ is Σ_2^0-complete.

(4) $\{e : T_e \text{ is bounded and} [T_e] \text{is nonempty}\}$ is Π_3^0-complete.

(5) $\{e : T_e \text{ is a.a.r.b. and} [T_e] \text{is nonempty}\}$ and
$\{e : T_e \text{ is a.a.r.b. and} [T_e] \text{is empty}\}$ are Σ_3^0-complete.

(6) $\{e : T_e \text{ is a.a.b. and} [T_e] \text{is nonempty}\}$ and
$\{e : T_e \text{ is a.a.b. and} [T_e] \text{is empty}\}$ are Σ_4^0-complete.

(8) $\{e : [T_e] \text{ is nonempty}\}$ is Σ_1^1-complete and
$\{e : [T_e] \text{ is empty}\}$ is Π_1^1-complete.

For any positive integer c,

(9) $\{e : T_e \text{ is r.b. and } Card([T_e]) > c\}$, $\{e : T_e \text{ is r.b. and } Card([T_e]) \leq c\}$, and $\{e : T_e \text{ is r.b. and } Card([T_e]) = c\}$ are all Σ_3^0-complete.

(10) $\{e : T_e$ is a.a.r.b. and $Card([T_e]) > c\}$,
$\{e : T_e$ is a.a.r.b. and $Card([T_e]) \le c\}$, and
$\{e : T_e$ is a.a.r.b. and $Card([T_e]) = c\}$ are all Σ_3^0-complete.
(11) $\{e : T_e$ is bounded and $Card([T_e]) \le c\}$ and
$\{e : T_e$ is bounded and $Card([T_e]) = 1\}$ are both Π_3^0-complete.
(12) $\{e : T_e$ is bounded and $Card([T_e]) > c\}$ and
$\{e : T_e$ is bounded and $Card([T_e]) = c + 1\}$ are both D_3^0-complete.
(13) $\{e : T_e$ is a.a.b. and $Card([T_e]) > c\}$,
$\{e : T_e$ is $a.a.$ bounded and $Card([T_e]) \le c\}$, and
$\{e : T_e$ is $a.a.$ bounded and $Card([T_e]) = c\}$ are all Σ_4^0-complete.
(14) $\{e : T_e$ is $r.b$, decidable, and $Card([T_e]) > c\}$,
$\{e : T_e$ is $r.b.$, dec. and $Card([T_e]) \le c\}$, and
$\{e : T_e$ is $r.b.$, dec. and $Card([T_e]) = c\}$ are all Σ_3^0-complete.
(15) $(\{e : Card([T_e]) > c\})$ is Σ_1^1-complete, $\{e : Card([T_e]) \le c\}$ is Π_1^1-complete
and $\{e : Card([T_e]) = c\}$ is Π_1^1-complete.

This is only a sample of the index set results that have been established for primitive recursive trees. For example, there are similar results when one replaces $[T_e]$ by $[T_e]_r$ in each of these statements. For each of the properties Pr in **(I)**-**(IX)** of trees, our goal is to find a corresponding property Pr' of finite normal predicate logic programs such that the complexity of the set of finite normal predicate logic programs P satisfying property Pr' refined by the cardinality of the stable models (recursive stable models) of P has the corresponding complexity of as the set of primitive recursive trees T satisfying property Pr refined by the cardinality of the set of infinite paths (recursive infinite paths) through T.

The outline of this paper is as follows. In Sect. 2, we shall define various properties on finite normal predicate logic programs which correspond to the properties **(I)**-**(IX)** described above. Many of the properties such as the finite support property and the recursive finite support property which correspond to bounded trees and recursively bounded trees have appeared in the literature. However, other properties such as a program being decidable, which correspond to decidable trees, are new. In Sect. 3, we shall state our main results. In Sect. 4, we state a number of results which classify the complexity of I_P for various properties P of finite normal predicate logic programs.

2   Properties of Finite Normal Logic Programs

In this section, we give the necessary background on normal logic programs.

We shall fix a recursive language $\mathcal{L}$ which has infinitely many constant symbols, infinitely many propositional letters, and infinitely many n-ary relation symbols and n-ary function symbols for each $n \ge 1$. A literal is an atomic formula or its negation. A ground literal is a literal which has no free variables. The Herbrand base of $\mathcal{L}$ is the set $H_{\mathcal{L}}$ of all ground atoms (atomic statements) of the language.

A (normal) logic programming clause C is of the form

$$c \leftarrow a_1, \ldots, a_n, \neg b_1, \ldots, \neg b_m \tag{1}$$

where $c, a_1, \ldots, a_n, b_1, \ldots, b_m$ are atoms of $\mathcal{L}$. Here we allow either n or m to be zero. In such a situation, we call c the *conclusion* of C, $a_1, \ldots, a_n$ the *premises* of C, $b_1, \ldots, b_n$ the *constraints* of C and $a_1, \ldots, a_n, \neg b_1, \ldots, \neg b_m$ the *body* of C and write $concl(C) = c$, $prem(C) = \{a_1, \ldots, a_n\}$, $constr(C) = \{b_1, \ldots, b_m\}$. A ground clause is a clause with no free variables. C is called a Horn clause if $constr(C) = \emptyset$, i.e., if C has no negated atoms in its body.

A finite normal predicate logic program P is a finite set of clauses of the form (1). P is said to be a Horn program if all its clauses are Horn clauses. A ground instance of a clause C is a clause obtained by substituting ground terms (terms without free variables) for all the free variables in C. The set of all ground instances of the program P is called *ground*(P). The Herbrand base of P, $H(P)$, is the set of all ground atoms that are instances of atoms that appear in P. For any set S, we let 2^S denote the set of all subsets of S.

Given a Horn program P, we let $T_P : 2^{H(P)} \to 2^{H(P)}$ be the one-step provability operator [15] associated with *ground*(P). That is, for $S \subseteq H(P)$,
$$T_P(S) = \{c : \exists_{C \in ground(P)}((C = c \leftarrow a_1, \ldots, a_n) \wedge (a_1, \ldots, a_n \in S))\}.$$
Then P has a least model $M = T_P \uparrow_\omega (\emptyset) = \bigcup_{n \geq 0} T_P^n(\emptyset)$ where for any $S \subseteq H(P)$, $T_P^0(S) = S$ and $T_P^{n+1}(S) = T_P(T_P^n(S))$. We denote the least model of a Horn program P by $lm(P)$.

Given a normal predicate logic program P and $M \subseteq H(P)$, we define the *Gelfond-Lifschitz reduct* [12] of P, P_M, via the following two step process. In Step 1, we eliminate all clauses $C = p \leftarrow q_1, \ldots, q_n, \neg r_1, \ldots, \neg r_m$ of *ground*(P) such that there exists an atom $r_i \in M$. In Step 2, for each remaining clause $C = p \leftarrow q_1, \ldots, q_n, \neg r_1, \ldots, \neg r_m$ of *ground*(P), we replace C by the Horn clause $C = p \leftarrow q_1, \ldots, q_n$. The resulting program P_M is a Horn propositional program and, hence, has a least model. If that least model of P_M coincides with M, then M is called a *stable model* for P.

Next, we define the notion of P-proof scheme of a normal *propositional* logic program P. Given a normal propositional logic program P, a P-proof scheme is defined by induction on its length n. Specifically, the set of P-proof schemes is defined inductively by declaring that

(I) $\langle \langle C_1, p_1 \rangle, U \rangle$ is a P-proof scheme of length 1 if $C_1 \in P$, $p_1 = concl(C_1)$, $prem(C_1) = \emptyset$, and $U = constr(C_1)$ and

(II) for $n > 1$, $\langle \langle C_1, p_1 \rangle, \ldots, \langle C_n, p_n \rangle, U \rangle$ is a P-proof scheme of length n if $\langle \langle C_1, p_1 \rangle, \ldots, \langle C_{n-1}, p_{n-1} \rangle, \bar{U} \rangle$ is a P-proof scheme of length $n - 1$ and C_n is a clause in P such that $concl(C_n) = p_n$, $prem(C_n) \subseteq \{p_1, \ldots, p_{n-1}\}$ and $U = \bar{U} \cup constr(C_n)$

If $\mathbb{S} = \langle \langle C_1, p_1 \rangle, \ldots, \langle C_n, p_n \rangle, U \rangle$ is a P-proof scheme of length n, then we let $supp(\mathbb{S}) = U$ and $concl(\mathbb{S}) = p_n$.

Example 1. Let P be the normal propositional logic program consisting of the following four clauses:
$C_1 = p \leftarrow$, $C_2 = q \leftarrow p, \neg r$, $C_3 = r \leftarrow \neg q$, and $C_4 = s \leftarrow \neg t$.
Then we have the following useful examples of P-proof schemes:

(a) $\langle \langle C_1, p \rangle, \emptyset \rangle$ is a P-proof scheme of length 1 with conclusion p and empty support.
(b) $\langle \langle C_1, p \rangle, \langle C_2, q \rangle, \{r\} \rangle$ is a P-proof scheme of length 2 with conclusion q and support $\{r\}$.
(c) $\langle \langle C_1, p \rangle, \langle C_3, r \rangle, \{q\} \rangle$ is a P-proof scheme of length 2 with conclusion r and support $\{q\}$.
(d) $\langle \langle C_1, p \rangle, \langle C_2, q \rangle, \langle C_3, r \rangle, \{q, r\} \rangle$ is a P-proof scheme of length 3 with conclusion r and support $\{q, r\}$.

In this example we see that the proof scheme in (c) had an unnecessary item, the first term, while in (d) the proof scheme was supported by a set containing q, one of atoms that were proved on the way to r.   □

A P-proof scheme differs from the usual Hilbert-style proofs in that it carries within itself its own applicability condition. In effect, a P-proof scheme is a *conditional* proof of its conclusion. It becomes applicable when all the constraints collected in the support are satisfied. Formally, for a set M of atoms, we say that a P-proof scheme $\mathbb{S}$ is *M-applicable* or that M *admits* $\mathbb{S}$ if $M \cap supp(\mathbb{S}) = \emptyset$. The fundamental connection between proof schemes and stable models is given by the following proposition which is proved in [17].

Proposition 1. *For every normal propositional logic program P and every set M of atoms, M is a stable model of P if and only if*

(i) *for every $p \in M$, there is a P-proof scheme $\mathbb{S}$ with conclusion p such that M admits $\mathbb{S}$ and*
(ii) *for every $p \notin M$, there is no P-proof scheme $\mathbb{S}$ with conclusion p such that M admits $\mathbb{S}$.*

A P-proof scheme may not need all its clauses to prove its conclusion. It may be possible to omit some clauses and still have a proof scheme with the same conclusion. Thus we define a pre-order on P-proof schemes $\mathbb{S}, \mathbb{T}$ by declaring that $\mathbb{S} \prec \mathbb{T}$ if (1) $\mathbb{S}, \mathbb{T}$ have the same conclusion and (2) Every clause in $\mathbb{S}$ is also a clause of $\mathbb{T}$. The relation $\prec$ is reflexive, transitive, and well-founded. Minimal elements of $\prec$ are minimal proof schemes. A given atom may be the conclusion of no, one, finitely many, or infinitely many different minimal P-proof schemes. These differences are clearly computationally significant if one is searching for a justification of a conclusion.

If P is a finite normal predicate logic program, then we define a P-proof scheme to be a *$ground(P)$*-proof scheme. Since we are considering finite normal programs over our fixed recursive language $\mathcal{L}$, we can use standard Gödel numbering techniques to assign code numbers to atomic formulas, clauses, proof schemes, and programs. It is then not difficult to verify that for any given finite normal predicate logic program P, the questions of whether a given n is the code of a ground atom, a ground instance of a clause in P, or a P-proof-scheme are primitive recursive predicates. The key observation to make is that since P is finite and the usual unification algorithm is effective, we can explicitly test whether a given number m is the code of a ground atom or a ground instance of a clause in P without doing any unbounded searches.

The set of Gödel numbers of well-formed programs is well-known to be primitive recursive (see Lloyd [15]). We let Q_e be the program with Gödel number e when this exists and let Q_e be the empty program otherwise. For any property $\mathcal{P}$ of finite normal predicate logic programs, let $I(\mathcal{P})$ be the set of indices e such that Q_e has property $\mathcal{P}$.

We say that a finite normal predicate logic program P over $\mathcal{L}$ has the *finite support (FS) property* if for every atom $a \in H(P)$, there are only finitely many inclusion-minimal supports of minimal $ground(P)$-proof schemes for a. We say that P has the *almost always finite support (a.a.FSP) property* if for all but finitely many atoms $a \in H(P)$, there are only finitely many inclusion-minimal supports of minimal $ground(P)$-proof schemes for a. We say that P has the *recursive finite support (rec.FSP) property* if it has the finite support property and there is an effective procedure which, given any atom $a \in H(P)$, produces the code of the set of the inclusion-minimal supports of $ground(P)$-proof schemes for a. We say that P has the *almost always recursive finite support (a.a.rec.FSP) property* if it has the a.a.FSP property and there is an effective procedure which, for all but a finite set of atoms $a \in H(P)$, produces the code of the set of the inclusion-minimal supports of $ground(P)$-proof schemes for a.

Next, we define two additional properties of recursive normal propositional logic programs that have not been previously defined in the literature. Suppose that P is a recursive normal propositional logic program consisting of ground clauses in $\mathcal{L}$ and M is a stable model of P. Then for any atom $p \in M$, we say that a minimal P-proof scheme $\mathbb{S}$ is the *smallest minimal P-proof for p relative to M* if $concl(\mathbb{S}) = p$ and $supp(\mathbb{S}) \cap M = \emptyset$ and there is no minimal P-proof scheme $\mathbb{S}'$ such that $concl(\mathbb{S}') = p$ and $supp(\mathbb{S}') \cap M = \emptyset$ and the Gödel number of $\mathbb{S}'$ is less than the Gödel number of $\mathbb{S}$.

We say that P is *decidable* if for all $N > 0$ and any finite (possibly empty) set of ground atoms $\{a_1, \ldots, a_n\} \subseteq H(P)$ such that the code of each a_i is less than or equal to N, and any finite set of minimal P-proof schemes $\{\mathbb{S}_1, \ldots, \mathbb{S}_n\}$ such that $concl(\mathbb{S}_i) = a_i$, we can effectively decide whether there is a stable model of M of P such that

(a) $a_i \in M$ and $\mathbb{S}_i$ is the smallest minimal P-proof scheme for a_i such that $supp(\mathbb{S}_i) \cap M = \emptyset$ and
(b) for any ground atom $b \notin \{a_1, \ldots, a_n\}$ such that the code of b is less than or equal to N, $b \notin M$.

We say that a finite normal predicate logic program is decidable if $ground(P)$ is decidable.

It will turn out that under our coding of trees into finite predicate logic programs, decidable trees induce decidable programs and under our coding of finite predicate logic programs into trees, decidable programs induce decidable trees. Moreover, decidability combined with the property of having the recursive finite support property ensures that there exists processable stable models when there are stable models. That is, we have the following theorem.

Theorem 1. *Suppose that P is a recursive normal logic program which has the recursive finite support property and is decidable. Then if P has a stable model, we can effectively find a recursive stable model of P.*

Proof. Let $a_0, a_1, \ldots$ be a list of all elements of $H(P)$ by increasing code numbers. That is, if c_i is the code of a_i, then $c_0 < c_1 < \ldots$. We will effectively construct a list of pairs of sets (A_i, R_i) for $i \geq 0$ such that for all i, $A_i \cap R_i = \emptyset$, $\{a_0, \ldots, a_i\} \subseteq A_i \cup R_i$, $A_i \subseteq \{a_0, \ldots, a_I\}$, $A_i \subseteq A_{i+1}$, and $R_i \subseteq R_{i+1}$. Then $A = \bigcup_{i \geq 0} A_i$ will be our desired recursive stable model. Thus we shall think of A_i as being the set of atoms that we have accepted to be in the stable model at stage i and R_i as being the set of atoms that have been rejected from being in A at stage i. Our construction will proceed in stages.

Stage 0. Consider a_o. Since P has the recursive finite support property, we can effectively find the supports of all the minimal P-proof schemes with conclusion a_0. If U is the support of a minimal proof scheme with conclusion a_0, then the fact that the set of minimal proofs schemes of P is r.e. means that we can search through the list of minimal proof schemes of P until we find the minimal proof scheme $\mathbb{S}_U$ with the smallest possible code such the conclusion of $\mathbb{S}_U$ is a_0 and the support of $\mathbb{S}_U$ is U. Thus if $U_1, \ldots, U_k$ is the set of all supports of minimal proof schemes with conclusion a_0, then we can effectively find proof schemes $\mathbb{S}_1, \ldots, \mathbb{S}_k$ such that for each i, $\mathbb{S}_i$ is the smallest minimal proof scheme such that the conclusion of $\mathbb{S}_i$ is a_0 and the support of $\mathbb{S}_i$ is U_i. Then, since P is decidable, we use our effective procedure with $N = c_0$ to determine whether there is a stable model M for which $\mathbb{S}_i$ is the smallest minimal proof scheme such that $supp(\mathbb{S}_i) \cap M = \emptyset$. If there is no such i, then a_0 is not in any stable model so we set $A_0 = \emptyset$ and $R_0 = \{a_0\}$. If there is such an i, then we let t_0 be the least such i and we set $A_0 = \{a_0\}$ and $R_0 = supp(\mathbb{S}_{t_0})$.

Stage $s + 1$. Assume that at stage s, we have constructed A_s and R_s such that $A_s \cap R_s = \emptyset$, $\{a_0, \ldots, a_s\} \subseteq A_s \cup R_s$, $A_s \subseteq \{a_0, \ldots, a_s\}$, and for each $a \in A_s$, we have constructed a proof scheme $\mathbb{S}_a$ such that if $A_s = \{d_1, \ldots, d_k\}$ and $N_s = c_{s+1}$. Then our decision procedure associated with the decidability of P will answer yes when we give it N_S, the set $\{d_1, \ldots, d_k\}$ and the corresponding proof schemes $\mathbb{S}_{d_1}, \ldots, \mathbb{S}_{d_k}$. Moreover, we assume that

$$R_s = \{a_i : i \leq s \ \& \ a_i \notin A_s\} \cup \bigcup_{i=1}^{k} supp(\mathbb{S}_{d_i}).$$

This means that there is at least one stable model M such that for each i, $\mathbb{S}_{d_i}$ is the least proof scheme that witnesses that d_i is in M and $(\{a_0, \ldots, a_s\} - A_s) \cap M = \emptyset$.

Now consider a_{s+1}. By the fact that P has the recursive support property, we can effectively find the finite set of supports $V_1, \ldots, V_r$ of the minimal P-proof schemes of a_{s+1} and we can find P-proof schemes $\mathbb{T}_1, \ldots, \mathbb{T}_r$ such that for each $1 \leq i \leq r$, $\mathbb{T}_i$ is the smallest possible proof scheme with conclusion a_{s+1} and support V_i. Then for each $i < r$, we can query the decision procedure associated with the decidability of P on the set $\{d_1, \ldots, d_k, a_{s+1}\}$ and the corresponding proof schemes $\mathbb{S}_{d_1}, \ldots, \mathbb{S}_{d_k}, \mathbb{T}_i$. If we get an answer yes for any i, then we let t_{s+1} be the least such i and we set $A_{s+1} = A_s \cup \{a_{s+1}\}$ and $R_{s+1} = R_s \cup supp(\mathbb{T}_{t_{s+1}})$. Note that since $\mathbb{S}_{d_1}, \ldots, \mathbb{S}_{d_k}, \mathbb{T}_i$ are the smallest minimal P proof schemes that witness that $d_1, \ldots, d_k, a_{s+1}$ are in some fixed stable model M such that $(\{a_0, \ldots, a_{s+1}\} - A_{s+1}) \cap M = \emptyset$, we must have that $A_{s+1} \cap R_{s+1} = \emptyset$. If there is no such i, then there is no stable model M which contains a_{s+1} and is such that for each i, $\mathbb{S}_{d_i}$ is the least proof scheme that witnesses that d_i is in M and $(\{a_0, \ldots, a_s\} - A_s) \cap M = \emptyset$. In that case, we let $A_{s+1} = A_s$ and $R_{s+1} = R_s \cup \{a_{s+1}\}$. It easily follows that our inductive assumption will hold at stage $s + 1$.

This completes the construction. It is easy to see that if $A = \bigcup_{s \geq 0} A_s$ and $R = \bigcup_{s \geq 0} R_s$, then $A \cap R = \emptyset$ and $\{a_0, a_1, \ldots\} \subseteq A \cup R$. Thus A and R partition $H(P)$. It is also easy to see that A is recursive since our construction is effective and at stage s, we have determined whether $a_s \in A$. We claim that A is stable model. That is, if A is not a stable model, then either there exists an a_s such that $a_s \in A$ and a_s has no P-proof scheme admitted by A or there is an $a_t \notin A$ such that a_t has an P-proof scheme which is admitted by A. Our construction ensures that if a_s is in A, then a_s has an P-proof scheme admitted by A. Thus suppose that $a_t \notin A$. Then let $W_1, \ldots, W_k$ be the supports of the minimal proof schemes of a_t. Let a_r be the largest element in $W_1 \cup \ldots \cup W_k$. Then consider what happens at stage r. Suppose $A_r = \{e_1, \ldots, e_k\}$. Then our construction also specifies minimal P-proof schemes $\mathbb{S}_1, \ldots, \mathbb{S}_k$ such that there is a stable model M such that for $1 \leq i \leq k$, $\mathbb{S}_i$ is the smallest proof scheme which witnesses that e_i is in M, $supp(\mathbb{S}_1) \cup \ldots \cup supp(\mathbb{S}_k) \subseteq R_r$, and $\{a_0, \ldots, a_r\} - A_r \cap M = \emptyset$. Thus a_t is not in M. Let $V_1, \ldots, V_b$ be the supports of the minimal P-proof schemes of a_{t+1}. This means that $M \cap supp(V_i) \neq \emptyset$ for each i. But for each i, $supp(V_i) \subseteq \{a_0, \ldots, a_r\}$ and $M \cap \{a_0, \ldots, a_r\} = A_r$. Thus it must be that case that $supp(V_i) \cap A_r \neq \emptyset$ for all i and hence a_t does not have a P-proof scheme admitted by A. Thus A is a stable model of P.

We now introduce and illustrate a technical concept that will be useful for our later considerations. At first glance, there are some obvious differences between stable models of normal propositional logic programs and models of sets of sentences in a propositional logic. For example, if T is a set of sentences in a propositional logic and $S \subseteq T$, then it is certainly the case that every model of T is a model of S. Thus a set of propositional sentences T has the property that if T has a model, then every subset of T has a model. This is not true for normal propositional logic programs. That is, suppose that P_0 is a normal propositional logic program which has a stable model and a is atom which is not in the Herbrand base of P_0, $H(P_0)$. Let P be the normal propositional logic program consisting of P_0 plus the clause $C = a \leftarrow \neg a$. Then P automatically does not have a stable model. That is, consider a potential stable model M of P. If $a \in M$, then C does not contribute to P_M so that there will be no clause of P_M with a in the head. Hence, a is not in the least model of P_M so that M is not a stable model of P. On the other hand, if $a \notin M$, then C will contribute the clause $a \leftarrow$ to P_M so that a must be in the least model of P_M. It follows that $P_0 \cup \{a \leftarrow \neg a, a \leftarrow\}$ has a stable model but $P_0 \cup \{a \leftarrow \neg a\}$ does not.

One can see from the example above that there may be a finite set of clauses in a normal propositional or predicate logic program P which prevent P from having a stable model. Our next definition captures the key property which ensures that the T_P which corresponds to a given finite normal predicate logic program is forced to be finite. We say that a finite normal predicate logic program Q_e over $\mathcal{L}$ has an *explicit initial blocking set* if there is an m such that

1. for every $i \leq m$, either i is not the code of an atom of $ground(P)$ or the atom a coded by i has the finite support property relative to P and there is at least one atom a in $H(P)$ whose code is less than or equal to m and
2. for all $S \subseteq \{0, \ldots, m\}$, either
   (a) there exists an $i \in S$ such that i is not the code of an atom in $H(P)$, or
   (b) there is an $i \notin S$ such that there exists a minimal P-proof scheme p such that $concl(p) = a$ where a is the atom of $H(P)$ with code i and $supp(p) \subseteq \{0, \ldots, m\} - S$, or
   (c) there is an $i \in S$ such that every minimal P-proof scheme $\mathbb{S}$ of the atom a of $H(P)$ with code i has $supp(\mathbb{S}) \cap S \neq \emptyset$.

The definition of a finite normal predicate logic program Q_e over $\mathcal{L}$ having an *initial blocking set* is the same as Q_e having an explicit initial blocking set, except that we drop the condition that for every $i \leq m$ which is the code of an atom $a \in H(P)$, a must have the finite support property relative to P.

3   Main Results

Next we state the main results, some of which were first discussed in [7], which reduce the problem of computing index sets for finite normal predicate logic programs to the problem of computing index sets for primitive recursive trees. We shall only give a sketch of the proofs of our main results. The full proofs are long and technical and can be found in [8].

Theorem 2. *There is a uniform effective procedure which given any recursive tree $T \subseteq \omega^{<\omega}$ produces a finite normal predicate logic program P_T such that the following hold.*

1. *There is an effective one-to-one degree preserving correspondence between the set of stable models of P_T and the set of infinite paths through T.*
2. *T is bounded if and only if P_T has the FS property.*
3. *T is recursively bounded if and only if P_T has the rec.FS property.*
4. *T is decidable and recursively bounded if and only if P_T is decidable and has the rec.FS property.*

Proof Sketch. Let T be a recursive tree contained in $\omega^{<\omega}$. Note that the empty sequence, whose code is 0, is in T. Below we shall only describe the program P_T. The details that P_T has the desired properties can be found in [8].

A classical result, first explicit in [1,23], but known earlier in equational form, is that every r.e. relation can be computed by a suitably chosen predicate over the least model of a finite predicate logic Horn program. An elegant method of proof due to Shepherdson [22] uses the representation of recursive functions by means of finite register machines. When such machines are represented by Horn programs in the natural way, we get programs in which every atom can be proved in only finitely many ways; see also [19]. Thus we have the following.

Proposition 2. *Let $r(\cdot, \cdot)$ be a recursive relation. Then there is a finite predicate logic program P_r computing $r(\cdot, \cdot)$ such that every atom in the least model M_r of P_r has only finitely many minimal proof schemes and there is a recursive procedure such that given an atom a in Herbrand base of P_r produces the code of the set of P_r-proof schemes for a. Moreover, the least model of P_r is recursive.*

It follows that there exists the following three normal finite predicate logic programs such that the set of ground terms in their underlying language are all of the form 0 or $s^n(0)$ for $n \geq 1$ where 0 is a constant symbol and s is a unary function symbol. We shall use n as an abbreviation for the term $s^n(0)$ for $n \geq 1$.

(I) There is a finite predicate logic Horn program P_0 such that for a predicate $tree(\cdot)$ of the language of P_0, the atom $tree(n)$ belongs to the least Herbrand model of P_0 if and only if n is a code for a finite sequence σ and $\sigma \in T$.

(II) There is a finite predicate logic Horn program P_1 such that for a predicate $seq(\cdot)$ of the language of P_1, the atom $seq(n)$ belongs to the least Herbrand model of P_1 if and only if n is the code of a finite sequence $\alpha \in \omega^{<\omega}$.

(III) There is a finite predicate logic Horn program P_2 which correctly computes the following recursive predicates on codes of sequences.

  (a) $samelength(\cdot, \cdot)$. This succeeds if and only if both arguments are the codes of sequences of the same length.

  (b) $diff(\cdot, \cdot)$. This succeeds if and only if the arguments are codes of sequences which are different.

  (c) $shorter(\cdot, \cdot)$. This succeeds if and only both arguments are codes of sequences and the first sequence is shorter than the second sequence.

  (d) $length(\cdot, \cdot)$. This succeeds when the first argument is a code of a sequence and the second argument is the length of that sequence.

  (e) $notincluded(\cdot, \cdot)$. This succeeds if and only if both arguments are codes of sequences and the first sequence is not an initial segment of the second.

  (f) $num(\cdot)$. This succeeds if and only if the argument is either 0 or $s^n(0)$ for some $n \geq 1$.

Now let P^- be the finite predicate logic program $P_0 \cup P_1 \cup P_2$. We denote its language by $\mathcal{L}^-$ and we let M^- be the least model of P^-. By Proposition 2, P^- is a Horn program, M^- is recursive, and for each ground atom a in the Herbrand base of P^-, we can explicitly construct the set of all P^--proof schemes of a. In particular, $tree(n) \in M^-$ if and only if n is the code of node in T.

Our final program P_T will consist of P^- plus clauses (1)-(7) given below. We assume no predicate that appears in the head of any of these clauses is in the language $\mathcal{L}^-$. However, we do allow predicates from P^- to appear in the body of clauses (1) to (7). It follows that for any stable model of the extended program, its intersection with the set of ground atoms of $\mathcal{L}^-$ will be M^-. In particular, the meaning of the predicates listed above will always be the same. We can now write the additional clauses which, together with P^-, will form the desired program P_T. First of all, we select three new unary predicates:

(i) $path(\cdot)$, whose intended interpretation in any given stable model M of P_T is that it holds only on the set of codes of sequences that lie on infinite path through T. This path will correspond to the path encoded by the stable model of M,

(ii) $notpath(\cdot)$, whose intended interpretation in any stable model M of P_T is the set of all codes of sequences which are in T but do not satisfy $path(\cdot)$, and

(iii) $control(\cdot)$, which will be used to ensure that $path(\cdot)$ always encodes an infinite path through T.

This given, the final 7 clauses of our program are the following.

(1) $path(X) \longleftarrow tree(X), \neg notpath(X)$

(2) $notpath(X) \longleftarrow tree(X), \neg path(X)$

(3) $path(0) \longleftarrow$

(4) $notpath(X) \longleftarrow tree(X), path(Y), tree(Y), samelength(X, Y), diff(X, Y)$

(5) $notpath(X) \longleftarrow tree(X), tree(Y), path(Y), shorter(Y, X), notincluded(Y, X)$

(6) $control(X) \longleftarrow path(Y), length(Y, X)$

(7) $control(X) \longleftarrow \neg control(X), num(X)$

Clearly, $P_T = P^- \cup \{(1), \ldots, (7)\}$ is a finite program.

Theorem 3. *There is a uniform recursive procedure which given any finite normal predicate logic program P produces a primitive recursive tree T_P such that the following hold.*

1. *There is an effective one-to-one degree-preserving correspondence between the set of stable models of P and the set of infinite paths through T_P.*

2. *P has the FS property or P has an explicit initial blocking set if and only if T_P is bounded.*

3. *If P has a stable model, then P has the FS property if and only if T_P is bounded.*

4. *P has the rec.FS property or an explicit initial blocking set if and only if T_P is recursively bounded.*

5. *If P has a stable model, then P has the rec.FS property if and only if T_P is recursively bounded.*

6. P has the a.a.FS property or P has an explicit initial blocking set if and only if T_P is nearly bounded.
7. If P has a stable model, then P has the a.a.FS property if and only if T_P is nearly bounded.
8. P has the a.a.rec.FS property or an explicit initial blocking set if and only if T_P is nearly recursively bounded.
9. If P has a stable model, then P has the a.a.rec.FS property if and only if T_P is nearly recursively bounded.
10. If P has a stable model, then P is decidable if and only if T_P is decidable.

Proof Sketch.

Our basic strategy is to encode a stable model M of $ground(P)$ by a path $f_M = (f_0, f_1, \ldots)$ through the complete ω-branching tree $\omega^{<\omega}$ as follows.

1. First, for all $i \geq 0$, $f_{2i} = \chi_M(i)$. That is, at the stage $2i$, we encode the information about whether or not the atom encoded by i belongs to M. Thus, in particular, if i is not the code of ground atom in $H(P)$, then $f_{2i} = 0$.
2. If $f_{2i} = 0$, then we set $f_{2i+1} = 0$. But if $f_{2i} = 1$ so that $i \in M$ and i is the code of a ground atom in $H(P)$, then we let f_{2i+1} equal $q_M(i)$ where $q_M(i)$ is the least code for a minimal P-proof scheme $\mathbb{S}$ for i such that the support of $\mathbb{S}$ is disjoint from M. That is, we select a minimal P-proof scheme $\mathbb{S}$ for i, or to be precise for the atom encoded by i, such that $\mathbb{S}$ has the smallest possible code of any P-proof scheme $\mathbb{T}$ such that $supp(\mathbb{T}) \cap M = \emptyset$. If M is a stable model, then, by Proposition 1, at least one such P-proof scheme exists for i.

Clearly, $M \leq_T f_M$ since it is enough to look at the values of f_M at even places to read off M. Now given an M-oracle, it should be clear that for each $i \in M$, we can use an M-oracle to find $q_M(i)$ effectively. This means that $f_M \leq_T M$. Thus the correspondence $M \mapsto f_M$ is an effective degree-preserving correspondence.

Then, given a program P, we construct a primitive recursive tree $T_P \subseteq \omega^\omega$ such that $[T_P] = \{f_M : M \in stab(P)\}$. The details of this construction and the verification that it has the desired properties can be found in [8].

4   Corollaries, Conclusions and Further Work

Theorems 2 and 3 allow us to transfer many results about paths through recursive trees to stable models of finite normal predicate logic programs. We give a brief sample of some the results that we have proved in this manner.

(1) $\{e : Q_e$ has the rec.FSP$\}$ is Σ_3^0-complete.

This can be proved as follows. First it is a straightforward exercise to show that the property that Q_e has the rec.FSP is a Σ_3^0 predicate. Then Cenzer and Remmel proved that $\{e : T_e$ is rec. bounded$\}$ is Σ_3^0-complete. Next, it follows from Theorem 3.1 that $\{e : T_e$ is rec. bounded$\}$ is one-to-one reducible to the set $\{e : Q_e$ is rec.FSP$\}$. Hence the set of $\{e : Q_e$ is rec.FSP$\}$ is Σ_3^0-complete.

The following results can be proved by similar type reasoning.

(2) $\{e : Q_e$ has the FSP$\}$ is Π_3^0-complete.
(3) $\{e : Q_e$ has the rec.FSP and is dec.$\}$ is Σ_3^0-complete.
(4) $\{e : Q_e$ has the rec.FSP and $Stab(Q_e) \neq \emptyset\}$ is Σ_3^0-complete.
(5) $\{e : Q_e$ has the FSP and $Stab(Q_e) \neq \emptyset\}$ is Π_3^0-complete.
(6) $\{e : Stab(Q_e) \neq \emptyset\}$ is Σ_1^1-complete.

For any positive integer c,

(7) $\{e : Q_e$ has the rec.FSP and $Card(Stab(Q_e)) > c\}$,
    $\{e : Q_e$ has the rec.FSP and $Card(Stab(Q_e)) \leq c\}$,
    and $\{e : Q_e$ has the rec.FSP and $Card(Stab(Q_e)) = c\}$ are all Σ_3^0-complete.
(8) $\{e : Q_e$ has the FSP and $Card(Stab(Q_e)) \leq c\}$ and
    $\{e : Q_e$ has the FSP and $Card(Stab(Q_e)) = 1\}$ are both Π_3^0-complete.
(9) $\{e : Q_e$ has the FSP and $Card(Stab(Q_e)) > c\}$ and
    $\{e : Q_e$ is has the FSP and $Card(Stab(Q_e)) = c+1\}$ are both D_3^0-complete.
(10) $\{e : Q_e$ has the rec.FSP and is dec. and $Card(Stab(Q_e)) > c\}$,
    $\{e : Q_e$ has the rec.FSP and is dec. and $Card(Stab(Q_e)) \leq c\}$,
    and $\{e : Q_e$ has the rec.FSP and is dec. and $Card(Stab(Q_e)) = c\}$ are all
    Σ_3^0-complete.
(11) $\{e : Card(Stab(Q_e)) > c\}$ is Σ_1^1-complete and
    $\{e : Card(Stab(Q_e)) \leq c\}$ and $\{e : Card(Stab(Q_e)) = c\}$ are Π_1^1-complete.

In [8], we proved many more results of this type. The properties that we considered involve both whether a finite normal predicate logic program possesses a blocking set or has various properties related to finite support property as well has properties about the types and complexity of its stable models such the cardinality of its set of stable models or the cardinality of its set of recursive stable models. This required that we prove some new index type results for trees and to modify the constructions of Theorems 2 and 3.

We believe that the types of relationship established in this paper between the sets of stable models of a finite normal predicate logic programs and the sets of infinite paths through recursive trees is a technology which can be applied to study the complexity of other notions that have appeared in the Answer Set Programming literature. For example, Cenzer and Remmel [10] showed that there is an intimate connection between the well-founded semantics of logic programs and Cantor-Bendixson derivatives. One could also ask whether our correspondences can be extended to handle cases where one adds additional constructs to logic programs such as aggregates, and more generally non-monotone set-constraints.

References

1. Andreka, H., Nemeti, I.: The generalized completeness of horn predicate logic as a programming language. Acta Cybernetica **4**, 3–10 (1978)
2. Apt, K.R., Blair, H.A.: Arithmetic classification of perfect models of stratified programs. Fund. Inf. **14**, 339–343 (1991)

3. Apt, K.R., Blair, H.A., Walker, A.:Towards a theory of declarative knowledge. In: Foundations of Deductive Databases and Logic Programming, pp. 89–148 (1988)
4. Bonatti, P.: Reasoning with infinite stable models. Art. Int. J. **156**, 75–111 (2004)
5. Calautti, M., Greco, S., Spezzano, F., Trubitsyna, I.: Checking termination of bottom-up evaluation of logic programs with function symbols. Theor. Pract. Log. Program. **15**, 854–859 (2015)
6. Calimeri, F., Cozza, S., Ianni, G., Leone, N.: Computable functions in ASP: theory and implementation. In: Garcia de la Banda, M., Pontelli, E. (eds.) ICLP 2008. LNCS, vol. 5366, pp. 407–424. Springer, Heidelberg (2008)
7. Cenzer, D., Marek, V.W., Remmel, J.B.: Index sets for finite predicate logic programs. In: Eiter, T., Gottlob, G. (eds.) FLOC 1999 Workshop on Complexity-theoretic and Recursion-theoretic Methods in Databases, Artificial Intelligence and Finite Model Theory, pp. 72–80 (1999)
8. Cenzer, D., Marek, V.W., Remmel, J.B.: Index sets for finite predicate logic programs (in preparation 2016)
9. Cenzer, D., Remmel, J.B.: Index Sets for Π_1^0 classes. Ann. Pure Appl. Logic **93**, 3–61 (1998)
10. Cenzer, D., Remmel, J.B.: A connection between Cantor-Bendixson derivatives and the well-founded semantics of finite logic programs. Ann. Math. Art. Int. **65**, 1–24 (2012)
11. Gebser, M., Kaufmann, B., Neumann, A., Schaub, T. : Conflict-driven answer set solving. In: Veloso, M. (ed.) Proceedings of Joint International Conference on Artificial Intelligence, p. 386 (2007)
12. Gelfond, M., Lifschitz, V.: The stable semantics for logic programs. In: Proceedings of the 5th International Symposium on Logic Programming, pp. 1070–1080. MIT Press (1998)
13. Leone, N., Pfeifer, G., Faber, W., Eiter, T., Gottlob, G., Perri, S., Scarcello, F.: The DLV system for knowledge representation and reasoning. ACM Trans. Comput. Log. **7**, 499–562 (2006)
14. Lierler, Y., Lifschitz, V.: One more decidable class of finitely ground programs. In: Hill, P.M., Warren, D.S. (eds.) ICLP 2009. LNCS, vol. 5649, pp. 489–493. Springer, Heidelberg (2009)
15. Lloyd, J.: Foundations of Logic Programming. Springer, New York (1989)
16. Marek, V.W., Nerode, A., Remmel, J.B.: How complicated is the set of stable models of a recursive logic program. Ann. Pure App. Logic **56**, 119–136 (1992)
17. Marek, V.W., Nerode, A., Remmel, J.B.: The stable models of predicate logic programs. J. Log. Program **21**, 129–153 (1994)
18. Niemelä, I., Simons, P.: Smodels — an implementation of the stable model and well-founded semantics for normal logic programs. In: Fuhrbach, U., Dix, J., Nerode, A. (eds.) LPNMR 1997. LNCS, vol. 1265, pp. 420–429. Springer, Heidelberg (1997)
19. Nerode, A., Shore, R.: Logic for Applications. Springer, New York (1993)
20. Soare, R.: Recursively Enumerable Sets and Degrees. Springer, Heidelberg (1987)
21. Syrjänen, T.: Omega-restricted logic programs. In: Eiter, T., Faber, W., Truszczyński, M. (eds.) LPNMR 2001. LNCS (LNAI), vol. 2173, pp. 267–279. Springer, Heidelberg (2001)
22. Shepherdson, J.C.: Unsolvable problems for SLDNF-resolution. J. Log. Program. **10**, 19–22 (1991)
23. Smullyan, R.M.: First-Order Logic. Springer, Heidelberg (1968)

Multiple Conclusion Rules in Logics with the Disjunction Property

Alex Citkin[✉]

Metropolitan Telecommunications, New York, NY 10041, USA
acitkin@gmail.com

Abstract. We prove that for the intermediate logics with the disjunction property any basis of admissible rules can be reduced to a basis of admissible m-rules (multiple-conclusion rules), and every basis of admissible m-rules can be reduced to a basis of admissible rules. These results can be generalized to a broad class of logics including positive logic and its extensions, Johansson logic, normal extensions of S4, n-transitive logics and intuitionistic modal logics.

Keywords: Intermediate logic · Admissible rule · Multiple conclusion rule · Basis of admissible rules

1 Introduction

The notion of admissible rule evolved from the notion of auxiliary rule: if in a given calculus (deductive system) S a formula B can be derived from a set of formulas $A_1, \ldots, A_n$, one can shorten derivations by using a rule $A_1, \ldots, A_n/B$. The application of such a rule does not extend the set of theorems, i.e. such a rule is admissible (permissible). In [24, p. 19] P. Lorenzen called the rules not extending the class of the theorems "zulässing", and the latter term was translated as "admissible", the term we are using nowadays. In [25] Lorenzen also linked the admissibility of a rule to existence of an elimination procedure.

Independently, P.S. Novikov, in his lectures on mathematical logic, had introduced the notion of derived rule: a rule $\mathcal{A}_1, \ldots, \mathcal{A}_n/\mathcal{B}$, where $\mathcal{A}_1, \ldots, \mathcal{A}_n, \mathcal{B}$ are variable formulas of some type, is derived in a calculus S if $\vdash_S \mathcal{B}$ holds every time when $\vdash_S \mathcal{A}_1, \ldots, \vdash_S \mathcal{A}_n$ hold (see [28, p. 30][1]). And he distinguished between two types of derived rules: a derived rule is strong, if $\vdash_S \mathcal{A}_1 \rightarrow (\mathcal{A}_2 \rightarrow \ldots (\mathcal{A}_n \rightarrow \mathcal{B}) \ldots)$ holds, otherwise a derived rule is weak.

For classical propositional calculus (CPC), the use of admissible rules is merely a matter of convenience, for every admissible for CPC rule $A_1, \ldots, A_n/B$ is derivable, that is $A_1, \ldots, A_n \vdash B$ (see, for instance [1]). It was observed by R. Harrop in [14] that the rule $\neg p \rightarrow (q \vee r)/(\neg p \rightarrow q) \vee (\neg p \rightarrow r)$ is admissible for the intuitionistic propositional calculus (IPC), but is not derivable in

---

[1] This book was published in 1977, but it is based on the notes of a course that P.S. Novikov taught in 1950th; A.V. Kuznetsov was recalling that P.S. Novikov had used the notion of derivable rule much earlier, in this lectures in 1940th.

© Springer International Publishing Switzerland 2016
S. Artemov and A. Nerode (Eds.): LFCS 2016, LNCS 9537, pp. 76–89, 2016.
DOI: 10.1007/978-3-319-27683-0_6

IPC. Later, in mid 1960s, A.V. Kuznetsov observed that the rule $(\neg\neg p \to p) \to$ $(p \vee \neg p)/((\neg\neg p \to p) \to \neg p) \vee ((\neg\neg p \to p) \to \neg\neg p)$ is also admissible for IPC, but not derivable. Another example of an admissible for IPC not derivable rule was found in 1971 by G. Mints (see [26]).

In 1974 A.V. Kuznetsov asked whether admissible for IPC rules have a finite basis, that is, whether there is a finite set R of admissible for IPC rules such that every admissible for IPC rule can be derived from R. Independently, in [10, Problem 40] H. Friedman asked whether the problem of admissibility for IPC is decidable, that is, whether there is a decision procedure that by a given rule r decides whether r is admissible for IPC. Also, in [30] W. Pogorzelski introduced a notion of structural completeness: as deductive system S is structurally complete if every admissible for S structural rule is derivable in S. Thus, CPC is structurally complete, while IPC is not. Naturally, a question which intermediate logics are structurally complete has been posed. Thus, for intermediate logics, and, later, for modal and various types of propositional (and not only propositional) logics, for a given logic L, first, we ask (a) whether L is structurally complete, that is, whether there are admissible for L not derivable rules; if L is not structurally complete, we ask (b) whether admissible for L rules have a finite, or at least recursive[2], basis; or, at last, (c) whether a problem of admissibility for L is decidable[3].

It was established by V. Rybakov (see [33,34]) that there is no finite basis of admissible for Int (and S4) rules, i.e. Kuznetsov's question has a negative answer, but the problem of admissibility for Int (and S4) is decidable, i.e. Friedman's problem has a positive answer. Later, using ideas from [33,34], V. Rybakov has constructed a basis of admissible rules for S4 (see [39]). For Int, P. Roziére (see [32]) and R. Iemhoff (see [15]), using different techniques, have found a recursive basis of admissible rules. Using this technique, R. Iemhoff has found the bases of admissible rules for different intermediate logics (see [16,17]). Some very useful information on admissibility in intermediate logics as well as in modal logics can be found in the book [37] by V. Rybakov.

In the review [22] on aforementioned book [37], M. Kracht suggested to study admissibility of multiple-conclusion rules: a rule $A_1, \ldots, A_n/B_1, \ldots, B_n$ is admissible for a logic L if every substitution that makes all the premises valid in L, makes at least one conclusion valid in L (see also [23]). A natural example of multiple-conclusion rule (called m-rule for short) admissible for IPC is the following rule, representing the disjunction property (DP for short): $DP := p \vee q/p, q$. That is, if a formula $A \vee B$ is valid in IPC, then at least one of the formulas A, B is valid in IPC (for more on disjunction property see [5]). It was reasonable to ask the same questions regarding m-rules: whether a given logic has

---

[2] Using idea from [8], it is not hard to show that if an intermediate logic has a recursively enumerable explicit basis of admissible rules, it has a recursive basis.

[3] In [6] A. Chagrov has constructed a decidable modal logic having undecidable admissibility problem, and gave a negative answer to V.Rybakov's question [35, Problem (1)]. The problem whether there exists a decidable intermediate logic with undecidable admissibility problem remains open.

admissible, not derived m-rules, whether m-rules have a finite or recursively enumerable basis, or whether the admissibility of m-rules is decidable. The bases of m-rules for a variety of intermediate and normal modal logics were constructed in [11–13, 19, 20].

For logics with the DP, there is a close relation between m-rules and rules: with each m-rule $r := \Gamma/\Delta$ one can associate a rule $r^q := \bigwedge \Gamma \vee q / \bigvee \Delta \vee q$, where variable q does not occur in formulas from Γ, Δ. Our goal is to prove that if m-rules $r_i, i \in I$ form a basis of m-rules admissible for a given intermediate logic L with the DP, then rules $r^q, i \in I$ form a basis of rules admissible for L (comp. [19, Theorem 3.1]). To prove this, we will use the main theorem from [7]. As a consequence, we obtain that for intermediate logics with the DP, each of the mentioned above problems for the m-rules and rules are equivalent. In the last section, we will discuss how this result can be extended beyond intermediate logics. In order to extend the results from intermediate logics to normal extension of S4, we are not using Gödel-McKinsey-Tarski translation; instead, we make a use of some common properties of the algebraic models (Heyting algebras and S4-algebras), and this gives us an ability to extend the results even further.

2 Background

2.1 Multiple-Conclusion Rules

We consider (propositional) formulas built in a usual way from the propositional variables from a countable set $\mathcal{P}$ and connectives from a finite set $\mathcal{C}$. By Fm we denote the set of all formulas, and by Σ we denote the set of all substitutions, that is the set of all mappings $\sigma : \mathcal{P} \to$ Fm. In a natural way, every substitution σ can be extended to a mapping Fm $\to$ Fm.

A *multiple-conclusion rule* (m-rule for short) is an ordered pair of finite sets of formulas $\Gamma, \Delta \subseteq$ Fm written as Γ/Δ; Γ is a set of *premises*, and Δ is a set of *conclusions*. A *rule* is an m-rule, that has the set of conclusions consisting of a single formula.

A *structural multiple-conclusion consequence relation* (m-consequence for short) is a binary relation $\vdash$ between finite sets of formulas for which the following holds: for any formula $A \in$ Fm and any finite sets of formulas $\Gamma, \Gamma', \Delta, \Delta' \subseteq$ Fm

(R) $A \vdash A$;
(M) if $\Gamma \vdash \Delta$, then $\Gamma \cup \Gamma' \vdash \Delta \cup \Delta'$;
(T) if $\Gamma, A \vdash \Delta$ and $\Gamma' \vdash A, \Delta'$, then $\Gamma \cup \Gamma' \vdash \Delta \cup \Delta'$;
(S) if $\Gamma \vdash \Delta$, then $\sigma(\Gamma) \vdash \sigma(\Delta)$ for each substitution $\sigma \in \Sigma$.

The class of all m-consequences will be denoted by $\mathcal{M}$.

Let $\vdash$ be an m-consequence and $r := \Gamma/\Delta$ be an m-rule. An m-rule r is *derivable* w.r.t. $\vdash$ (in written $\vdash$ r), if $\Gamma \vdash \Delta$.

Every collection R of m-rules defines an m-consequence $\vdash_R$, namely, the least m-consequence relative to which every rule from R is derivable:

$$\vdash_R := \bigcap \{\vdash \in \mathcal{M} \mid \vdash r \text{ for every } r \in R\}.$$

m-relation $\vdash_R$ always exists, because arbitrary meets of m-relations preserve derivability of m-rules.

An m-rule r is said to be *derivable from a set of m-rules* R (in written $R \vdash r$), if $\vdash_R$ r.

Every m-consequence $\vdash$ defines a logic $L(\vdash) \coloneqq \{A \in \mathsf{Fm} \mid \vdash A\}$. If L is a logic, an m-rule Γ/Δ is said to be *admissible* for L if for every substitution $\sigma \in \Sigma$

$$\sigma(\Gamma) \subseteq L \text{ entails } \sigma(\Delta) \cap L \neq \emptyset.$$

If L is a logic, by $Adm(L)$ we denote the set of all m-rules admissible for L, and by $Adm^{(1)}(L)$ we denote the set of all rules admissible for L.

Given a logic L, a set of m-rules $R \subseteq Adm(L)$ forms an *basis of admissible m-rules* (m-basis for short), if every rule $r \in Adm(L)$ is derivable from R; and a set of rules $R \subseteq Adm^{(1)}(L)$ forms a *basis of admissible rules* (s-basis for short), if every rule $r \in Adm^{(1)}(L)$ is derivable from R.

2.2 Algebraic Semantics

Basic Definitions. Algebraic models for intermediate logics are Heyting algebras, that is algebras $\langle A; \wedge, \vee, \rightarrow, \neg, \mathbf{1}, \mathbf{0} \rangle$, where $\langle A; \wedge, \vee, \mathbf{1}, \mathbf{0} \rangle$ is a bounded distributive lattice, and $\rightarrow, \neg$ are respectively a relative pseudo-complement and a pseudo-complement. The class of all Heyting algebras forms a variety that is denoted by $\mathcal{H}$.

Let **A** be a (Heyting) algebra, A be a formula, $r' \coloneqq A_1, \ldots, A_n/B$ be a rule and $r \coloneqq A_1, \ldots, A_n/B_1, \ldots, B_m$ be an m-rule. *A formula A is valid in a (Heyting) algebra* **A** (in written, $\mathbf{A} \models A$) if for every assignment $\nu : \mathcal{P} \rightarrow \mathbf{A}$ the value $\nu(A)$, that is, the value obtained by interpreting the connectives by operations of **A**, is $\mathbf{1}$. Accordingly, *rule r' is valid in* **A** (in written, $\mathbf{A} \models r'$), for every assignment ν, if $\nu(A_1) = \cdots = \nu(A_n) = \mathbf{1}$ yields $\nu(B) = \mathbf{1}$. And m-rule r if for every assignment ν, $\nu(A_1) = \cdots = \nu(A_n) = \mathbf{1}$ yields that at least for some $j = 1, \ldots, m$, $\nu(B_j) = \mathbf{1}$.

Let $\mathcal{K}$ be a class of algebras. If F is a family of formulas (R' is a family of rules, or R is a family of m-rules), then by $\mathcal{K} \models F$ ($\mathcal{K} \models R'$ or $\mathcal{K} \models R$) we mean that every formula (rule of m-rule) is valid in each algebra $\mathbf{A} \in \mathcal{K}$.

Immediately from the definition of validity of rule and the fact that for each non-degenerate Heyting algebras $\mathbf{A}, \mathbf{B}$ there is a homomorphism of **A** to **B**, we have the following:

Proposition 1. *Let r be a rule and $\mathbf{A}_i, i \in I$ be a family of algebras. Then, $\mathbf{A}_i \models r$ for all $i \in I$ if and only if $\prod_{i \in I} \mathbf{A}_i \models r$.*

Let us observe that for m-rules the situation is quite different: if **A** is a two-element Boolean algebra, then $\mathbf{A} \models DP$, but $\mathbf{A}^2 \not\models DP$.

It is not hard to see that any set of formulas F defines a variety $\mathbb{V}(F) \coloneqq \{\mathbf{A} \mid \mathbf{A} \models F\}$; any set R' of rules defines a quasivariety $\mathbb{Q}(R') = \{\mathbf{A} \mid \mathbf{A} \models R'\}$; any set R of m-rules defines a universal class $\mathbb{U}(R) = \{\mathbf{A} \mid \mathbf{A} \models R\}$).

On the other hand, if $\mathcal{K}$ is a family of algebras, by $\mathbb{V}(\mathcal{K})$, $\mathbb{Q}(\mathcal{K})$ and $\mathbb{U}(\mathcal{K})$, we denote respectively a variety, quasivariety and universal class generated by algebras in $\mathcal{K}$.

There is 1-1-correspondence between intermediate logics and non-trivial varieties of Heyting algebras. Moreover, there is 1-1correspondence between consequence relations and subquasivarieties of $\mathcal{H}$, and between m-consequences and universal subclasses of $\mathcal{H}$ (see, for instance, [3]). If $\mathcal{V}$ is a variety corresponding to a logic L, then a formula A is valid in L (a rule r' is admissible for L, or an m-rule r is m-admissible for L) if and only if $\mathbf{F}_\mathcal{V} \models A$ (accordingly $\mathbf{F}_\mathcal{V} \models r'$, or $\mathbf{F}_\mathcal{V} \models r$).

Let us note the following important property: if R is a set of m-rules (or rules) and r is an m-rule (or a rule), then $R \vdash r$ if and only if

$$\mathbf{A} \models R \text{ entails } \mathbf{A} \models r \text{ for every algebra } \mathbf{A} \in \mathcal{H}. \tag{1}$$

Well-Connected Algebras. An algebra $\mathbf{A}$ is said to be *well-connected*, if for every $\mathbf{a}, \mathbf{b} \in \mathbf{A}$, if $\mathbf{a} \vee \mathbf{b} = 1$, then $\mathbf{a} = 1$ or $\mathbf{b} = 1$.

The finite well-connected algebras are exactly subdirectly irreducible algebras. On the other hand, the free algebras of variety $\mathcal{H}$ are well-connected and they are not subdirectly irreducible.

Proposition 2. *Let $\mathbf{A}$ be a well-connected algebra, Γ/Δ be an m-rule and q be a variable not occurring in Γ/Δ. Then the following are equivalent*

(a) $\mathbf{A} \models \Gamma/\Delta$;
(b) $\mathbf{A} \models \Gamma/\bigvee_{B\in\Delta} B$;
(c) $\mathbf{A} \models \bigwedge_{A\in\Gamma} A \vee q/\bigvee_{B\in\Delta} B \vee q$;

Proof. (a) $\Rightarrow$ (b) is trivial.

(b) $\Rightarrow$ (a) due to well-connectedness of $\mathbf{A}$.

(b) $\Rightarrow$ (c). Suppose $\mathbf{A} \not\models \bigwedge_{A\in\Gamma} A \vee q/\bigvee_{B\in\Delta} B \vee q$. We need to prove that $\mathbf{A} \not\models \Gamma/\bigvee_{B\in\Delta} B$.

Indeed, let ν be a refuting valuation, that is

$$\bigwedge_{A\in\Gamma} \nu(A) \vee \nu(q) = 1_\mathbf{A} \text{ while } \bigvee_{B\in\Delta} \nu(B) \vee \nu(q) \neq 1_\mathbf{A}. \tag{2}$$

Then, clearly,

$$\bigvee_{B\in\Delta} \nu(B) \neq 1_\mathbf{A} \tag{3}$$

and

$$\nu(q) \neq 1_\mathbf{A}. \tag{4}$$

Due to well-connectedness of $\mathbf{A}$, from (2) and (4) we have

$$\bigwedge_{A\in\Gamma} \nu(A) = 1_\mathbf{A}. \tag{5}$$

And (5) together with (3) mean that ν is a refuting valuation for $\Gamma / \bigvee_{B \in \Delta} B$, that is, $\mathbf{A} \not\models \Gamma / \bigvee_{B \in \Delta} B$.

(c) $\Rightarrow$ (b). Since q does not occur in the formulas from Γ, Δ, we can substitule q with $\mathbf{0_A}$ and reduce (c) to (b). $\qquad \square$

The above Proposition can be restated in the following way:

Corollary 3. *Let* $\mathbf{A}$ *be a well-connected algebra,* r *be an m-rule and* q *be a variable not occurring in* r. *Then* $\mathbf{A} \models$ r *if and only if* $\mathbf{A} \models$ r^q.

3 The Case of Intermediate Logics

In this section we prove that for the intermediate logics with the disjunction property, any basis of admissible rules can be reduced to a basis of admissible m-rules (multiple-conclusion rules), and every basis of admissible m-rules can be reduced to a basis of admissible rules.

3.1 Reductions

We consider formulas in the signature $\wedge, \vee, \rightarrow, \neg, \bot, \top$. *Intermediate logic* is understood as a set of formulas L such that $\mathsf{Int} \subseteq \mathsf{L} \subseteq \mathsf{Cl}$, where Cl is classical logic, and closed under Modus Ponens and substitution. Clearly, for each intermediate logic L there is an m-consequence defining it: one can take a consequence relation that is defined by L (viewed as a set of axiom schemata) and by Modus Ponens. By $\vdash_{\mathsf{Int}}$ we denote a consequences relation defined by intuitionistic axiom schemata and the rule Modus Ponens. **In Sect. 3 we consider only m-consequences extending** $\vdash_{\mathsf{Int}}$ **and defining intermediate logics.**

A (intermediate) logic L enjoy the *disjunction property* (DP for short) if $(A \vee B) \in \mathsf{L}$ yields $A \in \mathsf{L}$ or $B \in \mathsf{L}$ for any formulas A, B. It is clear that L has the DP if and only if m-rule

$$\mathsf{DP} := p \vee q / p, q$$

is admissible for L.

Definition 1. *Let* r $:= \Gamma / \Delta$ *be an m-rule. The following rule is called a reduction of rule* r:

$$\mathsf{r}^\circ \leftrightharpoons \bigwedge_{A \in \Gamma} A / \bigvee_{B \in \Delta} B, \tag{6}$$

where $\bigwedge_{A \in \Gamma} A = \top$, *if* $\Gamma = \emptyset$, *and* $\bigvee_{B \in \Delta} B = \bot$, *if* $\Delta = \emptyset$.

Note, that the rule r$^\circ$ is always a single-conclusion rule.

It is not hard to see that rule DP expresses the DP.

We also will use the following m-rule:

$$\mathsf{NT} := \bot / \emptyset, \tag{7}$$

that we will call a *Non-Theorem rule*. Let us observe that NT is valid in every non-degenerate (i.e. having more than one element) algebra. Hence, NT is admissible for every intermediate logic. Moreover, for any set of formulas Γ, if a rule $\Gamma/\emptyset$ admissible for an intermediate logic L, this rule is derived from NT, for if NT is admissible for L, then $\Gamma \vdash_L \bot$, and, applying NT, we get $\Gamma/\emptyset$. Let us note that rules $\Gamma/\emptyset$ correspond to non-negative clauses (see [3]).

Proposition 4. *Let* R *be a set of rules from which rules* DP *and* NT *are derived, and* $r := \Gamma/\Delta$ *be an m-rule. Then*

$$R \vdash r \text{ if and only if } R \vdash r°. \tag{8}$$

Proof ($\Rightarrow$). Suppose $R \vdash r$, that is, $\Gamma \vdash_R \Delta$. We need to prove $R \vdash r°$, that is, we need to show that $\bigwedge_{A \in \Gamma} A \vdash_R \bigvee_{B \in \Delta} B$.

If $\Delta = \emptyset$, then $\Gamma \vdash_R \emptyset$ yields $\Gamma \vdash_R \bot$, because $\vdash_R$ is closed under (M). In its turn, $\Gamma \vdash_R \bot$ entails $\bigwedge_{A \in \Gamma} A \vdash_R \bot$, for $\Gamma \vdash_{Int} \bigwedge_{A \in \Gamma} A$, and $\vdash_{Int} \subseteq \vdash_R$.

The case $\Delta = \{B\}$ is trivial.

Suppose $\Delta = \{B_1, \ldots, B_n, B_{n+1}\}$. Let us prove that for all Δ'

$$\Gamma \vdash_R B_1, B_2, \Delta' \text{ yields } \Gamma \vdash_R B_1 \vee B_2, \Delta' \tag{9}$$

and then one can complete the proof of ($\Rightarrow$) by induction on cardinality of Δ.

Assume

$$\Gamma \vdash_R B_1, B_2, \Delta'. \tag{10}$$

Let us observe that $B_1 \vdash_{Int} B_1 \vee B_2$. Since $\vdash_{Int} \subseteq \vdash_R$, we can conclude that

$$B_1 \vdash_R B_1 \vee B_2. \tag{11}$$

From (10) and (11) by (T) we have

$$\Gamma \vdash_R B_1 \vee B_2, B_2, \Delta'. \tag{12}$$

Now, we use $B_2 \vdash_{Int} B_1 \vee B_2$, and by (M) and $\vdash_{Int} \subseteq \vdash_R$ we get

$$B_2 \vdash_R B_1 \vee B_2, \Delta'. \tag{13}$$

And from (12) and (13) by (T) we obtain

$$\Gamma \vdash_R B_1 \vee B_2, \Delta', \tag{14}$$

and this completes the proof of $\Rightarrow$.

Proof of ($\Leftarrow$). Suppose $R \vdash r°$, i.e. $\bigwedge_{A \in \Gamma} A \vdash_R \bigvee_{B \in \Delta} B$. Then, due to $\Gamma \vdash_{Int} \bigwedge_{A \in \Gamma} A$, we get $\Gamma \vdash_R \bigvee_{B \in \Delta} B$. If $\Delta = \emptyset$, due to $R \vdash NT$, we have $R \vdash \Gamma/\emptyset$, that is $R \vdash r$. If $\Delta \neq \emptyset$, due to $R \vdash DP$, we have $\bigvee_{B \in \Delta} B \vdash_R \Delta$. Thus, $\Gamma \vdash_R \Delta$, that is, $R \vdash r$. $\qquad \square$

Let us note that $\Rightarrow$ part of Proposition 4 holds for any sets of rules.

3.2 q-Reductions

Definition 2. *With every m-rule* $r := \Gamma/\Delta$ *and a variable* q *we associate a rule*

$$r^q := \bigwedge_{A \in \Gamma} A \vee q / \bigvee_{B \in \Delta} B \vee q. \tag{15}$$

The rule r^q we call a *q-reduction* of the rule r. If R is a set of m-rules and q is a variable, we let $R^q \rightleftharpoons \{r^q \mid r \in R\}$.

Proposition 5. *If an m-rule* Γ/Δ *is admissible for a given logic* L, *then for every substitution* $\sigma \in \Sigma$ *the m-rule* $\sigma(\Gamma)/\sigma(\Delta)$ *is admissible for* L.

Proof. The proof follows immediately from the definition of admissible m-rule and from the observation that a composition of two substitutions is a substitution. $\qquad\square$

Proposition 6. *Let a logic* L *enjoys DP and* q *be a variable not occurring in an m-rule* r. *Then m-rule* r *is admissible for* L *if and only if the rule* r^q *is admissible for* L.

Proof. Let $r := \Gamma/\Delta$ be admissible for L. We need to prove that for every substitution $\sigma \in \Sigma$,

$$\text{if } \sigma(\bigwedge_{A \in \Gamma} A \vee q) \in \mathsf{L} \text{ then } \sigma(\bigvee_{B \in \Delta} B \vee q) \in \mathsf{L}. \tag{16}$$

Indeed, if $\sigma(\bigwedge_{A \in \Gamma} A \vee q) \in \mathsf{L}$, by DP, one of the following holds

(a) $\sigma(\bigwedge_{A \in \Gamma}) \in \mathsf{L}$;
(b) $\sigma(q) \in \mathsf{L}$.

In the case (b), $\sigma(q) \in \mathsf{L}$ and, clearly, $\sigma(\bigvee_{B \in \Delta} B \vee q) = \sigma(\bigvee_{B \in \Delta} B) \vee \sigma(q) \in \mathsf{L}$.

In the case (a), $\sigma(\bigwedge_{A \in \Gamma}) \in \mathsf{L}$, hence, due to r is admissible for L, we have that $\sigma(B) \in \mathsf{L}$ for some $B \in \Delta$ and, hence, $\sigma(\bigvee_{B \in \Delta} B) = \bigvee_{B \in \Delta} \sigma(B) \in \mathsf{L}$. Therefore $\sigma(\bigvee_{B \in \Delta} B \vee q) \in \mathsf{L}$.

Conversely, suppose that r^q is admissible for L. Recall that the variable q is not occurring in Γ, Δ, and let ψ be a substitution such that $\psi : q \mapsto \bot$ and $\psi : p \mapsto p$ for all variables $p \neq q$. By virtue of Proposition 5, the following rule, obtained from r^q by applying ψ,

$$\bigwedge_{A \in \Gamma} A \vee \bot / \bigvee_{B \in \Delta} B \vee \bot. \tag{17}$$

is admissible for L.

Assume that σ is such a substitution that $\sigma(A) \in \mathsf{L}$ for all $A \in \Gamma$. Then, $\sigma(\bigwedge_{A \in \Gamma} A \vee \bot) \in \mathsf{L}$, and, due to rule (17) is admissible for L, we have

$$\sigma(\bigvee_{B \in \Delta} B \vee \bot) \in \mathsf{L}.$$

Since $\sigma(\bigvee_{B\in\Delta} B \vee \perp) = \bigvee_{B\in\Delta} \sigma(B) \vee \perp$ and the right hand formula is equivalent in Int to $\bigvee_{B\in\Delta} \sigma(B)$, we have

$$\bigvee_{B\in\Delta} \sigma(B) \in \mathsf{L}.$$

Due to logic L enjoys DP, for one of the formulas $B \in \Delta$ we have $\sigma(B) \in \mathsf{L}$, and this, by the definition of admissibility, means that the rule r is admissible for L. $\square$

Proposition 7. *For any logic* L, *if a rule* Γ/Δ *is admissible for* L, *so is the rule* $\bigwedge_{A\in\Gamma} A / \bigvee_{B\in\Delta} B$.

Proof. Straightforward. $\square$

A *problem of m-admissibility* (of admissibility) for a logic L is a problem of recognizing by a given m-rule (by a given rule) r whether r is admissible for L, i.e. whether r $\in Adm(\mathsf{L})$ (respectively, whether r $\in Adm^{(1)}(\mathsf{L})$). Thus, the problem of m-admissibility (of admissibility) for L is decidable if and only if the set $Adm(\mathsf{L})$ (the set $Adm^{(1)}(\mathsf{L})$) is recursive. Recall that two decision problems are *equivalent*, if they are reducible to each other.

Since $Adm^{(1)}(\mathsf{L}) \subseteq Adm(\mathsf{L})$ for every L and for every m-rule r we can effectively recognize whether r has a single conclusion, or not, that is, we can effectively recognize whether r $\in Adm^{(1)}(\mathsf{L})$, the decidability of the problem of m-admissibility yields the decidability of the problem of admissibility. In case when L enjoys the DP, the converse also holds.

Corollary 8. *For every logic* L *enjoying DP, the problems of m-admissibility and admissibility are equivalent. That is, the set* $Adm(\mathsf{L})$ *is recursive if and only if the set* $Adm^{(1)}(\mathsf{L})$ *is recursive.*

For instance, it is well known that Int enjoys the DP, hence from decidability of the admissibility of rules for Int (see [33]) it follows that the problem of m-admissibility for Int is decidable (in algebraic terms, that the universal theory of the free Heyting algebras is decidable [33, Theorem 10]) .

Remark 1. It is known from [31] that Medvedev's Logic ML is structurally complete and enjoys DP. From Proposition 6 it immediately follows that the rules DP, NT form m-basis of ML. It is not hard to see that m-rule DP is not derivable in ML. In fact, for any intermediate logic L m-rule DP is not derivable from rules admissible for L: all rules admissible for L are valid in the four-element Boolean algebra, while m-rule DP is not. Rule NT is not derivable in any intermediate logic too: all s-rules are valid in the degenerate algebra, while m-rule NT is not.

3.3   Reduction of Basis

In Sect. 3.2 we saw that for the logics with the DP, the admissibility of m-rule and its reduction are equivalent. In this section we will prove that the m-rules and their q-reductions are related even closer. More precisely, we will prove that using any basis of m-rules, one can effectively construct a basis of rules, and, using any basis of rules, one can construct a basis of m-rules.

Theorem 9. *Let* L *be a logic enjoying DP. Then the following holds*

(a) If rules R *form an s-basis, then m-rules* R $\cup$ {DP, NT} *form an m-basis.*
(b) If a set of m-rules R *forms an m-basis and* q *is a variable not occurring in any rule from* R, *then the rules* R^q *form an s-basis.*

Proof of (a)

Suppose R is a basis for L. We need to prove that every rule $\Gamma/\Delta \in Adm(\mathsf{L})$ is derived from R $\cup$ {DP, NT}. We know that every admissible for L rule $\Gamma/\emptyset$ is derivable from NT. Thus, we only need to prove that for every admissible m-rule Γ/Δ, where $\Delta \neq \emptyset$, we have $\Gamma \vdash_{\mathsf{R}\cup\{\mathsf{DP}\}} \Delta$.

Indeed, if $r := \Gamma/\Delta$ is an admissible m-rule, by Proposition 7, the rule $r^{\circ} = \bigwedge_{A\in\Gamma} A/\bigvee_{B\in\Delta} B$ is admissible for L. By our assumption, R is a basis, hence, $\mathsf{R} \vdash r^{\circ}$, that is,

$$\bigwedge_{A\in\Gamma} A \vdash_{\mathsf{R}} \bigvee_{B\in\Delta} B. \tag{18}$$

Next, we apply Proposition 4 and we obtain

$$\Gamma \vdash_{\mathsf{R}\cup\{\mathsf{DP},\mathsf{NT}\}} \Delta, \tag{19}$$

i.e. the set of m-rules R $\cup$ {DP, NT} forms an m-basis.

Proof of (b)

Suppose R is a basis of admissible m-rules and q is a variable not occurring in the rules from R. We need to prove that the set R^q forms a basis of admissible rules. For this, we will demonstrate that quasivariety $\mathcal{Q} := \mathbb{Q}(\mathsf{R}^q)$ is generated by algebra $\mathbf{F}$ – a free algebra of countable rank of the variety $\mathbb{V}(\mathsf{L})$, that is, we will show that $\mathcal{Q} = \mathbb{Q}(\mathbf{F})$.

Let $\mathbf{F}$ be a free algebra of $\mathbb{V}(\mathsf{L})$. Since L enjoys DP, $\mathbf{F}$ is well-connected. Due to rules R are admissible for L, the rules from R are valid in $\mathbf{F}$. Hence, by Proposition 2, all rules from R^q are valid in $\mathbf{F}$, that is, $\mathbf{F} \in \mathcal{Q}$. Therefore, $\mathbb{Q}(\mathbf{F}) \subseteq \mathcal{Q}$, and we need only to prove that $\mathbb{Q}(\mathbf{F}) \supseteq \mathcal{Q}$.

For contradiction: assume that $\mathbb{Q}(\mathbf{F}) \subset \mathcal{Q}$. Then there is an algebra $\mathbf{A} \in \mathcal{Q} \setminus \mathbb{Q}(\mathbf{F})$ in which all rules from R^q are valid. By virtue of [7, Theorem 1], the quasivariety $\mathcal{Q}$ is generated by its well-connected members. Thus, we can assume that $\mathbf{A}$ is well-connected. So, $\mathbf{A}$ is a well-connected algebra in which all rules from R^q are valid. Hence, by Proposition 2, all m-rules from R are valid in $\mathbf{A}$, hence, $\mathbf{A} \in \mathcal{U}$, where $\mathcal{U} = \mathbb{U}(\mathsf{R})$ is a universal class defined by all rules from R. Recall, that R forms an m-basis and, therefore, $\mathcal{U} = \mathbb{U}(\mathbf{F}) \subseteq \mathbb{Q}(\mathbf{F})$. Thus,

$$\mathbf{A} \in \mathcal{U} \subseteq \mathbb{Q}(\mathbf{F}),$$

and this contradicts that $\mathbf{A} \in \mathcal{Q} \setminus \mathbb{Q}(\mathbf{F})$.

Corollary 10. *Let* L *be a logic with the DP. Then* L *has a finite (recursive, recursively enumerable) s-basis if and only if* L *has a finite (recursive, recursively enumerable) m-basis.*

For example, since Int does not have a finite basis of admissible rules (see [34, Corollary 2]), it also does not have a finite basis of admissible m-rules [34, Theorem 9]. On the other hand, by adding m-rules DP and NT to a basis of s-rules, we can obtain an m-basis for Int (see [3, p.4235]).

Corollary 11. *If* L *is a logic with the DP and* R *is an s-basis, then* R^q *is an s-basis too. In other words, every intermediate logic with the DP has an s-basis consisting of q-extended rules.*

The bases consisting of q-reductions of rules also have the following important property.

Theorem 12. *Let* L *be a logic with the DP. If* R^q *is an independent s-basis, then* $R^q \cup \{DP, NT\}$ *is an independent m-basis.*

Proof. Assume that R^q is an independent basis. First, we will prove that $R^q \cup NT \nvdash DP$ and $R^q \cup DP \nvdash NT$. Indeed, since L is an intermediate logic and, therefore, L is consistent, the corresponding variety $\mathcal{V} := \mathbb{V}(L)$ is not trivial. Hence, its free algebra $\mathbf{F}_\mathcal{V}$ is not degenerate. Since all rules from R^q are admissible for L, we have $\mathbf{F}_\mathcal{V} \models R^q$ and, therefore, $\mathbf{F}_\mathcal{V}^2 \models R^q$ and $\mathbf{F}_\mathcal{V}^2 \models NT$, for $\mathbf{F}_\mathcal{V}^2$ is not degenerate. But $\mathbf{F}_\mathcal{V}^2 \nvDash DP$. Thus $R^q \cup NT \nvdash DP$.

$R^q \cup DP \nvdash NT$ simply because NT is invalid in degenerate algebra in which all rules from $R^q \cup DP$ are valid.

Now, let us assume that $r^q \in R^q$. We need to prove that $R_0^q \cup \{DP, NT\} \nvdash r^q$, where $R_0^q := R^q \setminus \{r^q\}$. Let us recall that basis R^q is independent, that is, $R_0^q \nvdash r^q$. Hence, there is an algebra $\mathbf{A}$ such that $\mathbf{A} \models R_0^q$ and $\mathbf{A} \nvDash r^q$. Since R_0^q consists of q-extensions of rules from R_0, we can apply [7, Lemma 1] and conclude that $\mathbf{A}$ is a subdirect product of well-connected algebras $\mathbf{A}_i, i \in I$ in which all rules R_0 are valid. Let $\mathcal{A} := \{\mathbf{A}_i, i \in I\}$. Due to all algebras from $\mathcal{A}$ being well-connected, $\mathcal{A} \models R_0$ yields $\mathcal{A} \models R_0^q$. Since $\mathbf{A} \nvDash r^q$, there is an algebra $\mathbf{A}_j \in \mathcal{A}$ such that $\mathbf{A}_j \nvDash r^q$. Now, let us observe that the rule DP is valid in every well-connected algebra and the rule NT is valid in every non-degenerate algebra. Hence $\mathbf{A}_j \models R_0^q \cup \{DP, NT\}$, but $\mathbf{A}_j \nvDash r^q$. And this completes the proof of the theorem. $\qquad\square$

Example 1. The m-bases for Gabbay-de Jongh logics D_n have been constructed in [13]: the m-rules $J_i, i \le n+1$ (see [13, Definition 17]) form a basis of m-rules of D_n for all n. By Theorem 9, $J_j^q, j \le n+1$ is a basis of admissible rules of logic D_n for all n.

4   Beyond Intermediate Logics

Let us note that all our proofs are based either on general properties of quasivarieties and universal classes or on the results from [7]. It was observed in

[7, Section 4] that all results from [7] can be extended to the logics for which there is a formula $R(p)$ such that $R(A) \vee R(B) \in \mathsf{L}$ yields $R(A) \in \mathsf{L}$ or $R(B) \in \mathsf{L}$, that is to the logics enjoying the DP relative to some formula $R(p)$. In this case, the corresponding algebraic model $\mathbf{A}$ is called well-connected if $R(\mathbf{a}) \vee R(\mathbf{b}) = \mathbf{1}_\mathbf{A}$ entails $R(\mathbf{a} = \mathbf{1}_\mathbf{A})$ or $R(\mathbf{a} = \mathbf{1}_\mathbf{A})$.

Thus, Theorems 9 and 12 hold for the following classes of logics

1. positive logic and its extensions (regarding admissibility for positive and Johansson logoics see [29]);
2. minimal (Johansson) [21] logic and its extensions;
3. logic KM (see [27]) and its extensions
4. K4 and its normal extensions;
5. intuitionistic modal logic MIPC (e.g. [2]) and its normal extensions;
6. n-transitive logics (e.g. [4]).

For instance, for logics $\mathsf{K4}, \mathsf{S4}, \mathsf{Grz}$ or GL one can take the m-basis (see [18]) and convert it into a basis of rules (see [18, Theorem 6.4.] where the same reduction as in Theorem 9 was used). Or one can take an s-basis of $\mathsf{S4}$ (see [39]), and convert it into an m-basis. Let us note that the proofs in [18, 39] are based on certain properties of Kripke models. On the other hand, an m-basis for logic GL can be obtained simply by extending the s-basis constructed in [9] by m-rules NT and $\Box_0 p \vee \Box_0 q / \Box_0 p, \Box_0 q$, where $\Box_0 \alpha \leftrightharpoons \Box \alpha \wedge \alpha$. Taking into account that GL does not have finite s-basis (see [36, Theorem 17]), we can conclude that GL has no finite m-basis.

For logics that have negation - but not the constant $\perp$ - one can use the following version of m-rule NT:

$$p \wedge \neg p / \emptyset. \tag{20}$$

Let us observe that m-rule NT is related to passive rules (see [38]). It was observed in [38] that for any consistent normal extension of logic D4, Rybakov's rule

$$\mathsf{RR} := \Diamond p \wedge \Diamond \neg p / \perp \tag{21}$$

forms an s-basis of all admissible passive s-rules. Let us note that RR is a consequence of the following m-rule:

$$\mathsf{RR}' := \Diamond p \wedge \Diamond \neg p / \emptyset, \tag{22}$$

and that rules RR, NT and rule RR$'$ are interderivable.

Since no m-rule with an empty set of conclusions is admissible for positive logic and its consistent extensions, in the proofs for positive logic we need to omit all references to NT. The same applies for those versions of Johansson's logics in which formula $(p \to p) \to \perp$ is valid. For the rest of the Johansson's logic versions, we can use the following modification of m-rule NT:

$$\mathsf{NT}' := (p \to p) \to \perp / \emptyset.$$

References

1. Belnap Jr., N.D., Leblanc, H., Thomason, R.H.: On not strengthening intuitionistic logic. Notre Dame J. Formal Log. **4**, 313–320 (1963)
2. Bezhanishvili, G.: Varieties of monadic heyting algebras. I. Stud. Logica **61**(3), 367–402 (1998)
3. Cabrer, L., Metcalfe, G.: Admissibility via natural dualities. J. Pure Appl. Algebra **219**(9), 4229–4253 (2015)
4. Chagrov, A., Zakharyaschev, M.: Modal Logic. Oxford Logic Guides. The Clarendon Press, vol. 35. Oxford University Press, Oxford Science Publications, New York (1997)
5. Chagrov, A., Zakharyashchev, M.: The disjunction property of intermediate propositional logics. Stud. Log. **50**(2), 189–216 (1991)
6. Chagrov, A.V.: A decidable modal logic for which the admissibility of inference rules is an undecidable problem. Algebra i Logika **31**(1), 83–93 (1992)
7. Citkin, A.: A note on admissible rules and the disjunction property in intermediate logics. Arch. Math. Logic **51**(1–2), 1–14 (2012)
8. Craig, W.: On axiomatizability within a system. J. Symb. Log. **18**, 30–32 (1953)
9. Fedorishin, B.R.: An explicit basis for the admissible inference rules in the Gödel-Löb logic GL. Sibirsk. Mat. Zh. **48**, 423–430 (2007). (In Russian)
10. Friedman, H.: One hundred and two problems in mathematical logic. J. Symb. Log. **40**, 113–129 (1975)
11. Goudsmit, J.: A note on extensions: admissible rules via semantics. In: Artemov, S., Nerode, A. (eds.) LFCS 2013. LNCS, vol. 7734, pp. 206–218. Springer, Heidelberg (2013)
12. Goudsmit, J.: Intuitionistic Rules Admissible Rules of Intermediate Logics. Ph.D. thesis, Utrech University (2015)
13. Goudsmit, J.P., Iemhoff, R.: On unification and admissible rules in Gabbay-de Jongh logics. Ann. Pure Appl. Log. **165**(2), 652–672 (2014)
14. Harrop, R.: Concerning formulas of the types $A \to B \bigvee C$, $A \to (Ex)B(x)$ in intuitionistic formal systems. J. Symb. Log. **25**, 27–32 (1960)
15. Iemhoff, R.: On the admissible rules of intuitionistic propositional logic. J. Symb. Log. **66**(1), 281–294 (2001)
16. Iemhoff, R.: Intermediate logics and Visser's rules. Notre Dame J. Formal Log. **46**(1), 65–81 (2005)
17. Iemhoff, R.: On the rules of intermediate logics. Arch. Math. Log. **45**(5), 581–599 (2006)
18. Jeřábek, E.: Admissible rules of modal logics. J. Log. Comput. **15**(4), 411–431 (2005)
19. Jeřábek, E.: Independent bases of admissible rules. Log. J. IGPL **16**(3), 249–267 (2008)
20. Jeřábek, E.: Canonical rules. J. Symb. Log. **74**(4), 1171–1205 (2009)
21. Kleene, S.C.: Introduction to Metamathematics. D. Van Nostrand Co. Inc., New York (1952)
22. Kracht, M.: Book review of [37]. Notre Dame J. Form. Log. **40**(4), 578–587 (1999)
23. Kracht, M.: Modal consequence relations, chap. 8. In: Blackburn, P., et al. (eds.) Handbook of Modal Logic. Studies in Logic and Practical Reasonong, vol. 3, pp. 491–545. Elsevier, New York (2007)

24. Lorenzen, P.: Einführung in die operative Logik und Mathematik. Die Grundlehren der mathematischen Wissenschaften in Einzeldarstellungen mit besonderer Berücksichtigung der Anwendungsgebiete, Bd. LXXVIII. Springer, Berlin-Göttingen-Heidelberg (1955)

25. Lorenzen, P.: Protologik. Ein Beitrag zum Begrndungsproblem der Logik. Kant-Studien 47, 1–4, Januay 1956, pp. 350–358. Translated in Lorenzen, P., Constructive Philosophy, 59–70. Univerisity of Massachusettes Press, Amherst (1987)

26. Mints, G.: Derivability of admissible rules. J. Sov. Math. 6 (1976), 417–421. Translated from Mints, G. E. Derivability of admissible rules. (Russian) Investigations in constructive mathematics and mathematical logic, V. Zap. Nauchn. Sem. Leningrad. Otdel. Mat. Inst. Steklov. (LOMI) 32, pp. 85–89 (1972)

27. Muravitsky, A.: Logic KM: a biography. In: Bezhanishvili, G. (ed.) Leo Esakia on Duality in Modal and Intuitionistic Logics. Outstanding Contributions to Logic, vol. 4, pp. 155–185. Springer, Netherlands (2014)

28. Novikov, P. S.: Konstruktivnaya matematicheskaya logika s tochki zreniya klassicheskoi [Constructive mathematical logic from the point of view of classical logic]. Izdat. "Nauka", Moscow (1977) With a preface by S. I. Adjan, Matematicheskaya Logika i Osnovaniya Matematiki. [Monographs in Mathematical Logic and Foundations of Mathematics] (in Russian)

29. Odintsov, S., Rybakov, V.: Unification and admissible rules for paraconsistent minimal Johanssons' logic $\mathbf{J}$ and positive intuitionistic logic $\mathbf{IPC}^+$. Ann. Pure Appl. Logic $\mathbf{164}$(7–8), 771–784 (2013)

30. Pogorzelski, W.A.: Structural completeness of the propositional calculus. Bull. Acad. Polon. Sci. Sér. Sci. Math. Astronom. Phys. $\mathbf{19}$, 349–351 (1971)

31. Prucnal, T.: Structural completeness of Medvedev's propositional calculus. Rep. Math. Logic $\mathbf{6}$, 103–105 (1976)

32. Roziére, P.: Régles admissibles en calcul propositionnel intuitionniste. Ph.D. thesis, Université Paris VII (1992)

33. Rybakov, V.V.: A criterion for admissibility of rules in the modal system S4 and intuitionistic logic. Algebra i Logika $\mathbf{23}$(5), 546–572 (1984)

34. Rybakov, V. V.: Bases of admissible rules of the modal system Grz and intuitionistic logic. Mat. Sb. (N.S.) vol. 128(170), 3, pp. 321–338, 446 (1985)

35. Rybakov, V.V.: Problems of admissibility and substitution, logical equations and restricted theories of free algebras. In: Logic, methodology and philosophy of science, VIII (Moscow, 1987), vol. 126. Stud. Logic Found. Math., pp. 121–139. North-Holland, Amsterdam (1989)

36. Rybakov, V.V.: Admissibility of rules of inference, and logical equations in modal logics that axiomatize provability. Izv. Akad. Nauk SSSR Ser. Mat. $\mathbf{54}$, 357–377 (1990). (In Russian)

37. Rybakov, V.V.: Admissibility of Logical Inference Rules. Studies in Logic and the Foundations of Mathematics, vol. 136. North-Holland Publishing Co., Amsterdam (1997)

38. Rybakov, V.V., Terziler, M., Gencer, C.: Unification and passive inference rules for modal logics. J. Appl. Non-Class. Log. $\mathbf{10}$(3–4), 369–377 (2000)

39. Rybakov, V.V.: Construction of an explicit basis for rules admissible in modal system S4. MLQ Math. Log. Q. $\mathbf{47}$(4), 441–446 (2001)

Multiple Conclusion Linear Logic: Cut Elimination and More

Harley Eades III[1][(✉)] and Valeria de Paiva[2]

[1] Computer and Information Sciences, Georgia Regents University,
Augusta, GA, Georgia
heades@gru.edu
[2] Nuance Communications, Sunnyvale, CA, USA

Abstract. Full Intuitionistic Linear Logic (FILL) was first introduced
by Hyland and de Paiva, and went against current beliefs that it was not
possible to incorporate all of the linear connectives, e.g. tensor, par, and
implication, into an intuitionistic linear logic. It was shown that their
formalization of FILL did not enjoy cut-elimination by Bierman, but
Bellin proposed a change to the definition of FILL in the hope to regain
cut-elimination. In this note we adopt Bellin's proposed change and give a
direct proof of cut-elimination. Then we show that a categorical model of
FILL in the basic dialectica category is also a LNL model of Benton and
a full tensor model of Melliès' and Tabareau's tensorial logic. Lastly, we
give a double-negation translation of linear logic into FILL that explicitly
uses par in addition to tensor.

Keywords: Full intuitionistic linear logic · Classical linear logic · Dialectica category · Cut-elimination · Tensorial logic · Linear/non-linear models · Categorical model · Proof theory · Par

1 Introduction

A commonly held belief during the early history of linear logic was that the
linear-connective par could not be incorporated into an intuitionistic linear logic.
This belief was challenged when de Paiva gave a categorical understanding of
Gödel's Dialectica interpretation in terms of dialectica categories [8,9].

Dialectica categories were initially believed to be models of intuitionistic
logic, but they are actually models of intuitionistic linear logic, containing the
linear connectives: tensor, implication, the additives, and the exponentials. Further work improved de Paiva's models to capture both intuitionistic and classical
linear logic. Armed with this semantic insight de Paiva gave the first formalization of Full Intuitionistic Linear Logic (FILL) [8]. FILL is a sequent calculus
with multiple conclusions in addition to multiple hypotheses. Logics of this type
go back to Gentzen's work on the sequent calculus for classical logic LK and for
intuitionistic logic LJ, and Maehara's work on LJ' [16,24]. The sequents in these
types of logics usually have the form $\Gamma \vdash \Delta$ where Γ and Δ are multisets of

© Springer International Publishing Switzerland 2016
S. Artemov and A. Nerode (Eds.): LFCS 2016, LNCS 9537, pp. 90–105, 2016.
DOI: 10.1007/978-3-319-27683-0_7

formulas. Sequents such as these are read as "the conjunction of the formulas in Γ imply the disjunction of the formulas in Δ". For a brief, but more complete history of logics with multiple conclusions see the introduction to [11].

Gentzen showed that to obtain intuitionistic logic one could start with the logic LK and then place a cardinality restriction on the right-hand side of sequents, however, this is not the only means of enforcing intuitionism. Maehara showed that in the propositional case one could simply place the cardinality restriction on the premise of the implication right rule, and leave all of the other rules of LK unrestricted. This restriction is sometimes called the Dragalin restriction, as it appeared in his AMS textbook [12]. The classical implication right rule has the form:

$$\frac{\Gamma, A \vdash B, \Delta}{\Gamma \vdash A \multimap B, \Delta} \text{ ImpR}$$

By placing the Dragalin restriction on the previous rule we obtain:

$$\frac{\Gamma, A \vdash B}{\Gamma \vdash A \multimap B} \text{ ImpR}$$

de Paiva's first formalization of FILL used the Dragalin restriction, see [8] p. 58, but Schellinx showed that this restriction has the unfortunate consequence of breaking cut-elimination [22].

Later, Hyland and de Paiva gave an alternative formalization of FILL with the intention of regaining cut-elimination [13]. This new formalization lifted the Dragalin restriction by decorating sequents with a term assignment. Hypotheses were assigned variables, and the conclusions were assigned terms. Then using these terms one can track the use of hypotheses throughout a derivation. They proposed a new implication right rule:

$$\frac{\Gamma, x : A \vdash t : B, \Delta \qquad x \notin \mathsf{FV}(\Delta)}{\Gamma \vdash \lambda x.t : A \multimap B, \Delta} \text{ ImpR}$$

Intuitionism is enforced in this rule by requiring that the variable being discharged, x, is not free terms annotating other conclusions. Unfortunately, this formalization did not enjoy cut-elimination either.

Bierman was able to give a counterexample to cut-elimination [4]. As Bierman explains the problem was with the left rule for the multiplicative disjunction par. The original rule was as follows:

$$\frac{\Gamma, x : A \vdash \Delta \qquad \Gamma', y : B \vdash \Delta'}{\Gamma, \Gamma', z : A \,\mathregular{⅋}\, B \vdash \mathsf{let}\, z \,\mathsf{be}\, (x \mathregular{⅋} -) \,\mathsf{in}\, \Delta \mid \mathsf{let}\, z \,\mathsf{be}\, (-\mathregular{⅋} y) \,\mathsf{in}\, \Delta'} \text{ ParL}$$

In this rule the pattern variables x and y are bound in each term of Δ and Δ' respectively. Notice that the variable z becomes free in every term in Δ and Δ'. Bierman showed that this rule mixed with the restriction on implication right prevents the usual cut-elimination step that commutes cut with the left rule for par. The main idea behind the counterexample is that in the derivation before commuting the cut it is possible to discharge z using implication right, but after the cut is commuted past the left rule for par, the variable z becomes free in more than one conclusion, and thus, can no longer be discharged.

In the conclusion of Bierman's note he gives an alternate left rule for par that he attributes to Bellin. This new left-rule is as follows:

$$\frac{\Gamma, x : A \vdash \Delta \quad \Gamma', y : B \vdash \Delta'}{\Gamma, \Gamma', z : A \,\invamp\, B \vdash \text{let-pat } z \,(x\invamp-)\, \Delta \mid \text{let-pat } z \,(-\invamp y)\, \Delta'} \text{ Parl}$$

In this rule let-pat $z\,(x\invamp-)\,t$ and let-pat $z\,(-\invamp y)\,t'$ only let-bind z in t or t' if $x \in FV(t)$ or $y \in FV(t')$. Otherwise the terms are left unaltered. Bellin showed that by adopting this rule cut-elimination can be proven by reduction to the cut-elimination procedure for proof nets for multiplicative linear logic with the mix rule [1]. However, this is an indirect proof that requires the adoption of proof nets.

Contributions. In this paper our main contribution is to give a direct proof of cut-elimination for FILL with Bellin's proposed par-left rule (Sect. 3). A direct proof accomplishes two goals: the first is to complete the picture of FILL Hyland and de Paiva started, and the second is to view a direct proof of cut-elimination as a means of checking the correctness of the formulation of FILL given here. The latter point is important for future work. Following the proof of cut-elimination we show that the categorical model of FILL called $\mathsf{Dial}_2(\mathsf{Sets})$, the basic dialectica category, is also a linear/non-linear model of Benton (Sect. 4) and a full tensor model of Melliès' and Tabareau's tensor logic (Sect. 5). Finally, we give a double-negation translation of multi-conclusion classical linear logic into FILL (Sect. 5.1). Due to the complexities of working in $\mathsf{Dial}_2(\mathsf{Sets})$ we have formalized all of the constructions and proofs used in Sects. 4 and 5 – although our formal verification does not include the double-negation translation in Sect. 5.1 – in the Agda proof assistant[1].

Related Work. The first formalization of FILL with cut-elimination was due to Braüner and de Paiva [5]. Their formalization can be seen as a linear version of LK with a sophisticated meta-level dependency tracking system. A proof of a FILL sequent in their formalization amounts to a classical derivation, π, invariant in what they call the FILL property:

– The hypothesis discharged by an application of the implication right rule in π is a dependency of the conclusion of the implication being introduced.

They were able to show that their formalization is sound, complete, and enjoys cut-elimination. In favor of the term assignment formalization given here over Braüner and de Paiva's formalization we can say that the dependency tracking system complicates both the definition of the logic and its use. However, one might conjecture that their system is more fundamental and hence more generalizable. It might be possible to prove cut-elimination of the term assignment

---

[1] The Agda development can be found at https://github.com/heades/cut-fill-agda.

formalization of FILL relative to Braüner and de Paiva's dependency tracking system by erasing the terms on conclusions and then tracking which variable is free in which conclusion. However, as we stated above a direct proof is more desirable than a relative one.

de Paiva and Pereira used annotations on the sequents of LK to arrive at full intuitionistic logic (FIL) with multiple conclusion that enjoys cut-elimination [11]. They annotate hypothesis with natural number indices, and conclusions with finite sets of indices. The sets of indices on conclusions correspond to the collection of the hypotheses that the conclusion depends on. Then they have a similar property to that of Braüner and de Paiva's formalization. In fact, the dependency tracking system is very similar to this formalization, but the dependency tracking has been collapsed into the object language instead of being at the meta-level.

Clouston et al. give both a deep inference calculus and a display calculus for FILL that admits cut-elimination [6]. Both of these systems are refinements of a larger one called bi-intuitionistic linear logic (BiLL). This logic contains every logical connective of FILL with the addition of the exclusion (or subtraction) connective. This connective can be defined categorically as the left-adjoint to par. Thus, exclusion is the dual to implication. A positive aspect to this work is that the resulting systems are annotation free, but at a price of complexity. Deep inference and display calculi are harder to understand, and their system requires FILL to be defined as a refinement of a system with additional connectives. We show in this paper that such a refinement is unnecessary. In addition, a term assignment system is closer to traditional logic than deep inference and display calculi, and it is closer, through the lens of the Curry-Howard-Lambek correspondence, to a type theoretic understanding of FILL.

2   Full Intuitionistic Linear Logic (FILL)

In this section we give a brief description of FILL. We first give the syntax of formulas, patterns, terms, and contexts. Following the syntax we define several meta-functions that will be used when defining the inference rules of the logic.

Definition 1. *The syntax for FILL is as follows:*

$$
\begin{aligned}
\textit{(Formulas)} \quad & A, B, C, D, E ::= \top \mid \bot \mid A \multimap B \mid A \otimes B \mid A \,\invamp\, B \\
\textit{(Patterns)} \quad & p ::= * \mid - \mid x \mid p_1 \otimes p_2 \mid p_1 \,\invamp\, p_2 \\
\textit{(Terms)} \quad & t, e ::= x \mid * \mid \circ \mid t_1 \otimes t_2 \mid t_1 \,\invamp\, t_2 \mid \lambda x.t \mid \text{let } t \text{ be } p \text{ in } e \mid t_1\, t_2 \\
\textit{(Left Contexts)} \quad & \Gamma ::= \cdot \mid x : A \mid \Gamma_1, \Gamma_2 \\
\textit{(Right Contexts)} \quad & \Delta ::= \cdot \mid t : A \mid \Delta_1, \Delta_2
\end{aligned}
$$

The formulas of FILL are standard, but we denote the unit of tensor as $\top$ and the unit of par as $\bot$. Patterns are used to distinguish between the various let-expressions for tensor, par, and their units. There are three different let-expressions:

Tensor: Par: Tensor Unit:

let t be $p_1 \otimes p_2$ in e let t be $p_1 \parr p_2$ in e let t be $*$ in e

In addition, each of these will have their own equational rules, see Fig. 2. The role each term plays in the overall logic will become clear after we introduce the inference rules.

At this point we introduce some syntax and meta-level functions that will be used in the definition of the inference rules for FILL. Left contexts are multisets of formulas each labeled with a variable, and right contexts are multisets of formulas each labeled with a term. We will often write $\Delta_1 \mid \Delta_2$ as syntactic sugar for Δ_1, Δ_2. The former should be read as "Δ_1 or Δ_2." We denote the usual capture-avoiding substitution by $[t/x]t'$, and its straightforward extension to right contexts as $[t/x]\Delta$. Similarly, we find it convenient to be able to do this style of extension for the let-binding as well.

Definition 2. *We extend let-binding terms to right contexts as follows:*

let t be p in $\cdot = \cdot$
let t be p in $(t' : A) = ($let t be p in $t') : A$
let t be p in $(\Delta_1 \mid \Delta_2) = ($let t be p in $\Delta_1) \mid ($let t be p in $\Delta_2)$

Lastly, we denote the usual function that computes the set of free variables in a term by $\mathsf{FV}(t)$, and its straightforward extension to right contexts as $\mathsf{FV}(\Delta)$.

$$\frac{}{x : A \vdash x : A} \ \text{Ax} \qquad \frac{\Gamma \vdash t : A \mid \Delta \quad \Gamma', y : A \vdash \Delta'}{\Gamma, \Gamma' \vdash \Delta \mid [t/y]\Delta'} \ \text{Cut} \qquad \frac{\Gamma \vdash \Delta}{\Gamma, x : \top \vdash \text{let } x \text{ be } * \text{ in } \Delta} \ \text{TL}$$

$$\frac{}{\cdot \vdash * : \top} \ \text{TR} \qquad \frac{\Gamma, x : A, y : B \vdash \Delta}{\Gamma, z : A \otimes B \vdash \text{let } z \text{ be } x \otimes y \text{ in } \Delta} \ \text{TenL}$$

$$\frac{\Gamma \vdash e : A \mid \Delta \quad \Gamma' \vdash f : B \mid \Delta'}{\Gamma, \Gamma' \vdash e \otimes f : A \otimes B \mid \Delta \mid \Delta'} \ \text{TenR} \qquad \frac{}{x : \bot \vdash \cdot} \ \text{PL} \qquad \frac{\Gamma \vdash \Delta}{\Gamma \vdash \circ : \bot \mid \Delta} \ \text{PR}$$

$$\frac{\Gamma, x : A \vdash \Delta \quad \Gamma', y : B \vdash \Delta'}{\Gamma, \Gamma', z : A \parr B \vdash \text{let-pat } z \ (x \parr -) \ \Delta \mid \text{let-pat } z \ (- \parr y) \ \Delta'} \ \text{ParL}$$

$$\frac{\Gamma \vdash \Delta \mid e : A \mid f : B \mid \Delta'}{\Gamma \vdash \Delta \mid e \parr f : A \parr B \mid \Delta'} \ \text{ParR} \qquad \frac{\Gamma \vdash e : A \mid \Delta \quad \Gamma', x : B \vdash \Delta'}{\Gamma, y : A \multimap B, \Gamma' \vdash \Delta \mid [y \, e/x]\Delta'} \ \text{ImpL}$$

$$\frac{\Gamma, x : A \vdash e : B \mid \Delta \quad x \notin \mathsf{FV}(\Delta)}{\Gamma \vdash \lambda x.e : A \multimap B \mid \Delta} \ \text{ImpR} \qquad \frac{\Gamma, x : A, y : B \vdash \Delta}{\Gamma, y : B, x : A \vdash \Delta} \ \text{ExL}$$

$$\frac{\Gamma \vdash \Delta_1 \mid t_1 : A \mid t_2 : B \mid \Delta_2}{\Gamma \vdash \Delta_1 \mid t_2 : B \mid t_1 : A \mid \Delta_2} \ \text{ExR}$$

Fig. 1. Inference rules for FILL

The inference rules for FILL are defined in Fig. 1. The PARL rule depends on the function let-pat $z \ p \ \Delta$ which we define next.

Definition 3. *The function* let-pat z p t *is defined as follows:*

$$\text{let-pat } z \ (x \bindnasrepma -) \ t = t \quad \text{let-pat } z \ (- \bindnasrepma y) \ t = t \quad \text{let-pat } z \ p \ t = \text{let } z \text{ be } p \text{ in } t$$
$$\text{where } x \notin \mathsf{FV}(t) \qquad \text{where } y \notin \mathsf{FV}(t)$$

It is straightforward to extend the previous definition to right-contexts, and we denote this extension by let-pat z p Δ.

The motivation behind this function is that it only binds the pattern variables in $x \bindnasrepma-$ and $-\bindnasrepma y$ if and only if those pattern variables are free in the body of the let. This overcomes the counterexample given by Bierman in [4].

The terms of FILL are equipped with an equivalence relation defined in Fig. 2. There are a number of α, β, and η like rules as well as several rules we call naturality rules. These rules are similar to the rules presented in [13].

$$\frac{y \notin \mathsf{FV}(t)}{t = [y/x]t} \ \text{Alpha} \qquad \frac{x \notin \mathsf{FV}(f)}{(\lambda x.f \ x) = f} \ \text{EtaFun} \qquad \frac{}{(\lambda x.e) \ e' = [e'/x]e} \ \text{BetaFun}$$

$$\frac{}{\text{let } * \text{ be } * \text{ in } e = e} \ \text{Eta1I} \qquad \frac{}{\text{let } u \text{ be } * \text{ in } [*/z]f = [u/z]f} \ \text{BetaI}$$

$$\frac{}{[\text{let } u \text{ be } * \text{ in } e/y]f = \text{let } u \text{ be } * \text{ in } [e/y]f} \ \text{NatI}$$

$$\frac{}{\text{let } e \otimes t \text{ be } x \otimes y \text{ in } u = [e/x, t/y]u} \ \text{Beta1Ten}$$

$$\frac{}{\text{let } u \text{ be } x \otimes y \text{ in } [x \otimes y/z]f = [u/z]f} \ \text{Beta2Ten}$$

$$\frac{}{[\text{let } u \text{ be } x \otimes y \text{ in } g/w]f = \text{let } u \text{ be } x \otimes y \text{ in } [g/w]f} \ \text{NatTen} \qquad \frac{}{u = \circ} \ \text{EtaParU}$$

$$\frac{}{(\text{let } u \text{ be } x \ \bindnasrepma - \text{ in } x) \ \bindnasrepma \ (\text{let } u \text{ be } - \bindnasrepma y \text{ in } y) = u} \ \text{EtaPar}$$

$$\frac{}{\text{let } u \ \bindnasrepma t \text{ be } x \ \bindnasrepma - \text{ in } e = [u/x]e} \ \text{Beta1Par}$$

$$\frac{}{\text{let } u \ \bindnasrepma t \text{ be } - \bindnasrepma y \text{ in } e = [t/y]e} \ \text{Beta2Par}$$

$$\frac{}{\text{let } t \text{ be } x \ \bindnasrepma - \text{ in } [u/x]f = [\text{let } t \text{ be } x \ \bindnasrepma - \text{ in } u/x]f} \ \text{Nat1Par}$$

$$\frac{}{\text{let } t \text{ be } - \bindnasrepma y \text{ in } [v/y]f = [\text{let } t \text{ be } - \bindnasrepma y \text{ in } v/y]f} \ \text{Nat2Par}$$

Fig. 2. Equivalence on terms

3   Cut-Elimination

FILL can be viewed from two different angles: i. as an intuitionistic linear logic with par, or ii. as a restricted form of classical linear logic. Thus, to prove cut-elimination of FILL one only needs to start with the cut-elimination procedure for intuitionistic linear logic, and then dualize all of the steps in the procedure for tensor and its unit to obtain the steps for par and its unit. Similarly, one could just as easily start with the cut-elimination procedure for classical linear logic, and then apply the restriction on the implication right rule producing a cut-elimination procedure for FILL.

The major difference between proving cut-elimination of FILL from classical or intuitionistic linear logic is that we must prove an invariant across each step in the procedure. The invariant is that if a derivation π is transformed into a derivation π', then the terms in the conclusion of the final rule applied in π must be transformable, when the equivalences defined in Fig. 2 are taken as left-to-right rewrite rules, into the terms in the conclusion of the final rule applied in π'.

We finally arrive at cut-elimination.

Theorem 1. *If $\Gamma \vdash t_1 : A_1, \ldots, t_i : A_i$ steps to $\Gamma \vdash t_1' : A_1, \ldots, t_i' : A_i$ using the cut-elimination procedure, then $t_j = t_j'$ for $1 \leq j \leq i$.*

Proof. The cut-elimination procedure given here is the standard cut-elimination procedure for classical linear logic except the cases involving the implication right rule have the FILL restriction. The structure of our procedure follows the structure of the procedure found in [17]. Throughout this proof we treat the equivalences defined in Fig. 2 as left-to-right rewrite rules. For the entire proof see the companion report [14].

Corollary 1 (Cut-Elimination). *Cut-elimination holds for FILL.*

4   Full LNL Models

One of the difficult questions considering the categorical models of linear logic was how to model Girard's exponential, !, which is read "of course". The ! modality can be used to translate intuitionistic logic into intuitionistic linear logic, and so the correct categorical interpretation of ! should involve a relationship between a cartesian closed category, and the model of intuitionistic linear logic.

de Paiva gave some of the first categorical models of both classical and intuitionistic linear logic in her thesis [8]. She showed that a particular dialectica category called $\mathsf{Dial}_2(\mathsf{Sets})$ is a model of FILL where ! is interpreted as a comonad which produces natural comonoids, see page 76 of [8].

Definition 4. *The category $\mathsf{Dial}_2(\mathsf{Sets})$ consists of*

- *objects that are triples, $A = (U, X, \alpha)$, where U and X are sets, and $\alpha \subseteq U \times X$ is a relation, and*

– *maps that are pairs* $(f, F) : (U, X, \alpha) \rightarrow (V, Y, \beta)$ *where* $f : U \rightarrow V$ *and* $F : Y \rightarrow X$ *such that*
  • *For any* $u \in U$ *and* $y \in Y$, $\alpha(u, F(y))$ *implies* $\beta(f(u), y)$.

Suppose $A = (U, X, \alpha)$, $B = (V, Y, \beta)$, *and* $C = (W, Z, \gamma)$. *Then identities are given by* $(\mathsf{id}_U, \mathsf{id}_X) : A \rightarrow A$. *The composition of the maps* $(f, F) : A \rightarrow B$ *and* $(g, G) : B \rightarrow C$ *is defined as* $(f; g, G; F) : A \rightarrow C$.

In her thesis de Paiva defines a particular class of dialectica categories called GC over a base category C, see page 41 of [8]. The category $\mathsf{Dial}_2(\mathsf{Sets})$ defined above can be seen as an instantiation of GC by setting C to be the category Sets of sets and functions between them. This model is a non-trivial model (all four units of the multiplicative and additive disjunction are different objects in the category), and does not model classical logic; see [8] page 58.

Seely gave a different, syntactic categorical model that confirmed that the of-course exponential should be modeled by a comonad [23]. However, Seely's model turned out to be unsound, as pointed out by Bierman [3]. This then prompted Bierman, Hyland, de Paiva, and Benton to define another categorical model called linear categories (Definition 5) that are sound, and also model ! using a monoidal comonad [3].

Definition 5. *A **linear category**, $\mathcal{L}$, consists of:*

– *A symmetric monoidal closed category $\mathcal{L}$,*
– *A symmetric monoidal comonad* $(!, \epsilon, \delta, m_{A,B}, m_I)$ *such that*
  • *For every free !-coalgebra* $(!A, \delta_A)$ *there are two distinguished monoidal natural transformations* $e_A : !A \rightarrow I$ *and* $d_A : !A \rightarrow !A \otimes !A$ *which form a commutative comonoid and are coalgebra morphisms.*
  • *If* $f : (!A, \delta_A) \rightarrow (!B, \delta_B)$ *is a coalgebra morphism between free coalgebras, then it is also a comonoid morphism.*

This definition is the one given by Bierman in his thesis, see [3] for full definitions.

Intuitionistic logic can be interpreted in a linear category as a full subcategory of the category of !-coalgebras for the comonad, see proposition 17 of [3].

Benton gave a more balanced view of linear categories called LNL models.

Definition 6. *A **linear/non-linear model (LNL model)** consists of*

– *a cartesian closed category* $(\mathcal{C}, 1, \times, \Rightarrow)$,
– *a SMCC* $(\mathcal{L}, I, \otimes, \multimap)$, *and*
– *a pair of symmetric monoidal functors* $(G, n) : \mathcal{L} \rightarrow \mathcal{C}$ *and* $(F, m) : \mathcal{C} \rightarrow \mathcal{L}$ *between them that form a symmetric monoidal adjunction with* $F \dashv G$.

See Benton, [2], for the definitions of symmetric monoidal functors and adjunctions.

A non-trivial consequence of the definition of a LNL model is that the ! modality can indeed be interpreted as a monoidal comonad. Suppose $(\mathcal{L}, \mathcal{C}, F, G)$ is a LNL model. Then the comonad is given by $(!, \epsilon :\, ! \to \mathsf{Id}, \delta :\, ! \to\, !!)$ where $! = FG$, ϵ is the counit of the adjunction and δ is the natural transformation $\delta_A = F(\eta_{G(A)})$, see page 15 of [2]. We recall the following result from Benton [2]:

Theorem 2 (LNL Models and Linear Categories)

  i. (Sect. 2.2.1 of [2]) Every LNL model is a linear category.
  ii. (Sect. 2.2.2 of [2]) Every linear category is a LNL model.

Proof. The proof of part i. is a matter of checking that each part of the definition of a linear category can be constructed using the definition of a LNL model. See lemmata 3–7 of [2].

As for the proof of part ii. Given a linear category we have a SMCC and so the difficulty of proving this result is constructing the CCC and the adjunction between both parts of the model. Suppose $\mathcal{L}$ is a linear category. Benton constructs the CCC out of the full subcategory of Eilenberg-Moore category $\mathcal{L}^!$ whose objects are exponentiable coalgebras denoted $\mathsf{Exp}(\mathcal{L}^!)$. He shows that this subcategory is cartesian closed, and contains the (co)Kleisli category, $\mathcal{L}_!$, Lemma 11 on page 23 of [2]. The required adjunction $F : \mathsf{Exp}(\mathcal{L}^!) \to L : G$ can be defined using the adjunct functors $F(A, h_A) = A$ and $G(A) = (!A, \delta_A)$, see lemmata 13–16 of [2].

Next we show that the category $\mathsf{Dial}_2(\mathsf{Sets})$ is a full version of a linear category. First, we extend the definitions of linear categories and LNL models to be equipped with the necessary categorical structure to model par and its unit.

Definition 7. *A **full linear category**, $\mathcal{L}$, consists of a linear category $(\mathcal{L}, \top, \otimes, \multimap, !A, e_A, d_A)$, a symmetric monoidal structure on L, $(\bot, \mathbin{⅋})$, and distribution natural transformations* $\mathsf{dist}_1 : A \otimes (B \mathbin{⅋} C) \to (A \otimes B) \mathbin{⅋} C$ *and* $\mathsf{dist}_2 : (A \mathbin{⅋} B) \otimes C \to A \mathbin{⅋} (B \otimes C)$. *The distributors must satisfy several coherence conditions which can all be found in [7].*

Definition 8. *A **full linear/non-linear model (full LNL model)** consists of a LNL model $(\mathcal{L}, \mathcal{C}, F, G)$, and a symmetric monoidal structure on L, $(\bot, \mathbin{⅋})$, as above.*

First we show that $\mathsf{Dial}_2(\mathsf{Sets})$ is a full linear category, and then using the proof by Benton that linear categories are LNL models we obtain that $\mathsf{Dial}_2(\mathsf{Sets})$ is a full LNL model. In order for this to work we need to know that $\mathsf{Dial}_2(\mathsf{Sets})$ has a symmetric *monoidal* comonad $(!, \epsilon, \delta, m_{A,B}, m_I)$. At the time of de Paiva's thesis it was not known that the comonad modeling the of-course exponential needed to be monoidal. We were able to show that the maps $m_{A,B}$ and m_I exist in the more general setting of dialectica categories, and thus, these maps exist in $\mathsf{Dial}_2(\mathsf{Sets})$. Intuitively, given two objects $A = (X, U, \alpha)$ and $B = (V, Y, \beta)$ of $\mathsf{Dial}_2(\mathsf{Sets})$ the map $m_{A,B}$ is defined as the pair $(\mathsf{id}_{U \times V}, F)$, where $F = (F_1, F_2)$, $F_1 : (U \times V) \Rightarrow (V \Rightarrow X)^* \to V \Rightarrow (U \Rightarrow X^*)$ and $F_2 : (U \times V) \Rightarrow (U \Rightarrow Y)^* \to$

$U \Rightarrow (V \Rightarrow Y^*)$. The maps F_1 and F_2 build the sequence of all the results of applying each function in the input sequence to the input coordinate.

We can now show that the following holds.

Lemma 1. *The category* $\mathsf{Dial}_2(\mathsf{Sets})$ *is a full linear category.*

Proof. We only give a sketch of the proof here, but for the full details see that companion report [14][2]. First, we must show that $\mathsf{Dial}_2(\mathsf{Sets})$ is a linear category. The majority of the linear structure of $\mathsf{Dial}_2(\mathsf{Sets})$ is in de Paiva's thesis [8]. We had to extend her definitions to show that the comonad $(!A, \delta, \epsilon)$ is monoidal, however, this is straightforward.

After showing that $\mathsf{Dial}_2(\mathsf{Sets})$ is a linear category one must show that $\mathsf{Dial}_2(\mathsf{Sets})$ is a model of par and its unit. This easily follows from de Paiva's thesis. The bifunctor which models par is given by de Paiva in Definition 10 on page 47 of [8].

Finally, $\mathsf{Dial}_2(\mathsf{Sets})$ must be distributive. The natural transformations $dist_1$ and $dist_2$ can be defined in terms of the maps $k : (A \otimes A') \otimes (B \mathbin{⅋} C) \to (A \otimes B) \mathbin{⅋} (A' \otimes C)$ and $k':(A \mathbin{⅋} B) \otimes (C \otimes C') \to (A \otimes C) \mathbin{⅋} (B \otimes C')$ given on page 52 of [8]. Set $A' = \top$ in k and $C = \top$ in k' to obtain $dist_1$ and $dist_2$ respectively. They can also be shown to satisfy the coherence conditions given in [7].

Corollary 2. *The category* $\mathsf{Dial}_2(\mathsf{Sets})$ *is a full LNL model.*

Proof. This follows directly from the previous lemma and Theorem 2 which shows that linear categories are LNL models.

The point of these calculations is to show that the several different axiomatizations available for models for linear logic are consistent and that a model proved sound and complete according to Seely's definition (using the Seely isomorphisms $!(A \times B) \cong !A \otimes !B$ and $!1 \cong \top$ but adding to it monoidicicty of the comonad) is indeed sound and complete as a LNL model too.

5   Tensorial Logic

Melliès and Tabareau introduced tensorial logic as a means of generalizing linear logic to a theory of tensor and a non-involutive negation called tensorial negation. That is, instead of an isomorphism $A = \neg\neg A$ we have only a natural transformation $A \to \neg\neg A$ [18]. Tensorial logic makes the claim that tensor and tensorial negation are more fundamental than tensor and negation defined via implication. This is at odds with FILL where implication is considered to be fundamental. In this section we show that multiplicative tensorial logic can be modeled by $\mathsf{Dial}_2(\mathsf{Sets})$ (Lemma 3) by showing that tensorial negation arises as a simple property of the implication in any SMCC (Lemma 2). While this is expected (after all negation being defined in terms of implication into absurdity

---

[2] This proof was formalized in the Agda proof assistant see the file https://github. com/heades/cut-fill-agda/blob/master/FullLinCat.agda.

is one of the staples of intuitionism) we think it bolsters our claim that linear implication is a fundamental connective that should not be redefined in terms of the multiplicative disjunction par. In any case, any model of FILL can be seen as a model of multiplicative tensorial logic.

A categorical model of tensorial logic is a symmetric monoidal category with a tensorial negation.

Definition 9. *A **tensorial negation** on a symmetric monoidal category $(\mathcal{C}, \otimes, I)$ is defined as a functor $\neg : \mathcal{C} \to \mathcal{C}^{\mathrm{op}}$ together with a family of bijections $\phi_{A,B,C} : \mathrm{Hom}_{\mathcal{C}}(A \otimes B, \neg C) \cong \mathrm{Hom}_{\mathcal{C}}(A, \neg(B \otimes C))$ natural in A, B, and C. Furthermore, the following diagram must commute:*

$$
\begin{array}{ccc}
\mathrm{Hom}(A \otimes (B \otimes C), \neg D) & \xrightarrow{\mathrm{Hom}(\alpha_{A,B,C},\mathrm{id}_{\neg D})} & \mathrm{Hom}((A \otimes B) \otimes C, \neg D) \\
\phi_{A,B\otimes C,D} \downarrow & & \downarrow \phi_{A\otimes B,C,D} \\
& & \mathrm{Hom}(A \otimes B, \neg(C \otimes D)) \\
& & \downarrow \phi_{A,B,C\otimes D} \\
\mathrm{Hom}(A, \neg((B \otimes C) \otimes D)) & \xrightarrow{\mathrm{Hom}(\mathrm{id}_A, \neg \alpha_{B,C,D})} & \mathrm{Hom}(A, \neg(B \otimes (C \otimes D)))
\end{array}
$$

The most basic form of tensorial logic is called multiplicative tensorial logic and only consists of tensor and a tensorial negation. The model of multiplicative tensorial logic is called a dialogue category.

Definition 10. *A **dialogue category** is a symmetric monoidal category equipped with a tensorial negation.*

At this point we show that tensorial negation arises as a simple property of implication, as is traditional.

Lemma 2. *In any monoidal closed category, $\mathcal{C}$, there is a natural bijection $\phi_{A,B,C,D} : \mathrm{Hom}_{\mathcal{C}}(A \otimes B, C \multimap D) \cong \mathrm{Hom}_{\mathcal{C}}(A, (B \otimes C) \multimap D)$. Furthermore, the following diagram commutes:*

$$
\begin{array}{ccc}
\mathrm{Hom}(A \otimes (B \otimes C), D \multimap E) & \xrightarrow{\mathrm{Hom}(\alpha_{A,B,C},\mathrm{id}_{D \multimap E})} & \mathrm{Hom}((A \otimes B) \otimes C, D \multimap E) \\
\phi_{A,B\otimes C,D,E} \downarrow & & \downarrow \phi_{A\otimes B,C,D,E} \\
& & \mathrm{Hom}(A \otimes B, (C \otimes D) \multimap E) \\
& & \downarrow \phi_{A,B,C\otimes D,E} \\
\mathrm{Hom}(A, ((B \otimes C) \otimes D) \multimap E) & \xrightarrow{\mathrm{Hom}(\mathrm{id}_A, \alpha_{B,C,D} \multimap E)} & \mathrm{Hom}(A, (B \otimes (C \otimes D)) \multimap E)
\end{array}
$$

Proof. Suppose $\mathcal{C}$ is a monoidal closed category. Then we can define $\phi(f : A \otimes B \to C \multimap D) = \mathrm{cur}(\alpha^{-1}; \mathrm{cur}^{-1}(f))$ and $\phi^{-1}(g : A \to (B \otimes C) \multimap D) = \mathrm{cur}(\alpha; \mathrm{cur}^{-1}(g))$. Clearly, these are mutual inverses, and hence, ϕ is a bijection. Naturality of ϕ easily follows. Lastly, the diagram given above also commutes. For the complete proof see the companion report [14].

Any model of FILL contains the unit of par, $\perp$, and thus, can be used to define the negation function $\neg A := A \multimap \perp$. Now replacing D and E in the previous result with $\perp$ yields the definition of tensorial negation.

Lemma 3. $\mathsf{Dial}_2(\mathsf{Sets})$ *is a model of multiplicative tensorial logic.*

Proof. We have already shown $\mathsf{Dial}_2(\mathsf{Sets})$ to be a model of FILL, and thus, has a SMCC structure as well as the negation functor, and thus, by Lemma 2 has a tensorial negation[3].

Extending a model of multiplicative tensorial logic with an exponential resource modality yields a model of full tensorial logic.

Definition 11. *A* ***resource modality*** *on a symmetric monoidal category* $(\mathcal{C}, \otimes, I)$ *is an adjunction with a symmetric monoidal category* $(\mathcal{M}, \otimes', I')$:

A resource modality is called an ***exponential resource modality*** *if* $\mathcal{M}$ *is cartesian where* $\otimes'$ *is the product and* I' *is the terminal object.*

A model of full tensorial logic is defined to be a model of multiplicative tensorial logic with an exponential resource modality. We now know that $\mathsf{Dial}_2(\mathsf{Sets})$ is a model of multiplicative tensorial logic. By constructing the co-Kleisli category which consists of the !-coalgebras as objects, and happens to be cartesian, we can show that $\mathsf{Dial}_2(\mathsf{Sets})$ is a model of full tensorial logic. The adjunction with the co-Kleisli category naturally arises from the proof that $\mathsf{Dial}_2(\mathsf{Sets})$ is a full LNL model (Corollary 2).

Lemma 4. *The category* $\mathsf{Dial}_2(\mathsf{Sets})$ *is a model of full tensorial logic.*

Proof. It suffices to show that there is an adjunction between $\mathsf{Dial}_2(\mathsf{Sets})$ and a cartesian category. Define the category $\mathsf{Dial}_2(\mathsf{Sets})_!$ as follows:

- Take as objects $(U, (U \Rightarrow X^*), \alpha_!)$ where U and X are sets, and $\alpha \subseteq U \times (U \Rightarrow X^*)$.
- Take as morphisms $(f, F) : (U, (U \Rightarrow X^*), \alpha_!) \to (V, (V \Rightarrow Y^*), \beta_!)$ where $f : U \to V$ and $F : (V \Rightarrow Y^*) \to (U \Rightarrow X^*)$ subject to the same condition on morphisms as $\mathsf{Dial}_2(\mathsf{Sets})$. Composition and identities are defined similarly to $\mathsf{Dial}_2(\mathsf{Sets})$.

   Next we must show that $\mathsf{Dial}_2(\mathsf{Sets})_!$ is cartesian. Notice that $\mathsf{Dial}_2(\mathsf{Sets})_!$ is a subcategory of $\mathsf{Dial}_2(\mathsf{Sets})$, and there is a functor $J : \mathsf{Dial}_2(\mathsf{Sets}) \to \mathsf{Dial}_2(\mathsf{Sets})_!$ which is defined equivalently to the endofunctor ! from the proof of Lemma 1.

---

[3] We give a full proof in the formalization see the file https://github.com/heades/ cut-fill-agda/blob/master/Tensorial.agda.

In fact, $\mathsf{Dial}_2(\mathsf{Sets})_!$ is the co-Kleisli category with objects free !-coalgebras and is cartesian closed [9]. However, we only need the fact that it is cartesian.

To show that $\mathsf{Dial}_2(\mathsf{Sets})_!$ is cartesian it suffices to show that J preserves the cartesian structure of $\mathsf{Dial}_2(\mathsf{Sets})$ – the proof that $\mathsf{Dial}_2(\mathsf{Sets})$ is cartesian can be found on page 48 of [8]. This follows by straightforward reasoning. For the complete proof see the companion report [14].

5.1   Double Negation Translation

In this section we show that we can use intuitionistic negation – which we showed was tensorial in the previous section – to define a negative translation of multi-conclusion linear logic (Fig. 3) into FILL where implication plays a central role. Melliès and Tabareau give a negative translation of the multiplicative fragment of linear logic into tensorial logic [19] using tensor as the main connective. For example, they define $(A \otimes B)^N = \neg(\neg(A)^N \otimes \neg(B)^N)$, and thus, they simulate par using tensor and negation. This definition would cause some syntactic issues with FILL, because the left-rule to par requires the let-pattern term defined in Definition 3, thus, encoding par in terms of tensor would require the let-pattern term to be used in the left-rule for tensor. While simulating par, using tensor and negation, can be seen as useful, in applications where only the tensor product can be actually calculated, in other applications we do have an extra bifunctor like par. This is true in the case of FILL, so we can modify Melliès and Tabareau's translation into one that better fits the source logical system.

$$\frac{}{A \vdash A}\ \text{LL\_Ax} \qquad \frac{\Gamma \vdash A, \Delta \quad \Gamma', A \vdash \Delta'}{\Gamma, \Gamma' \vdash \Delta, \Delta'}\ \text{LL\_Cut} \qquad \frac{\Gamma \vdash \Delta}{\Gamma, \top \vdash \Delta}\ \text{LL\_Tl}$$

$$\frac{}{\cdot \vdash \top}\ \text{LL\_Tr} \qquad \frac{\Gamma, A, B \vdash \Delta}{\Gamma, A \otimes B \vdash \Delta}\ \text{LL\_Tenl} \qquad \frac{\Gamma \vdash A, \Delta \quad \Gamma' \vdash B, \Delta'}{\Gamma, \Gamma' \vdash A \otimes B, \Delta, \Delta'}\ \text{LL\_Tenr}$$

$$\frac{}{\bot \vdash \cdot}\ \text{LL\_Pl} \qquad \frac{\Gamma \vdash \Delta}{\Gamma \vdash \bot, \Delta}\ \text{LL\_Pr} \qquad \frac{\Gamma, A \vdash \Delta \quad \Gamma', B \vdash \Delta'}{\Gamma, \Gamma', A \,\reflectbox{\&}\, B \vdash \Delta, \Delta'}\ \text{LL\_Parl}$$

$$\frac{\Gamma \vdash \Delta, A, B, \Delta'}{\Gamma \vdash \Delta, A \,\reflectbox{\&}\, B, \Delta'}\ \text{LL\_Parr} \qquad \frac{\Gamma \vdash A, \Delta \quad \Gamma', B \vdash \Delta'}{\Gamma, A \multimap B, \Gamma' \vdash \Delta, \Delta'}\ \text{LL\_Impl}$$

$$\frac{\Gamma, A \vdash B, \Delta}{\Gamma \vdash A \multimap B, \Delta}\ \text{LL\_Impr} \qquad \frac{\Gamma, A, B \vdash \Delta}{\Gamma, B, A \vdash \Delta}\ \text{LL\_Exl} \qquad \frac{\Gamma \vdash \Delta_1, A, B, \Delta_2}{\Gamma \vdash \Delta_1, B, A, \Delta_2}\ \text{LL\_Exr}$$

Fig. 3. Multi-Conclusion Linear Logic

The following definition defines a translation of linear logic formulas into FILL formulas.

Definition 12. *The following is the double-negation translation of linear logic into FILL:*

$$(\top)^N \quad = \top$$
$$(\bot)^N \quad = \bot$$
$$(A^{\bot})^N \quad = \neg((A)^N)$$
$$(A \mathbin{⅋} B)^N = \neg\neg((A)^N) \mathbin{⅋} \neg\neg((B)^N)$$
$$(A \otimes B)^N = \neg\neg((A)^N) \otimes \neg\neg((B)^N)$$

The main point of the previous definition is that because FILL has intuitionistic versions of all of the operators of linear logic we can give a very natural translation that preserves these operators by double negating their arguments.

At this point we need to extend the translation of linear logic formulas to sequents. However, we must be careful, because in FILL implication has the FILL restriction, and thus, if we choose the wrong translation then we will run into problems trying to satisfy the FILL condition. The method we employ here is to first translate a linear logic sequent into a single-sided sequent, and then translate that to FILL using the well-known translation. That is, it is easy to see that any linear logic sequent $A_1, \ldots, A_i \vdash B_1, \ldots, B_j$ is logically equivalent to the sequent $\cdot \vdash A_1^{\bot}, \ldots, A_i^{\bot}, B_1, \ldots, B_j$. Then we translate the latter into FILL as $x_1 : \neg((A_1^{\bot})^N), \ldots, x_i : \neg((A_i^{\bot})^N), y_1 : \neg((B_1)^N), \ldots, y_j : \neg((B_j)^N) \vdash \cdot$ for any free variables $x_1, \ldots, x_i$ and $y_1, \ldots, y_j$, but this is equivalent to $x_1 : \neg\neg((A_1)^N), \ldots, x_i : \neg\neg((A_i)^N), y_1 : \neg((B_1)^N), \ldots, y_j : \neg((B_j)^N) \vdash \cdot$. The reader may realize that this is indeed the translation of single-sided classical linear logic into single-conclusion intuitionistic linear logic. This translation also has the benefit that we do not have to worry about mentioning terms in the statement of the result.

Lemma 5 (Negative Translation). *If $A_1, \ldots, A_i \vdash B_1, \ldots, B_j$ is derivable, then for any unique fresh variables $x_1, \ldots, x_i$, and $y_1, \ldots, y_j$, the sequent $x_1 : \neg\neg((A_1)^N), \ldots, x_i : \neg\neg((A_i)^N), y_1 : \neg((B_1)^N), \ldots, y_j : \neg((B_j)^N) \vdash \cdot$ is derivable.*

Proof. This can be shown by induction on the assumed sequent. For the complete proof see the companion report [14].

6 Conclusion and Future Work

We first recalled the definition of full intuitionistic linear logic using the left rule for par proposed by Bellin in Sect. 2, but using only proof-theoretic methods, no proof nets. We then directly proved cut-elimination for FILL in Sect. 3 by adapting the well-known cut-elimination procedure for classical linear logic to FILL.

In Sect. 4 we showed that the category $\mathsf{Dial}_2(\mathsf{Sets})$, a model of FILL, is a full LNL model by showing that it is a full linear category, and then replaying the proof that linear categories are LNL models by Benton. Then in Sect. 5 we showed that $\mathsf{Dial}_2(\mathsf{Sets})$ is a model of full tensorial logic. The point of this exercise in categorical logic is to show that, despite linear logicians infatuation with linear negation, there is value in keeping all your connectives independent of each other. Only making them definable in terms of others, for specific applications.

Games, especially programming language games are the main motivation for Tensorial Logic and have been one of the sources of intuitions in linear logic all along. Since we are interested in the applications of tensorial logic to concurrency, we would like to see if our slightly more general framework can be applied to this task, just as well as tensorial logic.

Independently of the envisaged applications to programming, we are also interested in developing a "man in the street" game-like explanation for the finer-grained connectives of FILL, especially for par, the multiplicative disjunction. The second author has talked about games for FILL in the style of Lorenzen [10], building up on the work of Rahman [15,21]. Rahman showed that Lorenzen games could be defined for classical linear logic [20] and was able to define a sound and complete semantics in Lorenzen games for classical linear logic. While Rahman does mention that one could adopt a particular structural rule that enforces intuitionism, we have not seen a complete proof of soundness and completeness for this semantics. We plan to show that by adopting this rule we actually obtain a sound and complete semantics in Lorenzen games for FILL.

References

1. Bellin, G.: Subnets of proof-nets in multiplicative linear logic with MIX. J. Math. Struct. in CS. **7**, 663–669 (1997)
2. Benton, P.N.: A mixed linear and non-linear logic: Proofs, terms and models (preliminary report). Technical Report, UCAM-CL-TR-352, University of Cambridge Computer Laboratory (1994)
3. Bierman, G.M.: On Intuitionistic Linear Logic. Ph.D. thesis, Wolfson College, Cambridge (1993)
4. Bierman, G.M.: A note on full intuitionistic linear logic. J. Ann. Pure Appl. Logic **79**(3), 281–287 (1996)
5. Braüner, T., de Paiva, V.: A formulation of linear logic based on dependency-relations. In: Nielsen, M., Thomas, W. (eds.) CSL 1998, LNCS, vol. 1414, pp. 129–148. Springer, Heidelberg (1998)
6. Clouston, R., Dawson, J., Gore, R., Tiu, A.: Annotation-Free Sequent Calculi for Full Intuitionistic Linear Logic - extended version. CoRR, abs/1307.0289 (2013)
7. Cockett, J.R.B., Seely, R.A.G.: Weakly distributive categories. J. Pure Appl. Algebra **114**(2), 133–173 (1997)
8. de Paiva, V.: The Dialectica Categories. Ph.D. thesis, University of Cambridge (1988)
9. de Paiva, V.: Dialectica categories. In: Gray, J., Scedrov, A. (eds.) CSL 1989, AMS, vol. 92, pp. 47–62 (1989)
10. de Paiva, V.: Abstract: Lorenzen games for full intuitionistic linear logic. In: 14th International Congress of Logic, Methodology and Philosophy of Science (CLMPS) (2011)
11. de Paiva, V., Carlos, L.P.: A short note on intuitionistic propositional logic with multiple conclusions. MANUSCRITO - Rev. Int. Fil. **28**(2), 317–329 (2005)

12. Dragalin, A.G.: Mathematical Intuitionism. Introduction to Proof Theory, Translations of Mathematical Monographs. 67, AMS (1988)
13. Hyland, M., de Paiva, V.: Full intuitionistic linear logic (extended abstract). J. Ann. Pure Appl. Logic **64**(3), 273–291 (1993)
14. Eades III, H., de Paiva, V.: Compainion report: multiple conclusion intuitionistic linear logic and cut elimination. http://metatheorem.org/papers/FILL-report.pdf
15. Keiff, L.: Dialogical logic. In: Zalta, E.N. (ed.) The Stanford Encyclopedia of Philosophy. Summer 2011 edition (2011)
16. Maehara, S.: Eine darstellung der intuitionistischen logik in der klassischen. J. Nagoya Math. **7**, 45–64 (1954)
17. Melliés, P.-A.: Categorical semantics of linear logic. In: Curien, P.-L., Herbelin, H., Krivine, J.-L., Melliés, P.-A. (eds.) Interactive Models of Computation and Program Behaviour, Panoramas et Synthéses 27, Société Mathématique de France (2009)
18. Melliés, P., Tabareau, N.: Linear continuations and duality. Technical report, Université Paris VII (2008)
19. Melliès, P., Tabareau, N.: Resource modalities in tensor logic. J. Ann. Pure Appl. Logic **161**(5), 632–653 (2010)
20. Rahman, S.: Un desafio para las teorias cognitivas de la competencia log- ICA: los fundamentos pragmaticos de la semantica de la logica linear. Manuscrito **25**(2), 381–432 (2002)
21. Rahman, S., Keiff, L.: On how to be a dialogician. In: Vanderveken, D. (ed.) Logic, Thought and Action. Logic, Epistemology, and the Unity of Science, vol. 2, pp. 359–408. Springer, Netherlands (2005)
22. Schellinx, H.: Some syntactical observations on linear logic. J. Logic Comput. **1**(4), 537–559 (1991)
23. Seely, R.: Linear logic, *-autonomous categories and cofree coalgebras. CSL, AMS **92**, 371–381 (1989)
24. Takeuti, G.: Proof Theory. North-Holland, Amsterdam (1975)

The Online Space Complexity of Probabilistic Languages

Nathanaël Fijalkow[1,2,3](✉)

[1] LIAFA, Université Paris Diderot - Paris 7, Paris, France
[2] University of Warsaw, Warsaw, Poland,
[3] University of Oxford, Oxford, UK
nath@liafa.univ-paris-diderot.fr

Abstract. In this paper, we define the online space complexity of languages, as the size of the smallest abstract machine processing words sequentially and able to determine at every point whether the word read so far belongs to the language or not. The first part of this paper motivates this model and provides examples and preliminary results.

One source of inspiration for introducing the online space complexity of languages comes from a seminal paper of Rabin from 1963, introducing probabilistic automata, which suggests studying the online space complexity of probabilistic languages. This is the purpose of the second part of the current paper.

Keywords: Online complexity · Probabilistic languages · Automata · Online algorithms · Complexity theory

1 The Online Space Complexity

We introduce and study a new *complexity measure*, called *online space complexity*. The purpose of a complexity measure is to quantify the complexity of solving a given problem, focusing on a particular aspect. For instance, the most classical complexity measures are the *time and space complexity*, defined as the amount of time and space used by a Turing machine, while the *circuit complexity* counts the number of gates in a circuit; the *communication complexity* quantifies the amount of communication required when the input is spread among different agents.

The online complexity deals with the difficulty of observing the instance in an online fashion, *i.e.* one letter at a time. We consider deterministic abstract machines, which perform an action each time a letter is read. The task of the machine is to maintain enough information about the word read so far to answer boolean queries. The online space complexity focuses exclusively on space, *i.e.* the amount of information maintained.

A typical example is a machine presented with a sequence of a's and b's, which should at any point determine whether the sequence read so far contains

---

Work supported by the ANR STOCH-MC.

S. Artemov and A. Nerode (Eds.): LFCS 2016, LNCS 9537, pp. 106–116, 2016.
DOI: 10.1007/978-3-319-27083-0_8

exactly as many a's as b's. The canonical machine solving this problem uses as memory one counter taking integer values, the difference between the number of a's and the number of b's. This machine is of linear size, because after reading (up to) n letters it can be in at most $2n + 1$ different states; as we shall see, it is optimal, meaning that this problem has linear online space complexity.

Although, to the best of our knowledge, the following definition of the online space complexity is new, the concept of online computing is old and has been investigated in various scenarios and under various names. After giving the formal definitions, we will discuss further the relations between our framework and the existing ones.

1.1   Definitions

We fix an *alphabet* A, which is a finite set of letters. An instance of a problem is given by a *word*, which is a finite sequence of letters, often denoted $w = a_0 a_1 \cdots a_{n-1}$, where a_i's are letters from the alphabet A, *i.e.* $a_i \in A$. We say that w has length n. We denote A^* the set of all words and $A^{\leq n}$ the set of words of length at most n.

A computational problem is given by a set of words L, called a *language*; *i.e.* $L \subseteq A^*$. The online space complexity of a language measures the size of an abstract machine able to recognise the language in an online way: the machine processes words letter by letter, and must at any point be able to determine whether the word read so far belongs to the language or not.

The first definition that we give, that of a machine, is not new; it matches the classical definition of *deterministic automata*, except that the set of states is not assumed to be finite.

Definition 1 (Machine). *A machine is given by a (potentially infinite) set C of states, an initial state $c_0 \in C$, a transition function $\delta : C \times A \to C$ and a set of accepting states $\mathcal{A} \subseteq C$.*

When processing a word $w = a_0 a_1 \cdots a_{n-1}$, the machine assumes a sequence of states $c_0 c_1 \cdots c_n$, that we call the run on w, defined inductively by $c_{i+1} = \delta(c_i, a_i)$. It is unique, since we assume the machine to be deterministic and the transition function to be a total function. The last state of the run on w is denoted $c(w)$; the word w is accepted if $c(w)$ is accepting, *i.e.* if $c(w) \in \mathcal{A}$, and rejected otherwise. The language recognised by a machine is the set of words accepted by this machine.

Definition 2 (Size of a Machine). *The size of a machine is the function $s : \mathbb{N} \to \mathbb{N}$ defined by $s(n)$ being the number of different states reached by all words of length at most n. Formally:*

$$s(n) = \mathrm{Card} \left\{ c(w) \mid w \in A^{\leq n} \right\}.$$

Note that in complexity theory, it is usual to count the size of an object by the size of its description. Here, it would be natural, instead of counting of many different states are reached, how many bits are necessary to describe these states; this amounts to consider the logarithm of the number of states. We do not use this definition as it would erase the differences between, for instance, linearly many and quadratically many states.

For two functions $f, g : \mathbb{N} \to \mathbb{N}$, we say that f is smaller than g, denoted $f \leq g$, if it is true component wise: for all n, $f(n) \leq g(n)$.

Definition 3 (Online Space Complexity Class). *For a function $f : \mathbb{N} \to \mathbb{N}$, the class of languages* Online(f) *consists of all languages which are recognised by a machine of size smaller than f.*

1.2   Related Works

The definitions of online space complexity that we gave belong to the research area concerned with *online computing*. Unlike an *offline algorithm*, which has access to the whole input, an *online algorithm* is presented with its input in a restricted online way: it has to process it letter by letter. Various frameworks emerged from this versatile concept: we discuss three of them, *dynamic algorithms*, *streaming algorithms* and *competitive analysis of online algorithms*.

The field of dynamic algorithms, initiated by Patnaik and Immerman [6], focuses on the complexity of maintaining solutions to problems with online inputs. In this setting, the input can go through a series of changes, and the challenge is to store enough information to be able to solve the problem on the modified input. There are two differences between our approach and dynamic algorithms. The first is that whereas in our setting, the changes are only insertions, dynamic algorithms also consider deletions, and sometimes more complicated operations. The second is that the focus of dynamic algorithms is on the time complexity of the machines maintaining the information, whereas we consider instead the state space complexity of these machines, *i.e.* the number of different states they use.

The field of streaming algorithms, initiated by a series of papers (Munro and Paterson [5], then Flajolet and Martin [3], followed by the foundational paper of Alon, Matias and Szegedy [1]), focuses on algorithms having very limited available memory, much smaller than the input size. The challenge there is to use these constrained ressources to compute relevant information about the processed input, such as for instance statistics on frequency distributions. In this setting the input is a word, read letter by letter, and the focus is put on measuring the memory consumed by the machines processing the word, exactly is in our setting. The only difference is that streaming algorithms also have limited processing time per letter, whereas we abstract away this information, and only measure the state space complexity.

The field of competitive analysis of online algorithms, initiated by Sleator and Tarjan [8], and by Karp [4], compares the performances between offline and online algorithms. In this setting, each solution is assigned a real value, assessing its quality. An offline algorithm, having access to the whole input, can select the best solution. An online algorithm, however, has to make choices ignoring part of the input that is still to be read. The question is then whether there exists online algorithms that can perform nearly as good as offline algorithms, up to a competitive ratio. In this setting, the complexity of the machines is ignored, and the only question is what is the cost of making online choices rather than processing the whole input.

2   Preliminary Results

2.1   Remarks and Examples

We begin with a few simple remarks. The first remark is that the size of a machine satisfies the following inequality, for all n:

$$s(n) \leq 1 + \mathrm{Card}(A) + \mathrm{Card}(A^2) + \cdots + \mathrm{Card}(A^n) = \frac{\mathrm{Card}(A)^{n+1} - 1}{\mathrm{Card}(A) - 1}.$$

It follows that the maximal complexity of a language is exponential, and the online space complexity classes are relevant for functions smaller than exponential.

Definition 4 (Usual Online Space Complexity Classes).

- Online($Const$) *is the class of languages of constant online space complexity, defined as* $\bigcup_{K \in \mathbb{N}} \mathrm{Online}(n \mapsto K)$.
- Online(Lin) *is the class of languages of linear online space complexity, defined as* $\bigcup_{a \in \mathbb{N}} \mathrm{Online}(n \mapsto an)$.
- Online($Quad$) *is the class of languages of quadratic online space complexity, defined as* $\bigcup_{a \in \mathbb{N}} \mathrm{Online}(n \mapsto an^2)$.
- Online($Poly$) *is the class of languages of polynomial online space complexity, defined as* $\bigcup_{k \in \mathbb{N}} \mathrm{Online}(n \mapsto n^k)$.

We define completeness with respect to a online space complexity class as follows.

Definition 5 (Completeness for Complexity Classes).
   We say that L has linear online space complexity if:

- *(upper bound)* $L \in \mathrm{Online}(Lin)$, *i.e. there exists $a \in \mathbb{N}$ such that $L \in \mathrm{Online}(n \mapsto an)$,*
- *(lower bound) if $L \in \mathrm{Online}(f)$, then there exists $a \in \mathbb{N}$ such that for all n, $f(n) \geq an$.*

The definitions of L having constant, quadratic, polynomial or exponential complexity are similar.

We denote Reg the class of regular languages, *i.e.* those recognised by automata.

Theorem 1.

- Online($Const$) = Reg, *i.e. a language has constant online space complexity if, and only if, it is regular.*
- Online $\left(n \mapsto \frac{\mathrm{Card}(A)^{n+1}-1}{\mathrm{Card}(A)-1}\right)$ *contains all languages.*

The first item follows from the observation that deterministic automata are exactly machines with finitely many states. For the second item, consider a language L, we construct a machine recognising L of exponential size. Its set of states is A^*, the initial state is ε and the transition function is defined by $\delta(w, a) = wa$. The set of accepting states is simply L itself!

The languages defined in the following example will be studied later on to illustrate the lower bound techniques we give.

Example 1.

- Define $\mathrm{MAJ}_2 = \left\{w \in \{a, b\}^* \mid |w|_a > |w|_b\right\}$, the majority language over two letters. Here $|w|_a$ denotes the number of occurrences of the letter a in w. We construct a machine of linear size recognising MAJ_2: its set of states is $\mathbb{Z}$, the integers, the letter a acts as $+1$ and the letter b as -1, and the set of accepting states is $\mathbb{N}$, the positive integers. After reading the word w, the state is $|w|_a - |w|_b$, implying that the machine has linear size. So $\mathrm{MAJ}_2 \in$ Online(Lin), and we will show in Subsect. 2.2 that this bound is tight: MAJ_2 has linear online space complexity.
- Define $\mathrm{EQ} = \left\{w \in \{a, b, c\}^* \mid |w|_a = |w|_b = |w|_c\right\}$. We construct a machine of quadratic size recognising EQ: its set of states is $\mathbb{Z}^2$, the letter a acts as $(+1, +1)$, the letter b as $(-1, 0)$, the letter c as $(0, -1)$, and the only accepting state is 0. After reading the word w, the state is $(|w|_a - |w|_b, |w|_a - |w|_c)$, implying that the machine has quadratic size. So $\mathrm{EQ} \in$ Online($Quad$), and we will show in Subsect. 2.2 that this bound is tight: L_2 has quadratic online space complexity.
- Define $\mathrm{SQUARES} = \{ww \mid w \in A^*\}$. We will show in Subsect. 2.3 that $\mathrm{SQUARES}$ has exponential online space complexity.

2.2   Lower Bounds Using Formal Language Theory

We present a first technique to give lower bounds on the online space complexity, relying on the notion of left quotients.

Let w be a finite word, define its left quotient with respect to L by

$$w^{-1}L = \{u \mid wu \in L\}.$$

A well known result from automata theory states that the existence of a minimal deterministic automaton, called the *syntactic automaton*, whose states of the left quotients.

This construction extends *mutatis mutandis* when dropping the assumption that the automaton has finitely many states, *i.e.* going from deterministic automata to machines as we defined them. The statement, however, is more involved, and gives precise lower bounds on the online space complexity of the language.

Formally, consider a language L, we define the syntactic machine of L, denoted $\mathcal{M}_L$, as follows. We define the set of states as the set of all left quotients: $\{w^{-1}L \mid w \in A^*\}$. The initial state is $\varepsilon^{-1}L$, and the transition function is defined by $\delta(w^{-1}L, a) = (wa)^{-1}L$. Finally, the set of accepting states is $\{w^{-1}L \mid w \in L\}$. Denote s_L the size of the syntactic machine of L.

Theorem 2.

– $\mathcal{M}_L$ *recognises* L, *so* $L \in \mathrm{Online}(s_L)$,
– *for all* f, *if* $L \in \mathrm{Online}(f)$, *then* $f \geq s_L$.

Note that the implies the existence of a minimal function f such that $L \in \mathrm{Online}(f)$; in other words $L \in \mathrm{Online}(f)$ if, and only if, $f \geq s_L$. This was not clear a priori, because the order on functions is partial.

The first item is routinely proved, as in the case of automata. For the second item, assume towards contradiction that there exists f such that $L \in \mathrm{Online}(f)$ and $f \not\geq s_L$, *i.e.* there exists n such that $f(n) < s_L(n)$. Consider a machine $\mathcal{M}$ recognising L of size f. Since $f(n) < s_L(n)$, there exists two words u and v of length n such that $u^{-1}L \neq v^{-1}L$ but $c(u) = c(v)$, *i.e.* in $\mathcal{M}$ the words u and v lead to the same state. The left quotients $u^{-1}L \neq v^{-1}L$ being different, there exists a word w such that $uw \in L$ and $vw \notin L$, or the other way around. But $c(uw) = c(vw)$ since $\mathcal{M}$ is deterministic, so this state must be both accepting and rejecting, contradiction.

The view using left quotients is very powerful, as it gives the exact online space complexity of a language. However, it is sometimes hard to deal with, as it may involve complicated word combinatorics. We illustrate it on some of the examples introduced in Subsect. 2.1.

To prove for instance that L has an online space complexity at least linear using this technique, one has to exhibit, for infinitely many n, a family of linearly many words (*i.e.*, at least an words, for some constant a) of length at most n that induce pairwise distinct left quotients.

Example 2.

– Fix n. The words $a^{n+k}b^{n-k}$, for $0 \leq k \leq n$, have length $2n$ and all induce pairwise distinct left quotients, since $a^{n+k}b^{n-k} \cdot b^p \in \mathrm{MAJ}_2$ if, and only if, $p < 2k$. It follows from Theorem 2 that MAJ_2 has linear online space complexity.
– Fix n. The words $a^{2n-p-q}b^p c^q$, for $0 \leq p, q \leq n$, have length $2n$ and all induce pairwise distinct left quotients, since $a^{2n-p-q}b^p c^q \cdot a^{k+\ell}b^{2n-k}c^{2n-\ell} \in \mathrm{EQ}$ if, and only if, $k = p$ and $\ell = q$. It follows from Theorem 2 that EQ has quadratic online space complexity.

2.3   Lower Bounds Using Communication Complexity

We present a second technique to give lower bounds on the online space complexity. This is inspired by the field of streaming algorithms, which made several use of this idea.

Rather than giving a generic lower bound technique, we illustrate the ideas by giving an exponential lower bound for SQUARES, the language of squares, *i.e.* words of the form ww for some word w. Note that we are here using a very big hammer for a very simple result; the first technique using left quotients would very easily give an exponential lower bound.

We consider the following communication problem: Alice receives a word u of length n, and Bob another word v of length n. The goal for Bob is to determine whether $u = v$, with the least amount of communication between them. It is well known that no protocol can solve this problem using less than n bits of communication; the optimal protocol is simply for Alice to send her whole input u to Bob.

The idea now is to use a machine recognising SQUARES to construct a problem solving the above communication problem, thereby obtaining lower bounds on the size of such a machine. Denote s the size of the machine, and fix n. Consideer an input of length $2n$, which we denote $w = uv$ where u and v have length n. We construct the following protocol: Alice runs the machine on her input u, and communicates to Bob the state reached. Bob takes over from there, running the machine on his input, starting from the state sent by Alice. The last state he obtains determines whether the whole input, *i.e.* w, belongs to SQUARES, or equivalently whether $u = v$.

Now, to communicate the state reached after reading u, Alice only needs $\log(s(n))$ (indeed, if there are $s(n)$ different states, then they can be all described using $\log(s(n))$ bits). The lower bound on the communication problem implies than $\log(s(n)) \geq n$, *i.e.* $s(n) \geq 2^n$. It follows that SQUARES has exponential online space complexity.

2.4   Comparison to Circuit Complexity

We conclude this section by observing that the examples studied above show that online space complexity and circuit complexity are orthogonal. Indeed:

- The language MAJ has linear online space complexity (*small*), but does not belong to AC^0, *i.e.* it has a rather big circuit complexity. Another example of a language having a small online space complexity and a big circuit complexity is Parity: it is regular, and recognised by a machine of size 2, but does not belong to AC^0.
- The language SQUARES has exponential online space complexity (*large*), but it has a very small circuit complexity: it is recognised by a family of circuits of linear size of constant depth.

3   The Online Space Complexity of Probabilistic Automata

In his seminal paper introducing probabilistic automata [7], Rabin devotes a section to "approximate calculation of matrix products", which is related to, and inspired, online space complexity. In the end of this section, Rabin states a result, without proof; the aim of this section is to substantiate this claim, *i.e.* formalising and proving the result.

We start by defining probabilistic automata, state Rabin's claim, and prove that it holds true.

3.1   Probabilistic Automata

Let Q be a finite set of states. A distribution over Q is a function $\delta : Q \to [0, 1]$ such that $\sum_{q \in Q} \delta(q) = 1$. We denote $\mathcal{D}(Q)$ the set of distributions over Q.

Definition 6 (Probabilistic Automaton). *A probabilistic automaton $\mathcal{A}$ is given by a finite set of states Q, a transition function $\phi : A \to (Q \to \mathcal{D}(Q))$, an initial state $q_0 \in Q$, and a set of final states $F \subseteq Q$.*

In a transition function ϕ, the quantity $\phi(a)(s, t)$ is the probability to go from the state $s \in Q$ to the state $t \in Q$ reading the letter a. A transition function naturally induces a morphism $\phi : A^* \to (Q \to \mathcal{D}(Q))$. We denote $\mathbb{P}_{\mathcal{A}}(s \xrightarrow{w} t)$ the probability to go from a state s to a state t reading w on the automaton $\mathcal{A}$, *i.e.* $\phi(w)(s, t)$.

The *acceptance probability* of a word $w \in A^*$ by $\mathcal{A}$ is $\sum_{t \in F} \phi(w)(q_0, t)$, which we denote $\mathbb{P}_{\mathcal{A}}(w)$.

The following threshold semantics was introduced by Rabin [7].

Definition 7 (Probabilistic Language). *Let $\mathcal{A}$ be a probabilistic automaton, it induces the probabilistic language*

$$L^{>\frac{1}{2}}(\mathcal{A}) = \left\{ w \in A^* \mid \mathbb{P}_{\mathcal{A}}(w) > \frac{1}{2} \right\}.$$

3.2   Substantiating the Claim

In a section called "approximate calculation of matrix products" in the paper introducing probabilistic automata [7], Rabin asks the following question: is it possible, given a probabilistic automaton, to construct an algorithm which reads words and compute the acceptance probability in an online fashion? He then shows that this is possible under some restrictions on the probabilistic automaton, and concludes the section by stating that *"an example due to R. E. Stearns shows that without assumptions, a computational procedure need not exist"*. The example is not given, and to the best of the author's knowledge, has never been published anywhere.

In this section, we substantiate this claim, in the framework of online space complexity as we defined it. Note that the formalisation of Rabin's claim is subject to discussions, as for instance Rabin asks whether the acceptance probability can be computed up to a given precision; in our setting, the acceptance probability is not actually computed, but only compared to a fixed threshold, following Rabin's definition of probabilistic languages.

The following result shows that there exists a probabilistic automaton defining a language of maximal (exponential) online space complexity.

Theorem 3. *There exists a probabilistic automaton $\mathcal{A}$ such that $L^{>\frac{1}{2}}(\mathcal{A})$ has exponential online space complexity.*

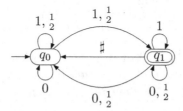

Fig. 1. The initial state is marked by an ingoing arrow and the accepting state by an outgoing arrow. The first symbol over a transition is a letter (either 0, 1, or ♯). The second symbol (if given) is the probability of this transition. If there is only one symbol, then the probability of the transition is 1.

In the original paper introducing probabilistic automata, Rabin [7] gave an example of a probabilistic automaton $\mathcal{A}$ computing the binary decomposition function (over the alphabet $\{0,1\}$), denoted bin, *i.e.* $\mathbb{P}_{\mathcal{A}}(u) = \text{bin}(u)$, defined by

$$\text{bin}(a_1 \dots a_n) = \frac{a_1}{2^n} + \dots + \frac{a_n}{2^1}$$

(*i.e.* $0.a_n \dots a_1$ in binary). We show that adding one letter and one transition to this probabilistic automaton gives an automaton with exponential online space complexity. This example appeared in [2].

The automaton $\mathcal{A}$ is represented in Fig. 1. The alphabet is $A = \{0, 1, ♯\}$. The only difference between the automaton proposed by Rabin [7] and this one is the transition over ♯ from q_1 to q_0. As observed by Rabin, a simple induction shows that for u in $\{0,1\}^*$, we have $\mathbb{P}_{\mathcal{A}}(u) = \text{bin}(u)$.

Let $w \in A^*$, it decomposes uniquely into $w = u_1 ♯ u_2 ♯ \cdots ♯ u_k$, where $u_i \in \{0,1\}^*$. Observe that $\mathbb{P}_{\mathcal{A}}(w) = \text{bin}(u_1) \cdot \text{bin}(u_2) \cdots \text{bin}(u_k)$.

Consider a machine recognising $L^{>\frac{1}{2}}(\mathcal{A})$ and fix n. The binary decomposition function maps words of length n to rationals of the form $\frac{a}{2^n}$, for $0 \le a < 2^n$. Consider two words u and v in $\{0,1\}^*$ of length n, we show that $(u1)^{-1}L^{>\frac{1}{2}}(\mathcal{A}) \neq (v1)^{-1}L^{>\frac{1}{2}}(\mathcal{A})$.

Without loss of generality assume $\text{bin}(u1) < \text{bin}(v1)$; observe that $\frac{1}{2} \leq \text{bin}(u1) < \text{bin}(v1)$. There exists w in $\{0,1\}^*$ such that $\text{bin}(u1) \cdot \text{bin}(w) < \frac{1}{2}$ and $\text{bin}(v1) \cdot \text{bin}(w) > \frac{1}{2}$: it suffices to choose w such that $\text{bin}(w)$ is in $\left(\frac{1}{2\text{bin}(v1)}, \frac{1}{2\text{bin}(u1)} \right)$, which exists by density of the dyadic numbers in $(0,1)$. Thus, $(u1)^{-1}L^{>\frac{1}{2}}(\mathcal{A}) \neq (v1)^{-1}L^{>\frac{1}{2}}(\mathcal{A})$, and we exhibited exponential many words having pairwise distinct left quotients. It follows from Theorem 2 that $L^{>\frac{1}{2}}(\mathcal{A})$ has exponential online space complexity.

4   Conclusion

We introduced a new complexity measure called the online space complexity, quantifying how hard it is to solve a problem when the input is given in an online fashion, focusing on the space consumption.

We considered the online space complexity of probabilistic automata, as hinted by Rabin in [7], and showed that probabilistic automata give rise to languages of high (maximal) online space complexity.

We mention some directions for future research about online space complexity.

The first is to give characterisations of the natural online space complexity classes (linear, quadratic, polynomial). Such characterisations could be in terms of logics, as it is done in descriptive complexity, or algebraic, as it is done in the automata theory. The canonical example is languages of constant online space complexity, which are exactly regular languages, defined by monadic second-order logic.

A second direction would be to extend the framework of online space complexity to quantitative queries. Indeed, we defined here the online space complexity of a language, $i.e.$ of qualitative queries: a word is either inside, or outside the language.

A third intriguing question is the existence of a dichotomy for the online space complexity of probabilistic automata. Is the following conjecture true: for every probabilistic language, its online space complexity is either polynomial or exponential?

References

1. Alon, N., Matias, Y., Szegedy, M.: The space complexity of approximating the frequency moments. In: STOC 1996, pp. 20–29 (1996). http://doi.acm.org/10.1145/237814.237823
2. Fijalkow, N., Skrzypczak, M.: Irregular behaviours for probabilistic automata. In: RP 2015, pp. 33–36 (2015)
3. Flajolet, P., Martin, G.N.: Probabilistic counting algorithms for data base applications. J. Comput. Syst. Sci. **31**(2), 182–209 (1985). http://dx.doi.org/10.1016/0022-0000(85)90041-8
4. Karp, R.M.: On-line algorithms versus off-line algorithms: how much is it worth to know the future? In: IFIP 1992, pp. 416–429 (1992)

5. Munro, J.I., Paterson, M.: Selection and sorting with limited storage. Theor. Comput. Sci. **12**, 315–323 (1980). http://dx.doi.org/10.1016/0304-3975(80)90061-4
6. Patnaik, S., Immerman, N.: Dyn-FO: A parallel, dynamic complexity class. In: PODS 1994, pp. 210–221 (1994). http://doi.acm.org/10.1145/182591.182614
7. Rabin, M.O.: Probabilistic automata. Inf. Control **6**(3), 230–245 (1963). http://dx.doi.org/10.1016/S0019-9958(63)90290-0
8. Sleator, D.D., Tarjan, R.E.: Amortized efficiency of list update and paging rules. Commun. ACM **28**(2), 202–208 (1985). http://doi.acm.org/10.1145/2786.2793

Type Theoretical Databases

Henrik Forssell[1], Håkon Robbestad Gylterud[2]([⊠]), and David I. Spivak[3]

[1] Department of Informatics, University of Oslo, Oslo, Norway
[2] Department of Mathematics, Stockholm University, Stockholm, Sweden
gylterud@math.su.se
[3] Department of Mathematics, MIT, Cambridge, MA, USA

Abstract. We show how the display-map category of finite simplicial complexes can be seen as representing the totality of database schemas and instances in a single mathematical structure. We give a sound interpretation of a certain dependent type theory in this model, and show how it allows for the syntactic specification of schemas and instances and the manipulation of the same with the usual type-theoretic operations. We indicate how it allows for the posing of queries. A novelty of the type theory is that it has non-trivial context constants.

Keywords: Dependent type theory · Simplicial sets · Relational databases

1 Introduction

Databases being, essentially, collections of (possibly interrelated) tables of data, a foundational question is how to best represent such collections of tables mathematically in order to study their properties and ways of manipulating them. The relational model, essentially treating tables as structures of first-order relational signatures, is a simple and powerful representation. Nevertheless, areas exist in which the relational model is less adequate than in others. One familiar example is the question of how to represent partially filled out rows or missing information. Another, more fundamental perhaps, is how to relate instances of different schemas, as opposed to the relatively well understood relations between instances of the same schema. Adding to this an increasing need to improve the ability to relate and map data structured in different ways suggests looking for alternative and supplemental ways of modelling tables more suitable to "dynamic" settings. It seems natural, in that case, to try to model tables of different shapes as living in a single mathematical structure, facilitating their manipulation across different schemas.

We investigate, here, a novel way of representing data structured in systems of tables which is based on simplicial sets and type theory rather than sets of relations and first-order logic. Formally, we present a soundness theorem (Theorem 1) for a certain dependent type theory with respect to a rather simple category of (finite, abstract) simplicial complexes. An interesting type-theoretic feature of this is that

S. Artemov and A. Nerode (Eds.): LFCS 2016, LNCS 9537, pp. 117–129, 2016.
DOI: 10.1007/978-3-319-27683-0_9

the type theory has context constants, mirroring that our choice of "display maps" does not include all maps to the terminal object. But from the database perspective the interesting aspect is that this category can in a natural way be seen as a category of tables; collecting in a single mathematical structure—an indexed or fibered category—the totality of schemas and instances.

This representation can be introduced as follows. Let a schema S be presented as a finite set $\mathbf{A}$ of attributes and a set of relation variables over those attributes. One way of allowing for missing information or partially filled out rows is to assume that whenever the schema has a relation variable R, say over attributes $A_0, \ldots, A_n$, it also has relation variables over all non-empty subsets of $\{A_0, \ldots, A_n\}$. So a partially filled out row over R is a full row over such a "partial relation" or "part-relation" of R. To this we add the requirement that the schema does not have two relation variables over exactly the same attributes[1]. This requirement means that a relation variable can be identified with the set of its attributes. Together with the first requirement, this means that the schema can be seen as a downward closed sub-poset of the positive power set of the set of attributes $\mathbf{A}$. Thus a schema is an (abstract) *simplicial complex*—a combinatorial and geometric object familiar from algebraic topology.

The key observation is now that an instance of the schema S can also be regarded as a simplicial complex, by regarding the data as attributes and the tuples as relation variables. Accordingly, an instance over S is a schema of its own, and the fact that it is an instance of S is "displayed" by a certain projection to S. Thus the category $\mathbb{S}$ of finite simplicial complexes and morphisms between them form a category of schemas which includes, at the same time, all instances of those schemas; where the connection between schema and instance is given by a collection D of maps in $\mathbb{S}$ called *display* maps.

We show, essentially, that $\mathbb{S}$ together with this collection D of maps form a so-called *display-map category* [7], a notion originally developed in connection with categorical models of dependent type theory. First, this means that the category $\mathbb{S}$ has a rich variety of ready-made operations that can be applied to schemas and instances. For example, the so-called dependent product operation can be used to model the natural join operation. Second, it is a model of dependent type theory. We specify a dependent type theory with context constants and a sound interpretation which interprets contexts as schemas and types as instances. This interpretation is with respect to the display-map category $(\mathbb{S}, D)$ in its equivalent form as an indexed category. The context constants are interpreted as distinguished single relation variable schemas (or *relation schemas* in the terminology of [1]), reflecting the special status of such schemas. The type theory

---

[1] Coming from reasons having to do with simplicity and wanting to stay close to the view of tables as relations, this requirement, and indeed the structure of the schemas we are considering, does mean that a certain care has to be taken with attribute names at the modeling level. For instance whether one should, when faced with two tables with exactly the same attributes, collect these into one table (possibly with an extra column), rename some attributes, or introduce new "dummy" or "relation name" attributes to keep the two tables apart. For reasons of space, we do not discuss these issues here.

allows for the syntactic specification of both schemas and instances, and formally derives the answers to queries; for instance, using the dependent product of the type theory, the elements of the natural join of two instances can be derived in the type theory.

We focus, in the space available here, on the basic presentation of the model and the type theory (Sects. 2 and 3, respectively). In Sect. 4 we then give a few brief indications of the use and further development of the model and the type theory: we give a suggestion of how the dependent structure of the model can be put to use to model e.g. updates; and we sketch the introduction and use of large types such as, and in particular, a *universe*. The universe allows reasoning generically about classes of instances of in the type theory itself, without having to resort to the metalanguage, and provides the basis for e.g. a precise, formal definition and analysis of the notion of *query* in this setting.

2   The Model

2.1   Complexes

We fix the following terminology and notation, adjusting the standard terminology somewhat for our purposes. More details can be found in [4]. A background on simplicial complexes and simplicial sets can be found in e.g. [5,6]. The question of whether vertices or attributes should be ordered is not essential for our presentation here, and is swept under the rug.

A *simplicial complex*, or just *complex*, X consists of the union of a finite set X_0 and a collection $X_{\geq 1}$ of non-empty, non-singleton subsets of X_0, satisfying the condition that if x is a set in $X_{\geq 1}$, then all non-empty, non-singleton subsets of x are in $X_{\geq 1}$. It is convenient to also allow singleton subsets, and identify them with the elements of X_0. The *natural order* on X is then given by subset inclusion. The elements of X_0 are referred to as *vertices*. The elements of $X_{\geq 1}$ as *faces*. For $n \geq 0$, we write X_n for the set of elements of X with size $n+1$, and refer to them as *faces of dimension n*. Accordingly, vertices are seen as having dimension 0. We use square rather than curly brackets for faces, e.g. $[A, B]$ rather than $\{A, B\}$.

A morphism $f : X \longrightarrow Y$ of complexes is a function $f_0 : X_0 \longrightarrow Y_0$ satisfying the condition that for all x in X the image $f_0(x)$ is in Y (again, with a singleton identified with its element). Thus a morphism of complexes can be seen as a morphism of posets by setting $f(x)$ to be the image $f_0(x)$.

A morphism $f : X \longrightarrow Y$ of complexes is said to be a *display map* if $x \in X_n$ implies $f(x) \in Y_n$ for all n. Thus display maps are the maps that "preserve dimension". Display maps are also the "local isomorphisms", in the sense that for $x \in X$ the restriction $f \restriction_{(\downarrow x)} : (\downarrow x) \to (\downarrow f(x))$ is an isomorphism. Note that a display map need not be a injection of vertices.

A poset P that is isomorphic to a complex can clearly be uniquely rewritten to a complex with the same set of vertices as P. For that reason, and as it is occasionally notationally convenient and can yield more intuitive examples, we allow ourselves to extend the notion of complex to any poset that is isomorphic to a complex. We say that it is a *strict* complex if we need to emphasize that it is of the form defined above.

2.2   Schemas and Instances

Schemas. A *simplicial schema* is a complex with the natural order reversed. We consider the resulting poset as a category, in the usual way. If $x \leq y$ in the natural order, we write $\delta_x^y : y \longrightarrow x$ for the corresponding arrow in the simplicial schema. In the context of simplicial schemas, we use "attribute" and "relation variable" synonomously with "vertex" and "face", respectively. Let $\mathbb{S}$ be the category of simplicial schemas and morphisms, and $\mathbb{S}_d$ be the category of simplicial schemas and display maps.

With respect to the traditional notion of schema, a simplicial schema X can be thought of as given in the usual way by a finite set of attributes $X_0 = \{A_0, \ldots, A_{n-1}\}$ and a set of relational variables $X_{\geq 1} = \{R_0, \ldots R_{m-1}\}$, each with a specification of column names in the form of a subset of X_0, but with the restrictions (1) that no two relation variables are over exactly the same attributes; and (2) for any non-empty subset of the attributes of a relation variable there exists a relation variable over (exactly) those attributes. As with "complex", we henceforth mostly drop the word "simplicial" and simply say "schema".

The category $\mathbb{S}_d$ contains in particular the *n-simplices* Δ_n and the *face maps*. Recall that the the n-simplex Δ_n is the complex given by the set $\{0, \ldots, n\}$ as vertices and all non-empty, non-singleton subsets as faces. For $0 \leq i \leq n + 1$, the face map $d_i^n : \Delta_n \longrightarrow \Delta_{n+1}$ is the morphism of complexes defined by the vertex function $k \mapsto k$, if $k < i$ and $k \mapsto k + 1$ else. These satisfy the simplicial identities $d_i^{n+1} \circ d_j^n = d_{j-1}^{n+1} \circ d_i^n$ if $i < j$. As a schema, Δ_n is the schema of a single relation on $n + 1$ attributes named by numbers $0, \ldots, n$ (and all its "generated" part-relations). The face map $d_i^n : \Delta_n \longrightarrow \Delta_{n+1}$ can be seen as the inclusion of the relation variable $[0, \ldots, i - 1, i + 1, \ldots, n + 1]$ in Δ_{n+1}. These schemas and morphisms play a special role in Sect. 3 where they are used to specify general schemas and instances syntactically.

Example 1. Let S be the schema the attributes of which are A, B, C and the relation variables $R : AB$ and $Q : BC$, with indicated column names. From a "simplicial" point of view, S is the category

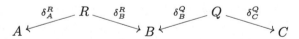

Replacing R with $[A, B]$ and Q with $[B, C]$ (and inverting the order) yields a strict complex. For another example, the 2-simplex Δ_2 can be seen as a schema on attributes 0,1, and 2, with relation variables $[0, 1, 2]$, and its part-relations. The function f_0 given by $A \mapsto 0$, $B \mapsto 1$, and $C \mapsto 2$ defines a morphism $f : S \longrightarrow \Delta_2$ of schemas/complexes. f is a display map. f_0^{-1} does not define a morphism of schemas.

Instances. Let X be a schema, say with attributes $X_0 = \{A_0, \ldots, A_{n-1}\}$. A functor $F : X \longrightarrow \mathbf{FinSet}$ from X to the category of finite sets and functions can be regarded as an instance of the schema X. For $x = [A_{i_0}, \ldots, A_{i_{m-1}}] \in X$, the set $F(x)$ can be regarded as a set of "keys" or "row-names". The "value" $k[A_{i_j}]$ of such a key $k \in F(x)$ at attribute A_{i_j} is then the element $k[A_{i_j}] := F(d^x_{A_{i_j}})(k)$. Accordingly, $F(x)$ maps to the set of tuples $F(A_{i_0}) \times \ldots \times F(A_{i_{m-1}})$ by $k \mapsto \langle k[A_{i_0}], \ldots, k[A_{i_{m-1}}] \rangle$. For arbitrary F, this function is not 1–1; that is, there can be distinct keys with the same values at all attributes. We say that F is a *relational instance* if this does not happen. That is, a relational instance is a functor $F : X \longrightarrow \mathbf{FinSet}$ such that for all $x \in X$ the functions $\{\delta^x_A \mid A \in x\}$ are jointly injective. Say that a relational instance is *strict* if the keys are actually tuples and the δ's are the expected projections.

Example 2. Let S be the schema of Example 1. Let an instance I be given by

R	A B		Q	B C
1	a b		1	b c
2	a' b		2	d e

Then I is the functor

$$I(\delta^R_A) \swarrow \quad I(R) = \{1,2\} \quad \searrow I(\delta^R_B) \qquad I(\delta^Q_B) \swarrow \quad I(Q) = \{1,2\} \quad \searrow I(\delta^Q_C)$$

$$I(A) = \{a, a'\} \qquad\qquad I(B) = \{b, d\} \qquad\qquad I(C) = \{c, e\}$$

with $I(\delta^R_A)(1) = a$, $I(\delta^R_B)(1) = b$ and so on.

Let J be the strict instance $J : \Delta_2 \longrightarrow \mathbf{FinSet}$ given in tabular form by

0 1 2	0 1	0 2	1 2	0	1	2
a b c	a b	a c	b c	a	b	c
	a' b	a' c	d e	a'	d	e

Explicitly, then, J is the functor which e.g. maps $[0,1]$ to $\{\langle a, b\rangle, \langle a', b\rangle\}$ and such that $J(\delta^{[0,1]}_1)(\langle a, b\rangle) = b$.

Substition, Strictification, and Induced Schemas. Let $f : X \longrightarrow Y$ be a morphism of schemas, and let $I : Y \longrightarrow \mathbf{FinSet}$ be a relational instance. Then it is easily seen that the composite $I \circ f : X \longrightarrow \mathbf{FinSet}$ is a relational instance. We write $I[f] := I \circ f$ and say it is the *substitution of I along f*.

It is clear that a relational instance is naturally isomorphic to exactly one strict relational instance with the same values. We say that the latter is the *strictification* of the former.

Example 3. Consider the morphism $f : S \longrightarrow \Delta_2$ of Example 1 and the instances I and J of Example 2. Then $J[f]$ is the strictification of I.

Let $\mathrm{Rel}(X)$ be the category of strict relational instances and natural transformations between them. For convenience and brevity (as with complexes) we often disregard the requirement that instances need to be strict in the sequel. However, working with relational instances up to strictification, or restricting to the strict ones, resolves the coherence issues so typical of categorical models of type theory. To have the "strict" instances be those "on tuple form" presents itself as a natural choice, both by the connection to the relational model and by the following formal connection between such instances and display maps.

Lemma 1. *Let $f : X \longrightarrow Y$ be a morphism of schemas. Then f is display if and only if for all strict instances J of Y the instance $J[f]$ is also strict.*

The connection between display maps, relational instances and simplicial schemas is given by the following. Let X be a schema and $F : X \longrightarrow \mathbf{FinSet}$ an arbitrary functor. Recall, e.g. from [8], that the category of elements $\int_X F$ has objects $\langle x, a \rangle$ with $x \in X$ and $a \in F(x)$. A morphism $\delta_{\langle y,b \rangle}^{\langle x,a \rangle} : \langle x, a \rangle \longrightarrow \langle y, b \rangle$ is a morphism $\delta_y^x : x \longrightarrow y$ with $F(\delta_y^x)(a) = b$. The projection $p : \int_X F \longrightarrow X$ is defined by $\langle x, a \rangle \mapsto x$ and $\delta_{\langle y,b \rangle}^{\langle x,a \rangle} \mapsto \delta_y^x$. We then have

Lemma 2. *Let X be a simplicial schema and $F : X \longrightarrow \mathbf{FinSet}$ be a functor. Then F is a relational instance if and only if $\int_X F$ is a simplicial schema and $p : \int_X F \to X$ is a display morphism.*

When F is a relational instance we write $X.F$ for $\int_X F$, and refer to it as the *canonical schema* corresponding to F. We refer to p as the *canonical projection*.

Example 4. The canonical schema of instance J of Example 2 has attribute set $\{\langle 0, a \rangle, \langle 0, a' \rangle, \langle 1, b \rangle, \langle 1, d \rangle, \langle 2, e \rangle, \langle 2, c \rangle\}$ and relation variables e.g. $\langle [0, 1, 2], \langle a, b, c \rangle \rangle$ (or, strictly, $[\langle 0, a \rangle, \langle 1, b \rangle, \langle 2, c \rangle]$).

Full Tuples. A schema X induces a canonical instance of itself by filling out the relations by a single row each, consisting of the attributes of the relation. This instance is terminal in the category of instances of X; that is, every other instance of X has a unique morphism to it. Accordingly, we define the *terminal instance* $1_X : X \longrightarrow \mathbf{FinSet}$ to be the functor defined by $x \mapsto \{x\}$.[2]

A *full* or *matching tuple* t of an instance I over schema X is a natural transformation $t : 1_X \Rightarrow I$. We write $\mathrm{Trm}_X(I)$ for the set of full tuples (indicating that we see them as terms type-theoretically).

Given a full tuple $t : 1_X \Rightarrow I$, the *induced section* is the morphism $\hat{t} : X \longrightarrow X.I$ defined by $x \mapsto \langle x, t_x(x) \rangle$. Notice that the induced section is always a display morphism.

Example 5. The instance I of Example 2 has precisely two full tuples. A full tuple can be seen as a tuple over the full attribute set of the schema with the

---

[2] Strictly speaking, we choose an isomorphic representation which is strict and stable under substitution. For current purposes, however, the current definition is notationally convenient.

property that for all relation variables the projection of the tuple is a row of that relation. The two full tuples of I are, then, $\langle a, b, c \rangle$ and $\langle a', b, c \rangle$. The instance J of Example 2 has precisely one full tuple $\langle a, b, c \rangle$.

2.3   Simplicial Databases

We have a functor $\mathrm{Rel}(-) : \mathbb{S}_d{}^{\mathrm{op}} \longrightarrow Cat$ which maps X to $\mathrm{Rel}(X)$ and $f : X \longrightarrow Y$ to $\mathrm{Rel}(f) = (-)[f] : \mathrm{Rel}(Y) \longrightarrow \mathrm{Rel}(X)$. We denote this indexed category by $\mathfrak{R}$, and think of it as a "category of databases" in which the totality of databases and schemas are collected. It is a model of a certain dependent type theory with context constants which we give in Sect. 3. We briefly outline some of the relevant structure available in $\mathfrak{R}$.

Definition 1. *For* $f : X \longrightarrow Y$ *in* $\mathbb{S}_d$ *and* $J \in \mathrm{Rel}(Y)$ *and* $t : 1_Y \Rightarrow J$ *in* $\mathrm{Trm}_Y(J)$:

1. *Define* $t[f] \in \mathrm{Trm}_X(J[f])$ *by* $x \mapsto t(f(x)) \in J[f](x)$. *Note that for* $g : Z \longrightarrow X$ *we have* $t[f][g] = t[f \circ g]$.
2. *With* $p_J : Y.J \longrightarrow Y$ *the canonical projection, let* $v_J : 1_{Y.J} \Rightarrow J[p_J]$ *be the full tuple defined by* $\langle y, a \rangle \mapsto a$. *(This term is needed for the type theory. We elsewhere leave subscripts on* v *and* p *determined by context.)*
3. *Denote by* $\tilde{f} : X.J[f] \longrightarrow Y.J$ *the schema morphism defined by* $\langle x, a \rangle \mapsto \langle f(x), a \rangle$. *Notice that since* f *is display, so is* $\tilde{f}$.

Lemma 3. *The following equations hold:*

1. *For* X *in* $\mathbb{S}_d$ *and* $I \in \mathrm{Rel}(X)$ *and* $t \in \mathrm{Trm}_X(I)$ *we have* $p \circ \hat{t} = \mathrm{id}_X$ *and* $t = v[\hat{t}]$.
2. *For* $f : X \longrightarrow Y$ *in* $\mathbb{S}_d$ *and* $J \in \mathrm{Rel}(Y)$ *and* $t \in \mathrm{Trm}_Y(J)$ *we have*
   (a) $p \circ \tilde{f} = f \circ p : X.J[f] \longrightarrow Y$;
   (b) $\tilde{f} \circ \widehat{t[f]} = \hat{t} \circ f : X \longrightarrow Y.J$; *and*
   (c) $v_J[\tilde{f}] = v_{J[f]} : 1_{X.J[f]} \Rightarrow J[f][p]$.
3. *For* $f : X \longrightarrow Y$ *and* $g : Y \longrightarrow Z$ *in* $\mathbb{S}_d$ *and* $J \in \mathrm{Rel}(Z)$ *we have* $\widetilde{g \circ f} = \tilde{g} \circ \tilde{f}$.
4. *For* $X \in \mathbb{S}_d$ *and* $I \in \mathrm{Rel}(X)$ *we have* $\tilde{p} \circ \hat{v} = \mathrm{Id}_{X.I}$.

The following instance-forming operations exist and commute with substitution.

0 and 1 instances: Given $X \in \mathbb{S}_d$ the terminal instance 1_X has already been defined. The *initial* instance 0_X is the constant empty functor, $x \mapsto \emptyset$.

Dependent Sum: Let $X \in \mathbb{S}_d$, $J \in \mathrm{Rel}(X)$, and $G \in \mathrm{Rel}(X.J)$. We define the instance $\Sigma_J G : X \longrightarrow \mathbf{FinSet}$ up to strictification by

$$x \mapsto \{\langle a, b \rangle \mid a \in J(x),\ b \in G(x, a)\}. \text{ For } \delta_y^x \text{ in } X, \text{ let } \Sigma_J G(\delta_y^x)(a, b) = \left\langle \delta_y^x(a), \delta_{y, \delta_y^x(a)}^{x, a}(b) \right\rangle.$$

Identity: Given $X \in \mathbb{S}_d$ and $J \in \mathrm{Rel}(X)$ the *Identity instance* $\mathrm{Id}_J \in \mathrm{Rel}(X.J.J[p])$ is defined, up to strictification, by $\langle \langle x, a \rangle, b \rangle \mapsto \star$ if $a = b$ and $\langle \langle x, a \rangle, b \rangle \mapsto \emptyset$ else.

$\star$ being e.g. the empty tuple. The full tuple refl $\in \text{Trm}_{(X.J)}(\text{Id}_J[\hat{v}])$ is defined by $\langle x, a \rangle \mapsto \star$.

Disjoint Union: Given $X \in \mathbb{S}_d$ and $I, J \in \text{Rel}(X)$, the instance $I + J \in \text{Rel}(X)$ is defined up to strictification by $x \mapsto \{\langle n, a \rangle \mid (n = 0 \wedge a \in I(x)) \vee (n = 1 \wedge a \in J(x))\}$. We have full tuples left $\in \text{Trm}_{X.I}((I + J)[p])$ defined by $\langle x, a \rangle \mapsto \langle 0, a \rangle$ and right $\in \text{Trm}_{X.J}((I + J)[p])$ defined by $\langle x, a \rangle \mapsto \langle 1, a \rangle$.

Dependent Product: Let $X \in \mathbb{S}_d$, $J \in \text{Rel}(X)$, and $G \in \text{Rel}(X.J)$. We define the instance $\Pi_J G : X \longrightarrow \textbf{FinSet}$ as strictification of the right Kan-extension (in the sense of e.g. [8]) of G along p. See [4] for an explicit construction. There are operations Ap and λ which for any full tuple $t \in \text{Trm}_{X.J}(G)$ yields a full tuple $\lambda t \in \text{Trm}_X(\Pi_J G)$, and for any full tuple $s \in \text{Trm}_X(\Pi_J G)$ yields a full tuple $\text{Ap}_s \in \text{Trm}_{X.J}(G)$. Moreover, $\text{Ap}_{\lambda t} = t$. We further indicate the relationship between the dependent product and full tuples, and the way in which the dependent product models the natural join operation, with the following example.

Example 6. Consider the schema S and instance I of Examples 1 and 2. Corresponding to the display map $f : S \longrightarrow \Delta_2$, we can present S an instance of Δ_2 as (ignoring strictification for readability) $S : \Delta_2 \longrightarrow \textbf{FinSet}$ by $S(0) = \{A\}$, $S(1) = \{B\}$, $S(2) = \{C\}$, $S(01) = \{R\}$, $S(12) = \{Q\}$, and $S(02) = S(012) = \emptyset$. Notice that, modulo the isomorphism between S as presented in Example 1 and $\Delta_2.S$, the morphism $f : S \longrightarrow \Delta_2$ is the canonical projection $p : \Delta_2.S \longrightarrow \Delta_2$. Similarly we have $I \in \Delta_2.S$ as (in tabular form, using subscript instead of pairing for elements in $\Delta_2.S$, and omitting the three single-column tables)

R_{01}	A_0	B_1
a	b
a'	b

Q_{12}	B_1	C_2
b	c
d	e

Then $\Pi_S I$ is, in tabular form (again omitting single column tables),

$0\ 1\ 2$	$0\ 1$	$0\ 2$	$1\ 2$
a b c	a b	a c	b c
a' b c	a' b	a' c	d e
		a e	
		a' e	

Notice that the three-column "top" table of $\Pi_S I$ is the natural join $R_{01} \bowtie Q_{12}$. The type theory of the next section will syntactically derive the rows of this table from the syntactic specification of S and I and the rules for the dependent product (see [4]).

3   The Type Theory

We introduce a Martin-Löf style type theory [9], with explicit substitutions (in the style of [3]), extended with context and substitution constants representing simplices and face maps. The type theory contains familiar constructs such as Σ- and Π-types. For this type theory we give an interpretation in the indexed category $\mathfrak{R}$ of the previous section. The goal is to use the type theory as a formal language for databases. We give examples how to specify instances and schemas formally in the theory. Further details can be found in [4].

3.1 The Type Theory $\mathcal{T}$

The type system has the following eight judgements, with intended interpretations.

Judgement	Interpretation
$? : \mathtt{Context}$	$[\![?]\!]$ is a schema
$? : \mathtt{Type}(\Gamma)$	$[\![?]\!]$ is an instance of the schema Γ
$? : \mathtt{Elem}(A)$	$[\![?]\!]$ is an full tuple in the instance A
$? : \Gamma \longrightarrow \Lambda$	$[\![?]\!]$ is a (display) schema morphism
$\Gamma \equiv \Lambda$	$[\![\Gamma]\!]$ and $[\![\Lambda]\!]$ are equal schemas
$A \equiv B : \mathtt{Type}(\Gamma)$	$[\![A]\!]$ and $[\![B]\!]$ are equal instances of $[\![\Gamma]\!]$
$t \equiv u : \mathtt{Elem}(A)$	$[\![t]\!]$ and $[\![u]\!]$ are equal full tuples in $[\![A]\!]$
$\sigma \equiv \tau : \Gamma \longrightarrow \Lambda$	the morphisms $[\![\sigma]\!]$ and $[\![\tau]\!]$ are equal

The type theory $\mathcal{T}$ has the rules listed in Figs. 1 and 2. The interpretation of these are given by the constructions in the previous section, and summarised in Fig. 3.

$$\sigma : \Gamma \longrightarrow \Delta,\ \tau : \Delta \longrightarrow \Theta \qquad\qquad \vdash \tau \circ \sigma : \Gamma \longrightarrow \Theta$$

$$\Gamma : \mathtt{Context} \qquad\qquad \vdash \mathrm{id}_\Gamma : \Gamma \longrightarrow \Gamma$$

$$A : \mathtt{Type}(\Gamma),\ \sigma : \Delta \longrightarrow \Gamma \qquad\qquad \vdash A[\sigma] : \mathtt{Type}(\Gamma)$$

$$a : \mathtt{Elem}(A),\ \sigma : \Delta \longrightarrow \Gamma \qquad\qquad \vdash a[\sigma] : \mathtt{Elem}(A[\sigma])$$

$$A : \mathtt{Type}(\Gamma) \qquad\qquad \vdash \Gamma.A : \mathtt{Context}$$

$$A : \mathtt{Type}(\Gamma) \qquad\qquad \vdash\ \downarrow_A\ : \Gamma.A \longrightarrow \Gamma$$

$$A : \mathtt{Type}(\Gamma) \qquad\qquad \vdash v : \mathtt{Elem}(A[\downarrow_A])$$

$$a : \mathtt{Elem}(A) \qquad\qquad \vdash a\uparrow\ : \Gamma \longrightarrow \Gamma.A$$

$$A : \mathtt{Type}(\Gamma),\ \sigma : \Delta \longrightarrow \Gamma \qquad\qquad \vdash \sigma.A : \Delta.A[\sigma] \longrightarrow \Gamma.A$$

$$\vdash \Delta_n : \mathtt{Context}$$

$$\vdash d_i^m : \Delta_n \longrightarrow \Delta_{n+1}$$

where where $n, i, j \in \mathbb{N}$ are such that $i < j \le n + 2$.

Fig. 1. Rules of the type theory: contexts and substitution

Each rule introduces a context, substitution, type or element. We will apply usual abbreviations such as $A \longrightarrow B$ for $\Pi_A B[\downarrow_A]$ and $A \times B$ for $\Sigma_A B[\downarrow_A]$. In addition to these term introducing rules there are a number of equalities which

$$
\begin{array}{ll}
A : \mathtt{Type}(\Gamma), \ B : \mathtt{Type}(\Gamma.A) & \vdash \Pi_A B : \mathtt{Type}(\Gamma) \\[4pt]
b : \mathtt{Elem}(B) & \vdash \lambda b : \mathtt{Elem}(\Pi_A B) \\[4pt]
f : \mathtt{Elem}(\Pi_A B) & \vdash \mathrm{apply}_f : \mathtt{Elem}(B) \\[4pt]
A : \mathtt{Type}(\Gamma), \ B : \mathtt{Type}(\Gamma.A) & \vdash \Sigma_A B : \mathtt{Type}(\Gamma) \\[4pt]
A : \mathtt{Type}(\Gamma), \ B : \mathtt{Type}(\Gamma.A) & \vdash \mathrm{pair} : \mathtt{Elem}(\Sigma_A B[\downarrow_A][\downarrow_B]) \\[4pt]
C : \mathtt{Type}(\Gamma.\Sigma_A B), & \\
c_0 : \mathtt{Elem}(C[(\downarrow_A \circ \downarrow_B).\Sigma_A B][\mathrm{pair}\uparrow]) & \vdash \mathrm{rec}_\Sigma \ c_0 : \mathtt{Elem}(C) \\[4pt]
A : \mathtt{Type}(\Gamma) & \vdash \mathrm{Id}_A : \mathtt{Type}(\Gamma.A.A[\downarrow_A]) \\[4pt]
A : \mathtt{Type}(\Gamma) & \vdash \mathrm{refl} : \mathtt{Elem}(\mathrm{Id}_A[v\uparrow]) \\[4pt]
C : \mathtt{Type}(\Gamma.A.A[\downarrow_A].\mathrm{Id}_A), & \\
c_0 : \mathtt{Elem}(C[(v\uparrow).\mathrm{Id}_A][\mathrm{refl}\uparrow]) & \vdash \mathrm{rec}_{\mathrm{Id}} \ c_0 : \mathtt{Elem}(C) \\[4pt]
A : \mathtt{Type}(\Gamma) \ B : \mathtt{Type}(\Gamma) & \vdash A + B : \mathtt{Type}(\Gamma) \\[4pt]
A : \mathtt{Type}(\Gamma) \ B : \mathtt{Type}(\Gamma) & \vdash l_{A,B} : \mathtt{Elem}((A+B)[\downarrow_A]) \\[4pt]
A : \mathtt{Type}(\Gamma) \ B : \mathtt{Type}(\Gamma) & \vdash r_{A,B} : \mathtt{Elem}((A+B)[\downarrow_B]) \\[4pt]
C : \mathtt{Type}(\Gamma.(A+B)), & \\
c_0 : \mathtt{Elem}(C[(\downarrow).(A+B)][l_{A,B}]) & \\
c_1 : \mathtt{Elem}(C[(\downarrow).(A+B)][r_{A,B}]) & \vdash \mathrm{rec}_+ \ c_0 \ c_1 : \mathtt{Type}(C) \\[4pt]
\Gamma : \mathtt{Context} & \vdash 0 : \mathtt{Type}(\Gamma) \\[4pt]
A : \mathtt{Type}(\Gamma.0) & \vdash \mathrm{rec}_0 : \mathtt{Elem}(A) \\[4pt]
\Gamma : \mathtt{Context} & \vdash 1 : \mathtt{Type}(\Gamma) \\[4pt]
\Gamma : \mathtt{Context} & \vdash * : \mathtt{Elem}(1) \\[4pt]
A : \mathtt{Type}(\Gamma.1), a : A[*\uparrow] & \vdash \mathrm{rec}_1 \ a : \mathtt{Elem}(A)
\end{array}
$$

Fig. 2. Rules of the type theory: Types

$$
\begin{array}{llll}
[\![\sigma \circ \tau]\!] = [\![\sigma]\!] \circ [\![\tau]\!] & [\![A[\sigma]]\!] = [\![A]\!] \, [\![\sigma]\!] & [\![id_\Gamma]\!] = id_{[\![\Gamma]\!]} & [\![a[\sigma]]\!] = [\![a]\!] \, [\![\sigma]\!] \\[4pt]
[\![\Gamma.A]\!] = [\![\Gamma]\!] . [\![A]\!] & [\![\downarrow_A]\!] = p & [\![v]\!] = v & [\![a\uparrow]\!] = \widehat{[\![a]\!]} \\[4pt]
[\![\sigma.A]\!] = \widehat{[\![\sigma]\!]} & [\![\Pi_A B]\!] = \Pi_{[\![A]\!]} \, [\![B]\!] & [\![\lambda f]\!] = \lambda \, [\![f]\!] & [\![\mathrm{apply}]\!] = Ap \\[4pt]
[\![\Sigma_A B]\!] = \Sigma_{[\![A]\!]} \, [\![B]\!] & [\![\mathrm{pair}]\!] = \mathrm{pair} & [\![\mathrm{rec}_\Sigma]\!] = \mathrm{rec}_\Sigma & [\![\mathrm{Id}_A]\!] = \mathrm{Id}_{[\![A]\!]} \\[4pt]
[\![\mathrm{refl}]\!] = \mathrm{refl} & [\![\mathrm{rec}_{\mathrm{Id}}]\!] = \mathrm{rec}_{\mathrm{Id}} & [\![\Delta_n]\!] = \Delta_n & [\![d_i^n]\!] = d_i^n
\end{array}
$$

Fig. 3. Interpretation of the type theory

should hold; such as the simplicial identities $d_i^{n+1} \circ d_j^n \equiv d_{j-1}^{n+1} \circ d_i^n : \Delta_n \longrightarrow \Delta_{n+2}$. We list the definitional equalities in Fig. 4.

$$(\upsilon \circ \tau) \circ \sigma \equiv \upsilon \circ (\tau \circ \sigma) \qquad id_\Gamma \circ \sigma \equiv \sigma \qquad \sigma \circ id_\Gamma \equiv \sigma$$

$$a[\sigma \circ \tau] \equiv a[\sigma][\tau] \qquad a[id_\Gamma] \equiv a \qquad \downarrow_A \circ (a{\uparrow}) \equiv id_\Gamma$$

$$v[a{\uparrow}] \equiv a \qquad \downarrow_A \circ (\sigma.A) \equiv \sigma \circ \downarrow_{A[\sigma]} \qquad v[f.A] \equiv v$$

$$(\sigma.A) \circ (a[\sigma]{\uparrow}) \equiv a{\uparrow} \circ \sigma \qquad (\sigma.A) \circ (\tau.A[\sigma]) \equiv (\sigma \circ \tau).A \qquad \downarrow.A \circ v{\uparrow} \equiv Id_{\Gamma.A}$$

$$\Pi_A B[\sigma] \equiv \Pi_{A[\Sigma]} B[\sigma.A] \qquad \mathrm{apply}(\lambda b) \equiv b$$

Fig. 4. Definitional equalities in the type theory

These all hold in our model. (The equalities for substitution are verified in Lemma 3. The remaining equations are mostly routine verifications.) We display this for reference.

Theorem 1. *The intended interpretation $[\![-]\!]$ yields a sound interpretation of the type theory $\mathcal{T}$ in $\mathfrak{R}$.*

3.2 Instance Specification as Type Introduction

The intended interpretation of $A : \mathtt{type}(\Gamma)$ is that A is an instance of the schema Γ. However, context extension allows us to view every instance as a schema in its own right; for every instance $A : \mathtt{type}(\Gamma)$, we get a schema $\Gamma.A$. It turns out that the most convenient way to specify a schema is by introducing a new type/instance over one of the simplex schemas Δ_n. To specify a schema, with a maximum of n attributes, may be seen as introducing a type in the context Δ_n. A relation variable with k attributes in the schema is introduced as an element of the schema substituted into Δ_k. Names of attributes are given as elements of the schema substituted down to Δ_0.

Example 7. We construct the rules of the schema S presented as an instance of Δ_2 as in Example 6. The introduction rules tells us the names of tables and attributes in S.

$$S : \mathtt{Type}(\Delta_2) \qquad\qquad A \equiv R[d_1] : \mathtt{Elem}(S[d_2 \circ d_1])$$
$$A : \mathtt{Elem}(S[d_2 \circ d_1]) \qquad\qquad B \equiv R[d_0] : \mathtt{Elem}(S[d_2 \circ d_0])$$
$$B : \mathtt{Elem}(S[d_2 \circ d_0]) \qquad\qquad B \equiv Q[d_1] : \mathtt{Elem}(S[d_0 \circ d_1])$$
$$C : \mathtt{Elem}(S[d_0 \circ d_0]) \qquad\qquad C \equiv Q[d_0] : \mathtt{Elem}(S[d_2 \circ d_0])$$
$$R : \mathtt{Elem}(S[d_2])$$
$$Q : \mathtt{Elem}(S[d_0])$$

From these introduction rules, we can generate an elimination rule. The elimination rule tells us how to construct full tuples in an instance over the schema S. Another interpretation of the elimination rule is that it formulates that the schema S contains only what is specified by the above introduction rules; it specifies the schema up to isomorphism.

An instance of a schema is a type depending in the context of the schema. Therefore instance specification is completely analoguous to schema specification. See [4] for an example. In [4] one can also find a derivation of the terms in $\mathcal{T}$ corresponding to the full tuples of the natural join in Example 6.

4 Dependent Structure, Large Types, and Queries

Most of this paper has been devoted to explaining the basic structure of the display map category of finite simplicial complexes seen as encoding systems of tables, and to stating the type theory which it models. In the space that remains, we briefly indicate some approaches of ongoing and future work, in particular emphasizing the definitions and roles of large types and universes. Before introducing additional types, however, we point to the use that can be made of the dependent structure itself. It is a cornerstone of the model that an instance over a schema can itself, by context extension, be seen as a schema over which new instances can be constructed. Thus context extension provides, for a given instance, the built in possibility to enter data related to the instance into tables formed by its rows. One immediate suggestion for the potential use of this feature is for updates; an update I' of an instance I over Γ is the instance over $\Gamma.I$ obtained by writing the new (or old or empty) row in the table formed by the row to be replaced (or kept or deleted). Adding new rows can be done by writing I' over $\Gamma.I + 1$ instead, as $I + 1$ has a copy of Γ over which new additions can be entered. (Multiple copies of Γ, and indeed of I, can be added if need be; notice that polynomial expressions over I such as $2I + 3$ yield meaningful instances over Γ). In this way a current update occurs in a context formed by a string of previous updates. Applying the dependent product operation gives an instance over the original schema Γ, if desired.

Returning to large types, Example 6 gives a glimpse of a "type-theoretic" operation and its relation to one of the standard queries of relational databases. A formal investigation of queries (and dependencies and views) in the setting of the type theory and model of the previous sections requires a formal understanding of what constitutes a query in this setting. For (partly) this purpose, we introduce a *universe*. This is a large type, corresponding to an infinite instance which encodes all finite instances over a fixed domain. Thus, given a schema X, the universe U_X of finite instances of X is an infinite instance of X where the full tuples encode the finite instances of X (over the fixed domain). The universe comes equipped with a *decoding instance* T_X over $X.U_X$ such that given a full tuple $t \in \mathrm{Trm}_X(U_X)$, the instance it encodes is decoded by $T_X[\hat{t}]$,

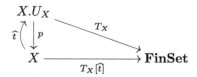

The universe and decoding instance are stable under substitution, and are closed under the other type constructions, such as Π- and Σ-types. We omit the details of the constructions.

The universe, U_Γ along with $T_\Gamma : \mathtt{Type}(\Gamma.U_\Gamma)$, allows reasoning generically about classes of instances of Γ in the type theory itself, without having to resort to the metalanguage. Since schemas can be though of as instances, they too can be constructed using the universe. In particular, given a schema Γ, the type $\Omega_\Gamma := \Sigma_{U_\Gamma} \Pi_{T_\Gamma} \Pi_{T_\Gamma[\downarrow_{T_\Gamma}]} \mathrm{Id}_{T_\Gamma}$ is the large type of subschemas of Γ. Its elements are decoded to instances by the family $\mathcal{O}_\Gamma := T[\downarrow_\Omega .U][\pi_0\uparrow]$. Given $t : \mathtt{Elem}(\Omega_\Gamma)$, the subschema it encodes is $\Gamma.\mathcal{O}[t\uparrow]$.

A query can then be seen as an operation which takes an instance of a schema to another instance of a related schema. Given codes for a source subschema $t : \mathtt{Elem}(\Omega)$ and a target subschema $u : \mathtt{Elem}(\Omega)$, the type of queries from t to u is thus $(\mathcal{O}[t\uparrow] \to U) \to (\mathcal{O}[u\uparrow] \to U)$. Having given a concrete type of queries leads the way to investigations as to exactly which queries can be expressed in the language. For illustration, we present an example query formulated in this way.

Example 8. In the spririt of Example 6, let $a : \mathtt{Elem}(\Omega)$ be the code for a subschema covering the schema Γ, in the sense that the set of attributes are the same. The query taking the dependent product, or natural join, of an instance of this subschema is expressed by the term

$$q := \lambda\lambda(\pi[(\downarrow_\Omega .U) \circ (\pi_0\uparrow)][a\uparrow.(T \to U)][\downarrow_1]) : \mathtt{Elem}((\mathcal{O}[a\uparrow] \to U) \to (1_\Gamma \to U)).$$

Further new types of interest and use are the type Λ_Γ of "tables", or faces, of a schema and the (large) type $\mathbb{N}_\Gamma$ of "finite" instances, or instances of the form $1 + 1 + \ldots + 1$. The former can be constructed so as to be both small and stable (whereas Ω_Γ seems to have to be large in order to be stable), and is sufficient for schema and instance specification (cf. Example 7). The latter is of relevance e.g. with respect to instances determined by their full tuples (or "with no nulls").

References

1. Abiteboul, S., Hull, R., Vianu, V.: Foundations of Databases. Addison-Wesley, Reading (1995)
2. Cartmell, J.: Formalising the network and hierarchical data models — an application of categorical logic. In: Pitt, D., Abramsky, S., Poigné, A., Rydeheard, D. (eds.) Category Theory and Computer Programming. LNCS, vol. 240, pp. 466–492. Springer, Heidelberg (1986)
3. Dybjer, P.: Internal type theory. In: Berardi, S., Coppo, M. (eds.) TYPES 1995. LNCS, vol. 1158. Springer, Heidelberg (1996)
4. Forssell, H., Gylterud, H.R., Spivak, D.I.: Type theoretical databases (2014). http://arxiv.org/abs/1406.6268
5. Friedman, G.: Survey article: an elementary illustrated introduction to simplicial sets. Rocky Mt. J. Math. **42**(2), 353–423 (2012)
6. Gabriel, P., Zisman, M.: Calculus of Fractions and Homotopy Theory. Springer, Heidelberg (1967)
7. Jacobs, B.: Categorical Logic and Type Theory. Elsevier, Amsterdam (1999)
8. Mac Lane, S.: Categories for the Working Mathematician. Springer, Heidelberg (1998)
9. Martin-Löf, P.: Intuitionistic Type Theory. Studies in Proof Theory, vol. 1. Bibliopolis, Naples (1984)

Augmenting Subset Spaces to Cope with Multi-agent Knowledge

Bernhard Heinemann[✉]

Faculty of Mathematics and Computer Science,
University of Hagen, 58084 Hagen, Germany
bernhard.heinemann@fernuni-hagen.de

Abstract. Subsequently, a particular extension of the bi-modal logic of subset spaces, LSS, to the case of several agents will be provided. The basic system, which originally was designed for revealing the intrinsic relationship between knowledge and topology, has been developed in several directions in recent years, not least towards a comprehensive knowledge-theoretic formalism. However, while subset spaces have been shown to be smoothly combinable with various epistemic concepts in the single-agent case, adjusting them to general multi-agent scenarios has brought about rather unsatisfactory results up to now. This is due to reasons inherent in the system so that one is led to consider more special cases. In the present paper, such a widening of LSS to the multi-agent setting is proposed. The peculiarity is here given by the case that the agents are supplied with certain knowledge-enabling functions allowing, in particular, for comparing their respective knowledge. It turns out that such circumstances can be modeled in corresponding logical terms to a considerable extent.

Keywords: Reasoning about knowledge of agents · Subset space semantics · Knowledge-enabling functions · Completeness · Decidability

1 Introduction

The starting point for this paper is *reasoning about knowledge.* This important foundational issue has been given a solid logical basis right from the beginning of the research into theoretical aspects of artificial intelligence, as can be seen, e.g., from the classic textbooks [5,14]. According to this, a binary *accessibility relation* R_A connecting *possible worlds* or *conceivable states of the world,* is associated with every instance A of a given finite group G of agents. The *knowledge of A* is then defined through the set of all *valid formulas,* where validity is understood with regard to every state the agent considers possible at the actual one. This widespread and well-established view of knowledge is complemented by Moss and Parikh's bi-modal *logic of subset spaces,* LSS (see [4,15], or Ch. 6 of [1]), of which the basic idea is reported in the following.

The *epistemic state* of an agent in question, i.e., the set of all those states that cannot be distinguished by what the agent topically knows, can be viewed

© Springer International Publishing Switzerland 2016
S. Artemov and A. Nerode (Eds.): LFCS 2016, LNCS 9537, pp. 130–145, 2016.
DOI: 10.1007/978-3-319-27683-0_10

as a *neighborhood* U of the actual state x of the world. Formulas are now interpreted with respect to the resulting pairs x, U called *neighborhood situations*. Thus, both the set of all states and the set of all epistemic states constitute the relevant semantic domains as particular subset structures. The two modalities involved, K and $\square$, quantify over all elements of U and 'downward' over all neighborhoods contained in U, respectively. This means that K captures the notion of knowledge as usual (see [5] again), and $\square$ reflects a kind of *effort to acquire knowledge* since gaining knowledge goes hand in hand with a shrinkage of the epistemic state. In fact, knowledge acquisition is this way reminiscent of a *topological procedure*. Thus, it was natural to ask for the appropriate logic of 'real' topological spaces, which could be determined by Georgatos shortly afterwards; see [6]. The subsequent research into subset and topological spaces, respectively, is quoted in the handbook [1], whereas more recent developments include, among others, the papers [2,16].

We focus on the knowledge-theoretic aspect of LSS as of now. Despite the fact that most treatises on this system deal with the single-agent case, a corresponding multi-agent version was proposed in the paper [9]. The key idea behind that approach is to implement the agents by means of additional modalities. This clearly leads to an essential modification of the logic, while the original semantics basically remains unchanged. On the contrary, if the agent structure shall be reflected in the neighborhood situations, then the scope of the modality K has to be restricted; see [10] for a detailed discussion on this topic. Anyway, it seems that a trade-off must be made between modifying the semantics and altering the logic in case of multiple agents.

For the scenarios considered in this paper, the additional semantic features will likewise appear *on top of* the subset space semantics. The variations of the basic logic, however, will be quite moderate. Such scenarios are constituted by, say, n agents whose knowledge need not be available at the actual situation instantaneously, but will be effective only after *enabling*. The process of enabling is formally described by agent-specific *functions* operating on neighborhood situations. When viewed 'from the outside', these functions quasi act as schedulers for individual reasoning. In the logic, they will be mirrored by additional modalities.[1]

It should be possible to model settings like this with the aid of the common logic of knowledge with incorporated time (cf. [5], Sect. 4.3.) as well, since we have just introduced a kind of *next step operator* (albeit for every agent). But sometimes it is unnecessary or even undesirable to make time explicit. For example, a particular ordering of the agents with regard to knowledge, or the effort spent on closing a knowledge gap between two agents, could be rated as more important than the factual distance of the agents in that sequence or the amount of time that trial costs in order to meet with success. We shall, therefore, define n-agent subset spaces in a way that such kind of qualitative weighting of

---

[1] If those knowledge-enabling functions shall depend on knowledge states alone, which is clearly worthy of discussion, then *topological nexttime logic,* see [8], would enter the field. This would lead to a somewhat more complicated but related system.

the agents can be reflected. For the sake of concision, however, just one sample application will actually be handled here, *leadership in knowledge*. This notion will be made precise below, with some new technical peculiarities coming along.

The rest of the paper is organized as follows. In the next section, we recapitulate the language and the logic of subset spaces for single agents. In Sect. 3, the ideas of both knowledge-enabling functions and leadership in knowledge are formalized. In Sect. 4, the soundness and completeness of the resulting logic is proved. In Sect. 5, the corresponding decidability problem is treated. Finally, we summarize and comment on some naturally arising questions.

All relevant facts from modal logic not explicitly introduced here can be found in the standard textbook [3].

2   The Language and the Logic of Subset Spaces Revisited

The purpose of this section is twofold: to clarify the starting point of our investigation on a technical level and to set up some concepts and results to be introduced and, respectively, proved later on.

First in this section, the language for (single-agent) subset spaces, $\mathcal{L}$, is defined precisely. Then, the semantics of $\mathcal{L}$ is linked with the common relational semantics of modal logic. Finally, the ensuing relationship is utilized after the most important facts on the logic of subset spaces have been recalled.

To begin with, we define the syntax of $\mathcal{L}$. Let $\mathsf{Prop} = \{p, q, \dots\}$ be a denumerably infinite set of symbols called *proposition variables* (which shall represent the basic facts about the states of the world). Then, the set SF of all *subset formulas* over Prop is defined by the rule

$$\alpha ::= \top \mid p \mid \neg\alpha \mid \alpha \wedge \alpha \mid \mathsf{K}\alpha \mid \Box\alpha.$$

The missing boolean connectives are treated as abbreviations, as needed. The operators which are dual to K and $\Box$ are denoted by L and $\Diamond$, respectively. In view of our remarks in the previous section, K is called the *knowledge operator* and $\Box$ the *effort operator*.

Second, we fix the semantics of $\mathcal{L}$. For a start, we single out the relevant domains. We let $\mathcal{P}(X)$ designate the powerset of a given set X.

Definition 1 (Semantic Domains).

1. *Let X be a non-empty set (of* states*) and $\mathcal{O} \subseteq \mathcal{P}(X)$ a set of subsets of X. Then, the pair $\mathcal{S} = (X, \mathcal{O})$ is called a* subset frame.
2. *Let $\mathcal{S} = (X, \mathcal{O})$ be a subset frame. The set*

$$\mathcal{N}_{\mathcal{S}} := \{(x, U) \mid x \in U \text{ and } U \in \mathcal{O}\}$$

   is then called the set of neighborhood situations *of $\mathcal{S}$.*
3. *Let $\mathcal{S} = (X, \mathcal{O})$ be a subset frame. Under an $\mathcal{S}$-valuation we understand a mapping $V : \mathsf{Prop} \to \mathcal{P}(X)$.*

4. *Let $\mathcal{S} = (X, \mathcal{O})$ be a subset frame and V an $\mathcal{S}$-valuation. Then, $\mathcal{M} := (X, \mathcal{O}, V)$ is called a* subset space *(based on $\mathcal{S}$).*

Note that neighborhood situations denominate the semantic atoms of the bi-modal language $\mathcal{L}$. The first component of such a situation indicates the actual state of the world, while the second reflects the uncertainty of the agent in question about it. Furthermore, Definition 1.3 shows that values of proposition variables depend on states only. This is in accordance with the common practice in epistemic logic; see [5] once more.

For a given subset space $\mathcal{M}$, we now define the relation of *satisfaction*, $\models_\mathcal{M}$, between neighborhood situations of the underlying frame and formulas from SF. Based on that, we define the notion of *validity* of formulas in subset spaces. In the following, neighborhood situations are often written without parentheses.

Definition 2 (Satisfaction and Validity). *Let $\mathcal{S} = (X, \mathcal{O})$ be a subset frame.*

1. *Let $\mathcal{M} = (X, \mathcal{O}, V)$ be a subset space based on $\mathcal{S}$, and let $x, U \in \mathcal{N}_\mathcal{S}$ be a neighborhood situation of $\mathcal{S}$. Then*

$$
\begin{aligned}
x, U &\models_\mathcal{M} \top && \text{is always true} \\
x, U &\models_\mathcal{M} p && :\iff x \in V(p) \\
x, U &\models_\mathcal{M} \neg\alpha && :\iff x, U \not\models_\mathcal{M} \alpha \\
x, U &\models_\mathcal{M} \alpha \wedge \beta && :\iff x, U \models_\mathcal{M} \alpha \text{ and } x, U \models_\mathcal{M} \beta \\
x, U &\models_\mathcal{M} \mathsf{K}\alpha && :\iff \forall y \in U : y, U \models_\mathcal{M} \alpha \\
x, U &\models_\mathcal{M} \Box\alpha && :\iff \forall U' \in \mathcal{O} : [x \in U' \subseteq U \Rightarrow x, U' \models_\mathcal{M} \alpha],
\end{aligned}
$$

where $p \in \mathsf{Prop}$ and $\alpha, \beta \in \mathsf{SF}$. In case $x, U \models_\mathcal{M} \alpha$ is true we say that α holds in $\mathcal{M}$ at the neighborhood situation x, U.

2. *Let $\mathcal{M} = (X, \mathcal{O}, V)$ be a subset space based on $\mathcal{S}$. A subset formula α is called* valid *in $\mathcal{M}$ iff it holds in $\mathcal{M}$ at every neighborhood situation of $\mathcal{S}$.*

Note that the idea of both knowledge and effort, as described in the introduction, is made precise by the first item of this definition. In particular, knowledge is here, too, defined as validity at all states that are indistinguishable to the agent.

Subset frames and subset spaces can be considered from a different perspective, as is known since [4] and reviewed in the following, for the reader's convenience. Let a subset frame $\mathcal{S} = (X, \mathcal{O})$ and a subset space $\mathcal{M} = (X, \mathcal{O}, V)$ based on it be given. Take $X_\mathcal{S} := \mathcal{N}_\mathcal{S}$ as a set of worlds, and define two accessibility relations $R_\mathcal{S}^\mathsf{K}$ and $R_\mathcal{S}^\Box$ on $X_\mathcal{S}$ by

$$
\begin{aligned}
(x, U) \, R_\mathcal{S}^\mathsf{K} \, (x', U') &:\iff U = U' \text{ and} \\
(x, U) \, R_\mathcal{S}^\Box \, (x', U') &:\iff (x = x' \text{ and } U' \subseteq U),
\end{aligned}
$$

for all $(x, U), (x', U') \in X_\mathcal{S}$. Moreover, let a valuation be defined by $V_\mathcal{M}(p) := \{(x, U) \in X_\mathcal{S} \mid x \in V(p)\}$, for all $p \in \mathsf{Prop}$. Then, bi-modal Kripke structures $\mathcal{S}_\mathcal{S} := (X_\mathcal{S}, \{R_\mathcal{S}^\mathsf{K}, R_\mathcal{S}^\Box\})$ and $\mathcal{M}_\mathcal{M} := (X_\mathcal{S}, \{R_\mathcal{S}^\mathsf{K}, R_\mathcal{S}^\Box\}, V_\mathcal{M})$ result in such a way that $\mathcal{M}_\mathcal{M}$ is equivalent to $\mathcal{M}$ in the following sense.

Proposition 1. *For all $\alpha \in$ SF and $(x, U) \in X_S$, we have that $x, U \models_{\mathcal{M}} \alpha$ iff $M_{\mathcal{M}}, (x, U) \models \alpha$.*

Here (and later on as well), the non-indexed symbol '$\models$' denotes the usual satisfaction relation of modal logic.

The proposition can easily be proved by structural induction on α. We call S_S and $M_{\mathcal{M}}$ the Kripke structures *induced* by the subset structures S and $\mathcal{M}$, respectively.

We now turn to the *logic* of subset spaces, LSS. The subsequent axiomatization from [4] was proved to be sound and complete in Sect. 1.2 and, respectively, Sect. 2.2 there.

1. All instances of propositional tautologies
2. $\mathsf{K}(\alpha \to \beta) \to (\mathsf{K}\alpha \to \mathsf{K}\beta)$
3. $\mathsf{K}\alpha \to (\alpha \wedge \mathsf{KK}\alpha)$
4. $\mathsf{L}\alpha \to \mathsf{KL}\alpha$
5. $(p \to \Box p) \wedge (\Diamond p \to p)$
6. $\Box(\alpha \to \beta) \to (\Box\alpha \to \Box\beta)$
7. $\Box\alpha \to (\alpha \wedge \Box\Box\alpha)$
8. $\mathsf{K}\Box\alpha \to \Box\mathsf{K}\alpha$,

where $p \in$ Prop and $\alpha, \beta \in$ SF. Note that LSS comprises the standard modal proof rules (only), i.e., modus ponens and necessitation with respect to each modality.

The last schema is by far the most interesting one, as it displays the interrelation between knowledge and effort. The members of this schema are called the *Cross Axioms* since [15]. Note that the schema involving only proposition variables is in accordance with the remark on Definition 1.3 above. (In other words, it is expressed by the latter schema that the language $\mathcal{L}$ essentially speaks about the development of *knowledge*.)

As the next step, let us take a brief look at the effect of the axioms from the above list within the framework of common modal logic. To this end, we consider bi-modal Kripke models $M = (W, R, R', V)$ satisfying the following four properties:

- the accessibility relation R of M belonging to the knowledge operator K is an equivalence,
- the accessibility relation R' of M belonging to the effort operator $\Box$ is reflexive and transitive,
- the composite relation $R' \circ R$ is contained in $R \circ R'$ (this is usually called the *cross property*), and
- the valuation V of M is constant along every R'-path, for all proposition variables.

Such a model M is called a *cross axiom model* (and the frame underlying M a *cross axiom frame*). Now, it can be verified without difficulty that LSS is sound with respect to the class of all cross axiom models. And it is also easy to see that every induced Kripke model is a cross axiom model (and every induced Kripke frame a cross axiom frame). Thus, the completeness of LSS for cross axiom models follows from that of LSS for subset spaces (which is Theorem 2.4 in [4]) by means of Proposition 1. This inferred completeness result can be used for proving the decidability of LSS; see [4], Sect. 2.3. We shall proceed in a similar way below, in Sect. 5.

3   Subset Spaces with Knowledge-Enabling Functions

The formalisms from the previous section will now be extended to the case of n agents, where $n \geq 2$ is a natural number. We again start with the logical language, which comprises n new operators $C_1, \ldots, C_n$ as of now. Thus, the set nSF of all n-subset formulas over Prop is defined by the rule

$$\alpha ::= \top \mid p \mid \neg\alpha \mid \alpha \wedge \alpha \mid K\alpha \mid \Box\alpha \mid C_1\alpha \mid \cdots \mid C_n\alpha.$$

Note that SF $\subseteq$ nSF. For $i = 1, \ldots, n$, the modality C_i is called the *knowledge-enabling operator associated with agent* i. The syntactic conventions from Sect. 2 apply correspondingly here. Note that there is no need to consider the dual to C_i separately since C_i will turn out to be self-dual.

Concerning semantics, the crucial modifications follow right now. We directly turn to the case that there is a leader in knowledge.

Definition 3 (Augmented n-Agent Subset Structures).

1. *Let $n \in \mathbb{N}$ be as above, and let $j \in \{1, \ldots, n\}$. Furthermore, let $\mathcal{S} = (X, \mathcal{O})$ be a subset frame. For all agents $i \in \{1, \ldots, n\}$ and states $x \in X$, let $f_{i,x} : \mathcal{O} \to \mathcal{O}$ be a partial function satisfying the following two conditions for every $U \in \mathcal{O}$.*
   (a) *The value $f_{i,x}(U)$ exists iff $x \in U$, and*
   (b) *if $f_{i,x}(U)$ exists, then $x \in f_{i,x}(U) \subseteq U$. (In this case, we also say that $f_{i,x}$ is* contracting.*)*
   Moreover, assume that, for all $i \in \{1, \ldots, n\}$, the set $f_{j,x}(U)$ is contained in $f_{i,x}(U)$ whenever $x \in U$. Then, the quadruple

$$\mathcal{S} = (X, \mathcal{O}, \{f_{i,x}\}_{1 \leq i \leq n, x \in X}, j)$$

   is called an augmented n-agent subset frame *(or an* aa-subset frame *for short), the mappings $f_{i,x}$, where $x \in X$, are called the* knowledge-enabling functions *for agent i, and j is called a* leader in knowledge.
2. *The notions of* neighborhood situation, $\mathcal{S}$-valuation *and* augmented-n-agent subset space *(aa-subset space) are completely analogous to those introduced in Definition 1.*

A detailed comment on this definition seems to be appropriate. For a start, note that the just introduced structures obviously do not correspond to the most general n-agent scenarios, but have already been adjusted to those indicated in the introduction. In fact, not only is an arbitrary agent capable of enabling its knowledge at any situation (according to (a) of the first item of the previous definition), but a particular one (namely j) is also doing better than all the others in this respect (because of the final assumption there). Note that an agent's enabling is always a kind of improvement since it is given as a shrinkage of a knowledge state (due to (b) above). The enabling functions obviously depend on agents *and* states of the world.[2] Furthermore, the 'distance' to the leader can be measured by the set inclusion relation as well. This will make it possible for us to express that distance with the aid of the 'global' effort operator $\Box$, quasi as a *missing effort*, later on. (Thus, $\Box$ may be called the *operator closing knowledge gaps*.)

With regard to satisfaction and validity, we need not completely present the analogue of Definition 2 at this place, but may confine ourselves to the clause for the new operators.

Definition 4 (Satisfaction). *Let $\mathcal{S} = (X, \mathcal{O}, \{f_{i,x}\}_{1 \leq i \leq n, x \in X}, j)$ be an aa-subset frame, $\mathcal{M}$ an aa-agent subset space based on $\mathcal{S}$, and $x, U \in \mathcal{N}_{\mathcal{S}}$ a neighborhood situation of $\mathcal{S}$. Then, for every $i \in \{1, \ldots, n\}$ and $\alpha \in n\mathsf{SF}$,*

$$x, U \models_{\mathcal{M}} \mathsf{C}_i \alpha : \iff x, f_{i,x}(U) \models_{\mathcal{M}} \alpha.$$

Since the operator K can no longer be assigned to a particular agent unambiguously, the knowledge of the agents involved in an aa-scenario must still be defined. We now are in a position to do this, viz through the validity of the K-prefixed formulas at the *respective* neighborhood situations. The latter are understood as those arising from the associated enabling functions as images. Thus we let, for $i \in \{1, \ldots, n\}$, *agent i know α at x, U* by definition, iff $x, U \models_{\mathcal{M}} \mathsf{K}\alpha$ *and* $U \in \mathrm{Im}(f_{i,x})$; in other words, K represents factual knowledge after enabling. (This also concerns the knowledge about another agent's knowledge.)

This fixing clearly requires a justification. To this end, note that the link between the relevant knowledge formulas and the semantic structures is accomplished 'externally' here, i.e., by means of an additional condition having no direct counterpart in the object language, namely the requirement that the subset component U of the actual neighborhood situation be contained in the image set of the enabling function for the agent in question. Relating to this, it should be mentioned that all the knowledge of agents we talk about in this paper is an 'ascribed' one (cf. [5], p. 8), in fact, by the system designer utilizing epistemic logic as a formal tool for specifying certain multi-agent scenarios. This gives us a kind of freedom regarding the choice of the system properties, which is only limited by the suitability of the approach for the intended applications. These are clearly limited to some extent by the lesser expressiveness of formulas here, but the knowledge development of the involved agents can just as well be described as the leadership in knowledge of a particular agent; see below for some examples.

---

[2] See footnote 1 above for an alternative way of modeling.

The final semantic issue to be mentioned is that of *induced* Kripke structures. Letting $\mathcal{S} = (X, \mathcal{O}, \{f_{i,x}\}_{1 \leq i \leq n, x \in X}, j)$ be any aa-subset frame, the following definition suggests itself.

$$(x, U) \, R_{\mathcal{S}}^{\mathsf{C}_i} \, (x', U') : \iff (x = x' \text{ and } U' = f_{i,x}(U)),$$

where $i \in \{1, \ldots, n\}$, $x, x' \in X$, and $U, U' \in \mathcal{O}$. With that, the corresponding analogue of Proposition 1 is obviously valid.

The *augmented logic of subset spaces*, ALSS, is given by the following list of axioms and the standard proof rules.[3]

1. All instances of propositional tautologies
2. $\mathsf{K}(\alpha \to \beta) \to (\mathsf{K}\alpha \to \mathsf{K}\beta)$
3. $\mathsf{K}\alpha \to (\alpha \wedge \mathsf{KK}\alpha)$
4. $\mathsf{L}\alpha \to \mathsf{KL}\alpha$
5. $(p \to \Box p) \wedge (\Diamond p \to p)$
6. $\Box (\alpha \to \beta) \to (\Box\alpha \to \Box\beta)$
7. $\Box\alpha \to (\alpha \wedge \Box\Box\alpha)$
8. $\mathsf{K}\Box\alpha \to \Box\mathsf{K}\alpha$
9. $\mathsf{C}_i(\alpha \to \beta) \to (\mathsf{C}_i\alpha \to \mathsf{C}_i\beta)$
10. $\mathsf{C}_i\neg\alpha \leftrightarrow \neg\mathsf{C}_i\alpha$
11. $\mathsf{K}\mathsf{C}_i\alpha \to \mathsf{C}_i\mathsf{K}\alpha$
12. $\Box\alpha \to \mathsf{C}_i\alpha$
13. $\mathsf{C}_i\Box\alpha \to \mathsf{C}_j\alpha,$

where j is the preassigned leader, $i \in \{1, \ldots, n\}$, $p \in$ Prop, and $\alpha, \beta \in n$SF.

Obviously, the first eight schemata of this list coincide with the LSS-axioms presented in Sect. 2. Thus we only comment on the others here, which are exactly those involving C_i for $i \in \{1, \ldots, n\}$. Axiom 9 is the usual distribution schema of modal logic, this time formulated for C_i. The next axiom captures the functionality of the accessibility relation associated with C_i; see, e.g., [7], Sect. 9 (for the operator *next*). In the present context, it comes along with the fact that we have assigned knowledge-enabling *functions* to the agents. The schema 11 is formally similar to the eighth one, thus comprising the Cross Axioms for K and C_i. The last but one schema mirrors the fact that the enabling functions, when defined, are contracting. With regard to the relational semantics, it says that the accessibility relation for C_i is contained in that for $\Box$. This schema, together with Axiom 10, is as well responsible for the fact that the counterpart of Axiom 5 is not needed for C_i. The most interesting new schema is the last one. In case all the involved modalities were equal, we would have the axioms capturing the *weak density* of the corresponding accessibility relation; see [7], Sect. 1. However, regarding augmented n-agent scenarios, the leadership of agent j in knowledge is thereby expressed.

As to an example of a schema of derived ALSS-sentences, let us recall the *reliably known* formulas $\alpha \in$ SF from [4], which satisfy $\mathsf{K}\alpha \to \Box\mathsf{K}\alpha \in$ LSS

---

[3] That is to say, the necessitation rule for each of the C_i's is added to the proof rules for LSS.

by definition. We now define *accessibly known* formulas $\alpha \in n\mathsf{SF}$ by analogy with that through the condition that $\mathsf{K}\alpha \to \mathsf{C}_i\mathsf{K}\alpha$ be in ALSS. Then, for every $\alpha \in n\mathsf{SF}$, the formula $\Box\alpha$ is of this type, as expected. In fact, $\mathsf{K}\Box\Box\alpha$ can be deduced from $\mathsf{K}\Box\alpha$ because of Axiom 7, from which we obtain $\Box\mathsf{K}\Box\alpha$ with the aid of Axiom 8. Now, Axiom 12 implies $\mathsf{C}_i\mathsf{K}\Box\alpha$.

Finally in this section, it is proved that the logic ALSS is *sound* with respect to the class of all aa-subset spaces.

Proposition 2. *Let* $\mathcal{M} = (X, \mathcal{O}, \{f_{i,x}\}_{1 \le i \le n, x \in X}, j, V)$ *be an aa-subset space. Then, every axiom from the above list is valid in* $\mathcal{M}$ *and every rule preserves validity.*

Proof. We confine ourselves to the last schema of the axioms. Let $\mathsf{C}_i\Box\alpha \to \mathsf{C}_j\alpha$ be an instance of this, and let $x, U \models_{\mathcal{M}} \mathsf{C}_i\Box\alpha$ be satisfied for any neighbourhood situation of the frame underlying $\mathcal{M}$. According to Definition 4, this means that $x, f_{i,x}(U) \models_{\mathcal{M}} \Box\alpha$. Thus, $x, f_{i,x}(U) \models_{\mathcal{M}} \alpha$ for all $U' \in \mathcal{O}$ such that $x \in U' \subseteq f_{i,x}(U)$. It follows that $x, f_{j,x}(U) \models_{\mathcal{M}} \alpha$ holds in particular, because of the leader-in-knowledge condition from Definition 3.1. Consequently, $x, U \models_{\mathcal{M}} \mathsf{C}_j\alpha$. This proves (the particular case of) the proposition.

Regarding the relational semantics, it will be seen that a certain property of *lying in between* corresponds to the schema treated in the preceding proof, as it is the case with the related axioms for weak density. We, therefore, call that schema ad hoc the *lying-in-between axioms*. These will crucially be utilized in the next section.

4   Completeness

In this section, we present the peculiarities required for proving the semantic completeness of ALSS on the class of all aa-subset spaces. As it is mostly the case with subset space logics, the overall structure of such a proof consists of an infinite step-by-step model construction.[4] Using such a procedure seems to be necessary, since subset spaces in a sense do not harmonize with the main modal means supporting completeness, viz *canonical models*.

The canonical model of ALSS will come into play nevertheless. So let us fix some notations concerning that model first. Let $\mathcal{C}$ be the set of all maximal ALSS-consistent sets of formulas. Furthermore, let $\xrightarrow{\mathsf{K}}$, $\xrightarrow{\Box}$, and $\xrightarrow{\mathsf{C}_i}$ be the accessibility relations induced on $\mathcal{C}$ by the modalities K, $\Box$, and C_i, respectively, where $i \in \{1, \ldots, n\}$. And finally, let $\alpha \in n\mathsf{SF}$ be a formula which is *not* contained in ALSS. Then, we have to find a model for $\neg\alpha$.

This model is constructed stepwise and incrementally in such a way that better and better intermediary structures are obtained (which means that more and

---

[4] One or another proof of that kind can be found in the literature; see, e.g., [4] for a fully completed proof regarding LSS and [9] for a particular multi-agent variation. Thus, it is really sufficient to confine ourselves to the case-specific issues here (which are quite difficult enough in themselves).

more existential formulas are realized). In order to ensure that the finally resulting limit structure behaves as desired, several requirements on those approximations have to be met at every stage. This makes up the technical core of the
proof, of which the specific features are described reasonably accurately in a
minute.

First, however, we need a lemma.

Lemma 1. *Let $n, j \in \mathbb{N}$ be fixed as in the previous section. Suppose that $s, t \in \mathcal{C}$
are maximal ALSS-consistent sets of formulas satisfying $s \xrightarrow{C_j} t$. Then, for all
$i \in \{1, \ldots, n\}$, there is some $u \in \mathcal{C}$ such that $s \xrightarrow{C_i} u \xrightarrow{\Box} t$.*

Proof. One has to apply the lying-in-between axioms and can argue in a similar
way as in the case of weak density in doing so; cf. [7], p. 26.[5]

We now describe the main ingredients of the above mentioned approximation
structures. Their possible worlds are successively taken from a denumerably
infinite set of points, Y, chosen in advance. Also, another denumerably infinite
set, Q, is chosen such that $Y \cap Q = \emptyset$. The latter set shall gradually contribute
to a partially ordered set representing the subset space structure of the desired
limit model. Finally, we fix particular 'starting elements' $x_0 \in Y$, $\bot \in Q$, and
$\Gamma \in \mathcal{C}$ containing the formula $\neg \alpha$ from above. Then, a sequence of quintuples
$(X_m, P_m, j_m, \{g_i^m\}_{1 \leq i \leq n}, t_m)$ has to be defined inductively such that, for all $m \in
\mathbb{N}$ and $i \in \{1, \ldots, n\}$,

- X_m is a finite subset of Y containing x_0,
- P_m is a finite subset of Q containing $\bot$ and carrying a partial order $\leq_m$ with
  least element $\bot$,
- $j_m : P_m \to \mathcal{P}(X_m)$ is a function satisfying $(\pi \leq_m \rho \iff j_m(\pi) \supseteq j_m(\rho))$, for
  all $\pi, \rho \in P_m$,
- $g_i^m : X_m \times P_m \to P_m$ is a partial function such that, for all $x \in X_m$ and
  $\pi \in P_m$,
    - if $g_i^m(x, \pi)$ exists, then $(x \in j_m(g_i^m(x, \pi))$ and $\pi \leq_m g_i^m(x, \pi))$
    - if $g_i^m(x, \pi)$ and $g_j^m(x, \pi)$ exist, then $g_i^m(x, \pi) \leq_m g_j^m(x, \pi)$,
- $t_m : X_m \times P_m \to \mathcal{C}$ is a partial function such that, for all $x, y \in X_m$ and
  $\pi, \rho \in P_m$,
    - $t_m(x, \pi)$ is defined iff $x \in j_m(\pi)$; in this case it holds that
      $$* \text{ if } y \in j_m(\pi), \text{ then } t_m(x, \pi) \xrightarrow{K} t_m(y, \pi)$$
      $$* \text{ if } \pi \leq_m \rho, \text{ then } t_m(x, \pi) \xrightarrow{\Box} t_m(x, \rho)$$
      $$* \text{ if } g_i^m(x, \pi) = \rho, \text{ then } t_m(x, \pi) \xrightarrow{C_i} t_m(x, \rho)$$
    - $t_m(x_0, \bot) = \Gamma$.

It is now clear how to define the approximating partial functions $f_{i,x}^m : \text{Im}(j_m) \to
\text{Im}(j_m)$. For all $x \in X_m$ and $\pi, \rho \in P_m$, we let $f_{i,x}^m(j_m(\pi)) := j_m(\rho)$ iff $g_i^m(x, \pi) =
\rho$. Then, $f_{i,x}^m$ is contracting and satisfies, for every $\pi \in P_m$,

---

[5] Note that such a proof is necessary here, since a Sahlqvist argument is insufficient
because ALSS is a non-normal logic.

- $f_{i,x}^m(j_m(\pi))$ exists iff $g_i^m(x,\pi)$ exists,
- $f_{j,x}^m(j_m(\pi))$ is contained in $f_{i,x}^m(j_m(\pi))$ in case both sets exist.

The next five conditions reveal to what extent the final model is approximated by the structures $(X_m, P_m, j_m, \{g_i^m\}_{1 \le i \le n}, t_m)$. Actually, it will be ensured that, for all $m \in \mathbb{N}$ and $i \in \{1, \ldots, n\}$,

- $X_m \subseteq X_{m+1}$,
- P_{m+1} is an *almost end extension* of P_m, i.e., a superstructure of P_m such that, if an element $\pi \in P_{m+1} \setminus P_m$ is strictly smaller than some element of P_m, then there are uniquely determined $\rho, \sigma \in P_m$, $x \in X_m$, and $i \in \{1, \ldots, n\}$ satisfying $\rho \le_m \sigma$, $\sigma = g_j^m(x, \rho)$, $\pi \le_{m+1} \sigma$, and $\pi = g_i^{m+1}(x, \rho)$ (this means, in particular, that $g_j^m(x, \rho)$ and $g_i^{m+1}(x, \rho)$ are defined),
- $j_{m+1}(\pi) \cap X_m = j_m(\pi)$ for all $\pi \in P_m$,
- $g_i^{m+1}|_{X_m \times P_m} = g_i^m$,
- $t_{m+1}|_{X_m \times P_m} = t_m$.

Note that *end extensions* are usually dealt with at this point. In the present case, however, the new elements must suitably be edged in. This requires a different approach.

Finally, the construction complies with the following requirements on existential formulas: for all $m \in \mathbb{N}$, $x \in X_m$, $\pi \in P_m$, $\nabla \in \{\Diamond, \mathsf{C}_1, \ldots, \mathsf{C}_n\}$, and $\beta \in n\mathsf{SF}$,

- if $\mathsf{L}\beta \in t_m(x, \pi)$, then there are $m < k \in \mathbb{N}$ and $y \in j_k(\pi)$ such that $\beta \in t_k(y, \pi)$,
- if $\nabla\beta \in t_m(x, \pi)$, then there are $m < k \in \mathbb{N}$ and $\pi \le_k \rho \in P_k$ such that $\beta \in t_k(x, \rho)$.

With that, the final model refuting α can be defined easily. Furthermore, a relevant *Truth Lemma* (cf. [3], 4.21) can be proved for it, from which the completeness of ALSS with respect to the augmented n-agent semantics follows immediately. Thus, it remains to specify, for all $m \in \mathbb{N}$, the approximating structures $(X_m, P_m, j_m, \{g_i^m\}_{1 \le i \le n}, t_m)$ in a way that all the above requirements are met.

Since the case $m = 0$ is quite obvious, we only focus on the induction step. Here, some existential formula γ contained in some maximal ALSS-consistent set $t_m(x, \pi)$, where $x \in X_m$ and $\pi \in P_m$, must be made true according to the last group of the above requirements. We confine ourselves to the case of the enabling operator associated with agent i, where $i \in \{1, \ldots, n\}$. So let $\gamma = \mathsf{C}_i\beta \in t_m(x, \pi)$ be satisfied. Then, we distinguish two cases, each one of which having two subcases. First, let $i = j$. If $g_j^m(x, \pi)$ is undefined, then this case is less interesting because the same proceeding as in the $\Diamond$-case leads to success; cf. [4]. Note however that here is the place where the status of definiteness of the function g_j is changed, namely for the argument (x, π). On the other hand, if $g_j^m(x, \pi)$ is defined, then nothing has to be changed and everything goes well because of the functionality of C_j. Now, let $i \ne j$. Then, we may assume that $\rho = g_j^m(x, \pi)$ has already been defined; see the remark right before Theorem 1 below. The case that $g_i^m(x, \pi)$ as well has been defined in advance is easy, too. Thus suppose that

$g_i^m(x, \pi)$ is undefined. In this case, we choose both a new point $y \in Y$ and a fresh $\sigma \in Q$, and we let $X_{m+1} := X_m \cup \{y\}$ and $P_{m+1} := P_m \cup \{\sigma\}$. The partial order is extended to P_{m+1} by letting $\pi \leq_{m+1} \sigma \leq_{m+1} \rho$ and σ be not comparable with any other element not less than π. The function j_{m+1} is defined as follows. We let $j_{m+1}(\tau) := j_m(\tau) \cup \{y\}$ for all $\tau \leq_m \pi$ and $j_{m+1}(\sigma) := j_m(\rho) \cup \{y\}$; for all other arguments, j_m remains unchanged. Furthermore, the extension of the function g_i^m is defined by $g_i^{m+1}(x, \pi) := \sigma$. Finally, the mapping t_m has to be adjusted. From the *Existence Lemma* of modal logic (see [3], 4.20) we know that there is some point Γ_i of $\mathcal{C}$ such that $t_m(x, \pi) \xrightarrow{C_i} \Gamma_i$ and $\beta \in \Gamma_i$. Thus, we define the new part of t_{m+1} by $t_{m+1}(x, \sigma) = t_{m+1}(y, \sigma) := \Gamma_i$, and $t_{m+1}(y, \tau) := t_m(x, \tau)$ for all $\tau \leq_m \pi$; moreover, the maximal consistent set which is to be assigned to a pair (z, σ) where $x \neq z \in j_m(\rho)$, is obtained in the following way. We know from the induction hypothesis that $t_m(z, \pi) \xrightarrow{C_j} t_m(z, \rho)$ is valid. From Lemma 1 we therefore obtain the existence of a maximal ALSS-consistent set Δ satisfying $t_m(z, \pi) \xrightarrow{C_i} \Delta \xrightarrow{\square} t_m(z, \rho)$. Now we let $t_{m+1}(z, \sigma) := \Delta$. This completes the definition of t_{m+1} and thus that of $\left(X_{m+1}, P_{m+1}, j_{m+1}, \{g_i^{m+1}\}_{1 \leq i \leq n}, t_{m+1}\right)$ in the case under consideration.

We must now check that the properties stated in the second group of requirements are satisfied and that the validity of those stated in the first group is transferred from m to $m + 1$. Doing so, several items prove to be evident from the construction. In some cases, however, the particularities of the accessibility relations on $\mathcal{C}$ like the two cross properties have to be applied. Further details regarding this must be omitted here.

As to the realization of existential formulas, it has to be ensured that *all* possible cases are eventually exhausted. To this end, processing must suitably be scheduled with regard to each of the involved modalities. This can be done with the aid of appropriate enumerations. Concerning details relating to this, the reader is referred to [4] again, but not before we have mentioned that one should keep in mind to rearrange, if need be, those enumerations in such a way that some C_j-formula is treated *before* any C_i-formula at all times, i.e., for any pair (x, π) (which is clearly possible). All this finally yields the subsequent theorem.

Theorem 1 (Completeness). *Let $\alpha \in n\mathsf{SF}$ be a formula which is valid in all aa-subset spaces. Then α belongs to the logic* ALSS.

Concluding this section, we would like to stress that the functionality of the knowledge-enabling operators is a decisive prerequisite for the success of the above model construction.

5   Decidability

The standard method for proving the decidability of a given modal logic is *filtration*. By that method, inspection of the relevant models is restricted to those not exceeding a specified size, in this way making a decision procedure possible.

However, just as subset spaces do not harmonize with canonical models, they are incompatible with filtration. A detour is therefore required, which takes us back into the relational semantics. In the following, we shall single out a class of multi-modal Kripke structures for which ALSS is as well sound and complete, and which is closed under filtration in a suitable manner. This will give us the desired decidability result. Subsequently, K is supposed to correspond to R, $\square$ to R', and C_i to S_i for $i = 1, \ldots, n$. Furthermore, let $j \in \{1, \ldots, n\}$ be given in advance.

Definition 5 (AA-Model). *Let* $M := (W, R, R', S_1, \ldots, S_n, V)$ *be a multi-modal Kripke model, where* $R, R', S_1, \ldots, S_n \subseteq W \times W$ *are binary relations and* V *is a valuation. Then* M *is called an* aa-model *(with* j *the leader in knowledge), iff the following conditions are satisfied.*

1. R *is an equivalence relation,*
2. R' *is reflexive and transitive,*
3. S_i *is a functional relation contained in* R', *for every* $i \in \{1, \ldots, n\}$,
4. *each of the pairs* (R, R'), (R, S_1), $\ldots$, (R, S_n) *satisfies the cross property,*
5. $S_j \subseteq S_i \circ R'$ *for every* $i \in \{1, \ldots, n\}$, *and*
6. *for all proposition variables, the valuation* V *of* M *is constant along every* R'-path.

Note that the fifth item of the previous definition represents the relational version of the lying-in-between property.

The class of all Kripke models *induced* by an aa-subset space (see Sect. 2 and, respectively, Sect. 3 for this notion) is contained in the class of all aa-models, as can be seen easily. It follows that ALSS is (sound and) complete with respect to the latter class; see the remark at the end of Sect. 2. It suffices therefore, in order to prove the decidability of ALSS, to show that the class of all aa-models is closed under filtration.

To this end, let an ALSS-consistent formula $\alpha \in n\mathsf{SF}$ be given. Then, a *filter set* of formulas, involving the set $\mathrm{sf}(\alpha)$ of all subformulas of α, is defined as follows. We start off with $\Sigma_0 := \mathrm{sf}(\alpha) \cup \{\neg\beta \mid \beta \in \mathrm{sf}(\alpha)\}$. In the next step, we take the closure of Σ_0 under finite conjunctions of pairwise distinct elements of Σ_0. After that, we close under single applications of the operator L. And finally, we join the sets of subformulas of all the elements of the set obtained last. (This final step is necessary because L was introduced as an abbreviation.) The resulting set of formulas, denoted by Σ, is quite similar to the one used for LSS in [4] and will meet the case-specific requirements here. Note that Σ is a *finite* set.

Now, the canonical model of ALSS is filtered through Σ. As a filtration of the corresponding accessibility relations, we take the *smallest* one in each of the $n + 2$ cases. Let $M = (W, R, R', S_1, \ldots, S_n, V)$ be the resulting model, where the valuation V shall be in accordance with Definition 5.6 for the proposition variables outside of Σ. Then, the following lemma is crucial.

Lemma 2. *The structure* M *is a finite aa-model. Furthermore, the size of* M *can be computed from the length of* α.

Proof. The finiteness of W follows from that of Σ, and we must now show that the six conditions from Definition 5 are satisfied. Due to space limitations, we must be rather brief in doing so. Taking into account the way the filter set Σ was formed, the verification of 1 and 4 is lengthy, but not very difficult. The reflexivity of R' can easily be concluded from the mere fact that M is the result of a filtration. Moreover, establishing the transitivity of R' is covered by the proof of Lemma 2.10 from [4]. Both the inclusions $S_i \subseteq R'$, where $i \in \{1, \ldots, n\}$, and the validity of 6 for the proposition variables occurring in Σ arise from the circumstance that we have chosen the *smallest* filtration in each case. The same is true of the functionality of the relations S_i, but this is not completely obvious. In fact, the (suitably adapted) *Fun-Lemma* 9.9 from [7] must be applied for it, ensuring that every S_i-successor of an arbitrary point of W is of equal value in regard to the validity of the formulas from Σ so that any of them can be selected. (It will become clear in a minute which one will actually be the right one.) Thus the verification of the fifth condition only still requires an argument. Fortunately, the characteristic features of a smallest filtration again help so that we are done after mentioning the following. For any starting point $w \in W$, the relational lying-in-between property can be realized with the aid of an arbitrarily chosen S_j-successor of w, with thereby determining the appropriate S_i-successor of that point through the correspondingly utilized lying-in-between relation on the canonical model; as to the validity of the latter, see Lemma 1. In this manner, the lemma is proved.

The desired decidability result is now an immediate consequence of Lemma 2 and the facts stated at the beginning of this section.

Theorem 2 (Decidability). *The logic* ALSS *is a decidable set of formulas.*

This is what we can say about the effectiveness properties of the logic ALSS for the moment.

6   Conclusion

First in this section, the results obtained in this paper are summarized. Then, we comment on some further points, including possible extensions of our approach.

A special *subset space logic of n agents,* denoted by ALSS, has been introduced above. This system has been designed to cover *leadership in knowledge,* in particular. We proposed a corresponding *axiomatization,* which turned out to be *sound and complete* with respect to the intended class of models. This constitutes the first of our main results. The second assures the *decidability* of the new logic.

It is to be expected that the *complexity* of ALSS can be determined not until solving this problem for the usual logic of subset spaces. As to that, only partial results are known; see [2,12].

The main reason for the (relative) progress in multi-agent subset spaces achieved in this paper is the relaxation of the underlying idea of knowledge.

Accordingly, the agents come in rather indirectly, viz in terms of their enabling functions. To say it somewhat exaggeratedly, the absence of agent-specific knowledge operators even makes a multi-agent usage of subset spaces possible, at least for particular epistemic scenarios. Following that idea, a promising new field of research opens up, in which the issue of *other interesting agent interrelationships and their effects on knowledge* (not to forget the correspondingly adapted idea of *common knowledge*) could be tackled. Relating to this, both [13] (on the knowledge-theoretic side) and [9] (on multi-subset spaces) may serve as a starting point. (Contrasting the present approach which is new, the recent paper [11] is based on the latter article to some extent.)

Acknowledgement. I would like to take this opportunity to thank the anonymous referees very much for their detailed reviews which, among other things, contain valuable comments on the system presented here as well as suggestions for alternative approaches to multi-agent subset spaces being worth considering.

References

1. Aiello, M., Pratt-Hartmann, I.E., van Benthem, J.F.A.K.: Handbook of Spatial Logics. Springer, Dordrecht (2007)
2. Balbiani, P., van Ditmarsch, H., Kudinov, A.: Subset space logic with arbitrary announcements. In: Lodaya, K. (ed.) Logic and Its Applications. LNCS, vol. 7750, pp. 233–244. Springer, Heidelberg (2013)
3. Blackburn, P., de Rijke, M., Venema, Y.: Modal Logic, Cambridge Tracts in Theoretical Computer Science, vol. 53. Cambridge University Press, Cambridge (2001)
4. Dabrowski, A., Moss, L.S., Parikh, R.: Topological reasoning and the logic of knowledge. Ann. Pure Appl. Log. **78**, 73–110 (1996)
5. Fagin, R., Halpern, J.Y., Moses, Y., Vardi, M.Y.: Reasoning About Knowledge. MIT Press, Cambridge (1995)
6. Georgatos, K.: Knowledge theoretic properties of topological spaces. In: Masuch, M., Pólos, L. (eds.) Knowledge Representation and Uncertainty, Logic at Work. Lecture Notes in Artificial Intelligence, vol. 808, pp. 147–159. Springer, Heidelberg (1994)
7. Goldblatt, R.: Logics of Time and Computation. CSLI Lecture Notes, vol. 7, 2nd edn. Center for the Study of Language and Information, Stanford (1992)
8. Heinemann, B.: Topological nexttime logic. In: Kracht, M., de Rijke, M., Wansing, H., Zakharyaschev, M. (eds.) Advances in Modal Logic 1. CSLI Publications, vol. 87, pp. 99–113. Kluwer, Stanford (1998)
9. Heinemann, B.: Logics for multi-subset spaces. J. Appl. Non-Class. Log. **20**(3), 219–240 (2010)
10. Heinemann, B.: Coming upon the classic notion of implicit knowledge again. In: Buchmann, R., Kifor, C.V., Yu, J. (eds.) KSEM 2014. LNCS, vol. 8793, pp. 1–12. Springer, Heidelberg (2014)
11. Heinemann, B.: Subset spaces modeling knowledge-competitive agents. In: Zhang, S., Wirsing, M., Zhang, Z. (eds.) KSEM 2015. LNCS, vol. 9403, pp. 1–12. Springer, Heidelberg (2015). doi:10.1007/978-3-319-25159-2_1
12. Krommes, G.: A new proof of decidability for the modal logic of subset spaces. In: ten Cate, B. (ed.) Proceedings of the Eighth ESSLLI Student Session, pp. 137–147. Austria, Vienna, August 2003

13. Lomuscio, A., Ryan, M.: Ideal agents sharing (some!) knowledge. In: Prade, H. (ed.) ECAI 1998, 13th European Conference on Artificial Intelligence, pp. 557–561. John Wiley & Sons Ltd, Chichester (1998)
14. Meyer, J.J.C., van der Hoek, W.: Epistemic Logic for AI and Computer Science, Cambridge Tracts in Theoretical Computer Science, vol. 41. Cambridge University Press, Cambridge (1995)
15. Moss, L.S., Parikh, R.: Topological reasoning and the logic of knowledge. In: Moses, Y. (ed.) Theoretical Aspects of Reasoning about Knowledge (TARK 1992), pp. 95–105. Morgan Kaufmann, Los Altos, CA (1992)
16. Wáng, Y.N., Ågotnes, T.: Subset space public announcement logic. In: Lodaya, K. (ed.) Logic and Its Applications. LNCS, vol. 7750, pp. 245–257. Springer, Heidelberg (2013)

On Lambek's Restriction in the Presence of Exponential Modalities

Max Kanovich[1,2,3], Stepan Kuznetsov[4]([✉]), and Andre Scedrov[5,6]

[1] Queen Mary, University of London, London, UK
[2] University College London, London, UK
[3] National Research University Higher School of Economics, Moscow, Russia
`m.kanovich@qmul.ac.uk`
[4] Steklov Mathematical Institute, RAS, Moscow, Russia
`sk@mi.ras.ru`
[5] University of Pennsylvania, Philadelphia, USA
[6] National Research University Higher School of Economics, Moscow, Russia
`scedrov@math.upenn.edu`

Abstract. The Lambek calculus can be considered as a version of non-commutative intuitionistic linear logic. One of the interesting features of the Lambek calculus is the so-called "Lambek's restriction," that is, the antecedent of any provable sequent should be non-empty. In this paper we discuss ways of extending the Lambek calculus with the linear logic exponential modality while keeping Lambek's restriction. We present several versions of the Lambek calculus extended with exponential modalities and prove that those extensions are undecidable, even if we take only one of the two divisions provided by the Lambek calculus.

Keywords: Lambek calculus · Linear logic · Exponential modalities · Lambek's restriction · Undecidability

1 Introduction

The Lambek calculus was introduced by J. Lambek in [9] for mathematical description of natural language syntax by means of so-called *Lambek categorial (type-logical) grammars* (see, for example, [4,12,14]). In Lambek grammars, syntactic categories are represented by logical formulae involving three connectives: the *product* (corresponds to concatenation of words) and two *divisions* (left and right), and syntactic correctness of natural language expressions corresponds to derivability in the Lambek calculus.

For simplicity, in this paper we consider only the product-free fragment of the Lambek calculus. First we consider not the Lambek calculus **L** [9], but its variant **L*** [10]. The difference between **L** and **L*** is explained in the end of this introductory section (see "Lambek's Restriction").

L* is a substructural logic, and here we formulate it as a Gentzen-style sequent calculus. Formulae of **L*** are called *types* and are built from *variables,*

© Springer International Publishing Switzerland 2016
S. Artemov and A. Nerode (Eds.): LFCS 2016, LNCS 9537, pp. 146–158, 2016.
DOI: 10.1007/978-3-319-27683-0_11

or *primitive types* $(p, q, r, p_1, p_2, \dots)$ using two binary connectives: $\backslash$ *(left division)* and $/$ *(right division)*. Types are denoted by capital Latin letters, finite (possibly empty) sequences of types by capital Greek ones. Λ stands for the empty sequence. The Lambek calculus derives objects called *sequents* of the form $\Pi \to A$, where the *antecedent* Π is a linearly ordered sequence of types and *succedent* A is a type.

The axioms of $\mathbf{L}^*$ are all sequents $A \to A$, where A is a type, and the rules of inference are as follows:

$$\frac{A, \Pi \to B}{\Pi \to A \backslash B} \ (\to \backslash) \qquad \frac{\Pi \to A \quad \Delta_1, B, \Delta_2 \to C}{\Delta_1, \Pi, A \backslash B, \Delta_2 \to C} \ (\backslash \to)$$

$$\frac{\Pi, A \to B}{\Pi \to B / A} \ (\to /) \qquad \frac{\Pi \to A \quad \Delta_1, B, \Delta_2 \to C}{\Delta_1, B / A, \Pi, \Delta_2 \to C} \ (/ \to)$$

For $\mathbf{L}^*$ and other calculi introduced later in this paper, we do not include cut as an official rule of the system. However, the cut rule of the following non-commutative form

$$\frac{\Pi \to A \quad \Delta_1, A, \Delta_2 \to B}{\Delta_1, \Pi, \Delta_2 \to B} (\text{cut})$$

is admissible in $\mathbf{L}^*$ [10].

By $\mathbf{L}^*_/$ (resp., $\mathbf{L}^*_\backslash$) we denote the fragment of $\mathbf{L}^*$ with only the right (resp., left) division connective. Due to the subformula property, these fragments are obtained from the full calculus simply by restricting the set of rules.

We see that $\mathbf{L}^*$ lacks structural rules (except for the implicit rule of associativity).

$\mathbf{L}^*$ can be conservatively embedded [1,22] into a non-commutative, intuitionistic or cyclic, variant of Girard's [5] linear logic. In the spirit of linear logic connectives, the Lambek calculus can be extended with the *exponential* unary connective that enables structural rules (weakening, contraction, and commutativity) in a controlled way.

We'll denote this extended calculus by $\mathbf{EL}^*$. Types of $\mathbf{EL}^*$ are built from variables using two binary connectives ($\backslash$ and $/$) and a unary one, !, called the *exponential*, or, colloquially, *"bang."* If $\Gamma = A_1, \dots, A_k$, then by $!\Gamma$ we denote the sequence $!A_1, \dots, !A_k$. $\mathbf{EL}^*$ is obtained from $\mathbf{L}^*$ by adding the following rules:

$$\frac{\Delta_1, A, \Delta_2 \to B}{\Delta_1, !A, \Delta_2 \to B} \ (! \to) \qquad\qquad \frac{!\Gamma \to A}{!\Gamma \to !A} \ (\to !)$$

$$\frac{\Delta \to B}{!A, \Delta \to B} \ (\text{weak}) \qquad\qquad \frac{!A, !A, \Delta \to B}{!A, \Delta \to B} \ (\text{contr})$$

$$\frac{\Delta_1, B, !A, \Delta_2 \to C}{\Delta_1, !A, B, \Delta_2 \to C} \ (\text{perm}_1) \qquad \frac{\Delta_1, !B, A, \Delta_2 \to C}{\Delta_1, A, !B, \Delta_2 \to C} \ (\text{perm}_2)$$

The following theorem is proved in [6,8] and summarized in [7]. A weaker result that $\mathbf{EL}^*$ with the product and two divisions is undecidable follows from [11].

Theorem 1. *The derivability problem for* $\mathbf{EL}^*$ *is undecidable.*

Lambek's Restriction

However, the original Lambek calculus $\mathbf{L}$ [9] differs from the presented above in one detail: in $\mathbf{L}$, sequents with empty antecedents are not permitted. This restriction applies not only to the final sequent, but to all ones in the derivation. Thus, for example, the sequent $(q \backslash q) \backslash p \to p$ is derivable in $\mathbf{L}^*$, but not in $\mathbf{L}$, though its antecedent is not empty (but the $\mathbf{L}^*$-derivation involves the sequent $\to q \backslash q$ with an empty antecedent). Further we shall use the term *Lambek's restriction* for this special constraint. Actually, Lambek's restriction in $\mathbf{L}^*$ could potentially be violated only by application of the $(\to \backslash)$ and $(\to /)$ rules, therefore $\mathbf{L}$ can be obtained from $\mathbf{L}^*$ by adding the constraint "Π is non-empty" to these two rules.

At first glance, Lambek's restriction looks strange and formal, but it is highly motivated by linguistic applications. In syntactic formalisms based on the Lambek calculus, Lambek types denote syntactic categories.

Example 1. [12, 2.5] Let n stand for "noun phrase," then n / n is going to be a "noun modifier" (it can be combined with a noun phrase on the right producing a new, more complex noun phrase: $\mathbf{L} \vdash n / n, n \to n$), i.e. an adjective. Adverbs, as adjective modifiers, receive the type $(n / n) / (n / n)$. Now one can derive the sequent $(n / n) / (n / n), n / n, n \to n$ and therefore establish that, say, "very interesting book" is a valid noun phrase (belongs to syntactic category n). However, in $\mathbf{L}^*$ one can also derive $(n / n) / (n / n), n \to n$, where the antecedent describes syntactic constructions like "very book," that in fact aren't correct noun phrases.

This example shows that, for linguistic purposes, $\mathbf{L}$ is more appropriate than $\mathbf{L}^*$.

Suprisingly, however, it is not so straightforward to add the exponential to $\mathbf{L}$ or to impose Lambek's restriction on $\mathbf{EL}^*$. In Sect. 2 we discuss several ways how to do this and define a number of the corresponding calculi.

In Sect. 3 we state and prove undecidability results for calculi defined in Sect. 2. Finally, Sect. 4 contains general discussion of the results and possible directions of future work.

2   Imposing Lambek's Restriction on $\mathbf{EL}^*$

2.1   The First Approach: $\mathbf{EL}^{\mathrm{wk}}$

The **first**, naïve way of imposing Lambek's restriction on $\mathbf{EL}^*$ is to restrict only rules $(\to \backslash)$ and $(\to /)$. Notice that all other rules, including rules for the exponential, preserve the non-emptiness of the antecedent. Denote the calculus by $\mathbf{EL}^{\mathrm{wk}}$.

However, such a restriction doesn't change things significantly, since the following lemma provides the non-emptiness of the antecedent for free:

Lemma 1. *Let p be a variable not occurring in a sequent $\Gamma \to A$. Then*

$$\mathbf{EL}^* \vdash \Gamma \to A \iff \mathbf{EL}^{\mathrm{wk}} \vdash \,!p, \Gamma \to A.$$

This lemma shows that $\mathbf{EL}^*$-derivations can be enabled in $\mathbf{EL}^{\mathrm{wk}}$ by an easy technical trick. Therefore, Theorem 1 implies immediately that $\mathbf{EL}^{\mathrm{wk}}$ is undecidable.

Lemma 2. $\mathbf{EL}^{\mathrm{wk}} \vdash \,!B, \Gamma \to A \iff \mathbf{EL}^* \vdash \,!B, \Gamma \to A.$

These two lemmas are proved by induction on the derivations (recall that (cut) is not included in the calculi).

Thus, Lambek's restriction in $\mathbf{EL}^{\mathrm{wk}}$ vanishes as soon as the antecedent contains a formula with ! as the main connective. And, unfortunately, this acts non-locally: once $!A$ appears in the antecedent, one can derive unwanted things like "very book" (see Example 1 above).

2.2   The Second Approach: $\mathbf{EL}^-$

To overcome that $!B$ is able to mimic the empty antecedent, we impose more radical restrictions by constructing the following calculus $\mathbf{EL}^-$.

Any formula not of the form $!B$ is called a *non-bang-formula*. Now $\mathbf{EL}^-$ is defined by the following axioms and rules:

$$A \to A$$

$$\frac{A, \Pi \to B}{\Pi \to A \backslash B}(\to \backslash), \text{ where } \Pi \text{ contains a non-bang-formula}$$

$$\frac{\Pi, A \to B}{\Pi \to B \,/\, A}(\to /), \text{ where } \Pi \text{ contains a non-bang-formula}$$

$$\frac{\Pi \to A \quad \Delta_1, B, \Delta_2 \to C}{\Delta_1, \Pi, A \backslash B, \Delta_2 \to C}(\backslash \to) \qquad \frac{\Pi \to A \quad \Delta_1, B, \Delta_2 \to C}{\Delta_1, B \,/\, A, \Pi, \Delta_2 \to C}(/ \to)$$

$$\frac{\Delta_1, A, \Delta_2 \to B}{\Delta_1, !A, \Delta_2 \to B}(! \to), \text{ where } \Delta_1, \Delta_2 \text{ contains a non-bang-formula}$$

$$\frac{\Delta \to B}{!A, \Delta \to B}(\mathrm{weak}) \qquad \frac{!A, !A, \Delta \to B}{!A, \Delta \to B}(\mathrm{contr})$$

$$\frac{\Delta_1, B, !A, \Delta_2 \to C}{\Delta_1, !A, B, \Delta_2 \to C}(\mathrm{perm}_1) \qquad \frac{\Delta_1, !B, A, \Delta_2 \to C}{\Delta_1, A, !B, \Delta_2 \to C}(\mathrm{perm}_2)$$

Note that in the $(\to !)$ rule all the formulae in the antecedent are of the form $!B$. Therefore there is no $(\to !)$ rule in $\mathbf{EL}^-$. Also note that the cut rule is not included in $\mathbf{EL}^-$.

Lemma 3. *If $\Pi \to A$ is derivable in $\mathbf{EL}^-$ and doesn't contain non-bang-formula, then it is of the form $!\Gamma \to !B$, where B is a formula from Γ.*

Now Lambek's restriction in $\mathbf{EL}^-$ is stated in the following way: *in a non-trivial derivable sequent $\Pi \to A$ the antecedent Π should contain at least one non-bang-formula.*

2.3 The Third Approach: EL

Unfortunately, $\mathbf{EL}^-$ doesn't respect type substitution: e.g., $\mathbf{EL}^- \vdash p, !(p \backslash q) \to q$, but $\mathbf{EL}^- \not\vdash !r, !(!r \backslash q) \to q$. The original Lambek calculus has type substitution, and for a decent logical system it is a desired property.

In order to restore type substitution as much as possible we consider **the third** approach to imposing Lambek's restriction on $\mathbf{EL}^*$.

We present such a system in the form of *marked sequent calculus*. A marked sequent is an expression of the form $\Pi \to A$, where A is a type and Π is a sequence of pairs of the form $\langle B, m \rangle$, written as $B_{(m)}$, where B is a type and $m \in \{0, 1\}$ is the *marking bit*. A pair $B_{(0)}$ is called an *unmarked type*, and $B_{(1)}$ is called a *marked type*. The marking bits are utilized inside the derivation, and in the end they are forgotten, yielding a sequent in the original sense. If $\Gamma = B_{1(m_1)}, \ldots, B_{k(m_k)}$, then by $!\Gamma$ we denote the sequence $(!B_1)_{(m_1)}, \ldots, (!B_k)_{(m_k)}$.

Lambek's restriction is now formulated as follows: *every sequent should contain an unmarked type in the antecedent.*

The calculus $\mathbf{EL}$ is defined in the following way:

$$p_{(0)} \to p$$

$$\frac{\Pi, A_{(m)} \to B}{\Pi \to B \, / \, A}(\to /), \text{ where } \Pi \text{ contains an unmarked type}$$

$$\frac{A_{(m)}, \Pi \to B}{\Pi \to A \backslash B}(\to \backslash), \text{ where } \Pi \text{ contains an unmarked type}$$

$$\frac{\Pi \to A \quad \Delta_1, B_{(m)}, \Delta_2 \to C}{\Delta_1, (B \, / \, A)_{(m)}, \Pi, \Delta_2 \to C}(/ \to) \qquad \frac{\Pi \to A \quad \Delta_1, B_{(m)}, \Delta_2 \to C}{\Delta_1, \Pi, (A \backslash B)_{(m)}, \Delta_2 \to C}(\backslash \to)$$

$$\frac{\Delta_1, A_{(m)}, \Delta_2 \to B}{\Delta_1, (!A)_{(1)}, \Delta_2 \to B}(! \to), \text{ where } \Delta_1, \Delta_2 \text{ contains an unmarked type}$$

$$\frac{!\Gamma, \Delta \to A}{!\Gamma, !\Delta \to !A}(\to !) \qquad \frac{\Delta_1, \Delta_2 \to A}{\Delta_1, (!A)_{(1)}, \Delta_2 \to A}(\text{weak})$$

$$\frac{(!A)_{(m_1)}, (!A)_{(m_2)}, \Delta \to B}{(!A)_{(\min\{m_1, m_2\})}, \Delta \to B}(\text{contr})$$

$$\frac{\Delta_1, B_{(m_2)}, (!A)_{(m_1)}, \Delta_2 \to C}{\Delta_1, (!A)_{(m_1)}, B_{(m_2)}, \Delta_2 \to C}(\text{perm}_1) \qquad \frac{\Delta_1, (!B)_{(m_2)}, A_{(m_1)}, \Delta_2 \to C}{\Delta_1, A_{(m_1)}, (!B)_{(m_2)}, \Delta_2 \to C}(\text{perm}_2)$$

Recall that all proofs are cut-free. Also note that in **EL** we use a stronger form of the $(\to\,!)$ rule. In **EL*** this new rule could be simulated by applying the $(!\to)$ rule for all formulae in Δ and then using the original $(\to\,!)$ rule, but here the $(!\to)$ rule will fail to satisfy the restriction.

The substitution property is now formulated as follows:

Proposition 1. *Let $\widetilde{A}$ (resp., $\widetilde{\Pi}$) be the result of substituting D for p in type A (resp., marked sequence Π). Then* **EL** $\vdash \Pi \to A$ *implies* **EL** $\vdash \widetilde{\Pi} \to \widetilde{A}$.

Proof. By structural induction on D we prove that **EL** $\vdash D_{(0)} \to D$ for every type D. Then we just replace p with D everywhere in the proof.

Compare **EL** with **EL**$^-$. These two systems are not connected with any strong form of conservativity or equivalence: on one hand, the sequent $!r, r \setminus !p, !(p \setminus q) \to q$ is derivable in **EL**$^-$, but not in **EL**; on the other hand, for $!p, !(!p \setminus q) \to q$ the situation is opposite. Fortunately, the following holds:

Lemma 4. *If Γ, Π, and A do not contain $!$, then*

$$\textbf{EL} \vdash !\Gamma, \Pi \to A \iff \textbf{EL}^- \vdash !\Gamma, \Pi \to A.$$

Proof. Since for a sequent of the form $!\Gamma, \Pi \to A$ the rule $(\to\,!)$ can never appear in the proof, marked types in the antecedent are exactly the types starting with $!$, and the two versions of Lambek's restriction coincide.

2.4   Conservativity over L

The three approaches are conservative over **L**:

Proposition 2. *If Π and A do not contain $!$, then*

$$\textbf{L} \vdash \Pi \to A \iff \textbf{EL}^{\mathrm{wk}} \vdash \Pi \to A \iff \textbf{EL}^- \vdash \Pi \to A \iff \textbf{EL} \vdash \Pi \to A$$

(for **EL***, all types in Π get the 0 marking bit).*

Note that Π is necessarily non-empty.

Therefore, we guarantee that in all approaches the innovation affects only the new exponential connective, and keeps the original Lambek system intact. For **EL**$^-$ and **EL** adding fresh exponentials to the antecedent also doesn't affect Lambek's restriction:

Proposition 3. *If Π and A do not contain $!$, and p is a variable not occurring in Π and A, then* **EL**$^-$ $\vdash !p, \Pi \to A \iff$ **EL** $\vdash !p, \Pi \to A \iff$ **L** $\vdash \Pi \to A$ *(for the* **EL** *case, $!p$ gets marking bit 1 and types from Π get 0).*

For **EL**^{wk}, due to Lemma 1, the situation is different: if Π and A don't contain $!$, and p is a fresh variable, then

$$\textbf{EL}^{\mathrm{wk}} \vdash \Pi \to A \iff \textbf{EL}^* \vdash !p, \Pi \to A \iff \textbf{L}^* \vdash \Pi \to A.$$

Recall that, for example, $(q \setminus q) \setminus p \to p$ is derivable in **L***, but not in **L**.

3   Undecidability Results

3.1   Lambek Calculus with Non-Logical Axioms and Generative Grammars

In this subsection we introduce *axiomatic extensions* of the Lambek calculus **L**, following [3]. These extensions are going to be useful for proving undecidability results à la Theorem 1.

Let $\mathcal{A}$ be a set of sequents. Then by $\mathbf{L} + \mathcal{A}$ we denote **L** augmented with sequents from $\mathcal{A}$ as new axioms and also the cut rule (which is no longer eliminable). Elements of $\mathcal{A}$ are called *non-logical axioms*.

Further we consider non-logical axioms of a special form: either $p, q \to r$, or $p \mathbin{/} q \to r$, where p, q, r are variables. Buszkowski calls them *special* non-logical axioms. In this case, $\mathbf{L} + \mathcal{A}$ can be formulated in a cut-free way [3]: instead of non-logical axioms of the form $p, q \to r$ or $p \mathbin{/} q \to r$ we use rules

$$\frac{\Pi_1 \to p \quad \Pi_2 \to q}{\Pi_1, \Pi_2 \to r}(\mathrm{red}_1) \qquad \text{and} \qquad \frac{\Pi, q \to p}{\Pi \to r}(\mathrm{red}_2), \text{ where } \Pi \neq \Lambda$$

respectively. This calculus admits the cut rule [3]. Further we'll mean it when talking about $\mathbf{L} + \mathcal{A}$. We'll use the term *Buszkowski's rules* for (red_i).

Now we define two notions of *formal grammar*. The first one is the widely known formalism of *generative grammars* introduced by Chomsky. If Σ is an alphabet (i.e. a finite non-empty set), then by Σ^* we denote the set of all words over Σ (including the empty word). A generative grammar is a quadruple $G = \langle N, \Sigma, s, P \rangle$, where N and Σ are two disjoint alphabets, $s \in N$, and P is a set or *rules*. Here we consider only rules of two forms: $x \to y_1 y_2$ or $x_1 x_2 \to y$, where $x, y, x_i, y_i \in N \cup \Sigma$. If $v = u_1 \alpha u_2$, $w = u_1 \beta u_2$, and $(\alpha \to \beta) \in P$, then this rule can be *applied* to v yielding w: $v \Rightarrow w$. By $\Rightarrow^*$ we denote the reflexive and transitive closure of $\Rightarrow$. Finally, the *language generated by* G is the set of all words $w \in \Sigma^*$, such that $s \Rightarrow^* w$. Note that the empty word can't be produced by a generative grammar as defined above.

It is well known that the class of languages generated by generative grammars coincides with the class of all recursively enumerable (r. e.) languages without the empty word.

The second family of formal grammar we are going to consider is the class of *Lambek categorial grammars with non-logical axioms*. A Lambek grammar is a tuple $\mathcal{G} = \langle \Sigma, \mathcal{A}, H, \rhd \rangle$, where Σ is an alphabet, $\mathcal{A}$ is a set of non-logical axioms, H is a type, and $\rhd \subseteq \mathrm{Tp} \times \Sigma$ is a finite binary correspondence between types and letter, called *type assignment*. A word $w = a_1 \ldots a_n$ belongs to the language generated by $\mathcal{G}$ iff there exist such types $A_1, \ldots, A_n$ that $A_i \rhd a_i$ $(i = 1, \ldots, n)$ and $\mathbf{L} + \mathcal{A} \vdash A_1, \ldots, A_n \to H$.

If we use $\mathbf{L}_{/}$ instead of $\mathbf{L}$, we get the notion of $\mathbf{L}_{/}$-*grammar* with non-logical axioms. It's easy to see that all languages generated by Lambek grammars are r. e., therefore, they can be generated by generative grammars. Buszkowski [3] proves the converse:

Theorem 2. *Every language generated by a generative grammar can be generated by an* $\mathbf{L}_/$ *-grammar with special non-logical axioms.*

In comparison, for $\mathcal{A} = \varnothing$ Pentus' theorem [17] states that all languages generated are context-free. Thus, even simple (special) non-logical axioms dramatically increase the power (and complexity) of Lambek grammars.

Since there exist undecidable r. e. languages, Buszkowski obtains the following [3]:

Theorem 3. *There exists such* $\mathcal{A}$ *that the derivability problem for* $\mathbf{L}_/ + \mathcal{A}$ *is undecidable.*

3.2   Undecidability Proof for $\mathbf{EL}^-$ and $\mathbf{EL}$

Recall that $\mathbf{EL}^-$, defined in Sect. 2.2, involves two division operations, and no product. The calculus $\mathbf{EL}^-_/$ is the fragment of $\mathbf{EL}^-$, where we confine ourselves only to the right division.

Theorem 4. *The derivability problem for* $\mathbf{EL}^-$ *and even for* $\mathbf{EL}^-_/$ *is undecidable.*

We take a set $\mathcal{A}$ of non-logical axioms of non-logical axioms of the forms $p, q \to r$ or $p/q \to r$ and encode them in $\mathbf{EL}^-$ using the exponential. Let $\mathcal{G}_\mathcal{A} = \{(r/q)/p \mid (p,q \to r) \in \mathcal{A}\} \cup \{r/(p/q) \mid (p/q \to r) \in \mathcal{A}\}$ and let $\Gamma_\mathcal{A}$ be a sequence of all types from $\mathcal{G}_\mathcal{A}$ in any order. Then the following holds:

Lemma 5. $\mathbf{L}_/ + \mathcal{A} \vdash \Pi \to A \iff \mathbf{EL}^-_/ \vdash !\Gamma_\mathcal{A}, \Pi \to A.$

Proof. $\boxed{\Rightarrow}$ Proceed by induction on the derivation of $\Pi \to A$ in $\mathbf{L}_/ + \mathcal{A}$. If $\Pi \to A$ is an axiom of the form $A \to A$, then we get $\mathbf{EL}^-_/ \vdash !\Gamma_\mathcal{A}, A \to A$ by application of the (weak) rule.

If $A = B/C$, and $\Pi \to A$ is obtained using the $(\to /)$ rule, then $!\Gamma_\mathcal{A}, \Pi \to A$ is derived using the same rule:

$$\frac{!\Gamma_\mathcal{A}, \Pi, C \to B}{!\Gamma_\mathcal{A}, \Pi \to B/C}$$

Here Π is not empty, and consists of non-bang-formulae, therefore the application of this rule is eligible in $\mathbf{EL}^-_/$; $\mathbf{EL}^-_/ \vdash !\Gamma_\mathcal{A}, \Pi, C \to B$ by induction hypothesis.

If $\Pi = \Phi_1, (B/C), \Psi, \Phi_2$, and $\Pi \to A$ is obtained by $(/ \to)$ from $\Psi \to C$ and $\Phi_1, B, \Phi_2 \to A$, then for $!\Gamma_\mathcal{A}, \Pi \to A$ we have the following derivation in $\mathbf{EL}^-_/$, where * means several applications of the rules in any order.

$$\frac{\dfrac{\dfrac{!\Gamma_\mathcal{A}, \Psi \to C \quad !\Gamma_\mathcal{A}, \Phi_1, B, \Phi_2 \to A}{!\Gamma_\mathcal{A}, \Phi_1, (B/C), !\Gamma_\mathcal{A}, \Psi, \Phi_2 \to A} \, (/ \to)}{!\Gamma_\mathcal{A}, !\Gamma_\mathcal{A}, \Phi_1, (B/C), \Psi, \Phi_2 \to A} \, (\mathrm{perm}_1)^*}{!\Gamma_\mathcal{A}, \Phi_1, (B/C), \Psi, \Phi_2 \to A} \, (\mathrm{contr}, \mathrm{perm}_1)^*$$

Finally, $\Pi \to A$ can be obtained by application of Buszkowski's rules (red_1) or (red_2). In the first case, $A = r$, $\Pi = \Pi_1, \Pi_2$; $\mathbf{L} + \mathcal{A} \vdash \Pi_1 \to p$, and $\mathbf{L} + \mathcal{A} \vdash \Pi_2 \to q$. Furthermore, $\mathcal{G}_{\mathcal{A}} \ni (r\,/\,q)\,/\,p$, thus we get the following derivation in $\mathbf{EL}^{-}_{/}$:

$$
\cfrac{
  !\Gamma_{\mathcal{A}}, \Pi_1 \to p \qquad
  \cfrac{
    \cfrac{!\Gamma_{\mathcal{A}}, \Pi_2 \to q \quad r \to r}{r\,/\,q, !\Gamma_{\mathcal{A}}, \Pi_2 \to r}\ (/\to)
  }{}
}{}
$$

$$
\cfrac{
\cfrac{
\cfrac{
\cfrac{!\Gamma_{\mathcal{A}}, \Pi_1 \to p \quad \cfrac{!\Gamma_{\mathcal{A}}, \Pi_2 \to q \quad r \to r}{r\,/\,q, !\Gamma_{\mathcal{A}}, \Pi_2 \to r}\,(/\to)}{(r\,/\,q)\,/\,p, !\Gamma_{\mathcal{A}}, \Pi_1, !\Gamma_{\mathcal{A}}, \Pi_2 \to r}\,(/\to)}{!((r\,/\,q)\,/\,p), !\Gamma_{\mathcal{A}}, \Pi_1, !\Gamma_{\mathcal{A}}, \Pi_2 \to r}\,(!\to)}{!\Gamma_{\mathcal{A}}, !((r\,/\,q)\,/\,p), \Pi_1, \Pi_2 \to r}\,(\mathrm{contr}, \mathrm{perm}_1)^{*}}{!\Gamma_{\mathcal{A}}, \Pi_1, \Pi_2 \to r}\,(\mathrm{contr}, \mathrm{perm}_1)^{*}
$$

The application of $(!\to)$ here is legal, since Π_1 and Π_2 are non-empty and consist of non-bang-formulae.

In the (red_2) case, $A = r$, and we have $!\Gamma_{\mathcal{A}}, \Pi, q \to p$ in the induction hypothesis. Again, $\mathcal{G}_{\mathcal{A}} \ni r\,/(p\,/\,q)$, and we proceed like this:

$$
\cfrac{
\cfrac{
\cfrac{
\cfrac{!\Gamma_{\mathcal{A}}, \Pi, q \to p}{!\Gamma_{\mathcal{A}}, \Pi \to p\,/\,q}\,(\to/) \quad r \to r}{r\,/(p\,/\,q), !\Gamma_{\mathcal{A}}, \Pi \to r}\,(/\to)}{!(r\,/(p\,/\,q)), !\Gamma_{\mathcal{A}}, \Pi \to r}\,(!\to)}{!\Gamma_{\mathcal{A}}, \Pi \to r}\,(\mathrm{contr}, \mathrm{perm}_1)^{*}
$$

Here, again, Π is not empty and consists of non-bang-formulae, therefore we can legally apply $(!\to)$ and $(\to/)$.

$\boxed{\Leftarrow}$ For deriving sequents of the form $!\Gamma, \Pi \to A$, where Γ, Π, and A do not contain the exponential, one can use a simpler calculus than $\mathbf{EL}^{-}_{/}$:

$$
!\Gamma, p \to p
$$

$$
\cfrac{!\Gamma, \Pi, B \to A}{!\Gamma, \Pi \to A\,/\,B}\ (\to/), \text{ where } \Pi \neq \Lambda
\qquad
\cfrac{!\Gamma, \Pi \to B \quad !\Gamma, \Delta_1, A, \Delta_2 \to C}{!\Gamma, \Delta_1, A\,/\,B, \Pi, \Delta_2 \to C}\ (/\to)
$$

$$
\cfrac{!\Gamma, \Delta_1, B, \Delta_2 \to A}{!\Gamma, \Delta_1, \Delta_2 \to A}\ (!\to), \text{ where } B \text{ is a type from } \Gamma \text{ and } \Delta_1, \Delta_2 \neq \Lambda
$$

Here (weak) is hidden into the axiom, (contr) comes within $(\to/)$, and $(!\to)$ includes both (perm_i) and (contr) in the needed form. One can easily see that if $\mathbf{EL}^{-}_{/} \vdash !\Gamma, \Pi \to A$, where Γ, Π, and A do not contain $!$, then this sequent is derivable in the simplified calculus. Moreover, the $(!\to)$ rule is interchangeable with the others in the following ways:

$$
\cfrac{\cfrac{!\Gamma, \Delta_1, C, \Delta_2, B \to A}{!\Gamma, \Delta_1, C, \Delta_2 \to A\,/\,B}\,(\to/)}{!\Gamma, \Delta_1, \Delta_2 \to A\,/\,B}\,(!\to)
\qquad \rightsquigarrow \qquad
\cfrac{\cfrac{!\Gamma, \Delta_1, C, \Delta_2, B \to A}{!\Gamma, \Delta_1, \Delta_2, B \to A}\,(!\to)}{!\Gamma, \Delta_1, \Delta_2 \to A\,/\,B}\,(\to/)
$$

$$\frac{!\Gamma, \Pi \to B \quad !\Gamma, \Delta_1, A, \Delta_2', D, \Delta_2'' \to C}{\dfrac{!\Gamma, \Delta_1, A \,/\, B, \Pi, \Delta_2', D, \Delta_2'' \to C}{!\Gamma, \Delta_1, A \,/\, B, \Pi, \Delta_2', \Delta_2'' \to C} \,(! \to)} \,(/ \to)$$

$$\updownarrow$$

$$\frac{!\Gamma, \Pi \to B \quad \dfrac{!\Gamma, \Delta_1, A, \Delta_2', D, \Delta_2'' \to C}{!\Gamma, \Delta_1, A, \Delta_2', \Delta_2'' \to C} \,(! \to)}{!\Gamma, \Delta_1, A \,/\, B, \Pi, \Delta_2', \Delta_2'' \to C} \,(/ \to)$$

And the same, if D appears inside Δ_1 or Π. Finally, consecutive applications of $(! \to)$ are always interchangeable.

After applying these transformations, we achieve a derivation where $(! \to)$ is applied immediately after applying $(/ \to)$ with the same active type (the other case, when it is applied after the axiom to p, is impossible, since then it violates the non-emptiness condition). In other words, applications of $(! \to)$ appear only in the following two situations:

$$\frac{\dfrac{!\Gamma, \Pi \to p \quad !\Gamma, \Delta_1, r \,/\, q, \Delta_2 \to A}{!\Gamma, \Delta_1, (r \,/\, q) \,/\, p, \Pi, \Delta_2 \to A} \,(/ \to)}{!\Gamma, \Delta_1, \Pi, \Delta_2 \to A} \,(! \to)$$

and

$$\frac{\dfrac{!\Gamma, \Pi \to p \,/\, q \quad !\Gamma, \Delta_1, r, \Delta_2 \to A}{!\Gamma, \Delta_1, r \,/(p \,/\, q), \Pi, \Delta_2 \to A} \,(/ \to)}{!\Gamma, \Delta_1, \Pi, \Delta_2 \to A} \,(! \to)$$

Now we prove the statement $\mathbf{EL}^-_{/} \vdash \,!\Gamma_A, \Pi \to A \Rightarrow \mathbf{L} + \mathcal{A} \vdash \Pi \to A$ by induction on the above canonical derivation. If $!\Gamma_A, \Pi \to A$ is an axiom or is obtained by an application of $(/ \to)$ or $(\to /)$, we apply the corresponding rules in $\mathbf{L} + \mathcal{A}$, so the only interesting case is $(! \to)$. Consider the two possible situations.

In the $(r \,/\, q) \,/\, p$ case, by induction hypothesis we get $\mathbf{L} + \mathcal{A} \vdash \Pi \to p$ and $\mathbf{L} + \mathcal{A} \vdash \Delta_1, r \,/\, q, \Delta_2 \to A$, and then we develop the following derivation in $\mathbf{L} + \mathcal{A}$ (recall that (cut) is admissible there):

$$\frac{\Pi \to p \quad \dfrac{\dfrac{p, q \to r}{p \to r \,/\, q} \,(\to /) \quad \Delta_1, r \,/\, q, \Delta_2 \to A}{\Delta_1, p, \Delta_2 \to A} \,(\text{cut})}{\Delta_1, \Pi, \Delta_2 \to A} \,(\text{cut})$$

In the case of $r \,/(p \,/\, q)$, the derivation looks like this:

$$\frac{\Pi \to p \,/\, q \quad \dfrac{p \,/\, q \to r \quad \Delta_1, r, \Delta_2 \to A}{\Delta_1, p \,/\, q, \Delta_2 \to A} \,(\text{cut})}{\Delta_1, \Pi, \Delta_2 \to A} \,(\text{cut})$$

Note that in this proof we don't need any form of the cut rule for $\mathbf{EL}^-$.

Now Theorem 4 follows from Theorem 3.

Corollary 1. *The derivability problem for $\mathbf{EL}_/$ (and, thus, for $\mathbf{EL}$) is undecidable.*

Proof. Similary to Lemma 4 we have that if Γ, Π, and A do not contain !, then $\mathbf{EL}_/ \vdash\ !\Gamma, \Pi \to A \Longleftrightarrow \mathbf{EL}^-_/ \vdash\ !\Gamma, \Pi \to A$.

Of course, everything discussed above can be dually performed for $\backslash$ instead of $/$, yielding undecidability for $\mathbf{EL}_\backslash$ and $\mathbf{EL}^-_\backslash$.

Buszkowski's rules can be also emulated in $\mathbf{EL}^-_/$ without using the (weak) rule. Thus we get undecidability for the variant with only (perm$_1$), (perm$_2$), and (contr), and only one division.

4   Conclusion

The derivability problem for the original Lambek calculus, without exponential modalities, is decidable and belongs to the NP class. This happens because the cut-free proof of a sequent has linear size with respect to the sequent's length. For the full Lambek calculus [18] and for its fragments with any two of three connectives (two divisions [20] or one division and the product [21]) the derivability problem is NP-complete.

On the other hand, for derivability problem in $\mathbf{L}_/$ there exists a polynomial time algorithm [19]. Thus the one-division fragment of the Lambek calculus appears to be significantly simpler. Despite this, in our undecidability results for $\mathbf{EL}$ and $\mathbf{EL}^-$ we use only one of the two divisions.

Future Work

It appears that the technique used in the $\boxed{\Leftarrow}$ part of the proof of Lemma 5 is an instance of *focusing* [2,15] in the non-commutative situation, which should be investigated systematically and in detail.

We also plan to investigate calculi with modalities where not all of the structural rules ((weak), (perm$_i$), and (contr)) are kept. Once we remove (contr), the derivability problem becomes decidable and falls into the NP class. The interesting question is to determine precise complexity bounds (P or NP-hard) for the fragments with only one division and bang (if we have at least two of the three Lambek connectives, even the calculus without bang is NP-complete). The variants where we have only permutational rules (both (perm$_1$) and (perm$_2$), or only one of them) are particularly interesting for linguistic applications (see, for example, [12,13]). The other question here is (un)decidability of the variant without (perm$_i$) (only (weak) and (contr)).

In the case of commutative linear logic Nigam and Miller [16] consider calculi that have several modalities interacting with each other, and different modalities are controlled by different sets of structural rules. These modalities are called *subexponentials*. We plan to study subexponentials in the non-commutative case under the umbrella of the Lambek calculus.

Acknowledgments. Stepan Kuznetsov's research was supported by the Russian Foundation for Basic Research (grants 15-01-09218-a and 14-01-00127-a) and by the Presidential Council for Support of Leading Scientific Schools of Russia (grant NŠ 1423.2014.1). Max Kanovich's research was partly supported by EPSRC (grant EP/K040049/1).

This research was initiated during the visit by Stepan Kuznetsov to the University of Pennsylvania, which was supported in part by that institution. Further work was done during the stay of Kanovich and Scedrov at the National Research University Higher School of Economics, which was supported in part by that institution. We would like to thank Sergei O. Kuznetsov and Ilya A. Makarov for hosting us.

References

1. Abrusci, V.M.: A comparison between Lambek syntactic calculus and intuitionistic linear propositional logic. Zeitschr. für math. Logik and Grundl. der Math. (Math. Logic Quart.) **36**, 11–15 (1990)
2. Andreoli, J.-M.: Logical programming with focusing proofs in linear logic. J. Log. Comput. **2**, 297–347 (1992)
3. Buszkowski, W.: Some decision problems in the theory of syntactic categories. Zeitschr. für math. Logik und Grundl. der Math. (Math. Logic Quart.) **28**, 539–548 (1982)
4. Carpenter, B.: Type-logical semantics. MIT Press, Cambridge (1998)
5. Girard, J.-Y.: Linear logic. Theor. Comp. Sci. **50**(1), 1–102 (1987)
6. de Groote, P.: On the expressive power of the Lambek calculus extended with a structural modality. In: Casadio, C., et al. (eds.) Language and Grammar. CSLI Lect. Notes, vol. 168, pp. 95–111. Stanford University, Stanford (2005)
7. Kanazawa, M.: Lambek calculus: Recognizing power and complexity. In: Gerbrandy, J., et al. (eds.) JFAK. Essays dedicated to Johan van Benthem on the occasion of his 50th birthday. Vossiuspers, Amsterdam Univ. Press (1999)
8. Kanovich, M.: The expressive power of modalized purely implicational calculi. CSLI Report. Stanford University (1993)
9. Lambek, J.: The mathematics of sentence structure. Amer. Math. Monthly **65**(3), 154–170 (1958)
10. Lambek, J.: On the calculus of syntactic types. In: Jakobson (ed.) Structure of Language and Its Mathematical Aspects. Proc. Symposia Appl. Math., vol. 12, pp. 166–178. AMS, Providence, RI (1961)
11. Lincoln, P., Mitchell, J., Scedrov, A., Shankar, N.: Decision problems for propositional linear logic. Ann. Pure Appl. Log. **56**, 239–311 (1992)
12. Moot, R., Retoré, C. (eds.): The Logic of Categorial Grammars. LNCS, vol. 6850. Springer, Heidelberg (2012)
13. Morrill, G., Leslie, N., Hepple, M., Barry, G.: Categorial deductions and structural operations. In: Barry, G., Morrill, G. (eds.) Studies in Categorial Grammar. Edinburgh Working Papers in Cognitive Science, vol. 5, pp. 1–21. Centre for Cognitive Science, Edinburgh (1990)
14. Morril, G.: Categorial grammar. Logical syntax, semantics, and processing. Oxford University Press, Oxford (2011)
15. Nigam, V., Miller, D.: Focusing in linear meta-logic. In: Armando, A., Baumgartner, P., Dowek, G. (eds.) IJCAR 2008. LNCS (LNAI), vol. 5195, pp. 507–522. Springer, Heidelberg (2008)

16. Nigam, V., Miller, D.: Algorithmic specifications in linear logic with subexponentials. In: Proceedings of the 11th ACM SIGPLAN conference on Principles and practice of declarative programming, pp. 129–140 (2009)
17. Pentus, M.: Lambek grammars are context-free. In: Proceedings of the 8th Annual IEEE Symposium on Logic in Computer Science, pp. 429–433. IEEE Computer Society Press (1993)
18. Pentus, M.: Lambek calculus is NP-complete. Theor. Comput. Sci. **357**, 186–201 (2006)
19. Savateev, Yu.: Lambek grammars with one division are decidable in polynomial time. In: Hirsch, E.A., Razborov, A.A., Semenov, A., Slissenko, A. (eds.) Computer Science – Theory and Applications. LNCS, vol. 5010, pp. 273–282. Springer, Heidelberg (2008)
20. Savateev, Yu.: Product-free lambek calculus is NP-complete. In: Artemov, S., Nerode, A. (eds.) LFCS 2009. LNCS, vol. 5407, pp. 380–394. Springer, Heidelberg (2008)
21. Savateev, Yu.: Algorithmic complexity of fragments of the Lambek calculus (in Russian). Ph.D. thesis. Moscow State University (2009)
22. Yetter, D.N.: Quantales and (noncommutative) linear logic. J. Sym. Log. **55**(1), 41–64 (1990)

A Quest for Algorithmically Random Infinite Structures, II

Bakhadyr Khoussainov[(✉)]

The University of Auckland, Auckland, New Zealand
bmk@cs.auckland.ac.nz

Abstract. We present an axiomatic approach that introduces algorithmic randomness into various classes of structures. The central concept is the notion of a branching class. Through this technical yet simple notion we define measure, metric, and topology in many classes of graphs, trees, relational structures, and algebras. As a consequence we define algorithmically random structures. We prove the existence of algorithmically random structures with various computability-theoretic properties. We show that any nontrivial variety of algebras has an effective measure 0. We also prove a counter-intuitive result that there are algorithmically random yet computable structures. This establishes a connection between algorithmic randomness and computable model theory.

Keywords: Martin-Löf randomness · Halting problem · Measure · Computable infinite structure

1 Introduction

1.1 Background and Motivation

Algorithmic randomness of infinite strings has a captivating history going back to the work of Kolmogorov [7], Martin-Löf [13], Chaitin [3], Schnorr [18,19] and Levin [20]. In the last two decades the topic has attracted the attention of experts in complexity, computability, logic, philosophy, computational biology, and algorithms. Many algorithmic randomness and related notions for infinite strings have been introduced and investigated. These include Martin-Löf tests, Schnorr tests, prefix free complexity, 1-generic sets, K-triviality, martingales, connections to differentiability, Solovey and related reducibilities. The monographs by Downey and Hirschfeldt [4] and Nies [15] expose recent advances in the area. Standard textbooks in the field are Calude [2], Li and Vitanyi [11].

Martin-Löf tests (ML-tests) are central in defining algorithmic randomness in the setting of infinite strings. The ML-tests are roughly effective measure 0 sets in the Cantor space $\{0,1\}^\omega$. In spite of decades of work, research on algorithmic randomness of infinite strings has excluded the study of randomness for infinite

---

B. Khoussainov— The author also acknowledges support of Marsden Fund of Royal Society of New Zealand.

S. Artemov and A. Nerode (Eds.): LFCS 2016, LNCS 9537, pp. 159–173, 2016.
DOI: 10.1007/978-3-319-27683-0_12

structures such as graphs, trees, and algebras. The main reason is that it was unclear how one would define a meaningful measure in these classes. Through such a measure it would be possible to introduce algorithmic randomness for infinite structures.

The concept of algorithmic randomness has a strong intuitive underpinning. So, it is natural to ask what algebraic, model-theoretic, and computability-theoretic properties one expects from algorithmically random infinite structures. We list three of those desirable properties. First, algorithmic randomness should be an isomorphism invariant property; in other words, ML-randomness should not be a property of presentations but rather of structures. We call this the *absoluteness property* of randomness. Second, algorithmic randomness should be a property of a *collective* (*large class*), the idea that goes back to von Mises [14]. In particular, we would like to have continuum random structures, just like in the case of infinite strings. Third, there should be no effective way to describe the isomorphism types of ML-random structures in a formal language, e.g. through a finite (or effective) set of simple first order logic formulas. We might refer to this as *unpredictability* property. For instance, one would not like algorithmically random structures to be finitely presented (such as finitely presented group).

The three properties that we listed raise many questions. For instance, can an algorithmically random structure be computable (that is, isomorphic to a computable structure)? Recall that a computable structure (such as a graph) is one whose all atomic relations (the edge relation in the case of graphs) and the domain both are computable. For infinite strings, for instance, no computable string is algorithmically random; but the Rado graph (also known as the random graph) is computable. Another question is whether algorithmic randomness should be context dependent, e.g. can algorithmic randomness in one class imply algorithmic randomness in other class? Can a universally axiomatised structure (such as a group) be algorithmically random in the class of all structures? More immediate and direct questions concern the differences and similarities between algorithmic randomness of infinite strings and infinite structures such as graphs and trees. For example, an important property of ML-random strings is immunity; the property states that all attempts to effectively list an infinite subsequence in an ML-random string always fail. So, it is natural to ask if algorithmically random structures possess the immunity-like property. We address these fundamental questions in this paper.

A natural yet naive way to introduce algorithmic randomness for structures is to identify structures with binary strings coding the atomic diagrams of the structures. With this identification, call a structure *string-random* if the string coding it is ML-random. In [8] the author proves that all string-random structures are isomorphic to Fraïsse limit of finite structures. In particular, in the class of graphs, the string-random graph is thus the Rado graph. Hence, (1) all string-random structures are unique up to isomorphism; they are isomorphic to a computable structure; (2) string-random structures are described by extension axioms; (3) the theory of string-random structure is decidable [6]. These defy the intuitive notion of algorithmic randomness and the three properties discussed above, and call for an alternative approach.

1.2   Contributions

The novelty is that the paper develops an axiomatic approach through which we can reason about algorithmic randomness of infinite structures. This is based on introducing measures in many natural classes of structures. On particular contributions, we list the following four:

(1)   We introduce the notion of a *branching class* (or *B-class*) of finite structures through a list of axioms. This is a key technical yet simple concept that establishes a machinery for studying algorithmic randomness in various classes of structures. Section 2.2 gives many examples of branching classes including graphs, trees, and algebras. For instance, connected graphs of bounded degree form a branching class. Hence, one can study ML-randomness in this class; the class is algebraically rich as it includes Cayley graphs. For a branching class $\mathcal{K}$, we construct a computable tree $T(\mathcal{K})$ such that the nodes of $T(\mathcal{K})$ represent structures from $\mathcal{K}$ and edges of $T(\mathcal{K})$ represent "one-step" isomorphic embeddings. Importantly, the class $\mathcal{K}$ uniquely determines the class $\mathcal{K}_\omega$ of infinite structures. The structures from $\mathcal{K}_\omega$ are direct limits of the structures from $\mathcal{K}$ and can be viewed as paths through the tree $T(\mathcal{K})$; there is a bijective operator $\eta \to \mathcal{A}_\eta$ that associates paths η of $T(\mathcal{K})$ with the structures $\mathcal{A}_\eta$ from $\mathcal{K}_\omega$. Using the tree $T(\mathcal{K})$, we equip $\mathcal{K}_\omega$ with measure, topology, and metric. Hence, one can define ML-tests in the class $\mathcal{K}_\omega$, and prove that the number of ML-random structures in $\mathcal{K}_\omega$ is continuum. So, ML-randomness is a property of a *collective* as we desired above. The class $\mathcal{K}_\omega$, by Theorem 1, contains ML-random structures computable in the halting set. This part of the work extends [8] and our approach here is cleaner and refined. One technical issue is that the definition of B-classes assumes that the structures have constants in the signature, and the tree $T(\mathcal{K})$ depends on the constants. It turns out in all our examples, ML-randomness is independent on the choice of constants [8]. Hence, randomness introduced is a robust concept and has the *absoluteness property* discussed above.

(2)   Once the definition of branching classes $\mathcal{K}$ is given and a right machinery is developed, the existence of ML-random structures computable in the halting set is an expected phenomenon. Indeed, such structures correspond to the leftmost paths of computable finitely branching trees, and these paths are computable in the halting set. The mapping $\eta \to \mathcal{A}_\eta$ mentioned above is a computable operator; so, $\mathcal{A}_\eta$ is computable in any oracle that computes η. Sometimes the opposite is true. For instance, for the class $\mathcal{K}_\omega$ of finitely generated algebras, the path η can be computed from $\mathcal{A}_\eta$. Hence, no ML-random algebra is computable [8]. It has been long believed that no ML-random computable structures exist [16]. We, however, prove that the situation can be far from the intuition. Theorem 2 constructs a B-class $\mathcal{S}_\omega$ that contains a computable yet ML-random structure $\mathcal{A}$. The reason for this phenomenon is that building an ML-test for the structure $\mathcal{A}$ requires an access to the existential theory of $\mathcal{A}$. In our notation above, while building $\mathcal{A}$ we exploit the fact that constructing the path in the tree $T(\mathcal{S})$ that corresponds to $\mathcal{A}$ requires the jump of the open diagram of the structure. Theorem 2 has two important consequences. One is

that the theorem reveals intrinsic connections between ML-random structures and their existential first order theory. The other is that the theorem establishes an unexpected connection between algorithmic randomness and the theory of computable structures.

(3) Fix a functional signature σ. Structures of σ are called *algebras*. Consider the class of all finitely generated algebras. A variety is a class of algebras closed under sub-algebras, homomorphisms, and products, e.g. the class of groups is a variety. Any variety is axiomatised by a set of universally quantified equations. Theorem 3 proves that the class V of finitely generated algebras that belong to a nontrivial variety has effective measure 0. This result can be viewed as the *unpredictability property* of ML-random structures that we discussed above. The result confirms our intuition that any effective attempt to formally describe the isomorphism type of an algorithmically random structure fails. By this theorem, (1) the class of all finitely generated groups has effective measure 0, and (2) no set of universally quantified non-trivial equations can describe the isomorphism type of a ML-random algebra. In particular, no finitely presented algebra (e.g. group) in a variety is ML-random. For instance, no finitely presented algebra is ML-random.

(4) Existence of ML-random computable structures implies that no immunity-like property can be expected from ML-random structures. The underlying reason (for the existence of computable ML-random structure) is that, for the operator $\eta \rightarrow \mathcal{A}_\eta$ mentioned above, the construction of η from $\mathcal{A}_\eta$ requires the jump of the open diagram of $\mathcal{A}_\eta$. We consider B-classes, that we call *jumpless*, where the jump is not needed to compute η from $\mathcal{A}_\eta$. There are many examples of jumpless classes, e.g. the class of finitely generated algebras. For jumpless classes, no ML-random structure is computable. So, the question arises if ML-random structures in jumpless classes exhibit immunity-like property. The immunity property, for algebras, is formalised as follows. Call a finitely generated infinite algebra $\mathcal{A}$ *effectively infinite* if there is a computable infinite sequence t_1, t_2, ... of ground terms whose values in $\mathcal{A}$ are pairwise distinct. Otherwise, call $\mathcal{A}$ *immune*. Immunity of $\mathcal{A}$ implies that any effective attempt to list pairwise distinct elements of $\mathcal{A}$ fails. Thus, for the class of algebras the immunity property can be expressed as follows. Does there exist an effectively infinite ML-random algebra? Theorem 4 answers the question positively.

2   Branching Classes and Examples

2.1   Embedded Systems, Height Function, and Branching Classes

Fix a relational signature $\sigma = (R_1^{n_0}, \ldots, R_m^{n_m}, c_1, \ldots, c_k)$, where $R_i^{n_i}$ is a relational symbol of arity n_i and c_j is a constant symbol. We fix the signature and identify structures of the signature up to isomorphisms. Note that we have at least one constant symbol.

Definition 1. *An* embedded system *of structures is a sequence* $\{(\mathcal{A}_i, f_i)\}_{i \in \omega}$
such that each $\mathcal{A}_i$ *is a finite structure of the signature and each* f_i *is a strictly
into embedding from* $\mathcal{A}_i$ *into* $\mathcal{A}_{i+1}$. *We call the sequence* $\mathcal{A}_0, \mathcal{A}_1, \ldots$ *the* base *of
the embedded system.*

Structures with functional operations are turned into relational structures by
replacing operations with their graphs. We identify such structures with their
relational counterparts that we have just described.

The embedded system $\{(\mathcal{A}_i, f_i)\}_{i \in \omega}$ has a limit called the *direct limit* and
denoted by $\lim_i(\mathcal{A}_i, f_i)$. The direct limit $\lim_i(\mathcal{A}_i, f_i)$ is infinite as follows from
Definition 1.

Definition 2. *An embedded system* $\{(\mathcal{A}_i, f_i)\}_{i \in \omega}$ *is* strict *if its direct limit is
isomorphic to the direct limit of any embedded system with the same base.*

For instance, the sequence of finite successor structures $\mathcal{A}_n = (\{0, \ldots, n\}; S, 0)$
with distinguished element 0 is strict.

Let $\mathcal{K}$ be a decidable class of finite structures. Let $h : \mathcal{K} \to \omega$ be a computable
function; call h a *height function for* $\mathcal{K}$. When $h(\mathcal{A}) = i$ we say that $\mathcal{A}$ has
height i. We postulate that the number of structures from $\mathcal{K}$ of height i is finite
for all i. The function h is an isomorphism invariant. Assume that the height
function $h : \mathcal{K} \to \omega$ satisfies the following properties [1]:

1. We can compute the cardinality of $h^{-1}(i)$ for every i.
2. For every $\mathcal{A} \in \mathcal{K}$ of height i there is a substructure $\mathcal{A}[i-1]$ of height $i-1$
   such that all substructures of $\mathcal{A}$ of height $\leq i-1$ are contained in $\mathcal{A}[i-1]$.
   Hence, the substructure $\mathcal{A}[i-1]$ is the largest substructure of $\mathcal{A}$ of height
   $i-1$.
3. For all $\mathcal{A} \in \mathcal{K}$ of height i and $C \subseteq A \setminus A[i-1]$, the height of the substructure
   $C \cup A[i-1]$ is i in case the substructure belongs to $\mathcal{K}$.

The postulates on h imply that for all $\mathcal{A} \in \mathcal{K}$ of height i and $j \leq i$ there is a
substructure $\mathcal{A}[j]$ of height j such that all substructures of $\mathcal{A}$ of height $\leq j$ are
contained in $\mathcal{A}[j]$. Also, $\mathcal{A}[0] \subset \mathcal{A}[1] \subset \ldots \subset \mathcal{A}[i]$, where $\mathcal{A}[i] = \mathcal{A}$.

Lemma 1. *Let* $\mathcal{K}$ *and* h *be as above. Then for all* $\mathcal{A}, \mathcal{B} \in \mathcal{K}$, *the structures*
$\mathcal{A}$ *and* $\mathcal{B}$ *are isomorphic if and only if* $h(\mathcal{A}) = h(\mathcal{B})$ *and* $\mathcal{A}[j] = \mathcal{B}[j]$ *for all*
$j \leq h(\mathcal{A})$. □

Let $\mathcal{K}$ and h be as above. An important not so trivial lemma is the following:

Lemma 2. *All embedded systems of structures from* $\mathcal{K}$ *are strict.* □

---

[1] The next subsection provides many examples of classes with height function. For
now, for the reader a good example of a class with a height function is the class of
rooted finite binary trees.

As a consequence, every direct limit of embedded systems of structures $\{\mathcal{A}_i, f_i\}_{i \in \omega}$ from the class $\mathcal{K}$ is the direct limit of a "canonical" embedded system $\{\mathcal{B}_i, g_i\}_{i \in \omega}$ such that: (1) the direct limits $\lim_i(\mathcal{A}_i, f_i)$ and $\lim_i(\mathcal{B}_i, g_i)$ are isomorphic, (2) the height of each $\mathcal{B}_i$ is i, (3) the embeddings g_i are identity embeddings, and (4) for all $i \leq j$ we have $\mathcal{B}_j[i] = \mathcal{B}_i[i] = \mathcal{B}_i$.

We introduce branching classes of structures, a key concept of this paper.

Definition 3. *The class $\mathcal{K}$ with the height function h is a branching class, or a B-class for short, if for all $\mathcal{A} \in \mathcal{K}$ of height i there exist distinct structures $\mathcal{B}, \mathcal{C} \in \mathcal{K}$ such that $h(\mathcal{B}) = h(\mathcal{C}) > h(\mathcal{A})$ and $\mathcal{B}[i] = \mathcal{C}[i] = \mathcal{A}$.*

Consider the class $\mathcal{K}_\omega$ of all direct limits of structures from $\mathcal{K}$. Note that this class consist of infinite structures. We often refer to $\mathcal{K}_\omega$ as a B-class as well.

2.2 Examples of Branching Classes

We provide examples of B-classes for a better exposition. Some examples are taken from [8]. The proofs that these classes are branching are not too hard.

Lemma 3. *Each of the eight classes $PG(d)$, $Tree(d)$, $Str(d)$, $PO(d)$, $PAlg$, $OT(2)$, $Sparce(a,b)$, and $H(\delta, d)$ described below is a branching class.* □

Example 1 (Pointed Graphs). A *pointed graph of degree d*, where $d > 2$, is a connected finite graph G with a fixed tuple $\bar{c}$ such that the degree of every vertex of G is bounded by d. The edge relation is a symmetric relation E with no self-loops. Given a vertex v, one computes the shortest path-distance from c to v. For each pointed graph G, set $h(G)$ be the maximum of all path-distances from $\bar{c}$ to vertices $v \in G$. Consider the following class:

$$PG(d) = \{G \mid \text{there is a } v \text{ such that } distance(\bar{c}, v) = h(G(\mathcal{A})) \text{ and}$$
$$degree(v) < d\}.$$

Example 2 (Rooted Trees). Fix an integer $d > 1$. Consider the class $Tree(d)$ of all rooted trees T such that every node of T has not more than d immediate successors. The height $h(T)$ of T is the length of the longest path in the tree from the root.

Example 3 (Structures of Bounded Degree). Let $\mathcal{A}$ be a relational structure with exactly constant symbols $\bar{c}$. The Gaifman graph of $\mathcal{A}$ is the graph $G(\mathcal{A})$ with vertex set A and an edge between a and b, where $a \neq b$, if there is an atomic relation R and a tuple $\bar{x} \in R$ that contains a and b. Say that $\mathcal{A}$ is of *bounded degree d* if $G(\mathcal{A})$ is a connected graph of bounded degree d. Set $h(\mathcal{A})$ be the maximum of all path-distances from c to vertices $v \in G(\mathcal{A})$. Define:

$$Str(d) = \{\mathcal{A} \mid \text{there is a } v \text{ such that } distance(\bar{c}, v) = h(G(\mathcal{A})) \text{ and}$$
$$degree(v) < d\}.$$

Example 4 (Orders with the Least Element). Let $(P; \leq)$ be a partially ordered set. For $p, q \in P$, q *covers* p, written $C(p, q)$, if $p \leq q$ & $p \neq q$ & $\neg \exists x (p < x < q)$. Call the partially ordered set $\mathcal{P} = (P; \leq, C)$ of bounded degree d if every element in it has at most d covers. The height $h(\mathcal{P})$ of $\mathcal{P}$ is the length of the longest chain in it. Consider the following class:

$$PO(d) = \{\mathcal{P} \mid \mathcal{P} \text{ has the least element, has height } h \text{ and is of bounded degree } d\}$$

Example 5 (Algebras). Consider algebras $\mathcal{A}$ of signature $f_1, \ldots, f_n, c_1, \ldots, c_m$ consisting of function symbols and constants. The algebra $\mathcal{A}$ is *c-generated* if every element a of $\mathcal{A}$ is the interpretation of some ground term t; in other words, if every element of $\mathcal{A}$ is obtained from constants by a chain of atomic operations of $\mathcal{A}$. Call the term t a *representation* of a in $\mathcal{A}$.

The *height* $h(t)$ of a ground term t is defined by induction. If $t = c_i$ then $h(t) = 0$. If t is of the form $f_i(t_1, \ldots, t_{k_i})$, then $h(t) = \max\{h(t_1), \ldots, h(t_{k_i})\} + 1$. For a c-generated algebra $\mathcal{A}$ the *height* $h(a)$ of an element $a \in A$ is the minimal height among the heights of all the ground terms representing a. The *height* $h(\mathcal{A})$ of $\mathcal{A}$ is the supremum of all the heights of its elements.

A c-generated algebra is finite if and only if there exists an n such that all elements of $\mathcal{A}$ have height at most n. So, infinite c-generated algebras have height ω.

Let $\mathcal{A}$ be a c-generated algebra and $n \in \omega$. Set: $A(n) = \{a \in A \mid h(a) \leq n\}$. Each k_i-ary atomic operation f_i of $\mathcal{A}$ defines a *partial operation* $f_{i,n}$ on $A(n)$ as follows. For all $a_1, \ldots, a_{k_i} \in A(n)$ the value of $f_{i,n}(a_1, \ldots, a_{k_i})$ is $f_i(a_1, \ldots, a_{k_i})$ if $h(a_i) < n$ for $i = 1, \ldots, k_i$; and $f_{i,n}(a_1, \ldots, a_{k_i})$ is undefined otherwise. Thus, we have the partial algebra $\mathcal{A}(n)$ on the domain $A(n)$. Every infinite c-generated algebra $\mathcal{A}$ is the direct limit of the embedded system $\{\mathcal{A}(n)\}_{n \in \omega}$. For two c-generated algebras we have $\mathcal{A} \cong \mathcal{B}$ if and only if $\mathcal{A}(n) \cong \mathcal{B}(n)$ for all $n \in \omega$. Define:

$$PAlg = \{\mathcal{B} \mid \text{ there is an infinite algebra } \mathcal{A} \text{ and } n \in \omega \text{ such that } \mathcal{B} = \mathcal{A}(n)\}.$$

Example 6 (Binary Rooted Ordered Trees). The class $OT(2)$ consists of binary rooted trees where each node has either left or right-child. View these trees in the signature $\sigma = (L, R, c)$, where c is the root, $L(x, y)$ iff y is the left child of x, $R(u, v)$ iff v is the right child of u.

Example 7 (Sparse Graphs). A pointed graph G is (a, b)-*sparse* if every subgraph of G with n vertices has $\leq an + b$ edges, $a, b > 0$. Set: $Sparse(a, b) = \{G \mid G \text{ is (a,b) -sparse graph}\}$.

Example 8 (Hyperbolic Graphs). A pointed graph G of bounded degree d is δ-hyperbolic if for all $x, y, z \in G$ the distance from any shortest path from x to y to the union of the shortest paths from x to z and from z to y is at most δ. When $d = 0$, 0-hyperbolic graphs are just trees. The class of all δ-hyperbolic graphs is decidable. Set: $H(\delta, d) = \{G \mid G \text{ is } \delta \text{ -hyperbolic graph}\}$.

3   Martin-Löf Randomness in $\mathcal{K}_\omega$

3.1   Introducing Measure and Metric

Fix a B-class $\mathcal{K}$ with height function h. Consider the class $\mathcal{K}_\omega$ of all direct limits of structures from $\mathcal{K}$. Note that this class consist of infinite structures. We often refer to $\mathcal{K}_\omega$ as a B-class as well.

Let $r_\mathcal{K}(n)$ be the number of all structures from $\mathcal{K}$ of height n. This is a computable function as follows from the definition of B-class. Clearly, $r_\mathcal{K}(n) \leq r_\mathcal{K}(n+1)$ for all $n \in \omega$. We show that the structures from the class $\mathcal{K}_\omega$ can be viewed as paths through a finitely branching tree $T(\mathcal{K})$. This allows us to introduce measure and metric into $\mathcal{K}_\omega$. We now define the tree $T(\mathcal{K})$.

The tree $T(\mathcal{K})$ is defined as follows. The root of the tree $T(\mathcal{K})$ is the empty set. This is level -1 of $T(\mathcal{K})$. The nodes of the tree $T(\mathcal{K})$ at level $n \geq 0$ are all structures from $\mathcal{K}$ of height n. There are exactly $r_\mathcal{K}(n)$ of them. Let $\mathcal{B}$ be a structure from $\mathcal{K}$ of height n. Its successor on $T(\mathcal{K})$ is any relational structure $\mathcal{C}$ of height $n+1$ such that $\mathcal{B}$ and $\mathcal{C}[n]$ coincide at level n.

Lemma 4 (Computable Tree Lemma). *For the tree $T(\mathcal{K})$ we have the following:*

1. *Given a node x of $T(\mathcal{K})$, we can effectively compute the structure $\mathcal{B}_x$ associated with x. We identify the nodes x and the structures $\mathcal{B}_x$.*
2. *For each node x in $T(\mathcal{K})$, the structure $\mathcal{B}_x$ has an immediate successor. Moreover, we can compute the number of immediate successors of x.*
3. *For each path $\eta = \mathcal{B}_0, \mathcal{B}_1, \ldots$ in $T(\mathcal{K})$ we have: $\mathcal{B}_0 \subset \mathcal{B}_1 \subset \ldots$. Thus, the union of this chain determines the structure $\mathcal{B}_\eta = \cup_i \mathcal{B}_i \in \mathcal{K}_\omega$.*
4. *The map $\eta \to \mathcal{B}_\eta$ is a bijection between all infinite paths of $T(\mathcal{K})$ and $\mathcal{K}_\omega$.* □

For structure $\mathcal{A} \in \mathcal{K}_\omega$ in a B-class, set $\mathcal{A}[i]$ be the largest substructure of $\mathcal{A}$ of height i. The substructure $\mathcal{A}[i]$ is correctly defined and is unique.

Definition 4. *Define topology, measure and metric in $\mathcal{K}_\omega$ as follows:*

(Topology): *Let $\mathcal{B}$ be a structure of height n. The **cone** with base $\mathcal{B}$ is the set $Cone(\mathcal{B}) = \{\mathcal{A} \mid \mathcal{A} \in K_\omega, \text{ and } \mathcal{A}[n] = \mathcal{B}\}$ The cones $Cone(\mathcal{B})$ are form the base of the topology on $\mathcal{K}_\omega$.*

(Measure): *The measure of the cone based at the root is 1. Let $\mathcal{B}_x$ be a structure of height n. Assume that the measure $\mu(Cone(\mathcal{B}_x))$ has been defined. Let e_x be the number of structures of height $n+1$ that are immediate successors of $\mathcal{B}_x$ in the tree. Then for any immediate successor y of x we set $\mu(Cone(\mathcal{B}_y)) = \mu(Cone(\mathcal{B}_x))/e_x$.*

(Metric): *Let $\mathcal{A}$ and $\mathcal{C}$ be structures from $\mathcal{K}_\omega$. Let n be the maximal level at which $\mathcal{A}[n]$ and $\mathcal{C}[n]$ coincide. Let $\mathcal{B}$ be the node of the tree such that $\mathcal{A}[n] = \mathcal{B}$. The distance $d(\mathcal{A}, \mathcal{C})$ between $\mathcal{A}$ and $\mathcal{C}$ is then: $d(\mathcal{A}, \mathcal{C}) = \mu(Cone(\mathcal{B}))$.*

The next lemma shows that d is a metric in $\mathcal{K}_\omega$.

Lemma 5. *The function d is a metric on* $\mathcal{K}_\omega$. $\qquad\qquad\qquad\qquad$ □

The space $\mathcal{K}_\omega$ is compact. Finite unions of cones form clo-open sets in the topology. Furthermore, the set of all μ-measurable sets is a σ-algebra. We note that the metric d defined above for the class of pointed graphs is homeomorphic to Benjamin-Schramm metric in the class [1].

3.2 ML-randomness in $\mathcal{K}_\omega$

The set-up above allows us to define ML-random structures in the class $\mathcal{K}_\omega$ through definitions borrowed from algorithmic randomness. A class $C \subseteq \mathcal{K}_\omega$ is a Σ_1^0-*class* if there is computably enumerable (c.e.) sequence $\mathcal{B}_0, \mathcal{B}_1, \ldots$ of structures from $\mathcal{K}$ such that $C = \cup_{i \geq 1} Cone(\mathcal{B}_i)$. Computable enumerability of $\mathcal{B}_0, \mathcal{B}_1, \ldots$ implies that given i we can compute the open diagram of $\mathcal{B}_i$; in particular, we know the cardinality of $\mathcal{B}_i$.

Definition 5. *Let* $\mathcal{K}$ *be a B-class. Consider the class* $\mathcal{K}_\omega$ *of infinite structures.*

1. *A* Martin-Löf test *is a uniformly c.e. sequence* $\{G_n\}_{n \geq 1}$ *of* Σ_1^0-*classes such that*
   $G_{n+1} \subset G_n$ *and* $\mu(G_n) < 2^{-n}$ *for all* $n \geq 1$.
2. *A structure* $\mathcal{A}$ *from* $\mathcal{K}_\omega$ *fails a* Martin-Löf test $\{G_n\}_{n \geq 1}$ *if* $\mathcal{A}$ *belongs to* $\cap_n G_n$. *Otherwise, we say that the structure* $\mathcal{A}$ passes *the test.*
3. *A structure* $\mathcal{A}$ *from* $\mathcal{K}$ *is* ML-random *if it passes every Martin-Löf test.*

If a class $C \subset \mathcal{K}_\omega$ *is contained in a ML-test, then* C *has* effective measure 0.

It is standard to show that there exists a *universal* ML-test in the sense that passing that test is equivalent to passing all ML-tests. Formally, an ML-test $\{U_n\}_{n \geq 1}$ is *universal* if for any ML-test $\{G_m\}_{m \geq 1}$ we have $\cap_m G_m \subseteq \cap_n U_n$. A construction of a universal ML-test is the following. Enumerate all ML-tests $\{G_k^e\}_{k \geq 1}$, where $e \geq 1$, uniformly on e and k, and set $U_n = \cup_e G_{n+e+1}^e$. It is not hard to see that $\{U_n\}_{n \geq 1}$ is a universal ML-test. Hence, to prove that a structure $\mathcal{A} \in \mathcal{K}_\omega$ is ML-random it suffices to show that $\mathcal{A}$ passes the universal ML-test $\{U_n\}_{n \geq 1}$. The class of not random structures has effective measure 0. Thus, we have the following corollary:

Corollary 1. *Let* $\mathcal{K}$ *be a B-class. The number of ML-random structures in the class* $\mathcal{K}_\omega$ *is continuum. In particular, for all the examples of B-classes* $\mathcal{K}$ *from Sect. 2.2, each of the classes* $\mathcal{K}_\omega$ *contains continuum ML-random structures.* □

We would like to make two important comments. One is that in all the examples of B-classes $\mathcal{K}$ from Sect. 2.2, the tree $T(\mathcal{K})$ depends on the choice of constants. A natural question is if the constants matter. In [8] it is shown that for the class of graphs, trees, and finitely generated algebras ML-randomness does not depend on the constants. For instance, for any connected graph G of bounded degree, the pointed graph $(G, \bar{p})$ is ML-random iff $(G, \bar{q})$ is ML-random. The proof techniques from [8] can be applied (in a straightforward way) to prove the following:

Corollary 2. *For all the examples of B-classes $\mathcal{K}$ from Sect. 2.2, ML-randomness is independent on the choice of constants. Hence, ML-randomness is the isomorphism property of the structures (with constants removed from the signature).* □

The second comment concerns the measure μ. The measure depends on two parameters: the function h, and the number of "one-step" extensions of the structure $\mathcal{B}_x$ at node x. The measure μ is defined so that all "one-step" extensions of x are equally likely to occur. From this view point μ is a simple measure. The measure can obviously be redefined by taking into account algebraic properties of the structure $\mathcal{B}_x$ (e.g. automorphisms) and its "one-step" extensions. Every such new measure introduces its own ML-randomness, and this study is a future work.

3.3   Randomness in the Halting Set

We study computable aspects of ML-random structures. from $\mathcal{K}_\omega$, where $\mathcal{K}$ is a B-class. The following definition goes back to Malcev [12], Rabin [17].

Definition 6. *An infinite structure $\mathcal{A}$ is computable if it is isomorphic to a structure with domain ω such that all atomic relations of the structure are computable.*

Thus, computability is an isomorphism invariant. Clearly, a structure is computable if and only if it is isomorphic to a structure whose atomic diagram is a computable set. Next we provide a stronger definition that involves the height function of the class $\mathcal{K}$.

Definition 7. *A computable structure $\mathcal{A}$ from $\mathcal{K}_\omega$ is strictly computable if the size of the substructure $\mathcal{A}[i]$ can be effectively computed for all $i \in \omega$.*

The following proposition gives examples of strictly computable structures.

Proposition 1. *The following are true: (a) Every computable c-generated algebra is strictly computable. (b) A computable pointed graph G of bounded degree is strictly computable iff there is an algorithm that given a vertex v of G computes $degree(v)$. (c) A computable rooted tree T of bounded degree is strictly computable iff there is an algorithm that given a node $v \in T$ computes the number of immediate successors of v. (d) A computable d-bounded partial order with the least element is strictly computable iff there exists an algorithm that for every v element of the partial order computes all covers of v.* □

Strict computability implies non ML-randomness:

Proposition 2. *If $\mathcal{A}$ is strictly computable then $\mathcal{A}$ is not ML-random.* □

Corollary 3. *Let $\mathcal{A}$ be either an infinite pointed graph or tree or partial order of bounded degree. If $\mathcal{A}$ is computable and its $\exists$-diagram, that is the set*

$$\{\phi(\bar{a}) \mid \bar{a} \in A \text{ and } \mathcal{A} \models \phi(\bar{a}) \text{ and } \phi(\bar{x}) \text{ is an existential first-order formula}\},$$

is decidable then $\mathcal{A}$ is not ML-random.

Proof. Decidability of the existential diagrams of computable graphs and trees allows one to effectively compute the substructures of height i for given i.

A natural class that contains the class of all computable structures is the class of structures computable in the Halting set. For instance, finitely presented groups are computable in the halting set. We denote the halting set by $\mathcal{H}$. Here is a definition.

Definition 8. *A structure $\mathcal{A}$ is $\mathcal{H}$-computable if it is isomorphic to a structure with the domain ω whose all atomic relations and operations of $\mathcal{A}$ are computable in $\mathcal{H}$.*

Every computable structure is $\mathcal{H}$-computable. The next theorem shows that the proposition above can't be extended to $\mathcal{H}$-computable structures.

Theorem 1. *Every B-class contains an $\mathcal{H}$-computable ML-random structure.* □

Corollary 4. *For all the examples of B-classes $\mathcal{K}$ from Sect. 2.2, each of the classes $\mathcal{K}_\omega$ contains ML-random $\mathcal{H}$-computable structures.* □

4   Computable ML-random structures

Let $\mathcal{K}$ be a B-class and $\mathcal{A} \in \mathcal{K}_\omega$. There is a path $\eta \in T(\mathcal{K})$ such that the structure $\mathcal{B}_\eta$ is isomorphic $\mathcal{A}$. The path η can be constructed in the jump of the open diagram of $\mathcal{A}$. In particular, if $\mathcal{A}$ is computable then η is computable in the halting set $\mathcal{H}$. This observation suggests that some B-classes $\mathcal{K}_\omega$ might contain ML-random yet computable structures. This intuition is confirmed in the following theorem:

Theorem 2. *There exists a B-class $\mathcal{S}_\omega$ that has a computable ML-random structure.*

Proof. Consider the B-class $OT(2)$ of all finite ordered trees as in **Example 6**. Given a tree $\mathcal{B}$ from $OT(2)$ we order all the nodes of the tree at the same level (from left to right) in a natural way. Refer to this order as the level order of the nodes. We define the following subclass $\mathcal{S}$ of $OT(2)$. A binary tree $\mathcal{B} \in OT(2)$ belongs to $\mathcal{S}$ if the tree $\mathcal{B}$ has the following properties:

1. All leaves of $\mathcal{B}$ are of the same height,
2. If v in $\mathcal{B}$ has the right child then all nodes left of v on the v's level-order including v have the left and the right children, and
3. At every level i of $\mathcal{B}$ there exists at most one node such that it is the left child of its parent, and the parent does not have a right child in $\mathcal{B}$.

We provide several properties of the class $\mathcal{S}$.

Lemma 6. *If $\mathcal{B}$ belongs to $\mathcal{S}$ and has height n then there are exactly two non-isomorphic extensions of $\mathcal{B}$ of height $n + 1$ both in $\mathcal{S}$.*

The next lemma gives a nice algebraic property of trees in $\mathcal{S}$.

Lemma 7. *For $n \geq 0$, the set of all trees in $\mathcal{S}$ of height n form a chain of embedded structures.*

These lemmas imply that there is a natural bijection $x \to \mathcal{A}_x$ from the set of all binary strings to the class $\mathcal{S}$. Let $\preceq$ be the lexicographical order on binary strings. We have:

Corollary 5. *Assume that $x \preceq y$. Then: (1) If $|x| \leq |y|$ then $\mathcal{A}_x$ is embedded into $\mathcal{A}_y$; and (2) If $|x| > |y|$ then $\mathcal{A}_x$ is embedded into $\mathcal{A}_{yz}$ for all z such that $|x| \leq |yz|$.*

Thus, we identify the tree $T(\mathcal{S})$ with the full binary tree and use the mapping $x \to \mathcal{A}_x$ in our construction of ML-random computable tree in $\mathcal{S}_\omega$.

Let $\{U_n\}_{n \in \omega}$ be the universal ML-test. We construct a computable binary ordered tree $\mathcal{C}$ in $\mathcal{S}_\omega$ such that $\mathcal{C}$ passes the ML-test $\{U_n\}_{n \in \omega}$. It suffices to construct $\mathcal{C}$ so that $\mathcal{C} \notin U_1$. This follows a standard construction of a ML-random string. Since $\mu(U_1) < 1/2$, the complement $S_\omega \setminus U_1$ is the set of all infinite paths through a computable tree. The leftmost infinite path η of this tree determines the structure $\mathcal{C}$. One needs to note then the structure $\mathcal{C}$ that corresponds to η is a computable structure even if η isn't (in fact, η is a ML-random). The lemmas above guarantee that $\mathcal{C}$ is computable.

5   Measures of Varieties

Consider the class of finitely generated algebras $PAlg_\omega$ (Example 5, Sect. 2.2). A class of algebras V is a *variety* if its closed under sub-algebras, homomorphisms, and products. The class V is variety iff is axiomatised by a set E of universally quantified equations [5]. An equation is of the form $p(\bar{x}) = q(\bar{x})$ where p, q are terms whose variables are among $\bar{x}$. Call an equation $p(\bar{x}) = q(\bar{x})$ *non-trivial* if at least one of the terms contains a variable and $p \neq q$ syntactically. If E contains a non-trivial equation then we call the variety defined by E a *non-trivial variety*.

Let V be a variety defined by E and let R be a set of defining relations on generators $\bar{c}$. Defining relations are $t_1(\bar{c}) = t_2(\bar{c})$, where t_1 and t_2 are ground terms. The set of all c-generated algebras that satisfy E and R contains the free c-algebra $\mathcal{A}(E, R)$ in the variety V. This algebra is unique and any c-generated algebra from V that satisfies R is a homomorphic image of $\mathcal{A}(E, R)$. In this sense, we view the pair E and R as a description of $\mathcal{A}(E, R)$. Examples of such algebras are finitely presented groups. A natural question is if there are finitely presented yet ML-random algebras. We answer this question in a strongest form:

Theorem 3. *The class of all c-generated algebras that belong to a non-trivial variety has an effective measure 0. Hence, no finitely presented algebra of a non-trivial variety is ML-random.*

Proof (Sketch). For the proof we use notations and definitions from Example 4 of Sect. 2.2. We also denote the tree $T(PAlg)$ by T.

Let V be a nontrivial variety defined E. Let $p(\bar{x}) = q(\bar{x})$ be a nontrivial equation in E. Say that $p = f(t_1(\bar{x}), \ldots, t_k(\bar{x}))$ and $q = g(r_1(\bar{x}), \ldots, r_l(\bar{x}))$. It suffices to show that the variety V' defined by one equation $p(\bar{x}) = q(\bar{x})$ has effective measure 0.

For each node x at level n of the tree T consider the structure $\mathcal{A}_x$ that corresponds to node x. Define: $U_x = \{\bar{b} \mid$ either $p(\bar{b})$ or $q(\bar{b})$ is not defined$\}$. Call a node x *potential* if $p(\bar{b}) = q(\bar{b})$ in $\mathcal{A}_x$ for all $\bar{b} \notin U_x$. Potentiality indicates that $Cone(\mathcal{A}_x)$ might contain an algebra from V. If x is not potential then $Cone(\mathcal{A}_x)$ contains no algebra from V.

Roughly, probability that a partial algebra extends $\mathcal{A}_x$, where x is potential, and satisfies $p(\bar{b}) = q(\bar{b})$ for some $\bar{b} \in U_x$ is small. This is the underlying intuitive reason for V to have effective measure 0.

Corollary 6. *No finitely generated ML-random algebra exists that satisfies a nontrivial set of equations. Hence, no ML-random group, monoid, or lattice exists.* □

We describe finitely axiomatised varieties with non-zero measure. The last part of the corollary follows from the results in [10].

Corollary 7. *A measure of a finitely axiomatised variety V is either effectively 0 or a rational number > 0. The latter case occurs iff the variety is axiomatised by a trivial set of equations. Moreover, $\mu(V) > 0$ iff V is a finite union of cones.* □

6   ML-randomness and Immunity

Let $\mathcal{K}$ be a B-class. The bijective operator $\eta \to \mathcal{A}_\eta$ is such that the structure $\mathcal{A}_\eta$ is computable in η. In contrast, the inverse mapping $\mathcal{A}_\eta \to \eta$ requires the jump of the open diagram of $\mathcal{A}_\eta$ to compute η. The following definition removes this disparity:

Definition 9. *Call a B-class* jumpless *if for every structure A_η in $\mathcal{K}_\omega$ the path η can be constructed with an oracle for the open diagram of $\mathcal{A}_\eta$.*

For instance, the class of $PAlg$ is jumpless. Every computable structure $\mathcal{A}$ in a jumpless class $\mathcal{K}_\omega$ is strictly computable. Hence, by Proposition 2 we have:

Corollary 8. *No computable structure in a jumpless class is ML-random.* □

Every Martin-Löf random string α possesses *the immunity property*. The property states α contains no computable infinite substring. A natural question arises if such phenomenon occurs for ML-random structures in jumpless B-classes. We formalise this idea for the class of algebras:

Definition 10. *A an infinite c-generated algebra is* effectively infinite *if there exists an infinite computably enumerable sequence $t_0, t_1, \ldots$ of ground terms such that the values of the ground terms in the algebra are all pairwise distinct. Otherwise, we call the algebra* immune.

Immune algebras have undecidable word problem. It is also not too hard to provide examples of immune c-generated algebras. For instance in [9] the author, using Kolmogorov complexity, constructs an immune finitely generated algebra with computably enumerable word problem.

Thus, the question above becomes if there exists a jumpless B-class that contains an ML-random effectively infinite algebra. Here is our positive answer:

Theorem 4. *There exists a jumpless B-class of algebras such that every ML-random structure in the class is effectively infinite.*

Proof. We use the class $\mathcal{S}$ defined in Theorem 2. Our class $\mathcal{K}$ will be a class of algebras of signature (L, R, c) where L and R are unary function symbols, and c is a constant.

Let $\mathcal{B} \in \mathcal{S}$. We turn $\mathcal{B}$ into a partial algebra $\mathcal{B}'$. Assume that height of $\mathcal{B}$ is n. For every node $a \in B$ of height n, set $L(a)$ and $R(a)$ undefined. For every other $a \in B$, if b is the left child of a, we set $L(a) = b$. Note that every a in $\mathcal{B}$ of height $< n$ has a left child. If a has a right child c, we set $R(a) = c$. Otherwise, we set $R(a) = L(a)$. Define the following class $\mathcal{K}$: $\mathcal{K} = \{\mathcal{B}' \mid \mathcal{B} \in S\}$. The following is easy:

Lemma 8. *The class $\mathcal{K}$ is a jumpless B-class such that there is an effective one to one correspondence between ML-tests in $\mathcal{S}_\omega$ and $\mathcal{K}_\omega$.*

Thus, given $\mathcal{A} \in \mathcal{S}_\omega$, we can transform $\mathcal{A}$ into algebra $\mathcal{A}'$ just as we described above. In other words, $\mathcal{K}_\omega = \{\mathcal{A}' \mid \mathcal{A} \in \mathcal{S}_\omega\}$. However, the operator $\mathcal{A} \to \mathcal{A}'$ is such that constructing $\mathcal{A}'$ from $\mathcal{A}$ requires the haltings set $\mathcal{H}$.

The rest of the proof is clear. Indeed, every algebra $\mathcal{A}'$ in $\mathcal{K}_\omega$ is effectively infinite; in the algebra $\mathcal{A}'$ the sequence of terms r, $L(r)$, $LL(r)$, $LLL(r)$, $\ldots$, where r is the root, is an effective infinite sequence of ground terms whose values in $\mathcal{A}'$ are pairwise distinct. Hence, all ML-random algebras in $\mathcal{K}_\omega$ are effectively infinite.

Many questions are waiting to be investigated. This work calls for the study of ML-randomness for infinite structures in the setting of computational complexity. This paper shows that jumpless classes contain ML-random structures computable in the halting set. It would be interesting to sharpen this result by constructing jumpless classes that contain computably enumerable ML-random structures. There are other examples of B-classes where randomness can be studied, e.g. the class of planar graphs of bounded degree. Another interesting task is the study of ML-randomness in the class of transitive graphs due to connections to geometric group theory. For transitive graphs, the direct translation of the techniques of this paper do not work. We also conjecture that there are computable yet ML-random trees and graphs (in the classes of trees and graphs, respectively). Another possible direction of research is the study of connections with zero-one laws.

References

1. Benjamini, I., Schramm, O.: Recurrence of distributional limits of finite planar graphs. Electron. J. Probab. **6**(23), 1–13 (2001)
2. Calude, C.S.: Information and Randomness - An Algorithmic Perspective, 2nd edn. Springer, Heidelberg (2002). Revised and Extended
3. Chaitin, G.J.: On the length of programs for computing finite binary sequences: statistical considerations. J. ACM **16**, 145–159 (1969)
4. Downey, R., Hirschfeldt, D.: Algorithmic Randomness and Complexity. Theory and Applications of Computability. Springer, New York (2010)
5. Grätzer, G.: Universal Algebra. Springer, New York (2008). Revised reprint of the second edition
6. Hodges, W.: Model Theory. Encyclopaedia of Mathematics and Its Applications, vol. 42. Cambridge University Press, Cambridge (1993)
7. Kolmogorov, A.: Three approaches to the quantitative definition of information. Probl. Inf. Transm. **1**, 1–7 (1965)
8. Khoussainov, B.: A quest for algorithmically random infinite structures. In: Proceedings of LICS-CSL Conference Vienna, Austria (2014)
9. Khoussainov, B.: Randomness computability, and algebraic specifications. Anal. Pure Appl. Log. **91**(1), 1–15 (1998)
10. Kozen, D.: Complexity of finitely presented algebras. In: Proceedings of the 9th ACM Symposium on Theory of Computing, p. 164–177 (1977)
11. Li, M., Vitanyi, P.: An Introduction to Kolmogorov Complexity and its Applications. Springer, Heidelberg, 3rd Edn., p. xx+792 (2008)
12. Mal'cev, A.I.: Constructive algebras. I. Uspehi Mat. Nauk, 16(3(99)), 3–60 (1961)
13. Martin-Löf, P.: The definition of random sequences. Inf. Control **9**(6), 602–619 (1966)
14. Von Mises, R.: Grundlagen der Wahrscheinlichkeitsrechnung. Math. Z. **5**(191), 52–99 (1919)
15. Nies, A.: Computability and Randomness. Oxford University Press, New York (2009)
16. Nies, A., Stephan, F.: Personal communication
17. Rabin, M.O.: Computable algebra, general theory and theory of computable fields. Trans. Amer. Math. Soc. **95**, 341–360 (1960)
18. Schnorr, C.P.: A unified approach to the definition of random sequences. Math. Syst. Theory **5**(3), 246–258 (1971)
19. Schnorr, C.P. : The process complexity and effective random tests. In: Proceedings of the Fourth ACM Symposium of Theory of Computing, Denver, Colorado, 13 May (1972)
20. Zvonkin, A.K., Levin, L.A.: The complexity of finite objects and the development of the concepts of information and randomness by means of the theory of algoritheorems. Russ. Math. Surv. **25**(6), 83–124 (1970)

Probabilistic Justification Logic

Ioannis Kokkinis[1], Zoran Ognjanović[2], and Thomas Studer[1(✉)]

[1] Institute of Computer Science, University of Bern, Bern, Switzerland
{kokkinis,tstuder}@inf.unibe.ch
[2] Mathematical Institute SANU, Belgrade, Serbia
zorano@mi.sanu.ac.rs

Abstract. We present a probabilistic justification logic, PPJ, to study rational belief, degrees of belief and justifications. We establish soundness and completeness for PPJ and show that its satisfiability problem is decidable. In the last part we use PPJ to provide a solution to the lottery paradox.

Keywords: Justification logic · Probabilistic logic · Strong completeness · Decidability · Lottery paradox

1 Introduction

In epistemic modal logic, we use formulas of the form $\Box A$ to express that A *is believed*. Justification logic unfolds the $\Box$-modality into a family of so-called *justification terms* to represent evidence for an agent's belief. That is in justification logic we use $t : A$ to state that A *is believed for reason t*.

Originally, Artemov developed the first justification logic, the Logic of Proofs, to give a classical provability semantics for intuitionistic logic [1,2,15]. Later, Fitting [7] introduced epistemic models for justification logic. As it turned out this interpretation provides a very successful approach to study many epistemic puzzles and problems [3,5,14].

In this paper, we extend justification logic with probability operators in order to accommodate the idea that

different kinds of evidence for A lead to different degrees of belief in A. (1)

In [10] we have introduced a first probabilistic justification logic PJ, which features formulas of the form $P_{\geq s}(t : A)$ to state that *the probability of $t : A$ is greater than or equal to s*. The language of PJ, however, does neither include justification statements over probabilities (i.e. $t : (P_{\geq s}A)$) nor iterated probabilities (i.e. $P_{\geq r}(P_{\geq s}A)$).

In the present paper, we remedy these shortcomings and present the logic PPJ, which supports formulas of the form $t : (P_{\geq s}A)$ as well as $P_{\geq r}(P_{\geq s}A)$. This explains the name PPJ: the two Ps refer to iterated P-operators. We introduce syntax and semantics for PPJ and establish soundness and completeness. We also

S. Artemov and A. Nerode (Eds.): LFCS 2016, LNCS 9537, pp. 174–186, 2016.
DOI: 10.1007/978-3-319-27683-0_13

show that satisfiability for PPJ is decidable. In the final part we present an application of PPJ to the lottery paradox.

Related Work. The design of PPJ follows that of LPP_1, which is a probability logic over classical propositional logic [21,22]. The proofs that we present for PPJ are extensions of the corresponding proofs for LPP_1. Note, however, that these extensions are non-trivial due to the presence of formulas of the form $t : (P_{\geq s}A)$.

Our probability logics are not compact. Consider the set

$$T := \{\neg P_{=0}A\} \cup \{P_{<1/n}A \mid n \text{ is a positive integer}\}.$$

Although every finite subset of T is satisfiable, the set T is not. Hence in order to obtain a strong completeness result, we use an infinitary rule, which originates from [21,23].

Milnikel [19] proposes a logic with uncertain justifications. We thoroughly study the relationship between Milnikel's logic and our approach in [10] where we show that three of his four axioms are theorems in our logic and that the fourth axiom holds under an additional independence assumption.

In the preprint [9], Ghari presents fuzzy variants of justification logic, in which an agent can have a justification for a statement with certainty between 0 and 1. He introduces fuzzy Fitting models and establishes a graded completeness theorem. Ghari also shows that Milnikel's principles are valid in his fuzzy setting.

Recently, Fan and Liau [6] introduced a possibilistic justification logic, which is an explicit version of a graded modal logic. Their logic includes formulas $t :_r A$ to express that *according to evidence t, A is believed with certainty at least r*. However, the following principle holds in their logic:

$$s :_r A \wedge t :_q A \rightarrow s :_{\max(r,q)} A.$$

Hence all justifications for a belief yield the same (strongest) certainty, which is not in accordance with our guiding idea (1).

Funding. Ioannis Kokkinis and Thomas Studer are supported by the SNSF project 153169, *Structural Proof Theory and the Logic of Proofs*. Zoran Ognjanović is supported by the Serbian Ministry of Education, Science and Technological Development.

2   The Probabilistic Justification Logic PPJ

Justification terms are built from countably many constants and countably many variables according to the following grammar:

$$t ::= c \mid x \mid (t \cdot t) \mid (t + t) \mid !t$$

where c is a constant and x is a variable. Tm denotes the set of all terms and Con denotes the sets of all constants. For any term t and natural number n we define $!^0 t := t$ and $!^{n+1} t := ! \, (!^n t)$.

Let Prop be a countable set of atomic propositions. We denote the set of rational numbers by $\mathbb{Q}$. Further we set $\mathsf{S} := \mathbb{Q} \cap [0,1]$. The set of formulas $\mathcal{L}$ is defined by the following grammar[1]:

$$A ::= p \mid P_{\geq s}A \mid \neg A \mid A \wedge A \mid t : A$$

where $t \in \mathsf{Tm}, s \in \mathsf{S}$ and $p \in \mathsf{Prop}$. We employ the standard abbreviations for classical connectives. Additionally, we set

$$P_{<s}A \equiv \neg P_{\geq s}A \qquad\qquad P_{\leq s}A \equiv P_{\geq 1-s}\neg A$$
$$P_{>s}A \equiv \neg P_{\leq s}A \qquad\qquad P_{=s}A \equiv P_{\geq s}A \wedge P_{\leq s}A$$

The axiom schemes of PPJ are presented in Fig. 1.

(P)   finitely many schemes in the language of $\mathcal{L}$
axiomatizing classical propositional logic
(J) $\vdash u : (A \to B) \to (v : A \to u \cdot v : B)$
(+) $\vdash u : A \vee v : A \to u + v : A$
(PI) $\vdash P_{\geq 0}A$
(WE) $\vdash P_{\leq r}A \to P_{<s}A$, where $s > r$
(LE) $\vdash P_{<s}A \to P_{\leq s}A$
(DIS) $\vdash P_{\geq r}A \wedge P_{\geq s}B \wedge P_{\geq 1}\neg(A \wedge B) \to P_{\geq \min(1,r+s)}(A \vee B)$
(UN) $\vdash P_{\leq r}A \wedge P_{<s}B \to P_{<r+s}(A \vee B)$, where $r + s \leq 1$

Fig. 1. Axioms Schemes of PPJ

A *constant specification* is any set CS that satisfies

$$\mathsf{CS} \subseteq \{(c, A) \mid c \text{ is a constant and}$$
$$A \text{ is an instance of some axiom of PPJ}\}.$$

A constant specification CS is called:

Axiomatically Appropriate: if for every axiom instance A of PPJ, there exists a constant c such that $(c, A) \in \mathsf{CS}$;

Schematic: if for every constant c, the set $\{A \mid (c, A) \in \mathsf{CS}\}$ consists of all instances of several (possibly zero) axiom schemes;

Finite: if CS is a finite set;

Almost Schematic: if $\mathsf{CS} = \mathsf{CS}_1 \cup \mathsf{CS}_2$ where $\mathsf{CS}_1 \cap \mathsf{CS}_2 = \emptyset$, CS_1 is schematic and CS_2 is finite.

---

[1] In order to have a countable language and in order to obtain decidability we restrict our probabilistic operators to the rational numbers.

The notion of schematiness will be crucial for establishing decidability for PPJ.

Let CS be any constant specification. The deductive system PPJ$_{CS}$ is the Hilbert system obtained by adding to the axioms of PPJ the rules (MP), (CE), (ST) and (AN!) as given in Fig. 2.

---

axioms of PPJ

+

(AN!) $\vdash\ !^n c :\ !^{n-1} c : \cdots :\ !c : c : A$, where $(c, A) \in$ CS and $n \in \mathbb{N}$

(MP) if $T \vdash A$ and $T \vdash A \rightarrow B$ then $T \vdash B$

(CE) if $\vdash A$ then $\vdash P_{\geq 1} A$

(ST) if $T \vdash A \rightarrow P_{\geq s - \frac{1}{k}} B$ for every integer $k \geq \frac{1}{s}$ and $s > 0$

      then $T \vdash A \rightarrow P_{\geq s} B$

Fig. 2. System PPJ$_{CS}$

Note that (ST) is an infinitary rule, which we need to obtain strong completeness. Observe also the difference in the definitions of rules (MP), (ST) and (CE) in Fig. 2. Rule (CE) can only be applied to theorems of PPJ (i.e. formulas that are deducible from the empty set), whereas (MP) and (ST) can always be applied.

To introduce semantics for PPJ$_{CS}$, we begin with the notion of a basic evaluation, which is the cornerstone for many interpretations of justification logic [4,13]. In the following we use $\mathcal{P}(X)$ to denote the power set of a set X.

Definition 1 (Basic Evaluation). *Let* CS *be a constant specification. A* basic evaluation *for* CS, *or a basic* CS-*evaluation, is a function* $*$ *that maps atomic propositions to truth values and maps justification terms to subsets of* $\mathcal{L}$, *i.e.*

$$* : \mathsf{Prop} \rightarrow \{\mathsf{T}, \mathsf{F}\} \quad and \quad * : \mathsf{Tm} \rightarrow \mathcal{P}(\mathcal{L}),$$

such that for $u, v \in \mathsf{Tm}$, *for* $c \in \mathsf{Con}$ *and* $A, B \in \mathcal{L}$ *we have:*

1. $(A \rightarrow B \in u^* \ and \ A \in v^*) \implies B \in (u \cdot v)^*$
2. $u^* \cup v^* \subseteq (u + v)^*$
3. *if* $(c, A) \in$ CS *then for all* $n \in \mathbb{N}$ *we have[2]:*

$$!^{n-1} c :\ !^{n-2} c : \cdots :\!c : c : A \in (!^n c)^*$$

We usually write t^* *and* p^* *instead of* $*(t)$ *and* $*(p)$, *respectively.*

---

[2] We agree to the convention that the formula $!^{n-1} c :\ !^{n-2} c : \cdots :\ !c : c : A$ represents the formula A for $n = 0$.

Definition 2 (Algebra Over a Set). *Let W be a non-empty set and let H be a non-empty subset of $\mathcal{P}(W)$. We call H an algebra over W iff the following hold:*

- $W \in H$
- $U, V \in H \Longrightarrow U \cup V \in H$
- $U \in H \Longrightarrow W \setminus U \in H$

Definition 3 (Finitely Additive Measure). *Let H be an algebra over W and $\mu : H \to [0,1]$. We call μ a finitely additive measure iff the following hold:*

1. $\mu(W) = 1$
2. *for all $U, V \in H$:*

$$U \cap V = \emptyset \implies \mu(U \cup V) = \mu(U) + \mu(V)$$

Definition 4 (Probability Space). *A probability space is a triple $\mathsf{Prob} = \langle W, H, \mu \rangle$, where:*

- *W is a non-empty set*
- *H is an algebra over W*
- *$\mu : H \to [0,1]$ is a finitely additive measure*

Definition 5 (Model). *Let CS be a constant specification. A $\mathsf{PPJ_{CS}}$-model is a quintuple $M = \langle U, W, H, \mu, * \rangle$ where:*

1. *U is a non-empty set of objects called worlds*
2. *W, H, μ and $*$ are functions, which have U as their domain, such that for every $w \in U$:*
   - *$\langle W(w), H(w), \mu(w) \rangle$ is a probability space with $W(w) \subseteq U$*
   - *$*_w$ is a basic CS-evaluation[3]*

The ternary satisfaction relation $\models$ is defined between models, worlds, and formulas.

Definition 6 (Truth in a $\mathsf{PPJ_{CS}}$-model). *Let CS be a constant specification and let $M = \langle U, W, H, \mu, * \rangle$ be a $\mathsf{PPJ_{CS}}$-model. We define what it means for an $\mathcal{L}$-formula to hold in M at a world $w \in U$ inductively as follows:*

$$M, w \models p \quad :\Longleftrightarrow \quad p_w^* = \mathsf{T} \quad \text{for } p \in \mathsf{Prop}$$

$$M, w \models P_{\geq s} B \quad :\Longleftrightarrow \quad \left([B]_{M,w} \in H(w) \text{ and } \mu(w)([B]_{M,w}) \geq s \right)$$

$$\text{where } [B]_{M,w} = \{ x \in W(w) \mid M, x \models B \}$$

$$M, w \models \neg B \quad :\Longleftrightarrow \quad M, w \not\models B$$

$$M, w \models B \wedge C \quad :\Longleftrightarrow \quad (M, w \models B \text{ and } M, w \models C)$$

$$M, w \models t : B \quad :\Longleftrightarrow \quad B \in t_w^*$$

---

[3] We will usually write $*_w$ instead of $*(w)$.

Definition 7 (Measurable Model). *Let* CS *be a constant specification and let* $M = \langle U, W, H, \mu, * \rangle$ *be a* PPJ$_{CS}$*-model.* M *is called measurable iff for every* $w \in U$ *and for every* $A \in \mathcal{L}$:

$$[A]_{M,w} \in H(w)$$

PPJ$_{CS,Meas}$ *denotes the class of* PPJ$_{CS}$*-measurable models.*

For a model $M = \langle U, W, H, \mu, * \rangle$, $M \models A$ means that $M, w \models A$ for all $w \in U$. Let $T \subseteq \mathcal{L}$. Then $M \models T$ means that $M \models A$ for all $A \in T$. Further $T \models A$ means that for all $M \in$ PPJ$_{CS,Meas}$, $M \models T$ implies $M \models A$.

To be precise we should write $T \vdash_{CS} A$ and $T \models_{CS} A$ instead of $T \vdash A$ and $T \models A$, respectively, since these two notions depend on a given constant specification CS. However, CS will always be clear from the context and we thus omit it.

Definition 8 (Satisfiability). *We say a formula A of $\mathcal{L}$ is satisfiable if there exists a* PPJ$_{CS}$*-measurable model* $M = \langle U, W, H, \mu, * \rangle$ *and* $w \in U$ *with* $M, w \models A$.

We established the Deduction Theorem for PJ in [10]. Now we present the version for PPJ, which can be proved in the same way.

Theorem 1 (Deduction Theorem). *Let* $T \subseteq \mathcal{L}$ *and* $A, B \in \mathcal{L}$. *For any constant specification* CS *we have:*

$$T, A \vdash B \quad \Longleftrightarrow \quad T \vdash A \to B$$

3   Soundness and Completeness

As usual, we can establish soundness by induction on the depth of the derivation of a formula A.

Theorem 2 (Soundness). *For any constant specification* CS, PPJ$_{CS}$ *is sound with respect to the class of* PPJ$_{CS,Meas}$*-models. I.e. for any* $A \in \mathcal{L}$ *and* $T \subseteq \mathcal{L}$ *we have:*

$$T \vdash A \quad \Longrightarrow \quad T \models A.$$

The completeness proof for PPJ$_{CS}$ is a combination of the completeness proof for LPP$_1$ [22] and the completeness proof for PJ [10]. For lack of space, however, we cannot give a detailed completeness proof here. We will only present a series of definitions and lemmas (without proofs) that leads to the completeness result. First we need the notion of a PPJ$_{CS}$-consistent set.

Definition 9 (PPJ$_{CS}$-consistent Set). *Let* CS *be a constant specification and let* T *be a set of $\mathcal{L}$-formulas.*

– T *is said to be* PPJ$_{CS}$*-consistent iff* $T \nvdash \bot$. *Otherwise* T *is said to be* PPJ$_{CS}$*-inconsistent.*

– T *is said to be* maximal *iff for every* $A \in \mathcal{L}$ *either* $A \in T$ *or* $\neg A \in T$.
– T *is said to be* maximal $\mathsf{PPJ}_{\mathsf{CS}}$-consistent *iff it is maximal and* $\mathsf{PPJ}_{\mathsf{CS}}$-*consistent*.

The next lemma is shown for PJ in [10]. The proof for $\mathsf{PPJ}_{\mathsf{CS}}$ is similar.

Lemma 1 (Lindenbaum). *Let* CS *be a constant specification. Every* $\mathsf{PPJ}_{\mathsf{CS}}$-*consistent set can be extended to a maximal* $\mathsf{PPJ}_{\mathsf{CS}}$-*consistent set.*

Definition 10 (Canonical Model). *Let* CS *be a constant specification. The canonical model for* $\mathsf{PPJ}_{\mathsf{CS}}$ *is given by the quintuple* $M = \langle U, W, H, \mu, * \rangle$, *defined as follows:*

– $U = \{ w \mid w$ *is a maximal* $\mathsf{PPJ}_{\mathsf{CS}}$-*consistent set of* $\mathcal{L}$-*formulas* $\}$
– *for every* $w \in U$ *the probability space* $\langle W(w), H(w), \mu(w) \rangle$ *is defined as follows:*
  1. $W(w) = U$
  2. $H(w) = \{ (A)_M \mid A \in \mathcal{L} \}$ *where* $(A)_M = \{ x \mid x \in U, A \in x \}$
  3. *for all* $A \in \mathcal{L}$, $\mu(w)\big((A)_M\big) = \sup_s \{ P_{\geq s} A \in w \}$
– *for every* $w \in W$ *the basic* CS-*evaluation* $*_w$ *is defined as follows:*
  1. *for all* $p \in$ Prop:

$$p_w^* = \begin{cases} \mathsf{T} & \text{if } p \in w \\ \mathsf{F} & \text{if } \neg p \in w \end{cases}$$

  2. *for all* $t \in$ Tm:

$$t_w^* = \{ A \mid t : A \in w \}$$

Lemma 2. *Let* CS *be a constant specification. The canonical model for* $\mathsf{PPJ}_{\mathsf{CS}}$ *is a* $\mathsf{PPJ}_{\mathsf{CS}}$-*model.*

Lemma 3. *Let* $M = \langle U, W, H, \mu, * \rangle$ *be the canonical model for* $\mathsf{PPJ}_{\mathsf{CS}}$. *Then we have*

$$(\forall A \in \mathcal{L})(\forall w \in U)\big[[A]_{M,w} = (A)_M\big].$$

From Lemma 3 we get the following corollary.

Corollary 1. *Let* CS *be any constant specification. The canonical model for* $\mathsf{PPJ}_{\mathsf{CS}}$ *is a* $\mathsf{PPJ}_{\mathsf{CS,Meas}}$-*model.*

Making use of the properties of maximal consistent sets, we can establish the Truth Lemma.

Lemma 4 (Truth Lemma). *Let* CS *be some constant specification and let* $M = \langle U, W, H, \mu, * \rangle$ *be the canonical model for* $\mathsf{PPJ}_{\mathsf{CS}}$. *For every* $A \in \mathcal{L}$ *and any* $w \in U$ *we have:*

$$A \in w \quad \Longleftrightarrow \quad M, w \models A.$$

Finally, we get the completeness theorem as usual.

Theorem 3 (Strong Completeness for PPJ). *Let* CS *be a constant specification, let* $T \subseteq \mathcal{L}$ *and let* $A \in \mathcal{L}$. *Then we have:*

$$T \models A \quad \Longrightarrow \quad T \vdash A.$$

4   Decidability for a Fragment of $\mathcal{L}$

Before we can show that satisfiability is decidable for all $\mathcal{L}$-formulas, we have to show that satisfiability is decidable for a subset $\mathcal{L}^r \subseteq \mathcal{L}$ that is given by the following grammar:

$$A ::= p \mid \neg A \mid A \wedge A \mid t : B$$

where $t \in \mathsf{Tm}, p \in \mathsf{Prop}$, and $B \in \mathcal{L}$.

The key fact about $\mathcal{L}^r$ is that the truth of an $\mathcal{L}^r$-formula A at a world w in a $\mathsf{PPJ_{CS}}$-model $M = \langle U, W, H, \mu, * \rangle$ only depends on the basic CS-evaluation $*_w$.

Hence we can use the notation $* \models A$ if A is a formula of $\mathcal{L}^r$ and $*$ is a basic evaluation. We find that A is satisfiable (in the sense of $\mathsf{PPJ_{CS}}$) if and only if there exists a basic evaluation $*$ such that $* \models A$.

Therefore, we can use an extension of the usual decision procedure for the basic justification logic J, see [11,12,20], to decide satisfiability for formulas of $\mathcal{L}^r$.

Theorem 4. *Let CS be a decidable almost schematic constant specification. For any formula A of the restricted language $\mathcal{L}^r$, it is decidable whether A is satisfiable.*

For lack of space, we only give a proof sketch of the above theorem. As in the decidability proof for J, we make use of schematic variables so that we can represent a schematic constant specification in a finite way. A key step in the decidability proof is then to compute a most general unifier for schematic formulas. This is the step that needs some major adaptations for our probabilistic setting.

Consider, for example, the scheme (WE) given by $P_{\leq r} A \to P_{<s} A$. It has three schematic variables: A for formulas and r, s for rational numbers. Note that there is also a side condition, $s > r$, of which the unification algorithm has to take care. Hence in addition to constructing a substitution, the unification algorithm also has to build up a system of linear inequalities for the rational variables. For instance, in order to unify $P_{\geq r} A$ and $P_{\geq s} B$ the algorithm has to unify A and B and to equate r and s, i.e. it adds $r = s$ to the linear system. In the end, the constructed substitution only is a most general unifier if the linear system is satisfiable.

Of course, one has to take care of the syntactic abbreviations when representing axioms. That means, the scheme (WE) actually is $P_{\geq 1-r} \neg A \to \neg P_{\geq s} A$ with the side condition $s > 1 - r$ (note that the implication again is an abbreviation).

Another complication are constraints of the form

$$l = \min(1, r + s) \tag{2}$$

that originate from the scheme (DIS). Obviously, (2) is not linear. However, for a system C of linear inequalities, we find that

$$C \cup \{l = \min(1, r + s)\}$$

has a solution if and only if

$$C \cup \{l = r + s, r + s \leq 1\} \text{ or } C \cup \{l = 1, r + s > 1\}$$

has a solution. Thus we can reduce solving a system involving (2) to solving several linear systems.

5   Decidability of PPJ$_{CS}$

Definition 11 (Subformulas). *The set of subformulas of an $\mathcal{L}$-formula A, $\mathsf{subf}(A)$, is recursively defined by:*

$$\mathsf{subf}(p) := \{p\}$$
$$\mathsf{subf}(P_{\geq s}B) := \{P_{\geq s}B\} \cup \mathsf{subf}(B)$$
$$\mathsf{subf}(\neg B) := \{\neg B\} \cup \mathsf{subf}(B)$$
$$\mathsf{subf}(B \wedge C) := \{B \wedge C\} \cup \mathsf{subf}(B) \cup \mathsf{subf}(C)$$
$$\mathsf{subf}(t : B) := \{t : B\} \cup \mathsf{subf}(B)$$

Definition 12. *Let $A \in \mathcal{L}$ and assume that $\mathsf{subf}(A) = \{A_1, \ldots, A_k\}$. The set $\mathsf{subfCon}(A)$ contains all sets of the form $\{\pm A_1, \ldots, \pm A_k\}$, where $\pm A_i$ is either A_i or $\neg A_i$. Elements of $\mathsf{subfCon}(A)$ are interpreted conjunctively. That is for $C \in \mathsf{subfCon}(A)$, we simply write $M, w \models C$ instead of $M, w \models \bigwedge C$. Hence $M, w \models C$ means that all elements of C are true at w in M. Accordingly, we say that C is satisfiable if the formula $\bigwedge C$ is so.*

We define the mapping j on sets C of $\mathcal{L}$-formulas by:

$$\mathsf{j}(C) := C \cap \mathcal{L}^r.$$

Before proving that PPJ$_{CS}$ is decidable we need to establish some auxiliary lemmata.

Lemma 5. *Let $M = \langle U, W, H, \mu, * \rangle \in$ PPJ$_{CS,Meas}$ and let $A \in \mathcal{L}$. Let $B \in \mathsf{subf}(A)$, let $C \in \mathsf{subfCon}(A)$ and let $w \in U$. Assume that $M, w \models C$. Then we have:*

$$M, w \models B \quad \Longleftrightarrow \quad B \in C.$$

Proof. We prove the two directions of the lemma separately:

$\Longleftarrow$: From $B \in C$ and $M, w \models C$ we immediately get $M, w \models B$.

$\Longrightarrow$: Since B is a subformula of A, we have either $B \in C$ or $\neg B \in C$. If $\neg B \in C$, then we would have $M, w \models \neg B$, i.e. $M, w \not\models B$, which contradicts the fact that $M, w \models B$. Thus, we conclude $B \in C$. $\qquad\square$

Lemma 6. *Let CS be a constant specification and let $A \in \mathcal{L}$. Then A is satisfiable if and only if there exists a set $Y = \{B_1, \ldots, B_n\} \subseteq \mathsf{subfCon}(A)$ such that all of the following conditions holds:*

1. *for some* $i \in \{1, \ldots, n\}$, $A \in B_i$.
2. *for every* $1 \leq i \leq n$, $\mathsf{j}(B_i)$ *is satisfiable.*
3. *for every* $1 \leq i \leq n$, *there are variables* x_{ij} *with* $1 \leq j \leq n$, *such that the following system of linear inequalities is satisfiable:*

$$\sum_{j=1}^{n} x_{ij} = 1$$

$$(\forall 1 \leq j \leq n)\big[x_{ij} \geq 0\big]$$

$$\textit{for every } P_{\geq s}C \in B_i, \quad \sum_{\{j \mid C \in B_j\}} x_{ij} \geq s$$

$$\textit{for every } \neg P_{\geq s}C \in B_i, \quad \sum_{\{j \mid C \in B_j\}} x_{ij} < s$$

Proof. We prove the two directions of the lemma separately:

$\Longrightarrow$: Let $M = \langle U, W, H, \mu, * \rangle \in \mathsf{PPJ_{CS,Meas}}$. Assume that A is satisfiable in some world of M.

Let $\approx$ denote a binary relation over U such that for all $w, x \in U$ we have:

$$w \approx x \quad \text{if and only if} \quad \big(\forall B \in \mathsf{subf}(A)\big)\big[M, w \models B \Leftrightarrow M, x \models B\big].$$

It is easy to see that $\approx$ is an equivalence relation. Let $K_1, \ldots, K_n$ be the equivalence classes of $\approx$. For every $i \in \{1, \ldots, n\}$ we choose some $w_i \in K_i$. For every $i \in \{1, \ldots, n\}$ some subformulas of A hold in the world w_i and some do not. So for every $i \in \{1, \ldots, n\}$ there exists a $B_i \in \mathsf{subfCon}(A)$ such that $M, w_i \models B_i$. For $i \neq j$ we have $B_i \neq B_j$ since w_i and w_j belong to different equivalence classes. Let $Y = \{B_1, \ldots, B_n\}$. It remains to show that the conditions in the statement of the lemma hold:

1. Let $w \in U$ be such that $M, w \models A$. The world w belongs to some equivalence class of $\approx$, which is represented by w_i. Thus $M, w_i \models A$. By Lemma 5 we find $A \in B_i$, i.e. condition 1 holds.
2. For every $1 \leq i \leq n$ we have $M, w_i \models B_i$. Because of $\mathsf{j}(B_i) \subseteq B_i$ we immediately get $M, w_i \models \mathsf{j}(B_i)$. Hence condition 2 holds.
3. Let $i \in \{1, \ldots, n\}$. We set

$$y_{ij} = \mu(w_i)(K_j \cap W(w_i)), \text{ for every } 1 \leq j \leq n.$$

Some calculations show that these values y_{ij} satisfy the linear system in condition 3.

$\Longleftarrow$: Assume that there exists $Y = \{B_1, \ldots, B_n\} \subseteq \mathsf{subfCon}(A)$ such that conditions 1–3 hold. For every $1 \leq i \leq n$, let $*_i$ be a basic evaluation such that $*_i \models \mathsf{j}(B_i)$. We define the quintuple $M = \langle U, W, H, \mu, * \rangle$ by:

- $U = \{w_1, \ldots, w_n\}$ for some $w_1, \ldots, w_n$.
- For all $1 \leq i \leq n$ we set:

1. $W(w_i) = U$
2. $H(w_i) = \mathcal{P}(W(w_i))$
3. $\mu(w_i)(V) = \sum_{\{j \mid w_j \in V\}} x_{ij}$ for every $V \in H(w_i)$
4. $*_{w_i} = *_i$.

We can show that $M \in \mathsf{PPJ_{CS,Meas}}$. However, we have to omit the proof due to lack of space.

It remains to show $M, w_i \models A$ for some i. We first establish

$$(\forall D \in \mathsf{subf}(A))(\forall 1 \le i \le n)\big[D \in B_i \iff M, w_i \models D\big] \qquad (3)$$

by induction on the structure of D (again we have to omit the proof).

It holds that $A \in \mathsf{subf}(A)$. Thus, by (3) we find:

$$(\forall 1 \le i \le n)\big[A \in B_i \iff M, w_i \models A\big].$$

By condition 1, there exists an i such that $A \in B_i$. Thus, there exists an i such that $M, w_i \models A$. Hence, A is $\mathsf{PPJ_{CS,Meas}}$-satisfiable. □

In the proof of Lemma 6 we construct a model with at most $2^{|\mathsf{subf}(A)|}$ worlds that satisfies A. Hence a corollary of Lemma 6 is that any $A \in \mathcal{L}$ is $\mathsf{PPJ_{CS,Meas}}$-satisfiable if and only if it is satisfiable in a $\mathsf{PPJ_{CS,Meas}}$-model with at most $2^{|\mathsf{subf}(A)|}$ worlds. In other words, Lemma 6 implies a small model property for $\mathsf{PPJ_{CS}}$.

Moreover, Lemma 6 dictates a procedure to decide the satisfiability problem for $\mathsf{PPJ_{CS}}$.

Theorem 5. *Let* CS *be a decidable almost schematic constant specification. The* $\mathsf{PPJ_{CS,Meas}}$-*satisfiability problem is decidable.*

Proof. Let $A \in \mathcal{L}$. The formula A is satisfiable if and only if there exists some $Y \subseteq \mathsf{subfCon}(A)$, such that all conditions in the statement of Lemma 6 hold. Since $\mathsf{subfCon}(A)$ is finite, it suffices to show that for every $Y \subseteq \mathsf{subfCon}(A)$ the conditions 1–3 in the statement of Lemma 6 can be effectively checked:

- Decidability of condition 1 is trivial.
- Decidability of condition 2 follows from Theorem 4.
- In condition 3 we have to check for the satisfiability of a set of linear inequalities, which is a well-known decidable problem [18].

We conclude that the satisfiability problem for $\mathsf{PPJ_{CS}}$ is decidable. □

6   Application to the Lottery Paradox

Kyburg's famous lottery paradox [16] goes as follows. Consider a fair lottery with 1000 tickets that has exactly one winning ticket. Now assume a proposition is believed if and only if its degree of belief is greater than 0.99. In this setting it is rational to believe that ticket 1 does not win, it is rational to believe that ticket 2 does not win, and so on. However, this entails that it is rational to believe that no ticket wins because rational belief is closed under conjunction. Hence it is rational to believe that no ticket wins and that one ticket wins.

$\mathsf{PPJ_{CS}}$ makes the following analysis of the lottery paradox possible. First we need a principle to move from degrees of belief to rational belief (this formalizes what Foley [8] calls *the Lockean thesis*): we suppose that for each term t, there exists a term $\mathsf{pb}(t)$ such that

$$t : (P_{>0.99}A) \; \rightarrow \; \mathsf{pb}(t) : A. \tag{4}$$

pb stands for probabilistic belief. Let w_i be the proposition *ticket i wins*. For each $1 \leq i \leq 1000$, there is a term t_i such that $t_i : (P_{=\frac{999}{1000}} \neg w_i)$ holds. Hence by (4) we get

$$\mathsf{pb}(t_i) : \neg w_i \quad \text{for each } 1 \leq i \leq 1000. \tag{5}$$

Now if CS is axiomatically appropriate, then

$$s_1 : A \wedge s_2 : B \; \rightarrow \; \mathsf{con}(s_1, s_2) : (A \wedge B) \tag{6}$$

is a valid principle (for a suitable term $\mathsf{con}(s_1, s_2)$). Hence by (5) we conclude that

$$\text{there exists a term } t \text{ with } \; t \; : (\neg w_1 \wedge \cdots \wedge \neg w_{1000}), \tag{7}$$

which leads to a paradoxical situation since it is also believed that one of the tickets wins.

In $\mathsf{PPJ_{CS}}$ we can resolve this problem by restricting the constant specification such that (6) is valid only if $\mathsf{con}(s_1, s_2)$ does not contain two different subterms of the form $\mathsf{pb}(t)$. Then the step from (5) to (7) is no longer possible and we can avoid the paradoxical belief.

This analysis is inspired by Leitgeb's [17] solution to the lottery paradox and his *Stability Theory of Belief* according to which *it is not permissible to apply the conjunction rule for beliefs across different contexts*. Our proposed restriction of (6) is one way to achieve this in a formal system. A related and very interesting question is whether one can interpret the above justifications t_i as stable sets in Leitgeb's sense. Of course, our discussion of the lottery paradox is very sketchy but we think that probabilistic justification logic provides a promising approach to it that is worth further investigations.

Acknowledgements. We would like to thank the anonymous referees for many valuable comments that helped us improve the paper substantially.

References

1. Artemov, S.N.: Operational modal logic. Technical report MSI 95–29, Cornell University, December 1995
2. Artemov, S.N.: Explicit provability and constructive semantics. Bull. Symbolic Logic **7**(1), 1–36 (2001)

3. Artemov, S.N.: The logic of justification. Rev. Symbolic Logic **1**(4), 477–513 (2008)
4. Artemov, S.N.: The ontology of justifications in the logical setting. Studia Logica **100**(1–2), 17–30 (2012). published online February 2012
5. Bucheli, S., Kuznets, R., Studer, T.: Partial realization in dynamic justification logic. In: Beklemishev, L.D., de Queiroz, R. (eds.) WoLLIC 2011. LNCS, vol. 6642, pp. 35–51. Springer, Heidelberg (2011)
6. Fan, T., Liau, C.: A logic for reasoning about justified uncertain beliefs. In: Yang, Q., Wooldridge, M. (eds.) Proceedings of the IJCAI 2015. pp. 2948–2954. AAAI Press (2015)
7. Fitting, M.: The logic of proofs, semantically. Ann. Pure Appl. Logic **132**(1), 1–25 (2005)
8. Foley, R.: Beliefs, degrees of belief, and the lockean thesis. In: Huber, F., Schmidt-Petri, C. (eds.) Degrees of Belief, pp. 37–47. Springer, Heidelberg (2009)
9. Ghari, M.: Justification logics in a fuzzy setting. ArXiv e-prints (Jul 2014)
10. Kokkinis, I., Maksimović, P., Ognjanović, Z., Studer, T.: First steps towards probabilistic justification logic. Logic J. IGPL **23**(4), 662–687 (2015)
11. Kuznets, R.: On the complexity of explicit modal logics. In: Clote, P.G., Schwichtenberg, H. (eds.) CSL 2000. LNCS, vol. 1862, pp. 371–383. Springer, Heidelberg (2000)
12. Kuznets, R.: Complexity Issues in Justification Logic. Ph.D. thesis, City University of New York (May 2008). https://urldefense.proofpoint.com/v2/url?u=http-3A__gradworks.umi.com_33_10_3310747.html&d=BQIFAg&c=8v77JlHZOYsReeO xyYXDU39VUUzHxyfBUh7fw_ZfBDA&r=vtfGbdUbVrB70ksnRgFs_T_U9JW0Ku iKAy7FCssagUA&m=E2jdp6bvLx4F1tqhZdQcSOIFPo46MJsetU8juUkriqE&s=0 xOBGghqLpQLrHd4oCus0pdP3DATqtB-QgVJWT-1nBk&e=
13. Kuznets, R., Studer, T.: Justifications, ontology, and conservativity. In: Bolander, T., Braüner, T., Ghilardi, S., Moss, L. (eds.) AiML 9, pp. 437–458. College Publications (2012)
14. Kuznets, R., Studer, T.: Update as evidence: belief expansion. In: Artemov, S., Nerode, A. (eds.) LFCS 2013. LNCS, vol. 7734, pp. 266–279. Springer, Heidelberg (2013)
15. Kuznets, R., Studer, T.: Weak arithmetical interpretations for the logic of proofs. Logic Journal of IGPL (to appear)
16. Kyburg, H.E.J.: Probability and the Logic of Rational Belief. Wesleyan University Press, Middletown (1961)
17. Leitgeb, H.: The stability theory of belief. Philos. Rev. **123**(2), 131–171 (2014)
18. Luenberger, D.G., Ye, Y.: Linear and Nonlinear Programming. International Series in Operations Research and Management Science, vol. 116. Springer, Heidelberg (2008)
19. Milnikel, R.S.: The logic of uncertain justifications. Ann. Pure Appl. Logic **165**(1), 305–315 (2014)
20. Mkrtychev, A.: Models for the logic of proofs. In: Adian, S., Nerode, A. (eds.) LFCS 1997. LNCS, vol. 1234, pp. 266–275. Springer, Heidelberg (1997)
21. Ognjanović, Z., Rašković, M.: Some first order probability logics. Theoret. Comput. Sci. **247**, 191–212 (2000)
22. Ognjanović, Z., Rašković, M., Marković, Z.: Probability logics. Zbornik radova, subseries. Logic Comput. Sci. **12**(20), 35–111 (2009)
23. Rašković, M., Ognjanović, Z.: A first order probability logic, LP_Q. Puplications de L'Institut Mathèmatique **65**(79), 1–7 (1999)

Sequent Calculus for Intuitionistic Epistemic Logic IEL

Vladimir N. Krupski$^{(\boxtimes)}$ and Alexey Yatmanov

Faculty of Mechanics and Mathematics, Lomonosov Moscow State University,
Moscow 119992, Russia
krupski@lpcs.math.msu.su

Abstract. The formal system of intuitionistic epistemic logic IEL was
proposed by S. Artemov and T. Protopopescu. It provides the formal
foundation for the study of knowledge from an intuitionistic point of
view based on Brouwer-Hayting-Kolmogorov semantics of intuitionism.
We construct a cut-free sequent calculus for IEL and establish that polynomial space is sufficient for the proof search in it. We prove that IEL is
PSPACE-complete.

Keywords: Modal logic · Intuitionistic epistemic logic · Sequent calculus · Cut-elimination · PSPACE

1 Introduction

Modal logic IEL, the basic Intuitionistic Epistemic Logic, was proposed by
S. Artemov and T. Protopopescu in [1]. It was defined by the following Hilbert
system:

Axioms

1. Axioms of propositional intuitionistic logic,
2. $K(F \to G) \to (KF \to KG)$ (distribution),
3. $F \to KF$ (co-reflection),
4. $\neg K\bot$ (consistency).

Rule $F, F \to G \vdash G$ (Modus Ponens). Here knowledge modality K means
verified truth, as suggested by T. Williamson in [2]. According to the Brouwer-
Heyting-Kolmogorov semantics of intuitionistic logic, a proposition is true iff it
is proved. The co-reflection principle states that any such proof can be verified.

The intuitionistic meaning of implication provides an effective proof checking
procedure that produces a proof of KF given a proof of F. But the assumption
that its output always contains a proof of F is too restrictive. The procedure may
involve some trusted sources which do not necessarily produce explicit proofs of
what they verify[1]. So the backward implication which is the reflection principle

---

[1] For example, it can be some trusted database that stores true facts without proofs
or some zero-knowledge proof.

© Springer International Publishing Switzerland 2016
S. Artemov and A. Nerode (Eds.): LFCS 2016, LNCS 9537, pp. 187–201, 2016.
DOI: 10.1007/978-3-319-27683-0_14

$KF \to F$ used in the classical epistemic logic (see [3]) is wrong in the intuition-istic setting. In general, a proof of KF is less informative than a proof of F.

At the same time some instances of the reflection principle are true in IEL. In particular, it is the consistency principle which is equivalent to $K\bot \to \bot$. The proof of $K\bot$ contains the same information as the proof of $\bot$ because there is no such proof at all. The more general example is the reflection principle for negative formulas: $K\neg F \to \neg F$. It is provable in IEL (see [1]).

In this paper we develop the proof theory for IEL. Our main contributions are the cut-free sequent formulation and the complexity bound for this logic. It is established that polynomial space is sufficient for the proof search, so IEL is PSPACE-complete.

Some other sequent cut-free formalization of intuitionistic reasoning about knowledge was proposed in [4].[2] It is based on a bimodal logic with reflexive knowledge modality $\mathcal{K}$ that is different from the co-reflexive modality K used in IEL.

Our cut-elimination technique is syntactic (see [5]). We formulate a special cut-free sequent calculus IEL_G^- without structural rules (see Sect. 3) that is correct with respect to the natural translation into IEL. It has a specific K-introduction rule (KI_1) that also allows to contract a formula F in the presence of KF in antecedents. This choice makes it possible to prove the admissibility of the stan-dard contraction rule as well as the admissibility of all natural IEL-correct modal rules (Sects. 4, 5). The admissibility of the cut-rule is proved by the usual induc-tion on the cutrank (Sect. 6). As the result we obtain the equivalence between IEL_G^- and IEL_G^0. (The latter is the straightforwardly formulated sequent counter-part for IEL with the cut-rule). Finally we formulate a light cut-free variant of IEL_G^- with the contraction rule and with modal rules

$$\frac{\Gamma_1, \Gamma_2 \Rightarrow F}{\Gamma_1, K(\Gamma_2) \Rightarrow KF} \ (KI), \qquad \frac{\Gamma \Rightarrow K\bot}{\Gamma \Rightarrow F} \ (U).$$

It is equivalent to IEL_G^-.

The proof search for IEL can be reduced to the case of so-called minimal derivations (Sect. 7). We implement it as a game of polynomial complexity and use the characterization AP=PSPACE (see [7]) to prove the upper complexity bound for IEL. The matching lower bound follows from the same bound for intuitionistic propositional logic [8].

2   Sequent Formulation of IEL

The definition of intuitionistic sequents is standard (see [5]). Formulas are build from propositional variables and $\bot$ using $\wedge$, $\vee$, $\to$ and K; $\neg F$ means $F \to \bot$.

---

[2] Sequents in [4] are classical (multiconclusion) and contain special labels denoting worlds of a Kripke structure, so this formalization can be considered as a classical formulation of the theory of forcing relation in a Kripke structure that corresponds to the intuitionistic bimodal epistemic logic.

A sequent has the form $\Gamma \Rightarrow F$ where F is a formula and Γ is a multiset of formulas. $K(\Gamma)$ denotes $KF_1, \ldots, KF_n$ when $\Gamma = F_1, \ldots, F_n$.

Let $\mathsf{IEL}^0_\mathsf{G}$ be the following calculus:

Axioms

$$\Gamma, A \Rightarrow A \ \ (A \text{ is a variable}), \qquad \Gamma, \bot \Rightarrow F, \qquad \Gamma, K\bot \Rightarrow F.$$

Rules

$$\frac{\Gamma, F, F \Rightarrow G}{\Gamma, F \Rightarrow G} \ (Contraction) \qquad \frac{\Gamma_1 \Rightarrow F \quad \Gamma_2, F \Rightarrow G}{\Gamma_1, \Gamma_2 \Rightarrow G} \ (Cut)$$

$$\frac{\Gamma, F, G \Rightarrow H}{\Gamma, F \wedge G \Rightarrow H} \ (\wedge \Rightarrow) \qquad \frac{\Gamma \Rightarrow F \quad \Gamma \Rightarrow G}{\Gamma \Rightarrow F \wedge G} \ (\Rightarrow \wedge)$$

$$\frac{\Gamma, F \Rightarrow H \quad \Gamma, G \Rightarrow H}{\Gamma, F \vee G \Rightarrow H} \ (\vee \Rightarrow) \qquad \frac{\Gamma \Rightarrow F_i}{\Gamma \Rightarrow F_1 \vee F_2} \ (\Rightarrow \vee)_i \quad (i = 1, 2)$$

$$\frac{\Gamma \Rightarrow F \quad \Gamma, G \Rightarrow H}{\Gamma, F \rightarrow G \Rightarrow H} \ (\rightarrow \Rightarrow) \qquad \frac{\Gamma, F \Rightarrow G}{\Gamma \Rightarrow F \rightarrow G} \ (\Rightarrow \rightarrow)$$

$$\frac{\Gamma \Rightarrow F}{K(\Gamma) \Rightarrow KF} \ (KI_0) \qquad \frac{\Gamma, F, KF \Rightarrow G}{\Gamma, F \Rightarrow G} \ (KC)$$

Comment. $\mathsf{IEL}^0_\mathsf{G}$ is a straightforwardly formulated sequent counterpart of IEL. Axioms and rules without the modality correspond to the standard sequent formulation of the intuitionistic propositional logic (cf. system G2i from [5] with the cut-rule). The modal axiom corresponds to the consistency principle. Modal rules (KI_0) and (KC) reflect the distribution and co-reflection principles respectively. Instead of the K-contraction rule (KC) one can take the equivalent K-elimination rule:

$$\frac{\Gamma, KF \Rightarrow G}{\Gamma, F \Rightarrow G} \ (KE).$$

Theorem 1. $\mathsf{IEL}^0_\mathsf{G} \vdash \Gamma \Rightarrow F$ iff $\mathsf{IEL} \vdash \wedge \Gamma \rightarrow F$.

Proof. Straightforward induction on the derivations. □

Our goal is to eliminate the cut-rule. But the cut-elimination result for $\mathsf{IEL}^0_\mathsf{G}$ will not have the desirable consequences, namely, the subformula property and termination of the proof search procedure. Below we give a different formulation without these disadvantages.

3   Cut-Free Variant $\mathsf{IEL}^-_\mathsf{G}$ with Rules (KI_1) and (U)

Axioms

$$\Gamma, A \Rightarrow A, \quad A \text{ is a variable or } \bot.$$

Rules

$$\frac{\Gamma, F, G \Rightarrow H}{\Gamma, F \wedge G \Rightarrow H} \, (\wedge \Rightarrow) \qquad \frac{\Gamma \Rightarrow F \quad \Gamma \Rightarrow G}{\Gamma \Rightarrow F \wedge G} \, (\Rightarrow \wedge)$$

$$\frac{\Gamma, F \Rightarrow H \quad \Gamma, G \Rightarrow H}{\Gamma, F \vee G \Rightarrow H} \, (\vee \Rightarrow) \qquad \frac{\Gamma \Rightarrow F_i}{\Gamma \Rightarrow F_1 \vee F_2} \, (\Rightarrow \vee)_i \quad (i = 1, 2)$$

$$\frac{\Gamma, F \rightarrow G \Rightarrow F \quad \Gamma, G \Rightarrow H}{\Gamma, F \rightarrow G \Rightarrow H} \, (\rightarrow \Rightarrow) \qquad \frac{\Gamma, F \Rightarrow G}{\Gamma \Rightarrow F \rightarrow G} \, (\Rightarrow \rightarrow)$$

$$\frac{\Gamma, K(\Delta), \Delta \Rightarrow F}{\Gamma, K(\Delta) \Rightarrow KF} \, (KI_1) \qquad \frac{\Gamma \Rightarrow K\bot}{\Gamma \Rightarrow F} \, (U)$$

In the rule (KI_1) we additionally require that Γ does not contain formulas of the form KG. (This requirement is unessential, see Corollary 3).

We define the main (occurrences of) formulas for axioms and for all inference rules except (KI_1) as usual — they are the displayed formulas in the conclusions (not members of Γ, H). For the rule (KI_1) all members of $K(\Delta)$ and the formula KF are main.

Comment. The propositional part of $\mathsf{IEL}_\mathsf{G}^-$ is the same as in the system $\mathsf{G3m}$ from [5]. In the modal part we do not add (KE) or (KC), but modify (KI_0). In the presence of weakening (it is admissible, see Lemma 2) (KI_0) is derivable:

$$\frac{\dfrac{\Gamma \Rightarrow F}{K(\Gamma), \Gamma \Rightarrow F} \, (W)}{K(\Gamma) \Rightarrow KF} \, (KI_1).$$

So one can derive all sequents of the forms $F \Rightarrow F$ for complex F and $F \Rightarrow KF$. It can be shown by induction on the complexity of the formula F. The latter also requires weakening in the case of $F = KG$:

$$\frac{F \Rightarrow F}{F \Rightarrow KF} \, (KI_1), \; F \neq KG, \qquad \frac{\dfrac{KG \Rightarrow KG}{KG, G \Rightarrow KG} \, (W)}{KG \Rightarrow KKG} \, (KI_1).$$

Comment. (U) is necessary. There is no way to prove the sequent $K\bot \Rightarrow \bot$ in $\mathsf{IEL}_\mathsf{G}^-$ without the rule (U).

4   Structural Rules Are Admissible

We prove the depth-preserving admissibility of weakening and contraction. Our proof follows [5] except the case of the rule (KI_1). The corresponding inductive step in the proof of Lemma 6 does not require the inversion of the rule. Instead of it, some kind of contraction is build-in in the rule itself.[3]

We write $\vdash_n \Gamma \Rightarrow F$ for "$\Gamma \Rightarrow F$ has a $\mathsf{IEL}_\mathsf{G}^-$-proof of depth at most n".

---

[3] This method was introduced by Kleene in the construction of his G3 systems, see [6].

Lemma 2 (Weakening). *If $\vdash_n \Gamma \Rightarrow F$ then $\vdash_n \Gamma, G \Rightarrow F$.*

Proof. Induction on n, similar to the proof of Depth-preserving Weakening lemma from [5]. For example, consider the additional modal rule (KI_1). Consider a derivation

$$\frac{\begin{array}{c}\mathcal{D}\\ \hline \Gamma, K(\Delta), \Delta \Rightarrow F'\end{array}}{\Gamma, K(\Delta) \Rightarrow KF'} (KI_1)$$

with $\vdash_n \Gamma, K(\Delta), \Delta \Rightarrow F'$. If $G = KG'$ for some G' then, by the induction hypothesis, $\vdash_n \Gamma, K(\Delta), \Delta, KG', G' \Rightarrow F'$, so the rule (KI_1) can be applied to this premise and we have $\vdash_{n+1} \Gamma, K(\Delta), KG' \Rightarrow KF'$ by (KI_1). In the remaining case the rule (KI_1) can be applied to $\vdash_n \Gamma, K(\Delta), \Delta, G \Rightarrow F'$, so we have $\vdash_{n+1} \Gamma, K(\Delta), G \Rightarrow KF'$ too.   □

Corollary 3. *The extended K-introduction rule*

$$\frac{\Gamma_1, K(\Delta), \Delta, \Gamma_2 \Rightarrow F}{\Gamma_1, K(\Delta, \Gamma_2) \Rightarrow KF} (KI_{ext})$$

is admissible in $\mathsf{IEL}_\mathsf{G}^-$ *and* $\vdash_n \Gamma_1, K(\Delta), \Delta, \Gamma_2 \Rightarrow F$ *implies* $\vdash_{n+1} \Gamma_1, K(\Delta, \Gamma_2) \Rightarrow KF$.

Proof. Suppose $\vdash_n \Gamma_1, K(\Delta), \Delta, \Gamma_2 \Rightarrow F$ and $\Gamma_1 = \Gamma_1', K(\Gamma_1'')$ where Γ_1' does not contain formulas of the form KG. By Lemma 2,

$$\vdash_n \Gamma_1', K(\Gamma_1''), K(\Delta), K(\Gamma_2), \Gamma_1'', \Delta, \Gamma_2 \Rightarrow F.$$

So,

$$\vdash_{n+1} \Gamma_1', K(\Gamma_1'', \Delta, \Gamma_2) \Rightarrow KF$$

by (KI_1). But $\Gamma_1', K(\Gamma_1'', \Delta, \Gamma_2) = \Gamma_1, K(\Delta, \Gamma_2)$, so $\vdash_{n+1} \Gamma_1, K(\Delta, \Gamma_2) \Rightarrow KF$.   □

Corollary 4. *All axioms of* $\mathsf{IEL}_\mathsf{G}^0$ *are provable in* $\mathsf{IEL}_\mathsf{G}^-$.

Proof. It is sufficient to prove sequents $\Gamma, \bot \Rightarrow F$ and $\Gamma, K\bot \Rightarrow F$:

$$\frac{\dfrac{\Gamma, \bot \Rightarrow \bot}{\Gamma, \bot \Rightarrow K\bot} (KI_{ext})}{\Gamma, \bot \Rightarrow F} (U), \qquad \frac{\dfrac{\Gamma, \bot \Rightarrow \bot}{\Gamma, K\bot \Rightarrow K\bot} (KI_{ext})}{\Gamma, K\bot \Rightarrow F} (U).$$

□

Lemma 5 (Inversion lemma, cf. [5]). *Left rules are invertible in the following sense:*

   If $\vdash_n \Gamma, A \wedge B \Rightarrow C$ then $\vdash_n \Gamma, A, B \Rightarrow C$.

   If $\vdash_n \Gamma, A_1 \vee A_2 \Rightarrow C$ then $\vdash_n \Gamma, A_i \Rightarrow C$, $i = 1, 2$.

   If $\vdash_n \Gamma, A \rightarrow B \Rightarrow C$ then $\vdash_n \Gamma, B \Rightarrow C$.

Proof. The proof is essentially the same as the proof of Inversion lemma from [5]. The additional cases of modal rules are straightforward.   □

Lemma 6 (Contraction). *If* $\vdash_n \Gamma, F, F \Rightarrow G$ *then* $\vdash_n \Gamma, F \Rightarrow G$.

Proof. Induction on n. Case $n = 1$. When the first sequent is an axiom, the second one is an axiom too.

Case $n + 1$. When the displayed two occurrences of F in $\Gamma, F, F \Rightarrow G$ are not main for the last rule of the derivation, apply the induction hypothesis to the premises of the rule and contract F there.

Suppose one of the occurrences is main. Only axioms may have atomic main formulas, so we treat atomic F as in case $n = 1$.

When F has one of the forms $A \wedge B$, $A \vee B$ or $A \to B$, we use the same proof as in [5]. It is based on the items of Inversion lemma formulated in Lemma 5.

Case $F = KA$ is new. The derivation of $\Gamma, F, F \Rightarrow G$ of depth $n + 1$ has the form

$$\frac{\begin{array}{c}\mathcal{D}\\ \hline \Gamma', K(\Delta), \Delta \Rightarrow B \end{array}}{\Gamma', K(\Delta) \Rightarrow KB} (KI_1)$$

where $\Gamma, F, F = \Gamma', K(\Delta)$ and $G = KB$; the multiset Δ contains two copies of A. We have

$$\vdash_n \Gamma', K(\Delta), \Delta \Rightarrow B. \tag{1}$$

Let $(\ )^-$ means to remove one copy of A from a multiset. We apply the induction hypothesis to (1) and obtain $\vdash_n \Gamma, K(\Delta^-), \Delta^- \Rightarrow B$. Then, by (KI_1),

$$\vdash_{n+1} \Gamma, K(\Delta^-) \Rightarrow KB.$$

But $\Gamma, F = \Gamma', K(\Delta^-)$, so $\vdash_{n+1} \Gamma, F \Rightarrow G$.   □

5   Admissible Modal Rules

We have already seen that (KI_0) is admissible in $\mathsf{IEL}_\mathsf{G}^-$.

Lemma 7 (Depth-preserving K-elimination). *If* $\vdash_n \Gamma, KF \Rightarrow G$ *then* $\vdash_n \Gamma, F \Rightarrow G$.

Proof. Induction on n. Case $n = 1$. When the first sequent is an axiom, the second one is an axiom too.

Case $n + 1$. Consider a proof of depth $n + 1$ of a sequent $\Gamma, KF \Rightarrow G$. Let (R) be its last rule. When the displayed occurrence of KF is not main for (R), apply the induction hypothesis to its premises and then apply (R) to reduced premises. It will give $\vdash_n \Gamma, F \Rightarrow G$.

Suppose the occurrence of KF is main. The derivation has the form

$$\frac{\vdash_n \Gamma', K(\Delta), KF, \Delta, F \Rightarrow G'}{\vdash_{n+1} \Gamma', K(\Delta, F) \Rightarrow KG'} \ (KI_1) \ .$$

Apply the induction hypothesis to the premise and remove one copy of F. By Lemma 6, $\vdash_n \Gamma', K(\Delta), \Delta, F \Rightarrow G'$. Then apply an instance of (KI_{ext}) with $\Gamma_1 = \Gamma', F$ and empty Γ_2. By Corollary 3, $\vdash_{n+1} \Gamma', K(\Delta), F \Rightarrow KG'$. $\square$

Corollary 8 (Depth-preserving K-contraction). *If $\vdash_n \Gamma, KF, F \Rightarrow G$ then* $\vdash_n \Gamma, F \Rightarrow G$.

Proof. Apply (KE) and contraction. Both rules are admissible and preserve the depth (Lemmas 7, 6). $\square$

6   Cut Is Admissible

Consider an IEL_G^--derivation with additional cut-rule

$$\frac{\Gamma_1 \Rightarrow F \quad \Gamma_2, F \Rightarrow G}{\Gamma_1, \Gamma_2 \Rightarrow G} \ (Cut) \ . \tag{2}$$

and some instance of (Cut) in it. The level of the cut is the sum of the depths of its premises. The rank of the cut is the length of F.

Lemma 9. *Suppose the premises of (Cut) are provable in IEL_G^- without (Cut). Then the conclusion is also provable in IEL_G^- without (Cut).*

Proof. We define the following well-ordering on pairs of natural numbers: $(k_1, l_1) > (k_2, l_2)$ iff $k_1 > k_2$ or $k_1 = k_2$ and $l_1 > l_2$ simultaneously. By induction on this order we prove that a single cut of rank k and level l can be eliminated.

As in [5], we consider three possibilities:

I. One of the premises is an axiom. In this case the cut-rule can be eliminated. If the left premise of (2) is an axiom,

$$\frac{\Gamma_1', A \Rightarrow A \quad \Gamma_2, A \Rightarrow G}{\Gamma_1', A, \Gamma_2 \Rightarrow G} \ (Cut) \ ,$$

then (Cut) is unnecessary. The conclusion can be derived from the right premise by weakening (Lemma 2).

Now suppose that the right premise is an axiom. If the cutformula F is not main for the axiom $\Gamma_2, F \Rightarrow G$ then the conclusion $\Gamma_1, \Gamma_2 \Rightarrow G$ is also an axiom, so (Cut) can be eliminated. If F is main for the right premise then $F = G = A$ where A is atomic, so (2) has the form

$$\frac{\Gamma_1 \Rightarrow A \quad \Gamma_2, A \Rightarrow A}{\Gamma_1, \Gamma_2 \Rightarrow A} \ (Cut) \ .$$

The conclusion can be derived without (Cut) from the left premise by weakening (Lemma 2).

II. Both premises are not axioms and the cutformula is not main for the last rule in the derivation of at least one of the premises. In this case one can permute the cut upward and reduce the level of the cut. For example,

$$\cfrac{\cfrac{\Phi,F,K(\Psi),\Psi\Rightarrow B}{\Phi,F,K(\Psi)\Rightarrow KB}\quad\Gamma_1\Rightarrow F}{\Gamma_1,\Phi,K(\Psi)\Rightarrow KB}\,(Cut)\quad\rightsquigarrow\quad\cfrac{\cfrac{\Gamma_1\Rightarrow F\quad\Phi,F,K(\Psi),\Psi\Rightarrow B}{\Gamma_1,\Phi,K(\Psi),\Psi\Rightarrow B}\,(Cut)}{\Gamma_1,\Phi,K(\Psi)\Rightarrow KB}\,(KI_{ext})\ .$$

The cutformula remains the same, so the cut rule can be eliminated by induction hypothesis (see [5]).

III. The cutformula F is main for the last rules in the derivations of both premises. In this case we reduce the rank of cut and apply the induction hypothesis.

Note that F is not atomic. (The atomic case is considered in I.) If the last rule in the derivation of the left premise is (U) then (Cut) can be eliminated:

$$\cfrac{\cfrac{\Gamma_1\Rightarrow K\bot}{\Gamma_1\Rightarrow F}\,(U)\quad\Gamma_2,F\Rightarrow G}{\Gamma_1,\Gamma_2\Rightarrow G}\,(Cut)\quad\rightsquigarrow\quad\cfrac{\cfrac{\Gamma_1\Rightarrow K\bot}{\Gamma_1,\Gamma_2\Rightarrow K\bot}}{\Gamma_1,\Gamma_2\Rightarrow G}\,(U)\ .$$

Case $F=KA$, the last rule in the derivation of the left premise is (KI_1). Then the right premise is also derived by (KI_1):

$$\cfrac{\cfrac{\mathcal{D}}{\Gamma,K(\Delta),\Delta\Rightarrow A}{\Gamma,K(\Delta)\Rightarrow KA}\,(KI_1)\quad\cfrac{\cfrac{\mathcal{D'}}{\Gamma',K(\Delta',A),\Delta',A\Rightarrow B}}{\Gamma',K(\Delta'),KA\Rightarrow KB}\,(KI_1)}{\Gamma,K(\Delta),\Gamma',K(\Delta')\Rightarrow KB}\,(Cut)\quad\rightsquigarrow$$

From $\Gamma',K(\Delta',A),\Delta',A\Rightarrow B$ by K-contraction (Corollary 8) we obtain $\Gamma',K(\Delta'),\Delta',A\Rightarrow B$ and then reduce the rank:

$$\rightsquigarrow\quad\cfrac{\cfrac{\mathcal{D}}{\Gamma,K(\Delta),\Delta\Rightarrow A}\quad\cfrac{\mathcal{D''}}{\Gamma',K(\Delta'),\Delta',A\Rightarrow B}}{\cfrac{\Gamma,K(\Delta),\Delta,\Gamma',K(\Delta'),\Delta'\Rightarrow B}{\Gamma,K(\Delta),\Gamma',K(\Delta')\Rightarrow KB}\,(KI_1)}\ (Cut)\ .$$

In remaining cases (when F has one of the forms $A\wedge B$, $A\vee B$ or $A\rightarrow B$) we follow [5].   □

Comment. Our formulation of the rule (KI_1) combines K-introduction with contraction. It is done in order to eliminate the contraction rule and to avoid the case of contraction in the proof of Lemma 9. But the contraction rule remains admissible and can be added as a ground rule too, so we can simplify the formulation of the K-introduction rule. It results in a "light" cut-free version $\mathsf{IEL_G}$:

Axioms

$$\Gamma, A \Rightarrow A, \qquad A \text{ is a variable or } \bot.$$

Rules

$$\frac{\Gamma, \Delta, \Delta \Rightarrow G}{\Gamma, \Delta \Rightarrow G} \, (C)$$

$$\frac{\Gamma, F, G \Rightarrow H}{\Gamma, F \wedge G \Rightarrow H} \, (\wedge \Rightarrow) \qquad\qquad \frac{\Gamma \Rightarrow F \quad \Gamma \Rightarrow G}{\Gamma \Rightarrow F \wedge G} \, (\Rightarrow \wedge)$$

$$\frac{\Gamma, F \Rightarrow H \quad \Gamma, G \Rightarrow H}{\Gamma, F \vee G \Rightarrow H} \, (\vee \Rightarrow) \qquad \frac{\Gamma \Rightarrow F_i}{\Gamma \Rightarrow F_1 \vee F_2} \, (\Rightarrow \vee)_i \quad (i = 1, 2)$$

$$\frac{\Gamma \Rightarrow F \quad \Gamma, G \Rightarrow H}{\Gamma, F \to G \Rightarrow H} \, (\to \Rightarrow) \qquad\qquad \frac{\Gamma, F \Rightarrow G}{\Gamma \Rightarrow F \to G} \, (\Rightarrow \to)$$

$$\frac{\Gamma_1, \Gamma_2 \Rightarrow F}{\Gamma_1, K(\Gamma_2) \Rightarrow KF} \, (KI) \qquad\qquad \frac{\Gamma \Rightarrow K\bot}{\Gamma \Rightarrow F} \, (U)$$

Lemma 10. $\mathsf{IEL_G} \vdash \Gamma \Rightarrow F$ *iff* $\mathsf{IEL_G^-} \vdash \Gamma \Rightarrow F$.

Proof. Part "only if". The rule (KI) is a particular case of (KI_{ext}), so all rules of $\mathsf{IEL_G}$ are admissible in $\mathsf{IEL_G^-}$ (Lemmas 6, 2 and Corollary 3).

   Part "if". All missing rules are derivable in $\mathsf{IEL_G}$:

$$\frac{\dfrac{\Gamma, F \to G \Rightarrow F \quad \Gamma, G \Rightarrow H}{\Gamma, F \to G, F \to G \Rightarrow H} \, (\to \Rightarrow)}{\Gamma, F \to G \Rightarrow H} \, (C) \quad , \quad \frac{\dfrac{\Gamma, K(\Delta), \Delta \Rightarrow F}{\Gamma, K(\Delta), K(\Delta) \Rightarrow KF} \, (KI)}{\Gamma, K(\Delta) \Rightarrow KF} \, (C) \; .$$

□

Theorem 11. (Cut) *is admissible in* $\mathsf{IEL_G}$.

Proof. Lemma 9 implies the similar statement for the calculus $\mathsf{IEL_G}$. Indeed, one can convert $\mathsf{IEL_G}$-derivations into $\mathsf{IEL_G^-}$-derivations, eliminate a single cut in $\mathsf{IEL_G^-}$, and then convert the cut-free $\mathsf{IEL_G^-}$-derivation backward (Lemma 10). The statement implies the theorem.   □

Theorem 12. *The following are equivalent:*

1. $\mathsf{IEL_G^0} \vdash \Gamma \Rightarrow F$.
2. $\mathsf{IEL_G^-} \vdash \Gamma \Rightarrow F$.
3. $\mathsf{IEL_G} \vdash \Gamma \Rightarrow F$.
4. $\mathsf{IEL} \vdash \wedge \Gamma \to F$.

Proof. 1. $\Leftrightarrow$ 2. All rules of $\mathsf{IEL}^0_\mathsf{G}$ are admissible in $\mathsf{IEL}^-_\mathsf{G}$ (Lemmas 2, 6, 7, 9) and vice versa.

The equivalence of 2. and 3. is proved in Lemma 10, the equivalence of 1. and 4. – see Theorem 1. $\square$

7   Complexity of IEL

We prove that IEL is PSPACE-complete. The lower bound follows from the same lower bound for the intuitionistic propositional logic. To prove the upper bound we show that polynomial space is sufficient for the proof search. Our proof search technique is based on monotone derivations and is similar to one used by S.C. Kleene in his G3 systems (see [6]).

Definition 13. For a multiset Γ let $set(\Gamma)$ be the set of all its members. An instance of a rule

$$\frac{\Gamma_1 \Rightarrow F_1 \ \ldots \ \Gamma_n \Rightarrow F_n}{\Gamma \Rightarrow F}$$

is *monotone* if $set(\Gamma) \subseteq \bigcap_i set(\Gamma_i)$. A derivation is called monotone if it uses monotone instances of inference rules only.

Consider the extension IEL'_G of the calculus $\mathsf{IEL}^-_\mathsf{G}$ by the following rules: the contraction rule (C) and

$$\frac{\Gamma, F \wedge G, F, G \Rightarrow H}{\Gamma, F \wedge G, F \Rightarrow H} \ (\wedge^C_1 \Rightarrow), \qquad \frac{\Gamma, F \wedge G, F, G \Rightarrow H}{\Gamma, F \wedge G, G \Rightarrow H} \ (\wedge^C_2 \Rightarrow),$$

$$\frac{\Gamma, F \wedge G, F, G \Rightarrow H}{\Gamma, F \wedge G \Rightarrow H} \ (\wedge^C \Rightarrow), \qquad \frac{\Gamma, F \vee G, F \Rightarrow H \quad \Gamma, F \vee G, G \Rightarrow H}{\Gamma, F \vee G \Rightarrow H} \ (\vee^C \Rightarrow),$$

$$\frac{\Gamma, F \Rightarrow G}{\Gamma, F \Rightarrow F \rightarrow G} \ (\Rightarrow \rightarrow^W), \qquad \frac{\Gamma, F \rightarrow G \Rightarrow F \quad \Gamma, F \rightarrow G, G \Rightarrow H}{\Gamma, F \rightarrow G \Rightarrow H} \ (\rightarrow^C \Rightarrow),$$

$$\frac{\Gamma, K(\Delta_1, \Delta_2), \Delta_1, \Delta_2 \Rightarrow F}{\Gamma, \Delta_1, K(\Delta_1, \Delta_2) \Rightarrow KF} \ (KI^W_1).$$

In (KI^W_1) we require that the multiset Γ, Δ_1 does not contain formulas of the form KG.

Lemma 14. $\mathsf{IEL}'_\mathsf{G} \vdash \Gamma \Rightarrow F$ *iff* $\mathsf{IEL}^-_\mathsf{G} \vdash \Gamma \Rightarrow F$.

Proof. All new rules are some combinations of corresponding ground rules with structural rules. The latter are admissible in $\mathsf{IEL}^-_\mathsf{G}$ (Lemmas 6, 2). $\square$

Lemma 15. *Any derivation in* IEL'_G *can be converted into a monotone derivation of the same sequent.*

Proof. Consider a derivation which is not monotone. Choose the first (top-down) non-monotone instance (R) of a rule in it. (R) introduces a new formula A in the antecedent of its conclusion which is not present in antecedents of some of its premises. Add a copy of A to the antecedent of the conclusion and to antecedents of all sequents above it. When A has the form KB and is added to the antecedent of the conclusion of some instance of rules (KI_1) or (KI_1^W) above (R), add a copy of B to the antecedent of the premise of this rule and to antecedents of all sequents above it. When B has the form KC, do the same further up the derivation with C, etc. Finally, insert the contraction rule after (R):

$$\frac{\mathcal{D}}{A, \Gamma \Rightarrow F}(R) \qquad \rightsquigarrow \qquad \frac{\dfrac{\mathcal{D}'}{A, A, \Gamma \Rightarrow F}(R)}{A, \Gamma \Rightarrow F}(C).$$

The result is also a correct derivation with one non-monotone instance eliminated. Repeat the transformation until the derivation becomes monotone.   □

Lemma 16. *A monotone derivation of a sequent* $\Gamma \Rightarrow F$ *in* IEL'_G *can be converted into a monotone derivation of the sequent* $set(\Gamma) \Rightarrow F$ *that contains only sequents of the form* $set(\Gamma') \Rightarrow F'$. *The transformation does not increase the depth of the proof.*

Proof. Given a monotone derivation replace all sequents $\Gamma' \Rightarrow F'$ in it with $set(\Gamma') \Rightarrow F'$. This transformation converts axioms into axioms. We claim that an instance of an inference rule will be converted either into some other instance of a rule of IEL'_G or some premise of the converted instance will coincide with its conclusion, so the rule can be removed from the resulting proof. The depth of the proof does not increase.

Indeed, instances of $(\Rightarrow \wedge)$, $(\Rightarrow \vee)$ and (U) will be converted into some other instances of the same rule. An instance of (C) will be converted into the trivial rule that can be removed:

$$\frac{\Gamma, \Delta, \Delta \Rightarrow G}{\Gamma, \Delta \Rightarrow G}(C) \qquad \rightsquigarrow \qquad \frac{set(\Gamma, \Delta) \Rightarrow G}{set(\Gamma, \Delta) \Rightarrow G} \qquad \rightsquigarrow \qquad \text{remove.}$$

The remaining cases. Let $k, l, m, n, k', l', m', n' \geq 0$ and $F^k = \underbrace{F, \ldots, F}_{k}$.

All monotone instances of $(\wedge \Rightarrow)$, $(\wedge_1^C \Rightarrow)$, $(\wedge_2^C \Rightarrow)$, $(\wedge^C \Rightarrow)$ have the form

$$\frac{\Gamma, (F \wedge G)^{k+1}, F^{l+1}, G^{m+1} \Rightarrow H}{\Gamma, (F \wedge G)^{k'+1}, F^{l'}, G^{m'} \Rightarrow H}.$$

Contractions in antecedents will give

$$\frac{\Gamma', F \wedge G, F, G \Rightarrow H}{\Gamma', F \wedge G \Rightarrow H} \ (\wedge^C \Rightarrow), \ l' = m' = 0,$$

$$\frac{\Gamma', F \wedge G, F, G \Rightarrow H}{\Gamma', F \wedge G, F \Rightarrow H} \ (\wedge_1^C \Rightarrow), \ l' > 0, m' = 0,$$

$$\frac{\Gamma', F \wedge G, F, G \Rightarrow H}{\Gamma', F \wedge G, G \Rightarrow H} \ (\wedge_2^C \Rightarrow), \ l' = 0, m' > 0,$$

trivial rule (removed), $l', m' > 0.$

All monotone instances of $(\vee \Rightarrow)$, $(\vee^C \Rightarrow)$ have the form

$$\frac{\Gamma, (F \vee G)^{k+1}, F^{l+1}, G^m \Rightarrow H \quad \Gamma, (F \vee G)^{k+1}, F^l, G^{m+1} \Rightarrow H}{\Gamma, (F \vee G)^{k'+1}, F^l, G^m \Rightarrow H}.$$

Contractions in antecedents will give

$$\frac{\Gamma', F \vee G, F \Rightarrow H \quad \Gamma, F \vee G, G \Rightarrow H}{\Gamma', F \vee G \Rightarrow H} \ (\vee^C \Rightarrow), \ l = m = 0,$$

trivial rule (removed), $l > 0$ or $m > 0.$

All monotone instances of $(\Rightarrow \rightarrow)$, $(\Rightarrow \rightarrow^W)$ have the form

$$\frac{\Gamma, F^{k+1} \Rightarrow G}{\Gamma, F^{k'} \Rightarrow F \rightarrow G}.$$

Contractions in antecedents will give

$$\frac{\Gamma', F \Rightarrow G}{\Gamma', F \Rightarrow F \rightarrow G} \ (\Rightarrow \rightarrow^W), \ k' > 0,$$

$$\frac{\Gamma', F \Rightarrow G}{\Gamma' \Rightarrow F \rightarrow G} \ (\Rightarrow \rightarrow), \quad k' = 0.$$

All monotone instances of $(\rightarrow \Rightarrow)$, $(\rightarrow^C \Rightarrow)$ have the form

$$\frac{\Gamma, (F \rightarrow G)^{k+1}, G^l \Rightarrow F \quad \Gamma, (F \rightarrow G)^{k'+1}, G^{l+1} \Rightarrow H}{\Gamma, (F \rightarrow G)^{k+1}, G^l \Rightarrow H}.$$

Contractions in antecedents will give

$$\frac{\Gamma', (F \rightarrow G) \Rightarrow F \quad \Gamma', (F \rightarrow G), G \Rightarrow H}{\Gamma', (F \rightarrow G) \Rightarrow H} \ (\rightarrow^C \Rightarrow), \ l = 0,$$

trivial rule (removed), $l > 0.$

All monotone instances of (KI_1), (KI_1^W) have the form

$$\frac{\Gamma, G_1^{k_1+1}, (KG_1)^{l_1+1}, \ldots, G_n^{k_n+1}, (KG_n)^{l_n+1} \Rightarrow F}{\Gamma, G_1^{k_1'}, (KG_1)^{l_1'+1}, \ldots, G_n^{k_n'}, (KG_n)^{l_n'+1} \Rightarrow KF}$$

Contractions in antecedents will give

$$\frac{\Gamma', G_1, KG_1, \ldots, G_n, KG_n \Rightarrow F}{\Gamma', KG_1, \ldots, KG_n \Rightarrow KF} \ (KI_1), \quad \text{when } k_1' = \ldots = k_n' = 0,$$

an instance of (KI_1^W),   when $k_i' = 0, k_j' > 0$ for some i, j,

trivial rule (removed),   when $k_1', \ldots, k_n' > 0.$

$\square$

Lemma 17. (Subformula Property). *Consider a derivation of a sequent* $\Gamma \Rightarrow F$ *in* $\mathsf{IEL_G^-}$, $\mathsf{IEL_G}$ *or* $\mathsf{IEL_G'}$. *Any sequent in it is composed of subformulas of some formulas from the multiset* $\Gamma, F, K\bot$.

Proof. For any rule of these calculi, its premises are composed of subformulas of formulas occurring in its conclusion and, possibly, of $K\bot$. $\square$

Definition 18. A monotone $\mathsf{IEL_G'}$-derivation of a sequent $set(\Gamma) \Rightarrow F$ is called minimal if it contains only sequents of the form $set(\Gamma') \Rightarrow F'$ and has the minimal depth.

The size of a sequent $F_1, \ldots, F_k \Rightarrow F$ is the sum of the lengths of all formulas F_i and F.

Lemma 19. *Let* $\mathcal{M}_n$ *be the set of all minimal* $\mathsf{IEL_G'}$-*derivations of sequents of size* n. *There exist polynomials* p *and* q *such that for any derivation* $\mathcal{D} \in \mathcal{M}_n$, *its depth is bounded by* $p(n)$ *and the sizes of all sequents in* $\mathcal{D}$ *do not exceed* $q(n)$.

Proof. Consider a derivation $\mathcal{D} \in \mathcal{M}_n$ and a path from the root to some leaf in it:

$$\Gamma_0 \Rightarrow F_0, \ldots, \Gamma_N \Rightarrow F_N.$$

All sequents in it are distinct from each other, all of them composed of subformulas of the first sequent, $\bot$ and $K\bot$ (Lemma 17), and $\Gamma_i \subseteq \Gamma_{i+1}$ holds for $i < N$.

Divide the path into maximal intervals with the same Γ_i inside. The length of such interval is bounded by the number of possible formulas F_i, which is $O(n)$. The number of intervals is $O(n)$ too, because it does not exceed the maximal length of a strictly monotone sequence $\Delta_0 \subset \Delta_1 \subset \ldots \subset \Delta_k$ of subsets of S where S is the set of all subformulas of the first sequent extended by $\bot$ and $K\bot$. So, $|S| = O(n)$ and $N = O(n^2)$.

Any sequent $\Gamma_i \Rightarrow F_i$ consists of at most $|S| + 1$ formulas of length $O(n)$, so its size is $O(n^2)$. $\square$

Corollary 20. *The set of all* $\mathsf{IEL'_G}$*-derivable sequents belongs to* PSPACE.

Proof. The result follows from the known game characterization AP = PSPACE ([7], see also [9] or [10]). We reproduce here the argument from [11] where Kleene's technique is used in a similar way.

Let p, q be the polynomials from Lemma 19. Consider the following two-person game with players (P) and (V). The initial configuration b_0 is a sequent of the form $set(\Gamma) \Rightarrow F$ of size n. Player (P) moves the first. He writes down one or two sequents of sizes less than $q(n)$ and his opponent (V) chooses one of them, and so on. The game is over after $p(n)$ moves of (V) or when (V) chooses a sequent that is an axiom of $\mathsf{IEL'_G}$.

Let w_i and b_i for $i > 0$ denote the moves of players (P) and (V) respectively, so $b_0, w_1, b_1, w_2, b_2, \dots$ is a run of the game. The winning conditions for (P) are:

1. For every move of (P) the figure $\dfrac{w_i}{b_{i-1}}$ is a monotone instance of some inference rule of $\mathsf{IEL'_G}$.
2. All sequents written by (P) have the form $set(\Delta) \Rightarrow G$.
3. At his last move (V) is forced to choose an axiom of $\mathsf{IEL'_G}$.

The number and the sizes of moves are bounded by polynomials and the winning conditions are polynomial-time decidable, so the set M of initial configurations that admit a winning strategy for (P) belongs to PSPACE (see [7]).

By Lemma 19, a sequent belongs to M iff it has a minimal derivation. But it follows from Lemmas 15, 16, 2, that a sequent $\Gamma \Rightarrow F$ is $\mathsf{IEL'_G}$-derivable iff $set(\Gamma) \Rightarrow F$ has a minimal derivation. Thus, the general derivability problem for $\mathsf{IEL'_G}$ belongs to PSPACE too.   □

Theorem 21. *The derivability problems for* $\mathsf{IEL^0_G}$, $\mathsf{IEL^-_G}$, $\mathsf{IEL_G}$, $\mathsf{IEL'_G}$ *and* IEL *are* PSPACE-*complete.*

Proof. The lower bound PSPACE follows from the same lower bound for intuitionistic propositional logic [8]. The upper bound PSPACE for $\mathsf{IEL'_G}$ is established in Corollary 20. It can be extended to other calculi by Theorem 12 and Lemma 14.   □

Acknowledgements. We are grateful to Sergey Artemov for inspiring discussions of intuitionistic epistemic logic. We thank the anonymous referees for their comments.

The research described in this paper was partially supported by Russian Foundation for Basic Research (grant 14-01-00127).

References

1. Artemov, S., Protopopescu, T.: Intuitionistic Epistemic Logic (2014). arXiv:1406.1582v2
2. Williamson, T.: On intuitionistic modal epistemic logic. J. Philos. Log. **21**(1), 63–89 (1992)

3. Fagin, R., Halpern, J.Y., Moses, Y., Vardi, M.Y.: Reasoning About Knowledge. The MIT Press, Cambridge (1995)
4. Maffezioli, V., Naibo, A., Negri, S.: The Church-Fitch knowability paradox in the light of structural proof theory. Synthese **190**, 2677–2716 (2013)
5. Troelstra, A.S., Schwichtenberg, H.: Basic Proof Theory. Cambridge University Press, Cambridge (1996)
6. Kleene, S.C.: Introduction to Metamathematics. North-Holland Publ. Co., Amsterdam (1952)
7. Chandra, A.K., Kozen, D.C., Stockmeyer, L.J.: Alternation. J. Assoc. Comput. Mach. **28**, 114–133 (1981)
8. Statman, R.: Intuitionistic propositional logic is polynomial-space complete. Theor. Comput. Sci. **9**, 67–72 (1979)
9. Krupski, V.N.: Introduction to Computational Complexity. Factorial Press, Moscow (2006). (In Russian)
10. Kitaev, A., Shen, A., Vyalyi, M.: Classical and Quantum Computations. MCCME-CheRo, Moscow (1999). (In Russian)
11. Krupski, N.: Typing in reflective combinatory logic. Ann. Pure Appl. Log. **141**, 243–256 (2006)

Interpolation Method for Multicomponent Sequent Calculi

Roman Kuznets[(✉)]

Institut Für Computersprachen, Technische Universität Wien, Vienna, Austria
roman@logic.at

Abstract. The proof-theoretic method of proving the Craig interpolation property was recently extended from sequents to nested sequents and hypersequents. There the notations were formalism-specific, obscuring the underlying common idea, which is presented here in a general form applicable also to other similar formalisms, e.g., prefixed tableaus. It describes requirements sufficient for using the method on a proof system for a logic, as well as additional requirements for certain types of rules. The applicability of the method, however, does not imply its success. We also provide examples from common proof systems to highlight various types of interpolant manipulations that can be employed by the method. The new results are the application of the method to a recent formalism of grafted hypersequents (in their tableau version), the general treatment of external structural rules, including the analytic cut, and the method's extension to the Lyndon interpolation property.

Keywords: Craig interpolation · Lyndon interpolation · Sequent calculi · Hypersequent calculi · Nested sequent calculi

1 Introduction

Along with decidability and compactness, the Craig interpolation property (CIP) is one of the principal properties desired of any logic. One way of demonstrating it by **constructing** the interpolant is the so-called proof-theoretic method, which relies on an analytic proof system for the logic. Until recently, the scope of the method has been limited to logics that can be captured by analytic sequent (equivalently, tableau) proof formalisms, as well as by display and resolution calculi, the discussion of which is outside the scope of this paper.

In [6], it was shown how to prove the CIP using nested sequents. In [9], written and accepted before this paper but likely to be published after it, the same principles were successfully applied to hypersequents. An anonymous reviewer of [9] noted that the nested and hypersequent cases are essentially the same. The purpose of this paper, which is based on a talk given at the Logic Colloquium 2015, is to present a general formal method in uniform notation, of which

R. Kuznets—Supported by the Austrian Science Fund (FWF) Lise Meitner grant M 1770-N25.

S. Artemov and A. Nerode (Eds.): LFCS 2016, LNCS 9537, pp. 202–218, 2016.
DOI: 10.1007/978-3-319-27683-0_15

both prior applications are instances. Note that the applicability of the method does not imply that the CIP can be proved using this method or, indeed, at all. For instance, the method is applicable to the hypersequent calculi for S4.3 from [7,8,11], but S4.3 does not have the CIP [13].

Let us first outline how general our method is intended to be. We concentrate on internal calculi, which excludes display and labelled sequent calculi. We consider only calculi whose basic unit is what we call a *multisequent*, i.e., an object that can be viewed as a hierarchy of *components* with each component *containing* a(n ordinary) sequent, called a *sequent component*. Since a sequent is essentially a single-component multisequent, the method for multisequents (partially) subsumes the well-known method for sequents. Following [4], we restrict the type of sequents used to the so-called *symmetric sequents*, which are best suited for interpolation proofs.

Definition 1 (Symmetric Sequents). *Rules of a symmetric sequent system operate on 2-sided sequents with formulas in negation normal form*[1] *in such a way that formulas never move between the* antecedent *and the* consequent.

The use of symmetric sequent systems is not required by the method (neither of [6,9] used them). Rather, they are a contrivance used to avoid splitting sequents, first introduced by Maehara in [12] written in Japanese. Splitting a 2-sided sequent results in a rather counter-intuitive 4-sided contraption, whereas splitting a 1-sided sequent is more or less isomorphic to working with a 2-sided symmetric sequent. However, the restriction to symmetric sequents has an unfortunate side-effect: it rules out the application of our method to subclassical logics, which lack De Morgan laws. For such logics, so far we have not been able to use split 2-sided sequents either.

In this paper, by the *common language* invoked by the more general formulations of the CIP, we understand common propositional variables, with all examples taken from propositional modal logics.

Thus, the proof-theoretic part of our recipe requires a description of a **propositional modal logic** by a **symmetric multisequent proof system**. There is one more necessary (for now) ingredient: the modal logic in question needs to have a **Kripke semantics** (with the standard local interpretation of $\wedge$ and $\vee$). Although the algorithm is designed using semantic reasoning (in cases of a successful application of the method), the final algorithm for computing interpolants makes no mention of semantics, remaining fully *internal*.

2   Sufficient Criteria of Applicability

As discussed in the previous section, we assume that we are given a propositional modal logic L described by a symmetric multisequent proof system **SL** and complete with respect to a class $\mathcal{C}_\mathsf{L}$ of Kripke models. The logic L is formulated

---

[1] In negation normal form, formulas are built from $\wedge$, $\vee$, $\top$, $\bot$, propositional variables p, and their negations $\overline{p}$; $\neg$ is defined via De Morgan laws; $\rightarrow$ is defined via $\neg$.

in a language $\mathcal{L}$ in negation normal form. Each multisequent is (can be viewed as) a hierarchy of components, each containing some *sequent component* $\Gamma \Rightarrow \Delta$, where the *antecedent* Γ and *consequent* Δ are multisets (sets, sequences) of $\mathcal{L}$-formulas that are called *antecedent* and *consequent formulas* respectively.

Definition 2 (Craig Interpolation Property). *A* logic L *has the* CIP *iff whenever* $L \vdash A \to B$, *there is an* interpolant $C \in \mathcal{L}$ *such that each propositional variable of* C *occurs in both* A *and* B *and such that* $L \vdash A \to C$ *and* $L \vdash C \to B$.[2]

Our method requires relationships among L, **SL**, and $\mathcal{C}_L$ stronger than completeness. The first requirement is the completeness of **SL** w.r.t. implications.

Definition 3 (Singleton Multisequent). *A singleton multisequent is a multisequent with exactly one component.*

Requirement I. *If* $L \vdash A \to B$, *then* **SL** $\vdash \mathcal{G}$ *for some singleton multisequent* $\mathcal{G}$ *with sequent component* $A \Rightarrow B$.

The second requirement is semantical completeness w.r.t. implications:

Definition 4 (Logical Consequence). *For sets (multisets, sequences)* Γ *and* Δ *of* $\mathcal{L}$-*formulas,* $\Gamma \vDash_{\mathcal{C}_L} \Delta$ *if, for each model* $\mathcal{M} \in \mathcal{C}_L$ *and each world* w *of* $\mathcal{M}$,

$$\mathcal{M}, w \Vdash A \text{ for each } A \in \Gamma \qquad \Longrightarrow \qquad \mathcal{M}, w \Vdash B \text{ for some } B \in \Delta \ .$$

Requirement II. *If* $A \vDash_{\mathcal{C}_L} B$, *then* $L \vdash A \to B$.

Formulating the next requirement requires preparation. The idea of our method is to consider maps f from the components of a given multisequent $\mathcal{G}$ to the worlds of a given model $\mathcal{M} \in \mathcal{C}_L$ and to evaluate formulas from a component α of $\mathcal{G}$ (i.e., formulas from the sequent component $\Gamma \Rightarrow \Delta$ contained at α) at the world $f(\alpha) \in \mathcal{M}$. To faithfully represent the component hierarchy peculiar to the multisequent system **SL**, however, we need to restrict these maps. For each multisequent type and each class of models considered, we require that the notion of *good map* be defined for each pair of a multisequent and a model. After we formulate what is needed from such maps, we give examples of good maps for nested sequents and hypersequents.

Remark 5. By a slight abuse of notation, we write $f \colon \mathcal{G} \to \mathcal{M}$ for a mapping from the components of $\mathcal{G}$ to the worlds of $\mathcal{M}$. In the same vein, we write $\alpha \in \mathcal{G}$ to state that α is a component of $\mathcal{G}$ and $w \in \mathcal{M}$ to state that w is a world in $\mathcal{M}$.

Requirement III. *If* **SL** $\vdash \mathcal{G}$, *then for each model* $\mathcal{M} \in \mathcal{C}_L$ *and for each good map* $f \colon \mathcal{G} \to \mathcal{M}$, *there exists a component* $\alpha \in \mathcal{G}$ *containing* $\Gamma \Rightarrow \Delta$ *such that*

$$\mathcal{M}, f(\alpha) \nVdash A \text{ for some } A \in \Gamma \qquad or \qquad \mathcal{M}, f(\alpha) \Vdash B \text{ for some } B \in \Delta \ .$$

---

[2] Here $D \to E$ means $\overline{D} \vee E$, where $\overline{D}$ is the defined negation of D.

In other words, we understand a multisequent as a multiworld disjunction of its sequent components and require this disjunction to be valid with respect to all good maps, which direct where each sequent component is to be evaluated.

Example 6 (Nested sequents). Nested sequents are often described as trees of sequents (and are sometimes called *tree hypersequents*). To transfer this tree hierarchy of components into models, we define *good maps* from a given nested sequent $\mathcal{G}$ to a given model $\mathcal{M} = (W, R, V)$ to be those that satisfy the following condition: **if β is a child of α in $\mathcal{G}$, then** $f(\alpha)Rf(\beta)$. It has been shown in [6] that Req. III is satisfied for all such maps.

Example 7 (Hypersequents). The standard formula interpretation of a hypersequent $\Gamma_1 \Rightarrow \Delta_1 \mid \cdots \mid \Gamma_n \Rightarrow \Delta_n$ as $\bigvee_{i=1}^{n} \Box(\bigwedge \Gamma_i \to \bigvee \Delta_i)$ suggests that good maps send all components to worlds accessible from a single root: good maps from a given hypersequent $\mathcal{G}$ to a given model $\mathcal{M} = (W, R, V)$ are all maps satisfying the following condition: **there exists $w \in W$ such that $wRf(\alpha)$ for all $\alpha \in \mathcal{G}$.** For some classes of models, this formulation can be simplified, e.g., if R is an equivalence relation for all models from $\mathcal{C}_\mathsf{L}$ (as in the case of S5), it is sufficient[3] to require that $f(\alpha)Rf(\beta)$ for all $\alpha, \beta \in \mathcal{G}$.

Remark 8. Note that good maps are defined on components rather than on sequent components. This means that we must notationally distinguish occurrences of the same sequent component. The linear notation for hypersequents masks the problems when hypersequents are sets or multisets of sequents. We assume that in any multisequent system there is a way of distinguishing sequent components and rely on this, but we do not specify the details, which could involve converting sets/multisets to sequences as the underlying data structure for multisequent components and adding appropriate exchange rules or using explicit labels for sequent components.

Requirement IV. *For each singleton multisequent with component α, each model $\mathcal{M} \in \mathcal{C}_\mathsf{L}$, and each world $w \in \mathcal{M}$, the map $\{(\alpha, w)\}$ must be a good map.*

3   Reducing the CIP to the Componentwise Interpolation

Our aim is to generalize the CIP to multiple components. In particular, interpolants are to be evaluated via good maps and, hence, cannot be mere formulas.

Definition 9 (Uniformula). *A uniformula is obtained from a multisequent $\mathcal{G}$ by replacing all sequent components in $\mathcal{G}$ with such multisets of formulas that the union of these multisets contains exactly one formula.*

---

[3] Despite the homogeneity of the components of hypersequents, maps can only be used unrestrictedly if the worlds of the model are completely homogeneous too, as in the case of the class of all models with $R = W \times W$, another class of models used for S5.

In other words, a uniformula is a single formula C placed at a particular component α of a given multisequent $\mathcal{G}$. We call C the *formula contained in* the uniformula and the component α the *active component of* the uniformula. Let $\mathcal{G}(\underbrace{\Gamma_1 \Rightarrow \Delta_1}_{\alpha_1}; \ldots; \underbrace{\Gamma_n \Rightarrow \Delta_n}_{\alpha_n})$ for $n \geq 0$ denote a multisequent with displayed components α_i containing sequents $\Gamma_i \Rightarrow \Delta_i$. By $\mathcal{G}^\circ(\alpha_1; \ldots; \alpha_n)$ we denote the result of removing all sequent components from $\mathcal{G}(\Gamma_1 \Rightarrow \Delta_1; \ldots; \Gamma_n \Rightarrow \Delta_n)$ but keeping its components with the hierarchy intact. Further, we allow to insert new objects, such as formulas, into a displayed component α_i. Thus, each uniformula has the form $\mathcal{G}^\circ(C)$ for some multisequent $\mathcal{G}(\Gamma \Rightarrow \Delta)$ and some formula C.

Definition 10 (Multiformula). *A multiformula $\mho$ with structure $\mho^\circ$ is defined as follows. Each uniformula $\mathcal{G}^\circ(A)$ is a multiformula with structure $\mathcal{G}^\circ(\alpha)$. If $\mho_1$ and $\mho_2$ are multiformulas with $\mho_1^\circ = \mho_2^\circ$, then $\mho_1 \oslash \mho_2$ and $\mho_1 \oslash \mho_2$ are also multiformulas with the same structure.*

To be able to formulate a generalized interpolation statement, we need to define a satisfaction relation between good maps and multiformulas, which are used as interpolants. For any two multisequents/multiformulas with the same structure, there is a unique way of transferring a good map from one onto the other.

Definition 11 ($\Vdash$ on Multiformulas). *Let $f \colon \mathcal{G}^\circ(\alpha) \to \mathcal{M}$. For a uniformula $\mathcal{G}^\circ(C)$, we say that $f \Vdash \mathcal{G}^\circ(C)$ iff $\mathcal{M}, f(\alpha) \Vdash C$.*

If $\mho_1 \oslash \mho_2$ ($\mho_1 \oslash \mho_2$) is defined, then $\mho_1^\circ = \mho_2^\circ$. Let f be a map from this structure to a model $\mathcal{M}$. $f \Vdash \mho_1 \oslash \mho_2$ ($f \Vdash \mho_1 \oslash \mho_2$) iff $f \Vdash \mho_i$ for some (each) $i = 1, 2$.

In other words, a uniformula is forced by a map if the formula contained in it is forced at the world to which the active component is mapped. The external conjunction $\oslash$ and disjunction $\oslash$ on multiformulas behave classically.

To define the Componentwise Interpolation Property, we use abbreviations:

Definition 12. *For a good map f from a multisequent $\mathcal{G}$ to a model $\mathcal{M}$, we write $f \vDash \mathsf{Ant}(\mathcal{G})$ if $\mathcal{M}, f(\alpha) \Vdash A$ for each component $\alpha \in \mathcal{G}$ and each antecedent formula A contained in α. We write $f \vDash \mathsf{Cons}(\mathcal{G})$ if $\mathcal{M}, f(\beta) \Vdash B$ for some component $\beta \in \mathcal{G}$ and some consequent formula B contained in β.*

Definition 13 (Componentwise Interpolation Property, or CWIP). *A multiformula $\mho$ is a (componentwise) interpolant of a multisequent $\mathcal{G}$, written $\mathcal{G} \longleftarrow \mho$, if $\mho^\circ = \mathcal{G}^\circ$ and the following two conditions hold:*

1. *if a propositional variable occurs in $\mho$, it must occur both in some antecedent formula of $\mathcal{G}$ and in some consequent formula of $\mathcal{G}$;*
2. *for each model $\mathcal{M} \in \mathcal{C}_\mathsf{L}$ and each good map $f \colon \mathcal{G} \to \mathcal{M}$,*

$$f \vDash \mathsf{Ant}(\mathcal{G}) \implies f \Vdash \mho \quad and \quad f \Vdash \mho \implies f \vDash \mathsf{Cons}(\mathcal{G}) . \quad (1)$$

*A multisequent proof system **SL** has the CWIP iff every derivable multisequent has an interpolant.*

The CIP can be reduced to the CWIP if Reqs. I–IV are satisfied. The proof of the reduction requires another small piece of notation.

Definition 14 (Componentwise Equivalence). *Multiformulas $\mho_1$ and $\mho_2$ are called* componentwise equivalent, *written $\mho_1 \equiv \mho_2$, provided $\mho_1^\circ = \mho_2^\circ$ and $f \Vdash \mho_1 \Longleftrightarrow f \Vdash \mho_2$ for any good map f on the common structure of $\mho_1$ and $\mho_2$.*

Remark 15. The classical reading of $\oslash$ and $\oslash$ implies that each multiformula can be transformed to a componentwise equivalent multiformula both in the DNF and in the CNF. This will be used for some of the rules in the following section.

Lemma 16. *For singleton sequents, multiformulas and uniformulas are equi-expressive, i.e., for each multiformula $\mho$ with a structure $\mathcal{G}^\circ(\alpha)$ where α is the only component, there exists a uniformula $\mathcal{G}^\circ(C)$ such that $\mathcal{G}^\circ(C) \equiv \mho$ and it has the same propositional variables as $\mho$.*

Proof. By induction on the construction of $\mho$. The case when $\mho$ is a uniformula is trivial. Let $\mho_1 \equiv \mathcal{G}^\circ(C_1)$ and $\mho_2 \equiv \mathcal{G}^\circ(C_2)$ for some structure $\mathcal{G}^\circ(\alpha)$. Then it is easy to see that $\mho_1 \oslash \mho_2 \equiv \mathcal{G}^\circ(C_1 \vee C_2)$ and $\mho_1 \oslash \mho_2 \equiv \mathcal{G}^\circ(C_1 \wedge C_2)$ and that the condition on propositional variables is also satisfied. □

Theorem 17 (Reduction of CIP to CWIP). *Let a logic L, a multisequent proof system* **SL***, and a class of Kripke models $\mathcal{C}_L$ satisfy all Reqs. I–IV. If* **SL** *enjoys the CWIP, then* L *enjoys the CIP.*

Proof. Assume that **SL** satisfies the CWIP and that $L \vdash A \to B$. Then, by Req. I, $\mathbf{SL} \vdash \mathcal{G}(A \Rightarrow B)$ for some singleton multisequent $\mathcal{G}(A \Rightarrow B)$, which has a componentwise interpolant $\mho$ by the CWIP. By Lemma 16, $\mathcal{G}(A \Rightarrow B) \longleftarrow \mathcal{G}^\circ(C)$ for some uniformula $\mathcal{G}^\circ(C)$. Since A is the only antecedent and B is the only consequent formula of $\mathcal{G}(A \Rightarrow B)$, each propositional variable of C must occur in both A and B. For any model $\mathcal{M} \in \mathcal{C}_L$ and any world $w \in \mathcal{M}$, by Req. IV, $f := \{(\alpha, w)\}$ is a good map on $\mathcal{G}(A \Rightarrow B)$. In particular, $f \vDash \mathsf{Ant}(\mathcal{G}(A \Rightarrow B))$ implies $f \Vdash \mathcal{G}^\circ(C)$, i.e., $\mathcal{M}, w \Vdash A$ implies $\mathcal{M}, w \Vdash C$. Given the arbitrariness of $\mathcal{M}$ and w, we conclude that $A \vDash_{\mathcal{C}_L} C$. It now follows from Req. II that $L \vdash A \to C$. The proof of $L \vdash C \to B$ is analogous. □

Remark 18. An attentive reader would notice the absence of Req. III, the most complex one, from the proof of Theorem 17. While the reduction does not rely on Req. III, its violation renders the reduction vacuous by denying the possibility of the CWIP for **SL**. Indeed, if Req. III is violated, i.e., $\mathbf{SL} \vdash \mathcal{G}$ and $f \vDash \mathsf{Ant}(\mathcal{G})$ but $f \nvDash \mathsf{Cons}(\mathcal{G})$ for some $\mathcal{M} \in \mathcal{C}_L$ and some good map $f: \mathcal{G} \to \mathcal{M}$, then no multiformula $\mho$ could satisfy (1) for this f.

4   Demonstrating the CWIP

In this section, strategies for proving the CWIP for various types of multisequent rules are described. For many common types of rules, a general (but not universal) recipe for handling them is presented. Thus, every statement in this section is implicitly prefaced by the qualifier "normally".

Initial Sequents are interpolated by uniformulas. It is easy to see that the following are interpolants for the most popular initial sequents:

$$\mathcal{G}(\Gamma, A \Rightarrow A, \Delta) \longleftarrow \mathcal{G}^{\circ}(A) \qquad \mathcal{G}(\Gamma \Rightarrow A, \overline{A}, \Delta) \longleftarrow \mathcal{G}^{\circ}(\top)$$

$$\mathcal{G}(\Gamma, \overline{A} \Rightarrow \overline{A}, \Delta) \longleftarrow \mathcal{G}^{\circ}(\overline{A}) \qquad \mathcal{G}(\Gamma, \bot \Rightarrow \Delta) \longleftarrow \mathcal{G}^{\circ}(\bot)$$

$$\mathcal{G}(\Gamma, A, \overline{A} \Rightarrow \Delta) \longleftarrow \mathcal{G}^{\circ}(\bot) \qquad \mathcal{G}(\Gamma \Rightarrow \top, \Delta) \longleftarrow \mathcal{G}^{\circ}(\top)$$

Single-Premise Local Rules. By a *local* rule we mean a rule that does not affect the components and affects the sequent components mildly enough to use the same map for the conclusion and the premise(s) (cf. *component-shifting rules* on p. 13). (Normally,) single-premise local rules require no change to the interpolant. We formulate sufficient criteria for reusing the interpolant and then list common rules satisfying them.

Lemma 19. *Consider a single-premise rule* $\dfrac{\mathcal{G}}{\mathcal{H}}$ *such that* $\mathcal{G}^{\circ} = \mathcal{H}^{\circ}$ *and such that no antecedent and no consequent propositional variable from* $\mathcal{G}$ *disappears in* $\mathcal{H}$. *If for any good map* f *on the common structure of* $\mathcal{G}$ *and* $\mathcal{H}$,

$$f \vDash \mathsf{Ant}(\mathcal{H}) \implies f \vDash \mathsf{Ant}(\mathcal{G}) \qquad and \qquad f \vDash \mathsf{Cons}(\mathcal{G}) \implies f \vDash \mathsf{Cons}(\mathcal{H}) \ ,$$

then $\mathcal{H} \longleftarrow \mho$ *whenever* $\mathcal{G} \longleftarrow \mho$.

Proof. Follows directly from the definition of componentwise interpolation.   □

This almost trivial observation captures most of the common single-premise propositional rules, both logical and internal structural. We only provide a non-exhaustive list, leaving the proof to the reader: internal weakening IW, internal contraction IC, internal exchange IEx, and both internal-context sharing and splitting versions of the left conjunction and right disjunction rules; some modal rules can also be treated this way: e.g., the multisequent T rules for reflexive models or the multisequent (local) D rules for serial models; an example of such a rule with multiple active components is the hypersequent rule $\square\mathsf{L}^{\mathsf{s}}$ from [14] and its symmetric version $\Diamond\mathsf{L}^{\mathsf{s}}$ for equivalence models with good maps from Example 7 (see Fig. 1). The variants of these logical rules with embedded internal contraction are also local.

Multi-premise Local Rules are those for which any good map on the conclusion can be applied to any of the premises. It follows directly from the definition of CWIP:

Lemma 20 (Conjunctive Rules). *Consider a rule* $\dfrac{\mathcal{G}_1 \quad \cdots \quad \mathcal{G}_n}{\mathcal{H}}$ *such that* $\mathcal{G}_1^{\circ} = \cdots = \mathcal{G}_n^{\circ} = \mathcal{H}^{\circ}$ *and such that no antecedent and no consequent propositional variable from any* $\mathcal{G}_i$ *disappears in* $\mathcal{H}$. *If for any good map* f *on the common structure of* $\mathcal{G}_i$'s *and* $\mathcal{H}$,

$$f \vDash \mathsf{Ant}(\mathcal{H}) \implies (\forall i)\big(f \vDash \mathsf{Ant}(\mathcal{G}_i)\big) \quad and \quad (\forall i)\big(f \vDash \mathsf{Cons}(\mathcal{G}_i)\big) \implies f \vDash \mathsf{Cons}(\mathcal{H}),$$

then $\mathcal{H} \longleftarrow \mho_1 \otimes \ldots \otimes \mho_n$ *whenever* $\mathcal{G}_i \longleftarrow \mho_i$ *for each* $i = 1, \ldots, n$.

$$\frac{\mathcal{G}(\Gamma \Rightarrow \Delta)}{\mathcal{G}(\Gamma, \Pi \Rightarrow \Sigma, \Delta)} \; \mathsf{IW} \qquad \frac{\mathcal{G}(\Gamma, \Pi, \Pi \Rightarrow \Sigma, \Sigma, \Delta)}{\mathcal{G}(\Gamma, \Pi \Rightarrow \Sigma, \Delta)} \; \mathsf{IC} \qquad \frac{\mathcal{G}(\Gamma, A_i \Rightarrow \Delta)}{\mathcal{G}(\Gamma, A_1 \wedge A_2 \Rightarrow \Delta)} \; \wedge \Rightarrow$$

$$\frac{\mathcal{G}(\Gamma, A_1, A_2 \Rightarrow \Delta)}{\mathcal{G}(\Gamma, A_1 \wedge A_2 \Rightarrow \Delta)} \; \wedge \Rightarrow \qquad \frac{\mathcal{G}(\Gamma \Rightarrow A_i, \Delta)}{\mathcal{G}(\Gamma \Rightarrow A_1 \vee A_2, \Delta)} \Rightarrow \vee \qquad \frac{\mathcal{G}(\Gamma \Rightarrow A_1, A_2, \Delta)}{\mathcal{G}(\Gamma \Rightarrow A_1 \vee A_2, \Delta)} \Rightarrow \vee$$

$$\frac{\mathcal{G}(\Gamma, A \Rightarrow \Delta)}{\mathcal{G}(\Gamma, \Box A \Rightarrow \Delta)} \; \mathsf{T} \Rightarrow \qquad \frac{\mathcal{G}(\Gamma \Rightarrow A, \Delta)}{\mathcal{G}(\Gamma \Rightarrow \Diamond A, \Delta)} \Rightarrow \mathsf{T} \qquad \frac{\mathcal{G}(\overbrace{\Gamma, \Box A \Rightarrow \Delta}^{\alpha}; \; \overbrace{\Pi, A \Rightarrow \Sigma}^{\beta})}{\mathcal{G}(\underbrace{\Gamma, \Box A \Rightarrow \Delta}_{\alpha}; \; \underbrace{\Pi \Rightarrow \Sigma}_{\beta})} \; \Box \mathsf{L}^{\mathsf{s}}$$

$$\frac{\mathcal{G}(\Gamma, \Diamond A \Rightarrow \Delta)}{\mathcal{G}(\Gamma, \Box A \Rightarrow \Delta)} \; \mathsf{D} \Rightarrow \mathsf{loc} \qquad \frac{\mathcal{G}(\Gamma \Rightarrow \Box A, \Delta)}{\mathcal{G}(\Gamma \Rightarrow \Diamond A, \Delta)} \Rightarrow \mathsf{D \; loc} \qquad \frac{\mathcal{G}(\overbrace{\Gamma \Rightarrow \Diamond A, \Delta}^{\alpha}; \; \overbrace{\Pi \Rightarrow A, \Sigma}^{\beta})}{\mathcal{G}(\underbrace{\Gamma \Rightarrow \Diamond A, \Delta}_{\alpha}; \; \underbrace{\Pi \Rightarrow \Sigma}_{\beta})} \; \Diamond \mathsf{L}^{\mathsf{s}}$$

Fig. 1. Rules not requiring changes to the interpolant by Lemma 19

Lemma 21 (Disjunctive Rules). *Consider a rule* $\dfrac{\mathcal{G}_1 \; \cdots \; \mathcal{G}_n}{\mathcal{H}}$ *such that* $\mathcal{H}^{\circ} = \mathcal{G}_1^{\circ} = \cdots = \mathcal{G}_n^{\circ}$ *and such that no antecedent and no consequent propositional variable from any* $\mathcal{G}_i$ *disappears in* $\mathcal{H}$. *If for any good map* f *on the common structure of* $\mathcal{G}_i$*'s and* $\mathcal{H}$,

$$f \vDash \mathsf{Ant}(\mathcal{H}) \implies (\exists i)\big(f \vDash \mathsf{Ant}(\mathcal{G}_i)\big) \quad and \quad (\exists i)\big(f \vDash \mathsf{Cons}(\mathcal{G}_i)\big) \implies f \vDash \mathsf{Cons}(\mathcal{H}) \; ,$$

then $\mathcal{H} \longleftarrow \mho_1 \oslash \ldots \oslash \mho_n$ *whenever* $\mathcal{G}_i \longleftarrow \mho_i$ *for each* $i = 1, \ldots, n$.

The remaining propositional rules fall under the scope of these two lemmas: it is easy to see that both the internal-context splitting and sharing versions of the right conjunction rule $\Rightarrow \wedge$ (see Fig. 2) are conjunctive and both versions of the left disjunction rule $\vee \Rightarrow$ (see Fig. 3) are disjunctive rules in this sense.

$$\frac{\mathcal{G}(\Gamma \Rightarrow A, \Delta) \quad \mathcal{G}(\Gamma \Rightarrow B, \Delta)}{\mathcal{G}(\Gamma \Rightarrow A \wedge B, \Delta)} \Rightarrow \wedge \qquad \frac{\mathcal{G}(\Gamma_1 \Rightarrow A, \Delta_1) \quad \mathcal{G}(\Gamma_2 \Rightarrow B, \Delta_2)}{\mathcal{G}(\Gamma_1, \Gamma_2 \Rightarrow A \wedge B, \Delta_1, \Delta_2)} \Rightarrow \wedge$$

Fig. 2. Propositional conjunctive rules in the sense of Lemma 20

Analytic Cut. Another common local rule is cut. While the general cut rule is problematic even in the sequent case, it is well known that analytic cuts can be handled (see [4]). To extend this handling to the external-context sharing and internal-context splitting cuts on multisequents (see Fig. 4), we impose a condition that is both stronger and weaker than analyticity. While A need not

$$\frac{\mathcal{G}(\Gamma, A \Rightarrow \Delta) \quad \mathcal{G}(\Gamma, B \Rightarrow \Delta)}{\mathcal{G}(\Gamma, A \vee B \Rightarrow \Delta)} \vee \Rightarrow \qquad \frac{\mathcal{G}(\Gamma_1, A \Rightarrow \Delta_1) \quad \mathcal{G}(\Gamma_2, B \Rightarrow \Delta_2)}{\mathcal{G}(\Gamma_1, \Gamma_2, A \vee B \Rightarrow \Delta_1, \Delta_2)} \vee \Rightarrow$$

Fig. 3. Propositional disjunctive rules in the sense of Lemma 21

appear as a subformula in the conclusion as long as all its propositional variables occur there, these propositional variables must be on the same side of $\Rightarrow$ as in A or $\overline{A}$ displayed in the premises. This condition is necessary to use the interpolants of both premises and the formula $\overline{A}$ in constructing the interpolant for the conclusion.

Lemma 22 (Analytic Cut). *For the cut rules from Fig. 4, if no antecedent and no consequent propositional variable from any premise disappears in the conclusion, then* (Cut $\Rightarrow$) *and* ($\Rightarrow$ Cut) *are a disjunctive and conjunctive rule respectively and can be treated according to Lemmas 21 or 20 respectively. For the* (C $\overset{u}{\Rightarrow}$ t) *rule, we have* $\mathcal{G}(\Gamma_1, \Gamma_2 \Rightarrow \Delta_1, \Delta_2) \longleftarrow \mho_1 \oslash \left(\mathcal{G}^\circ(\overline{A}) \oslash \mho_2 \right)$ *whenever* $\mathcal{G}(\Gamma_1 \Rightarrow A, \Delta_1) \longleftarrow \mho_1$ *and* $\mathcal{G}(\Gamma_2, A \Rightarrow \Delta_2) \longleftarrow \mho_2$.

Proof. The common language requirement is clearly satisfied. Consider an arbitrary good map f from the common structure of the premises and the conclusion of (C $\overset{u}{\Rightarrow}$ t) to some $\mathcal{M} \in \mathcal{C}_\mathsf{L}$. Assume first that $f \vDash \mathsf{Ant}(\mathcal{G}(\Gamma_1, \Gamma_2 \Rightarrow \Delta_1, \Delta_2))$. It is immediate that $f \vDash \mathsf{Ant}(\mathcal{G}(\Gamma_1 \Rightarrow A, \Delta_1))$ and, hence, $f \Vdash \mho_1$. Further, either $\mathcal{M}, f(\alpha) \Vdash A$ or $\mathcal{M}, f(\alpha) \nVdash A$ for the active component α. In the latter case, $f \Vdash \mathcal{G}^\circ(\overline{A})$.[4] In the former case, $f \vDash \mathsf{Ant}(\mathcal{G}(\Gamma_2, A \Rightarrow \Delta_2))$ implying $f \Vdash \mho_2$. In either case, $f \Vdash \mathcal{G}^\circ(\overline{A}) \oslash \mho_2$ for the second conjunct of the proposed interpolant.

Assume now that $f \Vdash \mho_1 \oslash \left(\mathcal{G}^\circ(\overline{A}) \oslash \mho_2 \right)$. It follows from $f \Vdash \mho_1$ that $f \vDash \mathsf{Cons}(\mathcal{G}(\Gamma_1 \Rightarrow A, \Delta_1))$. If one of the forced formulas is not the displayed A, then $f \vDash \mathsf{Cons}(\mathcal{G}(\Gamma_1, \Gamma_2 \Rightarrow \Delta_1, \Delta_2))$, which is the desired result. Otherwise, we have $\mathcal{M}, f(\alpha) \Vdash A$. In this case, $f \nVdash \mathcal{G}^\circ(\overline{A})$ implying $f \Vdash \mho_2$. This, in turn, implies $f \vDash \mathsf{Cons}(\mathcal{G}(\Gamma_2, A \Rightarrow \Delta_2))$ and $f \vDash \mathsf{Cons}(\mathcal{G}(\Gamma_1, \Gamma_2 \Rightarrow \Delta_1, \Delta_2))$ again. □

Remark 23. Lemma 22 also applies to one-to-one multicut rules allowing multiple copies of the cut formula in both premises in Fig. 4.

External Structural Rules. From now on, interpolants for most rules rely on the specifics of goodness conditions. The guiding principle is that **any good map on the conclusion of the rule needs to be transformed in some natural and general way into a good map on the premise(s)**. We start with rule types that are reasonably common across various sequent types: external weakening EW, external contraction EC, external mix, and external exchange EEx.

---

[4] We assume the standard semantics, i.e., that exactly one of A or $\overline{A}$ holds at a world.

$$\frac{\mathcal{G}(\Gamma_1, A \Rightarrow \Delta_1) \quad \mathcal{G}(\Gamma_2, \overline{A} \Rightarrow \Delta_2)}{\mathcal{G}(\Gamma_1, \Gamma_2 \Rightarrow \Delta_1, \Delta_2)} \text{ Cut} \Rightarrow \qquad \frac{\mathcal{G}(\Gamma_1 \Rightarrow A, \Delta_1) \quad \mathcal{G}(\Gamma_2 \Rightarrow \overline{A}, \Delta_2)}{\mathcal{G}(\Gamma_1, \Gamma_2 \Rightarrow \Delta_1, \Delta_2)} \Rightarrow \text{Cut}$$

$$\frac{\mathcal{G}(\Gamma_1 \Rightarrow A, \Delta_1) \quad \mathcal{G}(\Gamma_2, A \Rightarrow \Delta_2)}{\mathcal{G}(\Gamma_1, \Gamma_2 \Rightarrow \Delta_1, \Delta_2)} \text{ C} \overset{u}{\Rightarrow} \text{t}$$

Fig. 4. Cut rules

$$\frac{\mathcal{G}()}{\mathcal{G}(\underbrace{\Gamma_1 \Rightarrow \Delta_1}_{\alpha_1}; \ldots; \underbrace{\Gamma_n \Rightarrow \Delta_n}_{\alpha_n})} \text{ EW} \qquad \frac{\mathcal{G}(\overbrace{\Gamma_1 \Rightarrow \Delta_1}^{\alpha_1}; \overbrace{\Pi_1 \Rightarrow \Sigma_1}^{\beta_1}; \ldots; \overbrace{\Gamma_n \Rightarrow \Delta_n}^{\alpha_n}; \overbrace{\Pi_n \Rightarrow \Sigma_n}^{\beta_n})}{\mathcal{G}(\underbrace{\Gamma_1, \Pi_1 \Rightarrow \Delta_1, \Sigma_1}_{\alpha_1}; \ldots; \underbrace{\Gamma_n, \Pi_n \Rightarrow \Delta_n, \Sigma_n}_{\alpha_n})} \text{ mix}$$

Fig. 5. External structural rules EW and mix

External weakening. By external weakening rules we understand rules EW from Fig. 5 where the conclusion is obtained by adding new components in such a way that all the sequent components already present in the premise, along with the hierarchical relationships among their components, remain intact.

Requirement V (For EW). *For each instance of* EW *from Fig. 5 and each good map f on its conclusion, the restriction $f \upharpoonright \mathcal{G}()$ of f onto the components of $\mathcal{G}()$ must be a good map on the premise.*

Lemma 24 (External Weakening). *Let $\mathcal{G}() \longleftarrow \mho$ for an instance of* EW *from Fig. 5 and $\mho'$ be the result of adding empty components $\alpha_1, \ldots, \alpha_n$ to each uniformula in $\mho$ in the same way they are added in the rule. Then Req. V implies $\mathcal{G}(\Gamma_1 \Rightarrow \Delta_1; \ldots; \Gamma_n \Rightarrow \Delta_n) \longleftarrow \mho'$.*

Proof. For a good map $f: \mathcal{G}(\Gamma_1 \Rightarrow \Delta_1; \ldots; \Gamma_n \Rightarrow \Delta_n) \to \mathcal{M}$, the map $f \upharpoonright \mathcal{G}()$ is good by Req. V. If $f \vDash \mathsf{Ant}(\mathcal{G}(\Gamma_1 \Rightarrow \Delta_1; \ldots; \Gamma_n \Rightarrow \Delta_n))$, then $f \upharpoonright \mathcal{G}() \vDash \mathsf{Ant}(\mathcal{G}())$. Thus, $f \upharpoonright \mathcal{G}() \Vdash \mho$. It is easy to show by induction on the construction of $\mho$ that $f \Vdash \mho'$ iff $f \upharpoonright \mathcal{G}() \Vdash \mho$. Thus, $f \Vdash \mho'$. The argument for the consequents is similar. The common language condition is also clearly fulfilled. $\qquad \square$

The external weakening rules of both hypersequents and nested sequents are covered by this lemma w.r.t. good maps from Examples 7 and 6 respectively.

Example 25. Consider symmetric nested sequents written in a hybrid Brünnler–Poggiolesi notation (a similar notation has been used in [5]). By Lemma 24,

$$\frac{B \Rightarrow B, [A \Rightarrow A] \quad \longleftarrow \quad [A] \oslash (B, [\,])}{B \Rightarrow B, [A \Rightarrow A, [C \Rightarrow D]], [E \Rightarrow F] \quad \longleftarrow \quad ([A, [\,], [\,]) \oslash (B, [[\,]], [\,])} \text{ EW} \; .$$

Mix and external contraction rules. By mix rules we understand rules mix from Fig. 5 where the conclusion is obtained by transferring all antecedent and consequent formulas contained in each β_i, $i = 1, \ldots, n$, to the antecedent and consequent respectively of α_i and removing the emptied components β_i.

Requirement VI (For mix **and** EC**).** *For each instance of* mix *from Fig. 5 and each good map f on its conclusion, $f[\alpha \Rightarrow \beta] := f \cup \{(\beta_1, f(\alpha_1)), \ldots, (\beta_n, f(\alpha_n))\}$ must be a good map on the premise.*

Lemma 26 (Mix). *Let $\mathcal{G}(\Gamma_1 \Rightarrow \Delta_1; \Pi_1 \Rightarrow \Sigma_1; \ldots; \Gamma_n \Rightarrow \Delta_n; \Pi_n \Rightarrow \Sigma_n) \longleftarrow \mho$ for an instance of* mix *from Fig. 5. Let $\mho'$ be the result of moving each formula within each β_i to α_i, leaving formulas contained in components other than β_i intact, and removing the emptied β_i's from each uniformula in $\mho$. Then $\mathcal{G}(\Gamma_1, \Pi_1 \Rightarrow \Delta_1, \Sigma_1; \ldots; \Gamma_n, \Pi_n \Rightarrow \Delta_n, \Sigma_n) \longleftarrow \mho'$ whenever Req. VI is fulfilled.*

Proof. If $f \colon \mathcal{G}(\Gamma_1, \Pi_1 \Rightarrow \Delta_1, \Sigma_1; \ldots; \Gamma_n, \Pi_n \Rightarrow \Delta_n, \Sigma_n) \to \mathcal{M}$ is good, so is $f[\alpha \Rightarrow \beta]$ by Req. VI. $f \vDash \mathsf{Ant}\big(\mathcal{G}(\Gamma_1, \Pi_1 \Rightarrow \Delta_1, \Sigma_1; \ldots; \Gamma_n, \Pi_n \Rightarrow \Delta_n, \Sigma_n)\big)$ implies $f[\alpha \Rightarrow \beta] \vDash \mathsf{Ant}\big(\mathcal{G}(\Gamma_1 \Rightarrow \Delta_1; \ldots; \Gamma_n \Rightarrow \Delta_n; \quad \Pi_1 \Rightarrow \Sigma_1; \ldots; \Pi_n \Rightarrow \Sigma_n)\big)$ because formulas from each Π_i are evaluated at $f[\alpha \Rightarrow \beta](\beta_i) = f(\alpha_i)$, same as in f. Thus, $f[\alpha \Rightarrow \beta] \Vdash \mho$. It is easy to show by induction on the construction of $\mho$ that $f \Vdash \mho'$ iff $f[\alpha \Rightarrow \beta] \Vdash \mho$. Thus, $f \Vdash \mho'$. The argument for the consequents is similar. The common language condition is also fulfilled. $\qquad\square$

For set-based sequent components, the external contraction EC is simply an instance of mix with $\Gamma_i = \Pi_i$ and $\Delta_i = \Sigma_i$ for each $i = 1, \ldots, n$. For multiset- and sequence-based ones, EC can be obtained from mix by internal contraction IC and internal exchange IEx. Since the definition of CWIP is not sensitive to multiplicities of formulas or their positions within the antecedent (consequent), Lemma 26 is equally applicable to EC (cf. also the application of Lemma 19 to IC and IEx).

Remark 27. Requirement VI does not yet guarantee that mix from Fig. 5 is a proper mix rule or that its variant with $\Gamma_i = \Pi_i$ and $\Delta_i = \Sigma_i$ is a proper contraction rule: that requires the α-components to have the same hierarchical relations as the β-components, both among themselves and as related to the rest of the multisequent. But this is not a problem of interpolation.

The external contraction rules of both hypersequents and nested sequents are covered by Lemma 26 w.r.t. good maps from Examples 7 and 6 respectively.

Example 28. An example of a nontrivial mix rule is *medial* from [2], represented here in the original nested-sequent notation: $\dfrac{\Gamma\{[\Delta_1], [\Delta_2]\}}{\Gamma\{[\Delta_1, \Delta_2]\}}$ med, with brackets used to represent the tree structure on the components. Thus, the root component of Δ_1 is mixed with that of Δ_2 and each child component of either root becomes a child of the mixed component. Below we present an example of the use of Lemma 26, where $C \Rightarrow D$ is mixed with $A \Rightarrow A$:

$$\frac{\Rightarrow, [C \Rightarrow D, [B \Rightarrow B]], [A \Rightarrow A, [\Rightarrow F]] \longleftarrow ([[]], [A, []]) \otimes ([[B]], [[]])}{\Rightarrow, [C, A \Rightarrow D, A, [B \Rightarrow B], [\Rightarrow F]] \longleftarrow ([A, [], []]) \otimes ([[B], []])} \text{ med} .$$

Clearly, Req. VI is satisfied for med w.r.t. the good maps from Example 6.

$$\frac{\mathcal{G}(\overbrace{\Gamma \Rightarrow \Delta}^{\alpha}; \overbrace{A \Rightarrow}^{\beta})}{\underbrace{\mathcal{G}(\Gamma, \Diamond A \Rightarrow \Delta)}_{\alpha}} \, \mathsf{K} \Rightarrow \qquad \frac{\mathcal{G}(\overbrace{\Gamma \Rightarrow \Delta}^{\alpha}; \overbrace{\Rightarrow A}^{\beta})}{\underbrace{\mathcal{G}(\Gamma \Rightarrow \Box A, \Delta)}_{\alpha}} \Rightarrow \mathsf{K}$$

Fig. 6. Modal K rules

External Exchange. These are the rules that change the structure of the multise-quent without changing a single sequent component. For them, it is sufficient to change the structures of each uniformula in the interpolant in the same way. It is required that good maps on the conclusion could be transferred to the premise without changing where each formula is evaluated.

Component-Removing Rules are modal rules that remove a component from the premise multisequent. Such rules can be highly logic-specific. We consider two most common ones that rely on the connection between the modality and the Kripke semantics and are likely to be present in one form or another in virtually every multisequent system. For these rules, the argument is almost the same as the one given in [6] for nested sequents. Hence, we only provide the proof for one. It should be noted that, to the best of our knowledge, these rules require the interpolant of the premise to be in the DNF or CNF, depending on the rule. We have not been able to extend the construction to arbitrary interpolants. For both rules in Fig. 6, the conclusion is obtained by removing the component β with a single formula and transferring this formula, prefaced with an appropriate modality, to the component α (as usual, copying the modalized formula to the premise makes no difference).

Requirement VII (For K Rules). *For each instance of each rule from Fig. 6 and each good map f from its conclusion to a model $\mathcal{M} = (W, R, V)$, it is required that $f \cup \{(\beta, w)\}$ be a good map on the premise of the rule whenever $f(\alpha)Rw$.*

Lemma 29 (K Rules). *Consider an instance of* (K $\Rightarrow$) *from Fig. 6 and let*

$$\mathcal{G}(\Gamma \Rightarrow \Delta; A \Rightarrow) \longleftarrow \bigotimes_{i=1}^{n} \left(\bigotimes_{j=1}^{m_i} \mho_{ij}(X_{ij}; \varnothing) \otimes \bigotimes_{k=1}^{l_i} \mathcal{G}^{\circ}(\varnothing; C_{ik}) \right) \qquad (2)$$

where β is not the active component of any uniformula $\mho_{ij}(X_{ij}; \varnothing)$. Then

$$\mathcal{G}(\Gamma, \Diamond A \Rightarrow \Delta) \longleftarrow \bigotimes_{i=1}^{n} \left(\bigotimes_{j=1}^{m_i} \mho_{ij}(X_{ij}) \otimes \mathcal{G}^{\circ} \left(\Diamond \bigwedge_{k=1}^{l_i} C_{ik} \right) \right) \qquad (3)$$

wherever Req. VII is fulfilled. Similarly, for the $(\Rightarrow \mathsf{K})$ *rule, if*

$$\mathcal{G}(\Gamma \Rightarrow \Delta; \Rightarrow A) \longleftarrow \bigotimes_{i=1}^{n} \left(\bigvee_{j=1}^{m_i} \mho_{ij}(X_{ij}; \varnothing) \otimes \bigvee_{k=1}^{l_i} \mathcal{G}^{\circ}(\varnothing; C_{ik}) \right),$$

then, in the presence of Req. VII,

$$\mathcal{G}(\Gamma \Rightarrow \Box A, \Delta) \longleftarrow \bigotimes_{i=1}^{n} \left(\bigvee_{j=1}^{m_i} \mho_{ij}(X_{ij}) \otimes \mathcal{G}^{\circ}\left(\Box \bigvee_{k=1}^{l_i} C_{ik} \right) \right).$$

Proof. We prove the statement for the $(\mathsf{K} \Rightarrow)$ rule. Let $f \vDash \mathsf{Ant}\big(\mathcal{G}(\Gamma, \Diamond A \Rightarrow \Delta)\big)$ for some good map $f \colon \mathcal{G}(\Gamma, \Diamond A \Rightarrow \Delta) \to \mathcal{M}$, where $\mathcal{M} = (W, R, V)$. Then $\mathcal{M}, f(\alpha) \Vdash \Diamond A$, so that there exists a world $w \in W$ such that $f(\alpha) R w$ and $\mathcal{M}, w \Vdash A$. By Req. VII, the good map $f \cup \{(\beta, w)\} \vDash \mathsf{Ant}\big(\mathcal{G}(\Gamma \Rightarrow \Delta; A \Rightarrow)\big)$. Assuming (2), the interpolant given there in the DNF is forced by $f \cup \{(\beta, w)\}$, i.e., for some i, the map $f \cup \{(\beta, w)\}$ forces one of the disjuncts of the DNF: in particular, $\mathcal{M}, w \Vdash C_{ik}$ for all $k = 1, \dots, l_i$ for this i. Given that $f(\alpha) R w$, we see that $\mathcal{M}, f(\alpha) \Vdash \Diamond \bigwedge_{k=1}^{l_i} C_{ik}$.[5] The removal of the empty β component from the remaining $\mho_{ij}(X_{ij}; \varnothing)$ works the same way as for mix in Lemma 26. Thus, after the removal, all these uniformulas remain forced by f for this i. It follows that f forces the interpolant from (3). The argument for the consequents is analogous.

It is crucial that only one diamond formula has to be satisfied. This is used to find one world to extend the good map with. To single out such diamond formulas, the interpolant of the premise needs to be in the DNF.   □

Composite Rules can be viewed as combinations of other rule types.

Component-shifting rules. Some rules seem local because the structure of the multisequent is unchanged, whereas in reality a new component is added to replace an old one. An example is the hypersequent $\Box \mathsf{R}$ rule from [14], which can be obtained from EW and $(\Rightarrow \mathsf{K})$ (see Fig. 7), necessitating both Reqs. V and VII.

Seriality rules. It was shown in [6] that of the modal nested rules from [2], only the basic K rules (Fig. 6) and the seriality D rules require changing interpolants, with changes for the D rules obtained from those for the K rules by swapping the antecedent and consequent versions. An explanation is depicted in Fig. 8. The $\neg$ rules do not fit into our paradigm: they are from split sequents. But in this example the second $\neg$ cancels the problem created by the first one. Thus, a transformation can be guessed and then proved to be correct independently.

---

[5] This is true also for $l_i = 0$: the empty conjunction is $\top$ and $\mathcal{M}, f(\alpha) \Vdash \Diamond \top$. However, $\mathcal{G}^{\circ}\left(\Diamond \bigwedge_{k=1}^{l_i} C_{ik} \right)$ cannot be dropped: the diamond formula in the disjunct that is forced ensures the existence of an accessible world and the possibility to use (2).

$$\frac{\mathcal{G}(\Rightarrow A)}{\mathcal{G}(\Rightarrow \Box A)}\Box R \qquad \rightsquigarrow \qquad \frac{\dfrac{\mathcal{G}(\qquad\Rightarrow A)}{\mathcal{G}(\Rightarrow\ ;\quad \Rightarrow A)}\,\mathsf{EW}}{\mathcal{G}(\Rightarrow \Box A\quad)}\Rightarrow \mathsf{K}$$

Fig. 7. A rule that looks local but should be treated as composite

$$\frac{\mathcal{G}(\Gamma \Rightarrow \Delta;\ \Rightarrow A)}{\mathcal{G}(\Gamma \Rightarrow \Diamond A, \Delta)}\Rightarrow \mathsf{D} \qquad \rightsquigarrow \qquad \frac{\dfrac{\dfrac{\dfrac{\dfrac{\mathcal{G}(\Gamma \Rightarrow \Delta;\ \Rightarrow A)}{\mathcal{G}(\Gamma \Rightarrow \Delta; \overline{A} \Rightarrow\ )}\,\neg}{\mathcal{G}(\Gamma, \Diamond\overline{A} \Rightarrow \Delta)}\,\mathsf{K}\Rightarrow}{\mathcal{G}(\Gamma, \Box\overline{A} \Rightarrow \Delta)}\,\mathsf{D}\Rightarrow\mathrm{loc}}{\mathcal{G}(\Gamma \Rightarrow \Diamond A, \Delta)}\,\neg$$

Fig. 8. A composite component-removing rule with illegal transitions.

Multicut rule. Unlike in Rem. 23, the multi-to-one multicut rule is external-context splitting and, hence, not local. In addition, one component is juxtaposed against many in the other premise. Fortunately, it can always be represented as a combination of local one-to-one multicuts and rules EW, making our method directly applicable. For the lack of space, we leave the details to the reader.

5   Grafted Hypersequents

To show the versatility of our general method, we apply it to the prefixed-tableau version of a new type of multisequents called *grafted hypersequents*, introduced in [10]. A grafted hypersequent itself is a (possibly empty) hypersequent with an additional *trunk* component, separated from the others by ∥. The formula interpretation for a hypersequent $\Gamma \Rightarrow \Delta \parallel \Pi_1 \Rightarrow \Sigma_1 \mid \cdots \mid \Pi_n \Rightarrow \Sigma_n$ is $\bigwedge \Gamma \to \bigvee \Delta \vee \bigvee_{i=1}^{n} \Box(\bigwedge \Pi_i \to \bigvee \Sigma_i)$. In [10], a prefixed tableau version equivalent to grafted hypersequents is developed for K5 and KD5. This system operates with signed prefixed formulas $\ell\colon SA$, where the sign $S \in \{T, F\}$ and the prefixes can be of three types: the *trunk prefix* •, countably many *limb prefixes* $1, 2, \ldots$ and countably many *twig prefixes* $1, 2, \ldots$. Twig prefixes do not appear in initial tableaus: they can only be introduced by tableau rules. Each branch of a tableau is considered to be a multisequent with each prefix ℓ on the branch determining the component ℓ that contains $\Gamma_\ell \Rightarrow \Delta_\ell$ where $\Gamma_\ell := \{A \mid \ell\colon TA$ occurs on the branch$\}$ and $\Delta_\ell := \{A \mid \ell\colon FA$ occurs on the branch$\}$. Since the prefix • is always present, the singleton multisequents contain no limb or twig prefixes. The interpolant is constructed beginning from a closed tableau and working backwards through the stages of the tableau derivation until the starting tableau whose only branch contains $\bullet\colon TA$ and $\bullet\colon FB$ is reached.

Example 30 (Grafted tableaus). A map from the prefixes occurring on a branch to worlds in a model $\mathcal{M} = (W, R, V)$ is called *good* if $f(\bullet)Rf(\mathsf{n})$ for any limb prefix n and $f(\bullet)R^k f(m)$ for some $k > 0$ for any twig prefix m.

Requirements I and II easily follow from the results of [10].[6] Given the formula interpretation of grafted hypersequents, Req. III follows from the equivalence of grafted tableaus and grafted hypersequents ([10]), from Def. 30, and from the fact that twig components do not occur in initial tableaus. Req. IV is also trivial.

$$R1 \frac{\bullet: \mathsf{F}\Box A}{\mathsf{n}: \mathsf{F}A} \; \mathsf{n} \text{ new} \qquad R2 \frac{\bullet: \mathsf{T}\Box A}{\mathsf{n}: \mathsf{T}A} \; \mathsf{n} \text{ occurs} \qquad R5 \frac{\mathsf{c}: \mathsf{F}\Box A}{m: \mathsf{F}A} \; m \text{ new} \qquad R6 \frac{\mathsf{c}: \mathsf{T}\Box A}{\mathsf{c}': \mathsf{T}A} \; \mathsf{c}' \text{ occurs}$$

$$R3 \frac{\bullet: \mathsf{T}\Diamond A}{\mathsf{n}: \mathsf{T}A} \; \mathsf{n} \text{ new} \qquad R4 \frac{\bullet: \mathsf{F}\Diamond A}{\mathsf{n}: \mathsf{F}A} \; \mathsf{n} \text{ occurs} \qquad R7 \frac{\mathsf{c}: \mathsf{T}\Diamond A}{m: \mathsf{T}A} \; m \text{ new} \qquad R8 \frac{\mathsf{c}: \mathsf{F}\Diamond A}{\mathsf{c}': \mathsf{F}A} \; \mathsf{c}' \text{ occurs}$$

Fig. 9. Grafted tableau rules for K5, where c and c′ are either limb or twig prefixes

The propositional logical and all structural rules fall into the categories discussed above. The cut rule is eliminable. Thus, to demonstrate the CIP for K5, it is sufficient to consider the modal rules from [10] and their symmetric variants, as presented in Fig. 9. Written in our general notation, R2 and R6 coincide with $\Box\mathsf{L}^s$ while R4 and R8 coincide with $\Diamond\mathsf{L}^s$ from Fig. 1. The locality of the rules R2 and R4 directly follows from Def. 30 as $f(\bullet)Rf(\mathsf{n})$ for any good map. The locality of the rules R6 and R8 relies on the fact that $f(\bullet)R^k f(\mathsf{c})$ and $f(\bullet)R^l f(\mathsf{c}')$ for $k, l > 0$ implies $f(\mathsf{c})Rf(\mathsf{c}')$ in Euclidean models. The rules R1 and R5 are variants of the $(\Rightarrow \mathsf{K})$ rule with the principal modal formula preserved in the premise, whereas R3 and R7 are such variants of the $(\mathsf{K} \Rightarrow)$ rule. Req. VII directly follows from Def. 30 for all four rules.

For the logic $\mathsf{KT}_\Box$ of shift reflexivity, the grafted hypersequents from [10] can be translated into grafted tableaus by replacing the modal rules R5–R8 from Fig. 9 with the modal rules S5–S8 from Fig. 10. Instead of Euclideanity, the semantic condition of shift reflexivity is imposed: wRw whenever vRw for some v. To prove the CIP for $\mathsf{KT}_\Box$, it is sufficient to note that S7 and S8 are local rules because $f(\bullet)R^k f(\mathsf{c})$ for some $k > 0$. Further, S5 and S6 can be represented as $(\Rightarrow \mathsf{K})$ and $(\mathsf{K} \Rightarrow)$ respectively, followed by a series of $\Box\mathsf{L}^s$ and $\Diamond\mathsf{L}^s$ rules. Req. VII is clearly fulfilled by Def. 30. Moreover, since the $\Box\mathsf{L}^s$ and $\Diamond\mathsf{L}^s$ rules are performed in one block with a K rule, we can assume $f(\mathsf{c})Rf(m)$ ensuring their locality. Note that Euclideanity was not used for the rules R1–R4 in K5.

Since the additional tableau rule for KD5 from [10] can be extended to SDL^+ and since both the rule and its symmetric version are variants of $(\mathsf{D} \Rightarrow)$ and $(\Rightarrow \mathsf{D})$ with embedded contraction, they can be dealt with in the manner described in Sect. 7.

Theorem 31. *The Craig interpolation property for K5, KD5, the logic of shift reflexivity $\mathsf{KT}_\Box$, and the extended standard deontic logic SDL^+ can be proved constructively using grafted tableau systems, based on [10].*

---

[6] While the tableaus presented in [10] are not symmetric, the necessary modifications are standard, and thus the completeness results from [10] can be applied here.

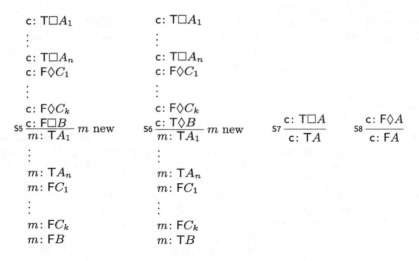

Fig. 10. Grafted tableau rules for $KT_\square$, where c is either a limb or twig prefix

6   Lyndon Interpolation

The CIP is often strengthened to the Lyndon interpolation property (LIP). Up to now, p and $\bar{p}$ have represented the same propositional variable. By contrast, for the LIP they are distinct: p $(\bar{p})$ can occur in the interpolant iff p $(\bar{p})$ occurs in both antecedent and consequent formulas. Thanks to the use of symmetric-type sequents, the interpolants constructed for the CIP can also be used to demonstrate the LIP for all the rules considered, with the exception of the analytic cut, which requires a strengthening of the condition on preservation of p and $\bar{p}$. The main condition for using our method to prove the LIP for custom-made rules is that no propositional letter, positive or negative, antecedent or consequent, disappears on the way from initial sequents to the endsequent.

Corollary 32. *The LIP for all 15 logics of the modal cube can be proved constructively using nested sequents from [2]. The LIP for S5 can be proved constructively using the hypersequent system from [1]. The LIP for K5, KD5, $KT_\square$, and SDL^+ can be proved constructively using grafted tableaus based on [10].*

7   Conclusion and Future Work

We have presented a general description of the constructive proof of Craig and Lyndon interpolation for hypersequents, nested sequents, and other multicomponent sequent formalisms such as grafted tableaus. This general description explains already existing results and facilitates the extension of the method to new rules, e.g., the analytic cut rule and the generalizations of the mix rule. We also provide a general formalism-independent treatment of external weakening and contraction rules. The natural next step is to apply this framework to new multisequent formalisms and to semantics other than Kripke models.

Acknowledgments. The author would like to thank Galina Savukova for editing the literary aspects of the paper, as well as the anonymous reviewers for their comments. The author is grateful to Rosalie Iemhoff and Sara Negri for insightful comments and engaging discussion at the Logic Colloquium 2015.

References

1. Avron, A.: The method of hypersequents in the proof theory of propositional non-classical logics. In: Hodges, W., Hyland, M., Steinhorn, C., Truss, J. (eds.) Logic: From Foundations to Applications: European Logic Colloquium, pp. 1–32. Clarendon Press (1996)
2. Brünnler, K.: Nested Sequents. Habilitation thesis, Institut für Informatik und angewandte Mathematik, Universität Bern (2010)
3. Craig, W.: Three uses of the herbrand-gentzen theorem in relating model theory and proof theory. J. Symbolic Log. **22**(3), 269–285 (1957)
4. Fitting, M.: Proof Methods for Modal and Intuitionistic Logics. D. Reidel Publishing Company, Dordrecht (1983)
5. Fitting, M.: Nested sequents for intuitionistic logics. Notre Dame J. Form. Log. **55**(1), 41–61 (2014)
6. Fitting, M., Kuznets, R.: Modal interpolation via nested sequents. Ann. Pure Appl. Logic **166**(3), 274–305 (2015)
7. Indrzejczak, A.: Cut-free hypersequent calculus for S4.3. Bull. Sect. Logic Univ. Łódź **41**(1–2), 89–104 (2012)
8. Kurokawa, H.: Hypersequent calculi for modal logics extending S4. In: Nakano, Y., Satoh, K., Bekki, D. (eds.) JSAI-isAI 2013. LNCS, vol. 8417, pp. 51–68. Springer, Heidelberg (2014)
9. Kuznets, R.: Craig interpolation via hypersequents. In: Concepts of Proof in Mathematics, Philosophy, and Computer Science. Ontos Verlag, Accepted (2016)
10. Kuznets, R., Lellmann, B.: Grafting Hypersequents onto Nested Sequents. E-print arXiv:1502.00814, August 2015
11. Lahav, O.: From frame properties to hypersequent rules in modal logics. In: 2013 28th Annual ACM/IEEE Symposium on Logic in Computer Science (LICS), pp. 408–417. IEEE Press (2013)
12. Maehara, S.: On the Interpolation Theorem of Craig (in Japanese). Sūgaku 12(4), 235–237 (1960–1961)
13. Maksimova, L.L.: Absence of the interpolation property in the consistent normal modal extensions of the dummett logic. Algebra Log. **21**(6), 460–463 (1982)
14. Restall, G.: Proofnets for S5: sequents and circuits for modal logic. In: Dimitracopoulos, C., Newelski, L., Normann, D., Steel, J.R. (eds.) Logic Colloquium 2005, pp. 151–172. Cambridge University Press (2007)

Adjoint Logic with a 2-Category of Modes

Daniel R. Licata[1]($\boxtimes$) and Michael Shulman[2]

[1] Wesleyan University, Middletown, CT, USA
dlicata@wesleyan.edu
[2] University of San Diego, San Diego, CA, USA
shulman@sandiego.edu

Abstract. We generalize the adjoint logics of Benton and Wadler [1994, 1996] and Reed [2009] to allow multiple different adjunctions between the same categories. This provides insight into the structural proof theory of cohesive homotopy type theory, which integrates the synthetic homotopy theory of homotopy type theory with the synthetic topology of Lawvere's axiomatic cohesion. Reed's calculus is parametrized by a preorder of modes, where each mode determines a category, and there is an adjunction between categories that are related by the preorder. Here, we consider a logic parametrized by a 2-category of modes, where each mode represents a category, each mode morphism represents an adjunction, and each mode 2-morphism represents a pair of conjugate natural transformations. Using this, we give mode theories that describe adjoint triples of the sort used in cohesive homotopy type theory. We give a sequent calculus for this logic, show that identity and cut are admissible, show that this syntax is sound and complete for pseudofunctors from the mode 2-category to the 2-category of adjunctions, and investigate some constructions in the example mode theories.

Keywords: Proof theory · Category theory · Homotopy type theory · Adjoint logic

1 Introduction

An adjunction $F \dashv U$ between categories $\mathscr{C}$ and $\mathscr{D}$ consists of a pair of functors $F : \mathscr{C} \to \mathscr{D}$ and $U : \mathscr{D} \to \mathscr{C}$ such that maps $FC \longrightarrow_{\mathscr{D}} D$ correspond naturally to maps $C \longrightarrow_{\mathscr{C}} UD$. A prototypical adjunction, which provides a mnemonic for the notation, is where U takes the underlying set of some algebraic structure

This material is based on research sponsored by The United States Air Force Research Laboratory under agreement number FA9550-15-1-0053. The U.S. Government is authorized to reproduce and distribute reprints for Governmental purposes notwithstanding any copyright notation thereon. The views and conclusions contained herein are those of the authors and should not be interpreted as necessarily representing the official policies or endorsements, either expressed or implied, of the United States Air Force Research Laboratory, the U.S. Government, or Carnegie Mellon University.

© Springer International Publishing Switzerland 2016
S. Artemov and A. Nerode (Eds.): LFCS 2016, LNCS 9537, pp. 219–235, 2016.
DOI: 10.1007/978-3-319-27683-0_16

such as a group, and F is the free structure on a set—the adjunction property says that a structure-preserving map from FC to D corresponds to a map of sets from C to UD (because the action on the structure is determined by being a homomorphism). Adjunctions are important to the proof theories and λ-calculi of modal logics, because the composite FU is a comonad on $\mathscr{D}$, while UF is a monad on $\mathscr{C}$. Benton and Wadler [2,3] describe an adjoint λ-calculus for mixing linear logic and structural/cartesian logic, with functors U from linear to cartesian and F from cartesian to linear; the $!A$ modality of linear logic arises as the comonad FU, while the monad of Moggi's metalanguage [16] arises as UF. Reed [19] describes a generalization of this idea to situations involving more than one category: the logic is parametrized by a preorder of *modes*, where every mode p determines a category, and there is an adjunction $F \dashv U$ between categories p and q (with $F : q \to p$) exactly when $q \geq p$. For example, the intuitionistic modal logics of Pfenning and Davies [18] can be encoded as follows: the necessitation modality $\square$ is the comonad FU for an adjunction between "truth" and "validity" categories, the lax modality $\bigcirc$ is the monad UF of an adjunction between "truth" and "lax truth" categories, while the possibility modality $\diamond$ requires a more complicated encoding involving four adjunctions between four categories. While specific adjunctions such as $(- \times A) \dashv (A \to -)$ arise in many logics, adjoint logic provides a formalism for abstract/uninterpreted adjunctions.

In Reed's logic, modes are specified by a preorder, which allows at most one adjunction between any two categories (more precisely, there can be two isomorphic adjunctions if both $p \geq q$ and $q \geq p$). However, it is sometimes useful to consider multiple different adjunctions between the same two categories. A motivating example is Lawvere's axiomatic cohesion [8], a general categorical interface that describes *cohesive spaces*, such as topological spaces, or manifolds with differentiable or smooth structures. The interface consists of two categories $\mathscr{C}$ and $\mathscr{S}$, and a quadruple of adjoint functors $\Pi_0, \Gamma : \mathscr{C} \to \mathscr{S}$ and $\Delta, \nabla : \mathscr{S} \to \mathscr{C}$ where $\Pi_0 \dashv \Delta \dashv \Gamma \dashv \nabla$. The idea is that $\mathscr{S}$ is some category of "sets" that provides a notion of "point", and $\mathscr{C}$ is some category of cohesive spaces built out of these sets, where points may be stuck together in some way (e.g. via topology). Γ takes the underlying set of points of a cohesive space, forgetting the cohesive structure. This forgetful functor's right adjoint $\Gamma \dashv \nabla$ equips a set with codiscrete cohesion, where all points are stuck together; the adjunction says that a map *into* a codiscrete space is the same as a map of sets. The forgetful functor's left adjoint $\Delta \dashv \Gamma$ equips a set with *discrete cohesion*, where no points are stuck together; the adjunction says that a map *from* a discrete space is the same as a map of sets. The further left adjoint $\Pi_0 \dashv \Delta$, gives the set of connected components—i.e. each element of $\Pi_0 C$ is an equivalence class of points of C that are stuck together. Π_0 is important because it translates some of the cohesive information about a space into a setting where we no longer need to care about the cohesion. These functors must satisfy some additional laws, such as Δ and ∇ being fully faithful (maps between discrete or codiscrete cohesive spaces should be the same as maps of sets).

A variation on axiomatic cohesion called *cohesive homotopy type theory* [20, 21, 23] is currently being explored in the setting of homotopy type theory and univalent foundations [25, 26]. Homotopy type theory uses Martin-Löf's intensional type theory as a logic of *homotopy spaces*: the identity type provides an ∞-groupoid structure on each type, and spaces such as the spheres can be defined by their universal properties using higher inductive types [13, 14, 22]. Theorems from homotopy theory can be proved *synthetically* in this logic [5, 6, 9–12], and these proofs can be interpreted in a variety of models [4, 7, 24]. However, an important but subtle distinction is that there is no *topology* in synthetic homotopy theory: the "homotopical circle" is defined as a higher inductive type, essentially "the free ∞-groupoid on a point and a loop," which a priori has nothing to do with the "topological circle," $\{(x, y) \in \mathbb{R}^2 \mid x^2 + y^2 = 1\}$, where $\mathbb{R}^2$ has the usual topology. This is both a blessing and a curse: on the one hand, proofs are not encumbered by topological details; but on the other, internally to homotopy type theory, we cannot use synthetic theorems to prove facts about topological spaces.

Cohesive homotopy type theory combines the synthetic homotopy theory of homotopy type theory with the synthetic topology of axiomatic cohesion, using an adjoint quadruple of $(\infty, 1)$-functors $\int \dashv \Delta \dashv \Gamma \dashv \nabla$. In this higher categorical generalization, $\mathscr{S}$ is an $(\infty, 1)$-category of homotopy spaces (e.g. ∞-groupoids), and $\mathscr{C}$ is an $(\infty, 1)$-category of cohesive homotopy spaces, which are additionally equipped with a topological or other cohesive structure at each level. The rules of type theory are now interpreted in $\mathscr{C}$, so that each type has an ∞-groupoid structure (given by the identity type) *as well as* a separate cohesive structure on its objects, morphisms, morphisms between morphisms, etc. For example, types have both morphisms, given by the identity type, and topological paths, given by maps that are continuous in the sense of the cohesion. In the ∞-categorical case, Δ's left adjoint $\int A$ (pronounced "shape of A") generalizes from the connected components to the *fundamental homotopy space* functor, which makes a homotopy space from the topological/cohesive paths, paths between paths, etc. of A. This captures the process by which homotopy spaces arise from cohesive spaces; for example, one can prove (using additional axioms) that the shape of the topological circle is the homotopy circle [23]. This allows synthetic homotopy theory to be used in proofs about topological spaces, and opens up possibilities for applications to other areas of mathematics and theoretical physics.

This paper begins an investigation into the structural proof theory of cohesive homotopy type theory, as a special case of generalizing Reed's adjoint logic to allow multiple adjunctions between the same categories. As one might expect, the first step is to generalize the mode preorder to a mode category, so that we can have multiple different morphisms $\alpha, \beta : p \geq q$. This allows the logic to talk about different but unrelated adjunctions between two categories. However, in order to describe an adjoint triple such as $\Delta \dashv \Gamma \dashv \nabla$, we need to know that the same functor Γ is both a left and right adjoint. To describe such a situation, we generalize to a 2-category of modes, and arrange the syntax of the logic to capture the following semantics. Each mode p determines a category (also denoted by p). Each morphism $\alpha : p \geq q$ determines adjoint functors $F_\alpha : p \to q$ and $U_\alpha : q \to p$ where $F_\alpha \dashv U_\alpha$. Each 2-cell $\alpha \Rightarrow \beta$ determines a morphism of adjunctions between $F_\alpha \dashv U_\alpha$ and $F_\beta \dashv U_\beta$, which consists of natural transformations $F_\beta \to F_\alpha$ and $U_\alpha \to U_\beta$ that are conjugate under the adjunction structure [15, Sect. IV.7]. For example, an adjoint triple is specified by the mode 2-category with

- objects c and s
- 1-cells $\mathsf{d} : s \geq c$ and $\mathsf{n} : c \geq s$
- 2-cells $1_c \Rightarrow \mathsf{n} \circ \mathsf{d}$ and $\mathsf{d} \circ \mathsf{n} \Rightarrow 1_s$ satisfying some equations

The 1-cells generate $F_\mathsf{d} \dashv U_\mathsf{d}$ and $F_\mathsf{n} \dashv U_\mathsf{n}$, while the 2-cells are sufficient to prove that U_d is naturally isomorphic to F_n, so we can define $\Delta := F_\mathsf{n}$, $\nabla := U_\mathsf{n}$, and $\Gamma := U_\mathsf{d} \cong F_\mathsf{n}$ and have the desired adjoint triple. Indeed, you may recognize this 2-category as the "walking adjunction" with $\mathsf{d} \dashv \mathsf{n}$—that is, we give an adjoint triple by saying that the mode morphism generating the adjunction $\Delta \dashv \Gamma$ is itself left adjoint to the mode morphism generating the adjunction $\Gamma \dashv \nabla$.

The main judgement of the logic is a "mixed-category" entailment judgement $A\,[\alpha] \vdash C$ where A has mode q and C has mode p and $\alpha : q \geq p$. Semantically, this judgement means a morphism from A to C "along" the adjunction determined by α—i.e. a map $F_\alpha\, A \longrightarrow C$ or $A \longrightarrow U_\alpha\, C$. However, taking the mixed-mode judgement as primitive makes for a nicer sequent calculus: U and F can be specified independently from each other, by left and right rules, in such a way that identity and cut (composition) are admissible, and the subformula property holds. While we do not consider focusing [1], we conjecture that the connectives can be given the same focusing behavior as in [19]: F is positive and U is negative (which, because limits are negative and colimits are positive, matches what left and right adjoints should preserve).

The resulting logic has a good definition-to-theorem ratio: from simple sequent calculus rules for F and U, we can prove a variety of general facts that are true for any mode 2-category (F_α and U_α are functors; $F_\alpha U_\alpha$ is a comonad and $U_\alpha F_\alpha$ is a monad; F_α preserves colimits and U_α preserves limits), as well as facts specific to a particular theory (e.g. for the adjoint triple above, Γ preserves both colimits and limits, because it can be written either has U_Δ or F_∇; and the comonad $\flat := \Delta\Gamma$ and monad $\sharp := \nabla\Gamma$ are themselves adjoint). Moreover, we can use different mode 2-categories to add additional structure; for example, moving from the walking adjunction to the walking reflection (taking $\Delta\nabla = 1$) additionally gives that Δ and ∇ are full and faithful and that $\flat$ and $\sharp$ are idempotent, which are some of the additional conditions for axiomatic cohesion.

We make a few simplifying restrictions for this paper. First, we consider only single-hypothesis, single-conclusion sequents, deferring an investigation of products and exponentials to future work. Second, on the semantic side, we consider only 1-categorical semantics of the derivations of the logic, rather than the ∞-groupoid semantics that we are ultimately interested in. More precisely, for any 2-category $\mathscr{M}$ of modes, we can interpret the logic using a pseudofunctor $S : \mathscr{M} \to \mathbf{Adj}$, where $\mathbf{Adj}$ is the 2-category of categories, adjunctions, and conjugate pairs of natural transformations [15, Sect. IV.7]. We show that the syntax forms such a pseudofunctor, and conjecture that the syntax is initial in some category or 2-category of pseudofunctors, but have not yet tried to make this precise. Third, we consider only a logic of simple-types, rather than a dependent type theory. Because many of the statements we would like to make require proving some equations between derivations (e.g. the monad laws), we give an equality judgement on sequent calculus derivations. This judgement is interpreted by actual equality of morphisms in the semantics above, but we intend some of these rules to be propositional equalities in an eventual adjoint type theory.

In Sect. 2, we define the rules of the logic, prove admissibility of identity and cut, and define an equational theory on derivations. In Sect. 3, we discuss the semantics of the logic in pseudofunctors $\mathcal{M} \to \mathbf{Adj}$. Finally, in Sect. 4, we examine some specific mode specifications for adjoint triples, which are related to the rules for spatial type theory used in [23]. All of the syntactic metatheory of the logic and the examples have been formalized in Agda [17].[1] An extended version of this paper, available from the authors' web sites, contains more discussion of the examples, definitions, and proofs.

2   Sequent Calculus and Equational Theory

2.1   Sequent Calculus

The logic is parametrized by a strict 2-category $\mathcal{M}$ of modes. We write p, q for the 0-cells (modes), $\alpha, \beta, \gamma, \delta : p \geq q$ for the 1-cells from q to p, and $e : \alpha \Rightarrow \beta$ for the 2-cells. The notation $p \geq q$ for the 1-cells follows [19], but in our case $\mathcal{M}$ is a general category, so there can be more than one morphism $p \geq q$. We use the notation $p \geq q$ for an arrow from q to p (an arrow points "lesser to greater") to match the sequent calculus, where the p-mode is on the left and the q-mode on the right ("validity is greater than truth"). We write $\beta \circ \alpha$ for 1-cell composition in function composition order (i.e. if $\beta : r \geq q$ and $\alpha : q \geq p$ then $\beta \circ \alpha : r \geq p$), $e_1 \cdot e_2$ for vertical composition of 2-cells in diagrammatic order, and $e_1 \circ_2 e_2$ for horizontal composition of 2-cells in "congruence of $\circ$" order (if $e_1 : \alpha \Rightarrow \alpha'$ and $e_2 : \beta \Rightarrow \beta'$ then $e_1 \circ_2 e_2 : \alpha \circ \beta \Rightarrow \alpha' \circ \beta'$). The equations for 2-cells say that $\cdot$ is associative with unit 1_α for any α, that $\circ_2$ is associative with unit 1_1, and that the interchange law $(e_1 \cdot e_2) \circ_2 (e_3 \cdot e_4) = (e_1 \circ_2 e_3) \cdot (e_2 \circ_2 e_4)$ holds. We think of the mode category as being fixed at the outset, and the syntax and judgements of the logic as being indexed by the 0/1/2-cells of this category.

In the pseudofunctor semantics, each object p of the mode category $\mathcal{M}$ will determine a category (also denoted by p). Syntactically, the judgement A type_p will mean that A determines an object of the category p. A morphism $\alpha : q \geq p$ in $\mathcal{M}$ determines an adjunction $F_\alpha \dashv U_\alpha$, with $F_\alpha : q \to p$ and $U_\alpha : p \to q$; note that the right adjoints are covariant and the left adjoints contravariant. Syntactically, the action on objects is given by $F_\alpha A$ type_p when A type_q and $U_\alpha A$ type_q when A type_p. The "pseudo" of the pseudofunctor means that, for

---

[1] See http://github.com/dlicata335/hott-agda/tree/master/metatheory/adjointlogic.

example, the types F_β (F_α A) and $F_{\alpha \circ \beta}$ A will be isomorphic but not definitionally equal. Finally, a 2-cell $e : \alpha \Rightarrow \beta$ in $\mathcal{M}$ determines natural transformations
$U_\alpha \to U_\beta$ and $F_\beta \to F_\alpha$ which are "conjugate" [15, Sect. IV.7]; again, the
right adjoints are covariant and the left adjoints are contravariant. Syntactically, these natural transformations will be definable using the sequent calculus
rules.

In addition to the connectives F_α A and U_α A, we allow an arbitrary collection of atomic propositions (denoted P), each of which has a designated mode;
these represent arbitrary objects of the corresponding categories. To add additional structure to a category or to all categories, we can add rules for additional
connectives; for example, a rule $A + B$ type_p if A type_p and B type_p (parametric
in p) says that any category p has a coproduct type constructor.

The sequent calculus judgement has the form $A [\alpha] \vdash C$ where A type_q and
C type_p and $\alpha : q \geq p$. The judgement represents a map from an object of some
category q to an object of another category p along the adjunction $F_\alpha \dashv U_\alpha$.
Semantically, this mixed-category map can be interpreted equivalently as an
arrow F_α $A \longrightarrow_p C$ or $A \longrightarrow_q U_\alpha$ C. In the rules, we write A_p to indicate an
elided premise A type_p. The rules for atomic propositions and for U and F are
as follows:

$$\frac{1 \Rightarrow \alpha}{P [\alpha] \vdash P} \; \text{hyp} \qquad \frac{A_r [\alpha \circ \beta] \vdash C_p}{F_{\alpha:r\geq q} \, A_r [\beta : q \geq p] \vdash C_p} \; \text{FL} \qquad \frac{\gamma : r \geq q \quad \gamma \circ \alpha \Rightarrow \beta \quad C_r [\gamma] \vdash A_q}{C_r [\beta : r \geq p] \vdash F_{\alpha:q\geq p} \, A_q} \; \text{FR}$$

$$\frac{\gamma : q \geq p \quad \alpha \circ \gamma \Rightarrow \beta \quad A_q [\gamma] \vdash C_p}{U_{\alpha:r\geq q} \, A_q [\beta : r \geq p] \vdash C_p} \; \text{UL} \qquad \frac{C_r [\beta \circ \alpha] \vdash A_p}{C_r [\beta : r \geq q] \vdash U_{\alpha:q\geq p} \, A_p} \; \text{UR}$$

The rules for other types do not change α—e.g., see the rules for coproducts in
Fig. 1.

$$\frac{C_q [\alpha] \vdash A_p}{C_q [\alpha] \vdash A_p + B_p} \; \text{Inl} \qquad \frac{C_q [\alpha] \vdash B_p}{C_q [\alpha] \vdash A_p + B_p} \; \text{Inr} \qquad \frac{A_q [\alpha] \vdash C_p \quad B_q [\alpha] \vdash C_p}{A_q + B_q [\alpha] \vdash C_p} \; \text{Case}$$

Fig. 1. Rules for coproducts; see the extended version for the definition of the admissible identity and cut principles, and for the equational theory extending the rules in
Sect. 2.3

These rules are guided by the usual design goals for sequent calculi: the only
rules are the left and right rules for each connective, the rules have the subformula
property (the premises only involve subformulas of the conclusion), and the
identity and cut rules are admissible. To achieve these goals, it is necessary to
treat the natural transformations $F_\beta \to F_\alpha$ and $U_\alpha \to U_\beta$ induced by a mode
2-cell $\alpha \Rightarrow \beta$ as an additional admissible structural rule: composing with such a
natural transformation transforms a derivation of $A [\alpha] \vdash C$ into a derivation of
$A [\beta] \vdash C$. The admissible rules are discussed further in Sect. 2.2 below.

Consider the rules FL and UR. When β is 1, these rules pass from $F_\alpha\,A\,[1] \vdash C$ and $A\,[1] \vdash U_\alpha\,C$ to $A\,[\alpha] \vdash C$, which makes sense because the judgement $A\,[\alpha] \vdash C$ is intended to mean either/both of these. When β is not 1, these rules compose the mode morphism in the connective with the mode morphism in the sequent. Semantically, this corresponds to one direction of the composition isomorphism between $F_{\alpha\circ\beta}\,A$ and $F_\beta\,F_\alpha\,A$ and similarly for U; see the derivation in Example 2 in Fig. 2. Though we do not consider focusing formally, we conjecture that these two rules are invertible.

Next, consider FR. The rule is a bit complex because it involves three different aspects of the pseudofunctor structure. First, in the case where γ is the identity 1-cell and $\beta = \alpha$ and the 2-cell is the identity, the rule gives functoriality of F (see Example 1 in Fig. 2). In the case where $\gamma = \beta$ and the 2-cell is the identity, the rule gives the other direction of the composition isomorphism between $F_{\alpha\circ\beta}\,A$ and $F_\beta\,F_\alpha\,A$ (see Example 2). In the case where γ is 1 and the rightmost premise is the identity sequent $A\,[1] \vdash A$, the rule gives a natural transformation $F_\beta \rightarrow F_\alpha$ induced by $e : \alpha \Rightarrow \beta$ (see Example 3). This is necessary because composition with such a natural transformation cannot always be pushed inside an application of functoriality, because a morphism from $\gamma \circ \alpha$ might not be constructed from a morphism from γ. In the general form of the rule, we combine these three ingredients: given $\alpha : q \geq p$ and $\beta : r \geq p$, to prove $F_\alpha\,A$ from C, choose a natural transformation that splits β as $\gamma \circ \alpha$ for some $\gamma : r \geq q$, and apply functoriality of α, which leaves proving $C\,[\gamma] \vdash A$. The UL rule is dual.

We give some additional examples in Fig. 2; these examples and many more like them are in the companion Agda code. The composites FU and UF should be a comonad and a monad respectively; define $\Box_\alpha\,A := F_\alpha\,(U_\alpha\,A)$ and $\bigcirc_\alpha\,A := U_\alpha\,(F_\alpha\,A)$ for any $\alpha : q \geq p$. As an example of the (co)monad structure, the comonad comultiplication is defined in the figure. An advantage of using a cut-free sequent calculus is that we can observe some non-provabilities. For example, there is not in general a map $P\,[1_p] \vdash \Box_\alpha\,P$ (unit for the comonad): by inversion, a derivation must begin with FR, but to apply this rule, we need a $\gamma : p \geq q$ and a 2-cell $\gamma \circ \alpha \Rightarrow 1$, which may not exist. Next, we give one half of the isomorphism showing that F preserves coproducts; this is a consequence of the left rule for $P + Q$ allowing an arbitrary α.

Because we are interested not only in provability, but also in the equational theory of proofs in this logic, one might think the next step would be to annotate the sequent judgement with a proof term, writing e.g. $x : A[\alpha] \vdash M : B$. However, the proof terms M would have exactly the same structure as the derivations of this typing judgement, so we instead use the derivations themselves as the proof terms. We sometimes write $D : A\,[\alpha] \vdash B$ to indicate "typing" in the metalanguage; i.e. this should be read "D is a derivation tree of the judgement $A\,[\alpha] \vdash B$."

Example 1: Functoriality, given $\alpha : q \geq p$ and $D : A\,[1_q] \vdash B$:

$$\dfrac{\dfrac{\overline{1_q : q \geq q} \quad \overline{1 : 1_q \circ \alpha \Rightarrow \alpha \circ 1_p} \quad D : A\,[1_q] \vdash B}{A\,[\alpha \circ 1_p] \vdash F_\alpha\,B} \text{ FR}}{F_\alpha\,A\,[1_p] \vdash F_\alpha\,B} \text{ FL}$$

Example 2: F on $\beta \circ \alpha$:

$$\dfrac{\overline{\beta : r \geq p} \quad \dfrac{\overline{1 : (\beta \circ \alpha) \circ 1 \Rightarrow \beta \circ \alpha}}{P\,[(\beta \circ \alpha) \circ 1] \vdash F_\alpha\,(F_\beta\,P)} \quad \dfrac{\dfrac{\overline{1 : r \geq r} \quad \overline{1 : 1 \circ \alpha \Rightarrow \alpha} \quad \overline{P\,[1] \vdash P}^{\text{hyp}\,1}}{P\,[\beta] \vdash F_\beta\,P} \text{ FR}}{} }{F_{\beta \circ \alpha}\,P\,[1] \vdash F_\alpha\,(F_\beta\,P)} \text{ FL}$$

$$\dfrac{\dfrac{\dfrac{\overline{1 : r \geq r} \quad \overline{1 : 1 \circ (\beta \circ \alpha) \Rightarrow \beta \circ (\alpha \circ 1)} \quad \overline{P\,[1] \vdash P}^{\text{hyp}\,1}}{P\,[\beta \circ (\alpha \circ 1)] \vdash F_{\beta \circ \alpha}\,P} \text{ FR}}{F_\beta\,P\,[\alpha \circ 1] \vdash F_{\beta \circ \alpha}\,P} \text{ FL}}{F_\alpha\,(F_\beta\,P)\,[1] \vdash F_{\beta \circ \alpha}\,P} \text{ FL}$$

Example 3: F/U on 2-cells:

$$\dfrac{\dfrac{\overline{1 : q \geq q} \quad \overline{e : 1 \circ \alpha \Rightarrow \beta} \quad \overline{P\,[1] \vdash P}^{\text{hyp}\,1}}{P\,[\beta] \vdash F_\alpha\,P} \text{ FR}}{F_\beta\,P\,[1] \vdash F_\alpha\,P} \text{ FL} \qquad \dfrac{\dfrac{\overline{1 : p \geq p} \quad \overline{e : \alpha \circ 1 \Rightarrow \beta} \quad \overline{P\,[1] \vdash P}^{\text{hyp}\,1}}{U_\alpha\,P\,[\beta] \vdash P} \text{ UL}}{U_\alpha\,P\,[1] \vdash U_\beta\,P} \text{ UR}$$

Example 4: Comonad comultiplication:

$$\dfrac{\overline{1 : q \geq q} \quad \overline{1 : \alpha \Rightarrow \alpha} \quad \dfrac{\dfrac{\dfrac{\overline{1 : q \geq q} \quad \overline{1 : \alpha \Rightarrow \alpha} \quad \dfrac{\overline{1 : p \geq p} \quad \overline{1 : \alpha \Rightarrow \alpha} \quad \dfrac{\overline{1 \Rightarrow 1} \quad \overline{P\,[1] \vdash P}}{}^{\text{hyp}}}{\dfrac{U_\alpha\,P\,[\alpha] \vdash P}{U_\alpha\,P\,[1] \vdash U_\alpha\,P} \text{ UR}} \text{ UL}}{U_\alpha\,P\,[\alpha] \vdash \Box_\alpha\,P} \text{ FR}}{U_\alpha\,P\,[1] \vdash U_\alpha\,\Box_\alpha\,P} \text{ UR}}{U_\alpha\,P\,[\alpha] \vdash \Box_\alpha\,\Box_\alpha\,P} \text{ FR}}{\dfrac{U_\alpha\,P\,[\alpha] \vdash \Box_\alpha\,\Box_\alpha\,P}{\Box_\alpha\,P\,[1] \vdash \Box_\alpha\,\Box_\alpha\,P} \text{ FL}}$$

Example 5: F preserves coproducts (one half of a natural isomorphism):

$$\dfrac{\dfrac{\dfrac{\dfrac{\overline{1 : q \geq q} \quad \overline{1 : \alpha \Rightarrow \alpha} \quad \overline{P\,[1] \vdash P}^{\text{hyp}\,1}}{P\,[\alpha] \vdash F_\alpha\,P} \text{ FR}}{P\,[\alpha] \vdash F_\alpha\,P + F_\alpha\,Q} \text{ Inl} \quad \dfrac{\dfrac{\overline{1 : q \geq q} \quad \overline{1 : \alpha \Rightarrow \alpha} \quad \overline{Q\,[1] \vdash Q}^{\text{hyp}\,1}}{Q\,[\alpha] \vdash F_\alpha\,Q} \text{ FR}}{Q\,[\alpha] \vdash F_\alpha\,P + F_\alpha\,Q} \text{ Inr}}{P + Q\,[\alpha] \vdash F_\alpha\,P + F_\alpha\,Q} \text{ Case}}{F_\alpha\,(P + Q)\,[1] \vdash F_\alpha\,P + F_\alpha\,Q} \text{ FL}$$

Fig. 2. Some examples

2.2   Admissible Rules

Adjunction morphisms As discussed above, composition with the natural transformations $F_\beta \to F_\alpha$ and $U_\alpha \to U_\beta$ induced by a 2-cell $e : \alpha \Rightarrow \beta$ is an admissible rule, which we write as $e_*(D) : A[\beta] \vdash B$:

$$\frac{\alpha \Rightarrow \beta \quad A[\alpha] \vdash C}{A[\beta] \vdash C} \;-_*(-)$$

The definition of this operation pushes the natural transformation into the premises of a derivation until it reaches a rule that builds in a transformation (FR,UL,hyp):

$$\begin{aligned}
e_*(\mathsf{hyp}\, e') &:= \mathsf{hyp}\,(e' \cdot e) \\
e_*(\mathsf{FR}^\gamma_{e'}(D)) &:= \mathsf{FR}^\gamma_{e'\cdot e}(D) \\
e_*(\mathsf{FL}(D)) &:= \mathsf{FL}((1 \circ_2 e)_*(D)) \\
e_*(\mathsf{UL}^\gamma_{e'}(D)) &:= \mathsf{UL}^\gamma_{e'\cdot e}(D) \\
e_*(\mathsf{UR}(D)) &:= \mathsf{UR}((e \circ_2 1)_*(D))
\end{aligned}$$

Identity The identity rule is admissible:

$$\frac{}{A_p\,[1] \vdash A_p}\;\mathsf{ident}$$

As a function from types to derivations, we have

$$\begin{aligned}
\mathsf{ident}_P &:= \mathsf{hyp}\,1 \\
\mathsf{ident}_{U_\alpha\,A} &:= \mathsf{UR}(\mathsf{UL}^1_1(\mathsf{ident}_A)) \\
\mathsf{ident}_{F_\alpha\,A} &:= \mathsf{FL}(\mathsf{FR}^1_1(\mathsf{ident}_A))
\end{aligned}$$

Cut The following cut rule is admissible:

$$\frac{A_r\,[\beta] \vdash B_q \quad B_q\,[\alpha] \vdash C_p}{A_r\,[\beta \circ \alpha] \vdash C_p}\;\mathsf{cut}$$

For example, consider the principal cut for F:

$$\frac{\dfrac{e : \gamma \circ \alpha_1 \Rightarrow \beta \quad D : A[\gamma] \vdash B}{A[\beta] \vdash F_{\alpha_1}\,B}\;\mathsf{FR} \quad \dfrac{E : B[\alpha_1 \circ \alpha] \vdash C}{F_{\alpha_1}\,B[\alpha] \vdash C}\;\mathsf{FL}}{A[\beta \circ \alpha] \vdash C}\;\mathsf{cut}$$

In this case the cut reduces to

$$\frac{e \circ_2 1 : (\gamma \circ \alpha_1) \circ \alpha \Rightarrow \beta \circ \alpha \quad \dfrac{D : A[\gamma] \vdash B \quad E : B[\alpha_1 \circ \alpha] \vdash C}{A[\gamma \circ \alpha_1 \circ \alpha] \vdash C}\;\mathsf{cut}}{A[\beta \circ \alpha] \vdash C}\;-_*(-)$$

As a transformation on derivations, we have

$$
\begin{aligned}
\mathsf{cut}\ (\mathsf{hyp}\,e)\ (\mathsf{hyp}\,e') &:= \mathsf{hyp}\,(e \circ_2 e') \\
\mathsf{cut}\ (\mathsf{FR}_e^\gamma(D))\ (\mathsf{FL}(E)) &:= (e \circ_2 1)_*(\mathsf{cut}\ D\ E) \\
\mathsf{cut}\ (\mathsf{UR}(D))\ (\mathsf{UL}_e^\gamma(E)) &:= (1 \circ_2 e)_*(\mathsf{cut}\ D\ E) \\
\mathsf{cut}\ D\ (\mathsf{FR}_e^\gamma(E)) &:= \mathsf{FR}_{1 \circ_2 e}^{\beta \circ \gamma}(\mathsf{cut}\ D\ E) \\
\mathsf{cut}\ D\ (\mathsf{UR}(E)) &:= \mathsf{UR}(\mathsf{cut}\ D\ E) \\
\mathsf{cut}\ (\mathsf{FL}(D))\ E &:= \mathsf{FL}(\mathsf{cut}\ D\ E) && \text{if } E \text{ is not a right rule} \\
\mathsf{cut}\ (\mathsf{UL}_e^\gamma(D))\ E &:= \mathsf{UL}_{e \circ_2 1}^{\gamma \circ \alpha}(\mathsf{cut}\ D\ E) && \text{if } E \text{ is not a right rule}
\end{aligned}
$$

The first case is for atomic propositions. The next two cases are the principal cuts, when a right rule meets a left rule; these correspond to β-reduction in natural deduction. The next two cases are right-commutative cuts, which push any D inside a right rule for E. The final two cases are left-commutative cuts, which push any E inside a left rule for D. The left-commutative and right-commutative cuts overlap when D is a left rule and E is a right rule; we give precedence to right-commutative cuts definitionally, but using the equational theory below, we will be able to prove the general left-commutative rules.

As an example using identity and cut, we give one of the maps from the bijection-on-hom-sets adjunction for F and U: given $\alpha : q \geq p$ we can transform $D : F_\alpha\ A\,[1] \vdash B$ into $A\,[1] \vdash U_\alpha\ B$:

$$
\cfrac{
\cfrac{
\cfrac{\overline{1 : q \geq q}\quad \overline{1 : \alpha \Rightarrow \alpha}\quad \cfrac{}{A\,[1] \vdash A}\ \text{ident}}
{A\,[\alpha] \vdash F_\alpha\ A}\ \text{FR}
\qquad D : F_\alpha\ A\,[1] \vdash B}
{A\,[\alpha] \vdash B}\ \text{cut}}
{A\,[1] \vdash U_\alpha\ B}\ \text{UR}
$$

2.3   Equations

When we construct proofs using the admissible rules $e_*(D)$ and ident_A and $\mathsf{cut}\ D\ E$, there is a natural notion of definitional equality induced by the above definitions of these operations—the cut- and identity-free proofs are normal forms, and a proof using cut or identity is equal to its normal form. However, to prove the desired equations in the examples below, we will need some additional "propositional" equations, which, because we are using derivations as proof terms, we represent by a judgement $D \approx D'$ on two derivations $D, D' : A\,[\alpha] \vdash C$. This judgement is the least congruence closed under the following rules. First, we have uniqueness/η rules. The rule for F says that any map from $F_\alpha\ A$ is equal to a derivation that begins with an application of the left rule and then cuts the original derivation with the right rule; the rule for U is dual.

$$
\cfrac{D : F_\alpha\ A\,[\beta] \vdash C}{D \approx \mathsf{FL}\,(\mathsf{cut}(\mathsf{FR}_1^1(\mathsf{ident}_A))\,D)}\ F\eta
\qquad
\cfrac{D : C\,[\beta] \vdash U_\alpha\ A}{D \approx \mathsf{UR}(\mathsf{cut}\,D(\mathsf{UL}_1^1(\mathsf{ident}_A)))}\ U\eta
$$

Second, we have rules arising from the 2-cell structure. For example, suppose we construct a derivation by $\mathsf{FR}_e^\gamma(D)$ for some $\gamma : r \geq q$ and $e : \gamma \circ \alpha \Rightarrow \beta$, but there is another morphism $\gamma' : r \geq q$ such that there is a 2-cell between γ and γ'. The following says that we can equally well pick γ' and suitably transformed e and D, using composition and $e_{2*}(-)$ to make the types match up.

$$\frac{e : \gamma \circ \alpha \Rightarrow \beta \quad D : C\,[\gamma'] \vdash A \quad e_2 : \gamma' \Rightarrow \gamma}{\mathsf{FR}_e^\gamma(e_{2*}(D')) \approx \mathsf{FR}_{((e_2 \circ_2 1)\cdot e)}^{\gamma'}(D')} \qquad \frac{e : \gamma \circ \alpha \Rightarrow \beta \quad D : C\,[\gamma'] \vdash A \quad e_2 : \gamma' \Rightarrow \gamma}{\mathsf{UL}_e^\gamma(e_{2*}(D')) \approx \mathsf{UL}_{((1 \circ_2 e_2)\cdot e)}^{\gamma'}(D')}$$

Semantically, these rules will be justified by some of the pseudofunctor laws.

The final rules say that left rules of negatives and right rules of positives commute. These are needed to prove the left-commutative cut equations in the case where E is a right rule, which seem necessary for showing that cut is unital and associative. For U and F, we have

$$\frac{(1 \circ_2 e_1) \cdot e_2 = (e_3 \circ_2 1) \cdot e_4}{\mathsf{UL}_{e_2}(\mathsf{FR}_{e_1}(D)) \approx \mathsf{FR}_{e_4}(\mathsf{UL}_{e_3}(D))}$$

The following additional equality rules are admissible for logic containing the U/F rules described above and the coproduct rules in Fig. 1. The rules in each line (except the first) are proved by mutual induction, and use the preceding lines:

$$\overline{1_*(D) = D} \qquad \overline{(e_1 \cdot e_2)_*(D) = e_{2*}(e_{1*}(D))}$$

$$\frac{D \approx D'}{e_*(D) \approx e_*(D')} \qquad \frac{e : \alpha \Rightarrow \alpha' \quad e' : \beta \Rightarrow \beta' \quad D : A\,[\alpha] \vdash B \quad D' : B\,[\beta] \vdash C}{(e \circ_2 e')_*(\mathsf{cut}\ D\ D') \approx \mathsf{cut}\ (e_*(D))\ (e'_*(D'))}$$

$$\overline{\mathsf{cut}\ D_1\ (\mathsf{cut}\ D_2\ D_3) \approx \mathsf{cut}\ (\mathsf{cut}\ D_1\ D_2)\ D_3}$$

$$\overline{\mathsf{cut}\ D\ \mathsf{ident} \approx D} \quad \overline{\mathsf{cut}\ \mathsf{ident}\ D \approx D} \quad \frac{D \approx D'}{\mathsf{cut}\ D\ E \approx \mathsf{cut}\ D'\ E} \quad \frac{E \approx E'}{\mathsf{cut}\ D\ E \approx \mathsf{cut}\ D\ E'}$$

$$\overline{\mathsf{cut}\ (\mathsf{FL}(D))\ E \approx \mathsf{FL}(\mathsf{cut}\ D\ E)} \quad \overline{\mathsf{cut}\ (\mathsf{UL}_e^\gamma(D))\ E \approx \mathsf{UL}_{e \circ_2 1}^{\gamma \circ \alpha}(\mathsf{cut}\ D\ E)}$$

3   Semantics

In the extended version of this paper, we give a detailed account of soundness and completeness results. Let **Adj** be the 2-category whose objects are categories, whose morphisms $\mathscr{C} \to \mathscr{D}$ are adjunctions $L \dashv R$ with $L : \mathscr{D} \to \mathscr{C}$ and $R : \mathscr{C} \to \mathscr{D}$, and whose 2-cells $(L^1 \dashv R^1) \to (L^2 \dashv R^2)$ are conjugate pairs of transformations $t^L : L^2 \to L^1$ and $t^R : R^1 \to R^2$. A pseudofunctor is a map between 2-categories that preserves identity and composition of 1-cells up to coherent isomorphism, rather than on the nose.

Theorem 1 (Soundness). *For any mode theory $\mathscr{M}$, the rules of adjoint logic can be interpreted in any pseudofunctor $\mathscr{M} \to$ **Adj**.*

The rules describe a pseudofunctor because $F_1 A \cong A \cong U_1 A$ and $F_{\beta \circ \alpha} A \cong F_\alpha F_\beta A$ and $U_{\beta \circ \alpha} A \cong U_\beta U_\alpha A$, but these are not equalities of types.

Theorem 2 (Completeness).   *The syntax of adjoint logic determines a pseudofunctor $\mathcal{M} \to \mathbf{Adj}$:*

1. *An object p of $\mathcal{M}$ is sent to the category whose objects are A type_p and morphisms are derivations of $A[1_p] \vdash B$ quotiented by $\approx$, with identities given by ident and composition given by cut.*
2. *For each q, p, there is a functor from the category of morphisms $q \geq p$ to the category of adjoint functors between q and p.*
   - *Each $\alpha : q \geq p$ is sent to $F_\alpha \dashv U_\alpha$ in $\mathbf{Adj}$—F_α and U_α are functors and they are adjoint.*
   - *Each 2-cell $e : \alpha \Rightarrow \beta$ is sent to a conjugate pair of transformations $(F(e), U(e)) : (F_\alpha \dashv U_\alpha) \to (F_\beta \dashv U_\beta)$, and this preserves 1 and $e_1 \cdot e_2$.*
3. *$F_1 A \cong A$ and $U_1 A \cong A$ naturally in A, and these are conjugate, so there is an adjunction isomorphism P^1 between $F_1 \dashv U_1$ and the identity adjunction.*
4. *$F_{\beta \circ \alpha} A \cong F_\alpha (F_\beta A)$ and $U_{\beta \circ \alpha} A \cong U_\beta (U_\alpha A)$ naturally in A, and these are conjugate, so there is an adjunction isomorphism $P^\circ(\alpha, \beta)$ between $F_{\beta \circ \alpha} \dashv U_{\beta \circ \alpha}$ and the composition of the adjunctions $F_\alpha \dashv U_\alpha$ and $F_\beta \dashv U_\beta$. Moreover, this family of adjunction isomorphisms is natural in α and β.*
5. *Three coherence conditions between these identity and composition isomorphisms are satisfied.*

Proof. We have given a flavor for some of the maps in the examples above; the complete construction is about 500 lines of Agda. There are many equations to verify—inverses, naturality, conjugation, and coherence—but they are all true for $\approx$.

Next, we summarize some constructions on $F_\alpha \dashv U_\alpha$ that can be made in the logic. We write $D \bullet E$ as an infix notation for $\mathsf{cut}\ D\ E$ (composition in diagrammatic order).

Lemma 1 (Some Constructions on Adjunctions).   *Let $\alpha : q \geq p$. Then:*

1. *The composite functor $\Box_\alpha A := F_\alpha U_\alpha A$ is a comonad:*
   $\mathsf{counit} : \Box_\alpha A[1] \vdash A$ *naturally in A*
   $\mathsf{comult} : \Box_\alpha A[1] \vdash \Box_\alpha \Box_\alpha A$ *naturally in A*
   $\mathsf{comult} \bullet (\Box\, \mathsf{comult}) \approx \mathsf{comult} \bullet \mathsf{comult}$ *and* $\mathsf{comult} \bullet \mathsf{counit} \approx \mathsf{ident}$
   and $\mathsf{comult} \bullet (\Box\, \mathsf{counit}) \approx \mathsf{ident}$.
2. *The composite functor $\bigcirc_\alpha A := U_\alpha F_\alpha A$ is a monad:*
   $\mathsf{unit} : A[1] \vdash \bigcirc_\alpha A$ *naturally in A*
   $\mathsf{mult} : \bigcirc_\alpha \bigcirc_\alpha A[1] \vdash \bigcirc_\alpha A$ *naturally in A*
   $(\bigcirc\, \mathsf{mult}) \bullet \mathsf{mult} \approx \mathsf{mult} \bullet \mathsf{mult}$ *and* $\mathsf{unit} \bullet \mathsf{mult} \approx \mathsf{ident}$
   and $(\bigcirc\, \mathsf{unit}) \bullet \mathsf{mult} \approx \mathsf{ident}$.
3. *F preserves coproducts: $F_\alpha (A + B) \cong F_\alpha A + F_\alpha B$ naturally in A and B.*

Proof. We showed some of the maps above; the (co)monad laws, naturality conditions, and inverse laws are all true for $\approx$; the construction is about 150 lines of Agda.

4   Example Mode Theories

4.1   Adjoint Triple

Consider the walking adjunction $d \dashv n$, which has

- objects c and s
- 1-cells $d : s \geq c$ and $n : c \geq s$
- 2-cells unit $: 1_c \Rightarrow n \circ d$ and counit $: d \circ n \Rightarrow 1_s$ satisfying
  $(1_d \circ_2 \text{unit}) \cdot (\text{counit} \circ_2 1_d) = 1$ and $(\text{unit} \circ_2 1_n) \cdot (1_n \circ_2 \text{counit}) = 1$.

The 1-cells specify two adjunctions $F_d \dashv U_d$ and $F_n \dashv U_n$. However, the natural transformations specified by the 2-cells also give adjunctions $F_d \dashv F_n$ and $U_d \dashv U_n$ (using the unit/counit definition of adjunction). Since a right or left adjoint of a given functor is unique up to isomorphism, it follows that the two functors $U_d, F_n : c \to s$ are isomorphic, resulting in an adjoint triple $F_d \dashv (U_d \cong F_n) \dashv U_n$. However, rather than proving $F_d \dashv F_n$ or $U_d \dashv U_n$ and then concluding $U_d \cong F_n$ from uniqueness of adjoints, we can construct the isomorphism directly:

Lemma 2. $U_d A \cong F_n A$ naturally in A.

Proof. We can write the maps as follows:

$$\cfrac{d : s \geq c \quad \text{counit} : d \circ n \Rightarrow 1 \quad \cfrac{1 : c \geq c \quad 1 : d \Rightarrow d \quad \cfrac{\overline{A\,[1] \vdash A}}{} \text{ident}}{\cfrac{U_d\,A\,[d] \vdash A}{} \text{FR}} \text{UL}}{U_d\,A\,[1] \vdash F_n\,A}$$

$$\cfrac{\cfrac{\cfrac{\text{unit} : 1 \Rightarrow n \circ d \quad \cfrac{\overline{A\,[1] \vdash A}}{} \text{ident}}{A\,[n \circ d] \vdash A} {-}_*(-)}{\cfrac{A\,[n] \vdash U_d\,A}{} \text{UR}}}{F_n\,A\,[1] \vdash U_d\,A} \text{FL}}{}$$

In the Agda code, we verify that these are inverse and natural.

We can develop some of the expected properties of an adjoint triple $L \dashv M \dashv R$, such as the fact that the "left" comonad LM is itself left adjoint to the "right" monad RM, and consequently, LM preserves colimits. In this case, we have $L = F_d$, $M = U_d \cong F_n$, and $R = U_n$, and we write $\Box_d A := F_d\,U_d\,A$ and $\bigcirc_n A := U_n\,F_n\,A$.

Theorem 3 (Properties of an Adjoint Triple)

1. $\Box_d \dashv \bigcirc_n$
2. $\Box_d (A + B) \cong \Box_d A + \Box_d A$

Proof. We can prove that $\square_d A$ and $\bigcirc_n A$ are isomorphic to a single F and U, respectively:

$$\square_d A = F_d\ U_d\ A \cong F_d\ F_n\ A \cong F_{nod}\ A$$
$$\bigcirc_n A = U_n\ F_n\ A \cong U_n\ U_d\ A \cong U_{nod}\ A$$

This implies the above properties because $F_{nod} \dashv U_{nod}$ and F_{nod} preserves coproducts and these facts respect natural isomorphism.

From a polarity point of view, it is unusual for a comonad $F\ U\ A$ to preserve positives, because the negative connective U interrupts focus/inversion phases. Here, this behavior is explained by the fact that $F_d\ U_d\ A$ is isomorphic to a single positive connective $F_{nod}\ A$. The ambipolar middle connective in an adjoint triple thus emerges from the presence of two isomorphic connectives, one positive and one negative.

4.2   Reflection

In our motivating example of axiomatic cohesion, the adjoint triple $\Delta \dashv \Gamma \dashv \nabla$ has some additional properties. We now write $\flat$ for the comonad $\Delta\Gamma$ and $\sharp$ for the monad $\nabla\Gamma$. $\flat$ takes a cohesive space and "retopologizes" it with the discrete cohesion, while $\sharp$ takes a cohesive space and retopoligizes it with the codiscrete cohesion. Intuitively, retopologizing twice should be the same as retopologizing once, because each retopologization forgets the existing cohesive structure; that is, we want $\flat\flat A \cong \flat A$ and $\sharp\sharp A \cong \sharp A$ and $\flat\sharp A \cong \flat A$ and $\sharp\flat A \cong \sharp A$. Moreover, Δ and ∇ should be full and faithful, because a map between discrete or codiscrete spaces is exactly a map of sets.

Recalling that a right (resp. left) adjoint is full and faithful exactly when the counit (resp. unit) of the adjunction is an isomorphism, we can capture these properties by considering a different mode 2-category, the "walking reflection". This has the same objects and morphisms as above, but we now take $d \circ n = 1$, with the counit being just the identity 2-cell, and the equations simplify to unit $\circ_2 1_n = 1$ and $1_d \circ_2$ unit $= 1$. Note that the only non-identity morphisms of this mode category are d, n, and n ∘ d.

We write $\Delta := F_d$, $\Gamma := (U_d \cong F_n)$, and $\nabla := U_n$, so $\flat = \square_d$ and $\sharp = \bigcirc_n$. Since in particular we still have an adjunction, this mode theory inherits all the theorems from the previous section; it also has the following additional properties:

Theorem 4 (Properties of the Walking Reflection)

1. $\flat\flat A \cong \flat A$ and $\sharp\sharp A \cong \sharp A$ naturally in A.
2. $\sharp\flat A \cong \sharp A$ and $\flat\sharp A \cong \flat A$ naturally in A.
3. F_d and U_n are full and faithful.

Proof. We discuss the first two parts. Using Theorem 2, the equality of morphisms $d \circ n = 1$ implies that

$$F_n\ F_d\ A \cong F_{don}\ A = F_1\ A \cong A$$
$$U_d\ U_n\ A \cong U_{don}\ A = U_1\ A \cong A$$

Consequently, by Lemma 2, the other (co)monads besides $\flat$ and $\sharp$ are trivial:

$$\bigcirc_{\mathsf{d}} A = U_{\mathsf{d}}\, F_{\mathsf{d}}\, A \cong F_{\mathsf{n}}\, F_{\mathsf{d}}\, A \cong A$$
$$\square_{\mathsf{n}} A = F_{\mathsf{n}}\, U_{\mathsf{n}}\, A \cong U_{\mathsf{d}}\, U_{\mathsf{n}}\, A \cong A$$

This gives idempotence and absorption:

$$\flat\,\flat\, A = F_{\mathsf{d}}\, (U_{\mathsf{d}}\, F_{\mathsf{d}}\, (U_{\mathsf{d}}\, A)) \cong F_{\mathsf{d}}\, U_{\mathsf{d}}\, A = \flat\, A$$
$$\sharp\,\sharp\, A = U_{\mathsf{n}}\, (F_{\mathsf{n}}\, U_{\mathsf{n}}\, (F_{\mathsf{n}}\, A)) \cong U_{\mathsf{n}}\, F_{\mathsf{n}}\, A = \sharp\, A$$
$$\flat\,\sharp\, A = F_{\mathsf{d}}\, (U_{\mathsf{d}}\, U_{\mathsf{n}}\, (F_{\mathsf{n}}\, A)) \cong F_{\mathsf{d}}\, F_{\mathsf{n}}\, A \cong F_{\mathsf{nod}}\, A \cong \flat\, A$$
$$\sharp\,\flat\, A = U_{\mathsf{n}}\, (F_{\mathsf{n}}\, F_{\mathsf{d}}\, (U_{\mathsf{d}}\, A)) \cong U_{\mathsf{n}}\, U_{\mathsf{d}}\, A \cong U_{\mathsf{nod}}\, A \cong \sharp\, A$$

4.3   Spatial Type Theory

The above mode theory allows us to work with cohesive types (which have mode c) and non-cohesive types (which have mode s). However, because Δ and ∇ are full and faithful, it is not strictly necessary to ever work in s itself—we could equivalently work in the image of Δ or ∇ in c. If we wish to restrict ourselves to constructions in c, we can simplify the mode theory to the (strictly) idempotent monad, which has one object c, one generating 1-cell $\mathsf{r} : \mathsf{c} \geq \mathsf{c}$ such that $\mathsf{r} \circ \mathsf{r} = 1$, and one generating 2-cell $\mathsf{unit} : 1 \Rightarrow \mathsf{r}$ satisfying $\mathsf{t} \circ_2 \mathsf{unit} = 1$ and $\mathsf{unit} \circ_2 \mathsf{r} = 1$. Observe that the only 1-cells are 1 and r and the only 2-cells are 1_1, 1_{r}, and unit. This mode theory embeds in the walking reflection, with $\mathsf{r} := \mathsf{n} \circ \mathsf{d}$, so we could equivalently work in the c-types above.

For this mode theory, we define $\flat := F_{\mathsf{r}}$ and $\sharp := U_{\mathsf{r}}$. In the walking reflection, we defined $\flat := \square_{\mathsf{d}}$ and $\sharp := \bigcirc_{\mathsf{n}}$ and then proved (in the proof of Theorem 3) that $\flat \cong F_{\mathsf{nod}}$ and $\sharp \cong U_{\mathsf{nod}}$. Here, we take the other side of this isomorphism as the definition, so we immediately have $\flat \dashv \sharp$ and $\flat$ preserves coproducts, but we must prove that they are (co)monads. A simple route to this is to prove absorption, because $\flat\,\sharp\, A = F_{\mathsf{r}}\, U_{\mathsf{r}}\, A$ is a comonad, and dually for $\sharp\,\flat\, A$.

Theorem 5 (Idempotence and Absorption). $\flat\,\flat\, A \cong \flat\, A$ and $\sharp\,\sharp\, A \cong \sharp\, A$ and $\sharp\,\flat\, A \cong \sharp\, A$ and $\flat\,\sharp\, A \cong \flat\, A$ naturally in A.

Proof. Because $\mathsf{r} \circ \mathsf{r} = \mathsf{r}$, idempotence is just the composition isomorphisms F° and U° from Theorem 2. The absorption isomorphisms are constructed directly.

In the extended version of this paper, we connect adjoint logic for this mode theory to the rules for spatial type theory used in Shulman [23].

5   Conclusion

In this paper, we have defined an adjoint logic that allows multiple different adjunctions between the same categories, shown soundness and completeness of the logic in pseudofunctors $\mathscr{M} \to \mathbf{Adj}$, and used some specific mode theories to model adjoint triples and the $\flat$ and $\sharp$ modalities of axiomatic cohesion. One direction for future work is to extend this adjoint logic with multiple assumptions and dependent types (we discuss some special cases in the extended version). This would provide a context for investigating the shape modality $\int \dashv \flat$. We could certainly give a mode theory with one mode and $\int \dashv \flat \dashv \sharp$, or with two modes and $\int \dashv \Delta \dashv \Gamma \dashv \nabla$, but it remains to be investigated whether this can provide the right properties for $\int$ beyond adjointness. On the one hand, too much might be true: $\int$ does not preserve identity types, and the general dependently typed rules for F might force it to. On the other, too little might be true: for applications such as relating the shape of the topological circle to the homotopical circle, extra properties are needed, such as $\int \mathbb{R} \cong 1$. Both of these issues can be addressed as in [23] by treating $\int$ not as an abstract adjoint, of the kind we can represent using the mode 2-category, but as a defined type (specifically, a higher inductive), which among other things has the property that it is adjoint to $\flat$ (adjoint logic is still essential for representing $\flat$ and $\sharp$ themselves). Another is to consider ∞-category semantics, rather than the 1-categorical semantics of derivations that we have considered here. A final direction for future work is to look for applications of other mode theories in our generalized adjoint logic beyond the motivating example of triple adjunctions and cohesive homotopy type theory.

Acknowledgments. We thank Jason Reed for helpful discussions about this paper and work, and we thank the anonymous reviewers for helpful feedback on a previous draft.

References

1. Andreoli, J.M.: Logic programming with focusing proofs in linear logic. J. Log. Comput. **2**(3), 297–347 (1992)
2. Benton, P.N.: A mixed linear and non-linear logic: Proofs, terms and models. In: Pacholski, L., Tiuryn, J. (eds.) CSL 1994. LNCS, vol. 933, pp. 121–135. Springer, Heidelberg (1995)
3. Benton, N., Wadler, P.: Linear logic, monads and the lambda calculus. In: IEEE Symposium on Logic in Computer Science. IEEE Computer Society Press (1996)
4. Bezem, M., Coquand, T., Huber, S.: A model of type theory in cubical sets. preprint, September 2013
5. Cavallo, E.: The Mayer-Vietoris sequence in HoTT. In: Talk at Oxford Workshop on Homotopy Type Theory, November 2014
6. Hou, K.B.: Covering spaces in homotopy type theory, talk at TYPES, May 2014
7. Kapulkin, C., Lumsdaine, P.L., Voevodsky, V.: The simplicial model of univalent foundations (2012). arXiv:1211.2851
8. Lawvere, F.W.: Axiomatic cohesion. Theory Appl. Categories **19**(3), 41–49 (2007)
9. Licata, D.R., Brunerie, G.: $\pi_n(S^n)$ in homotopy type theory. In: Certified Programs and Proofs (2013)
10. Licata, D.R., Brunerie, G.: A cubical approach to synthetic homotopy theory. In: IEEE Symposium on Logic in Computer Science (2015)
11. Licata, D.R., Finster, E.: Eilenberg-MacLane spaces in homotopy type theory. In: IEEE Symposium on Logic in Computer Science (2014)

12. Licata, D.R., Shulman, M.: Calculating the fundamental group of the circle in homotopy type theory. In: IEEE Symposium on Logic in Computer Science (2013)
13. Lumsdaine, P.L.: Higher inductive types: a tour of the menagerie, April 2011. https://homotopytypetheory.org/2011/04/24/higher-inductive-types-a-tour-of-th e-menagerie/
14. Lumsdaine, P.L., Shulman, M.: Higher inductive types, (in preparation 2015)
15. MacLane, S.: Categories for the Working Mathematician. Graduate Texts in Mathematics, vol. 5, 2nd edn. Springer, New York (1998)
16. Moggi, E.: Notions of computation and monads. Inf. Comput. **93**(1), 55–92 (1991)
17. Norell, U.: Towards a practical programming language based on dependent type theory. Ph.D. thesis, Chalmers University of Technology (2007)
18. Pfenning, F., Davies, R.: A judgmental reconstruction of modal logic. Math. Struct. Comput. Sci. **11**, 511–540 (2001)
19. Reed, J.: A judgemental deconstruction of modal logic (2009). Accessed www.cs. cmu.edu/jcreed/papers/jdml.pdf
20. Schreiber, U.: Differential cohomology in a cohesive ∞-topos (2013). arXiv:1310. 7930
21. Schreiber, U., Shulman, M.: Quantum gauge field theory in cohesive homotopy type theory. In: Workshop on Quantum Physics and Logic (2012)
22. Shulman, M.: Homotopy type theory VI: higher inductive types, April 2011. https://golem.ph.utexas.edu/category/2011/04/homotopy_type_theory_vi.html
23. Shulman, M.: Brouwer's fixed-point theorem in real-cohesive homotopy type theory (2015). arXiv:1509.07584
24. Shulman, M.: Univalence for inverse diagrams and homotopy canonicity. Math. Struct. Comput. Sci. **25**, 1203–1277 (2015) arXiv:1203.3253
25. Univalent Foundations Program: Homotopy Type Theory: Univalent Foundations of Mathematics (2013). Accessed homotopytypetheory.org/book
26. Voevodsky, V.: A very short note on homotopy λ-calculus, September 2006. https://www.math.ias.edu/vladimir/files/2006_09_Hlambda.pdf

Parallel Feedback Turing Computability

Robert S. Lubarsky[⊠]

Florida Atlantic University, Boca Raton, FL 33431, USA
Robert.Lubarsky@alum.mit.edu

Abstract. In contrast to most kinds of computability studied in mathematical logic, feedback computability has a non-degenerate notion of parallelism. Here we study parallelism for the most basic kind of feedback, namely that of Turing computability. We investigate several different possible definitions of parallelism in this context, with an eye toward specifying what is so computable. For the deterministic notions of parallelism identified we are successful in this analysis; for the non-deterministic notion, not completely.

Keywords: Parallel computation · Feedback · Determinism · Non-determinism · Reflection · Gap-reflection · Admissibility

AMS 2010 MSC: 03D10, 03D60, 03D70, 03E10, 03E75

1 Introduction

Parallelism, as far as the author is aware, has not been studied much in the kind of computability theory done by mathematical logicians (Turing degrees, arithmetic sets, admissibility). This is for a good reason: it can be mimicked, via dovetailing. Using a universal machine, a parallel computation can be simulated by a sequential computation. This is in stark contrast with complexity theory. For instance, an NP problem can be understood as a polynomial problem with parallelism, so the addition of parallelism to polynomial computation results in a new and quite important notion. Quantum computability can also be understood as a kind of parallelism, as can distributed computing.

This paper studies parallelism in an extension of Turing computability where it does make a difference, namely feedback. Feedback was first identified in [6], p. 406–407, even if not under that name, where some of the results of [1] were anticipated. Oddly enough, even though that was a very prominent text for decades, likely the best-known in (using the terminology of the day) recursion theory, no one ever picked up on those ideas. It was re-discovered independently for infinite time Turing machines in [2], where even parallelism was discussed, albeit briefly. Something was actually done with parallelism in [1], where it was

Thanks are due to Nate Ackerman, Cameron Freer, and Anil Nerode for their consultation during the preparation of this work.

© Springer International Publishing Switzerland 2016
S. Artemov and A. Nerode (Eds.): LFCS 2016, LNCS 9537, pp. 236–250, 2016.
DOI: 10.1007/978-3-319-27683-0_17

shown that its addition to feedback Turing computability is non-trivial, in the sense that it gets you strictly more than you had before. (It was also shown there that parallelism added to feedback primitive recursion is essentially trivial in the same sense.) It was left open there just what parallel feedback Turing computable (pfc in what follows) does compute.

In this paper, the issues around parallelism are clarified somewhat. Most fundamentally, as already discussed in [1], there are several different ways that parallelism can be included in this framework. It is not yet clear which are fruitful. By analyzing some of them, we hope to bring this issue along.

In the end, we discuss three. In the next section, we show one to provide no new computational power, and hence (presumably) to be uninteresting. In Sect. 3 we turn to the one from [1]; we expand upon its semantics, but despite that are still unable to characterize just what is so computable, although an upper bound is provided later. Finally, in the last section we define a semantics which is in a sense intermediate between these other two, and are successful in its characterization; we also provide here the upper bound for earlier.

We assume some familiarity with both feedback and parallelism, as presented in [1,2]. To summarize briefly, the oracle in a feedback computation contains the convergence and divergence facts about computations that call that very same oracle. So the oracle is a fixed point: whatever convergence and divergence facts which follow from using the oracle are already contained in the oracle. The particular fixed point we use here is a least fixed point. If a computation queries an oracle about a computation for which the oracle does not have an answer, that computation freezes dead in its tracks. For instance, if a computation ever asks the oracle about itself, it will freeze. This allows for parallelism, since a computation could ask the oracle for a program from a parameterized list of programs which does not freeze.

In the following, the notation $\langle e \rangle(n)$ will be used ambiguously to refer to any notion of feedback Turing computation (taking off of the standard notation $\{e\}(n)$ for regular Turing computation), the choice of which we hope is clear from the context.

2 Absolutely Deterministic Parallelism

If a parallel oracle call about $\langle e \rangle(\cdot)$ is to return an n such that $\langle e \rangle(n)$ does not freeze (if any), there is a clear invitation to non-determinism: which n? Indeed, in [1], a semantics for deterministic parallelism was offered and then quickly passed over, as it turns out for a good reason: it gets you nothing new. Here we show this, if for no other reason than to demonstrate that this definition should no longer be considered.

The idea is that the oracle is supposed to return the "least" n leading to non-freezing, by some measure. The measure to be used is primarily that of ordinal height of a computation. That is, n minimizes the height of the tree of sub-computations. To help keep this paper self-contained, this tree will be presented, albeit in a way different from in [2] or [1], tailored to the purpose at hand.

The tree $D_\alpha^{(e,n)}$ (D for determinism) is defined inductively on α, simultaneously for all e, n, as is whether rank$(e, n) = \alpha$. Assume this is known for all $\beta < \alpha$. Start the run of $\langle e \rangle (n)$, which is considered as taking place at the root of $D_\alpha^{(e,n)}$. Suppose at some stage of that computation, an oracle call e' is made. Then a child of the root is established, to the right of any previous children, for the outcome of this oracle call. Suppose there is an n' such that rank$(e', n') < \alpha$. Then let n' be chosen to minimize this rank; if there is more than one such, then among those pick the least in the natural ordering of ω. The tree $D^{(e',n')} = D_{\text{rank}(e',n')}^{(e',n')}$ is placed at the child, and the value $\langle e' \rangle (n')$ is returned to the main computation, which then continues. If there is no such n', then the computation pauses, and the construction of $D_\alpha^{(e,n)}$ is finished.

If no oracle calls pause, then by this stage α the computation $\langle e \rangle (n)$ is seen to be non-freezing; $D^{(e,n)}$ can be taken to be $D_\alpha^{(e,n)}$ and is the tree of sub-computations; rank$(e, n) \le \alpha$; and the value of $\langle e \rangle (n)$ is the content on the output tape if the main computation ever entered into a halting state, else $\uparrow$ if it did not.

It is not hard to show that the rank of a computation is the ordinal height of its tree of sub-computations. For a freezing computation, i.e. one that remains paused however big α is taken to be, I do not (yet) have a good notion of a tree of sub-computations. For the eternally paused node, which is trying to run, say, e' in parallel, it's paused because for each n' the trees $D_\beta^{(e',n')}$ remain paused, say at $e''_{n'}$. This could be viewed as countable branching from e', but of course this branching is different from that in $D^{(e,n)}$: in the latter tree, the branching shows the sequential computation, and the unsuccessful parallel runs are suppressed; from e', the branching represents all the parallel attempts. Of course, from $e''_{n'}$, the same story continues.

The problem with this notion is that it doesn't get us anything new.

Theorem 1. *If $\langle e \rangle (n)$ does not freeze, then $D^{(e,n)} \in L_{\omega_1^{CK}}$.*

Proof. By induction on the ordinal height of $D^{(e,n)}$. Consider the subtrees $D^{(e',n')}$ that occur on the top level (i.e. children of the root) of $D^{(e,n)}$. Inductively, they are all in $L_{\omega_1^{CK}}$. If there are only finitely many of them, then the ordinal α by which they all appear is easily less that ω_1^{CK}, since the latter is a limit ordinal. $D^{(e,n)}$ is then easily definable over L_α. If there are infinitely many, then the admissibility of ω_1^{CK} must be used to get α to be strictly less than ω_1^{CK}. The set of such $D^{(e',n')}$'s is the range of a Σ_1 definable function f with domain ω, since the run of $\langle e \rangle (n)$ is simply defined, and (mod the oracle calls) continues for ω-many steps; $f(k)$ is then the sub-tree $D^{(e',n')}$ for the kth oracle call.

3   Non-deterministic Parallelism

Since choosing one canonical output to a parallel call didn't work out so well, let's go to other extreme and allow all possible answers. So when a computation makes an oracle call $\langle e \rangle(\cdot)$, an acceptable answer is *any* choice of n such that $\langle e \rangle(n)$ does not freeze. But wait a minute – since computations are non-deterministic, it could be that some runs of $\langle e \rangle(n)$ freeze and others do not, depending on how the oracle calls made while running $\langle e \rangle(n)$ turn out. So what does it mean to say "$\langle e \rangle(n)$ does not freeze?" We take that to be that *some* run of that computation does not freeze, if for no other reason than that is the choice made in [1]. Since in the end we are not able to analyze this as we would like, perhaps it would have been better to say *all* runs do not freeze. Still, the question based on some run not freezing remains, and so we keep to that former notion, leaving the other for future work.

All of this can naturally be summarized in the *tree of runs*, defined below. This is not to be confused with the tree of sub-computations, so central in developing feedback. The tree of sub-computations summarized the sequential running of an algorithm, which can be viewed as traversing that tree, depth-first, from left to right. In contrast, the tree of runs captures the non-determinism. The splitting at a node is the many parallel runs of an oracle call. A single run of the algorithm is a path through the tree. There is no room in this tree for the sub-computations: if a node in the tree of runs represents $\langle e \rangle(n) = k$, the witness to that last computation is not contained in the tree, but rather must be found in the tree of runs for $\langle e \rangle(n)$.

3.1   The Tree of Runs

Definition 1. *The* **tree of runs** *is built from the root (thought of as being on the top) downwards, or, equivalently, as the computation proceeds, starting from the beginning, step 0. Each node has a start, meant to be the state of the computation when that node becomes active, and an end, meant as the state of the computation when the node becomes inactive. The start of the root is the program (e, n) being run. What the end of the root, or any other node for that matter, is, depends. If continuing the computation from the start of the node leads to an oracle call, say $\hat{e}$, then the end of the node is this $\hat{e}$; as need be, we may assume that the state of the computation at that point is also recorded in the node. If no such oracle call exists, then there are two possibilities. One is that after finitely many steps from the start of the node the computation has entered into a halting state. Then the end of the node is this halting state, and the content of the output tape is an output of the main computation. The other possibility is that the computation from the node's start never enters into a halting state, and so it diverges. Then the end of the node is this divergence, symbolically $\uparrow$, which is an output of the main computation.*

Nodes that end in a halting state or with divergence have no children. A node that ends with $\hat{e}$ may have children. For any natural number $\hat{n}$, and any output k of the computation $\langle \hat{e} \rangle(\hat{n})$, there is a child with start $(\hat{e}, \hat{n}, k)$, and which continues the computation of its parent with that start as the answer to the oracle call. Implicitly, and now explicitly, if there are no such $\hat{n}$ and k, then that node has no children, and the computation freezes there.

So a **run of a computation** is exactly a (maximal) path through its tree of runs. A finite output is given by a finite path, ending in a childless node in a halting state. A freezing computation is also given by a finite path, ending in a freezing node. A divergent computation can be given by a finite path, ending in $\uparrow$, and also by an infinite path.

The tree of sub-computations is absent from the tree of runs. It is hidden in the step from a node with end $\hat{e}$ to its children, or to its lack of children, which can be determined only by building $\hat{e}$'s own tree of runs. Of course, this latter tree might sub-contract out its own side-trees, and so on.

Because the semantics is given by a least fixed point, ordinal heights can be associated with these computations (when non-freezing). Ultimately, we will define the height of an output. But we must be careful here: because of the non-determinism, there could be wildly different ways to arrive at the same output. The simple solution to that would be to define the height of an output as the least ordinal among all the ordinals given by the different ways to get to that output. To do this right, one must define the height of a run of a computation, or, actually, the height of a *hereditary run*.

A **hereditary run** of a non-freezing computation is a run of that computation, along with an assignment, to each oracle call in the run (i.e. node in the run with end $\hat{e}$), with answer $(\hat{e}, \hat{n}, k)$ (i.e. the child in this run of that aforementioned node has start $(\hat{e}, \hat{n}, k)$), a hereditary run of $(\hat{e}, \hat{n})$ with output k.

The **height of a hereditary run** is defined inductively as the least ordinal greater than the heights of all of the sub-runs, meaning the hereditary runs assigned to oracle calls along the way.

The **height of a computation** $\langle e \rangle (n) = k$ is the smallest height of any hereditary run of such a computation. We will want to show that this is absolute among all transitive models.

Define $T_\alpha^{(e,n)}$, the sub-tree of the tree of runs of (e, n) which contains only those children of rank less than α, inductively on α.

For $\alpha = 0$, this tree contains only the root; if $\langle e \rangle (n)$ makes an oracle call then $T_0^{(e,n)}$ does not witness any output, else it witnesses either some finite k or $\uparrow$ as an output.

More generally, if $\beta < \alpha$, then $T_\beta^{(e,n)} \subseteq T_\alpha^{(e,n)}$. Furthermore, if a node in $T_\alpha^{(e,n)}$ ends with an oracle call $\hat{e}$, and there are $\beta < \alpha, \hat{n}$, and k (including $\uparrow$) such that $T_\beta^{(\hat{e}, \hat{n})}$ witnesses that k is an output, then the child with start $(\hat{e}, \hat{n}, k)$ is in $T_\alpha^{(e,n)}$.

The outputs witnessed by $T_\alpha^{(e,n)}$ are the outputs of any terminal node (i.e. k if a node ends in a halting state with output k, or $\uparrow$ if a node ends with $\uparrow$), and also $\uparrow$ if $T_\alpha^{(e,n)}$ is ill-founded.

Notice that the height of $\langle e \rangle (n) = k$ is at most α iff $T_\alpha^{(e,n)}$ witnesses k as an output.

Proposition 1. *The height of $\langle e \rangle(n) = k$ is absolute among all transitive models.*

Proof. Inductively on α, the trees $T_\alpha^{(e,n)}$ and the outputs they witness are absolute. The outputs witnessed by terminal nodes are clearly absolute, individual nodes being finite, and for divergence, well-foundedness is absolute for well-founded models.

3.2 Functions and Ordinal Notations

Ultimately we would like to characterize just what is parallel feedback computable. In the context of multi-valued functions, what this means should be clarified.

Definition 2. *A function f is **parallel feedback computable (pfc)** if there is an index e such that $\langle e \rangle(\cdot)$ is single valued and $\langle e \rangle(n) = f(n)$. A set is pfc if its characteristic function is.*

We would like to know what functions are pfc, and what relations are pfc.

While it should be no surprise that functions offer some benefits over relations, let's bring out a particular way that happens. Consider the index e which on any n returns both 0 and 1. (In more detail, let p be the parity function: $\{p\}(n)$ is 0 when n is even, 1 when odd. Let $\langle e \rangle(n)$ make a parallel call to p and return its output). Notice that the characteristic function of any set at all is given by some run of e. So if you're non-deterministically searching for, say, the truth set of some L_α, there may well be a pfc function that gives you what you want, but you can't distinguish that from this e. And it does you no good to pick one non-deterministically, because if you pick e, when you go to use it again later, you might get different answers.

Since we expect that the analysis of this will involve computing initial segments of L, we might have need of notation for ordinals, which can be defined à la Kleene's $\mathcal{O}$. In honor of this history, and since the current subject is parallelism, we will call it $\mathcal{P}$. Because of the non-determinism present, there are several options for how this can be defined (in the limit case).

Definition 3. *Functional $\mathcal{P}$ ($f\mathcal{P}$) is defined inductively:*

- *$0 \in f\mathcal{P}$ and $\mathrm{ord}(0) = 0$.*
- *If $a \in f\mathcal{P}$ then $2^a \in f\mathcal{P}$ and $\mathrm{ord}(2^a) = \mathrm{ord}(a) + 1$.*
- *If $\langle a \rangle(\cdot)$ is a function, and for all n we have $\langle a \rangle(n) \in f\mathcal{P}$, then $3 \cdot 5^a \in f\mathcal{P}$ and $\mathrm{ord}(3 \cdot 5^a) = \sup_n\{\mathrm{ord}\langle a \rangle(n)\}$.*

Definition 4. *Strict $\mathcal{P}$ ($s\mathcal{P}$) is defined inductively:*

- *$0 \in s\mathcal{P}$ and $\mathrm{ord}(0) = 0$.*
- *If $a \in s\mathcal{P}$ then $2^a \in s\mathcal{P}$ and $\mathrm{ord}(2^a) = \mathrm{ord}(a) + 1$.*
- *If $\langle a \rangle(\cdot)$ is a total relation, and for all n and any possible output k_n of $\langle a \rangle(n)$ we have $k_n \in s\mathcal{P}$, and moreover $\mathrm{ord}(k_n)$ is independent of the choice of k_n (for a fixed n), then $3 \cdot 5^a \in s\mathcal{P}$ and $\mathrm{ord}(3 \cdot 5^a) = \sup_n\{\mathrm{ord}(k_n)\}$, where k_n is any output for $\langle a \rangle(n)$.*

Definition 5. *Loose* $\mathcal{P}$ *($l\mathcal{P}$) is defined inductively:*

- $0 \in l\mathcal{P}$ *and ord(0) = 0.*
- *If $a \in l\mathcal{P}$ then $2^a \in l\mathcal{P}$ and ord(2^a) = ord(a) + 1.*
- *If $\langle a \rangle(\cdot)$ is a total relation, and for all n and any possible output k_n of $\langle a \rangle(n)$ we have $k_n \in l\mathcal{P}$, and $\sup_n\{\text{ord}(k_n)\}$ is independent of the choice of k_n's, then $3 \cdot 5^a \in l\mathcal{P}$ and ord($3 \cdot 5^a$) = $\sup_n\{\text{ord}(k_n)\}$, where k_n is any output for $\langle a \rangle(n)$.*

Clearly, $f\mathcal{P} \subseteq s\mathcal{P} \subseteq l\mathcal{P}$.

Proposition 2. *Every pfc well-ordering is isomorphic to one given by a functional ordinal notation.*

Proposition 3. *If X is pfc then $\mathcal{O}^X$ is pfc and ω_1^X has a functional ordinal notation.*

That $\mathcal{O}^X$ is pfc was proven in [1]. This is a slight extension of that argument.

Proposition 4. *If α has a loose ordinal notation then the Σ_1 truth set Tr_α of $L_{\omega_\alpha^{CK}}$ is pfc (where, as a function of α, ω_α^{CK} enumerates the closure of the set of admissible ordinals).*

Proof. Let $e \in \mathcal{P}^*$ be a fixed representation of α. By the recursion theorem, we can do this inductively on the ordinal height of $f <_{\mathcal{P}} e$.

If $f = 0$, then $Tr_f = \emptyset$.

If $f = 2^g$, then $Tr_f = \mathcal{O}^{Tr_g}$ from the previous proposition. (It is standard hyperarithmetic theory that $\mathcal{O}^X$ is Turing equivalent to the Σ_1 truth predicate of $L_{\omega_1^{CK}}$).

If $f = 3 \cdot 5^g$, then the truth or falsity of any Σ_1 assertion ϕ in the limit structure can be determined as follows. Let n run through ω, and see whether ϕ is true according to each $Tr_{g(n)}$ in turn. If you ever find such an n making ϕ true, halt, else continue. Using feedback, ask whether that computation halts. If so, then ϕ is true in the limit structure; else ϕ is false there.

Because of that last proposition, I bet the loose notations are ultimately the best, since they seem to capture the flavor of this kind of computation.

Proposition 5. *The characteristic function of $T_\alpha^{(e,n)}$ (along with the start and end of each node) is computable from a loose ordinal notation for α, as are the outputs witnessed by $T_\alpha^{(e,n)}$.*

With a bit of work, this could be presented as a corollary of the previous proposition, since $T_\alpha^{(e,n)}$ and its outputs are definable over $L_{\omega_\alpha^{CK}}$.

Proof. By a simultaneous induction on ordinal notations.

The only notation for the ordinal 0 is 0. To compute $T_0^{(e,n)}$, one first asks the oracle whether computing $\langle e \rangle(n)$ will ever lead to an oracle call. If so, one runs $\langle e \rangle(n)$ until that call, which becomes the end of the root, and then stops. If not, one asks the oracle whether computing $\langle e \rangle(n)$ will ever halt. If so, one runs it until it halts; if not, then the output is $\uparrow$.

Consider the ordinal notation $a = 2^b$ for $\alpha = \beta + 1$. Of course, the root of $T_\alpha^{(e,n)}$ is computable, as above. For any node in $T_\alpha^{(e,n)}$, to see whether a child is in $T_\alpha^{(e,n)}$, we may assume the node ends with $\hat{e}$. A child starting with $(\hat{e}, \hat{n}, k)$ is in $T_\alpha^{(e,n)}$ iff $T_\beta^{(\hat{e},\hat{n})}$ witnesses that k is an output, which inductively is computable from b. The end of such a node is deterministic in the start. To compute whether k is witnessed to be an output, one can use the oracle to see whether the search through $T_\alpha^{(e,n)}$ for a terminal node with output k will halt. In addition, when $k = \uparrow$, check whether $T_\alpha^{(e,n)}$ is well-founded, which is computable in its hyperjump (cf. the penultimate proposition).

Now consider the ordinal notation $a = 3 \cdot 5^b$. We must decide membership in $T_\alpha^{(e,n)}$ of children of nodes ending in $\hat{e}$. For the child starting with $(\hat{e}, \hat{n}, k)$, use the oracle to see whether the search for an i such that, with $\beta_i = \mathrm{ord}(b(i))$, the tree $T_{\beta_i}^{(\hat{e},\hat{n})}$ witnesses that k is an output, halts. The determination of $b(i)$ is, of course, non-deterministic, as is the value β_i, but as β_i is guaranteed to be cofinal in α, this makes no difference. The computation of the outputs witnessed is as above.

The hope is that the structure just identified will help in determining the pfc functions and relations, which we have not been able to do. Although the next section is dedicated to the study of a different kind of computation for its own sake, it also provides at least a coarse upper bound for those studied here.

4   Context-Dependent Determinism

4.1   Semantics

The problem of the first alternative offered is that it's too restrictive, and so gives you nothing new. The problem with the second is that it's too liberal, allowing for multi-valuedness, and so we couldn't analyze it. This time we're going for something in the middle. Any oracle call will return at most one value, but possibly a different value every time it's called.

The semantics begins just as in the non-deterministic case. Trees $C_\alpha^{(e,n)}$ (C for context) are defined inductively on α. The new intuition here is that these trees are built until an output is seen, and that first output is taken as the value of $\langle e \rangle(n)$. More precisely, $C_\alpha^{(e,n)}$ yields an output if it contains a halting node (with some integer output k) or a diverging node (with output $\uparrow$), or is ill-founded (with output $\uparrow$). Let α be the least ordinal such that $C_\alpha^{(e,n)}$ yields an output. If it yields more than one output, pick the left-most one. That is, starting at the root, traverse the tree downwards. Every non-terminal node ends with an oracle call $\hat{e}$. The child to be followed has start $(\hat{e}, \hat{n}, k)$, where $\hat{n}$ is the least natural number such that the tree beneath that node yields an outcome (and k is the value of $\langle \hat{e} \rangle(\hat{n})$).

As an example of this semantics in practice, the earlier proof that $\mathcal{O}^X$ is pfc from X [1] still works. The way that construction goes, given a non-well-founded order, and an n in the non-standard part, if k is in the standard part, it won't be chosen as a successor step after n, because that will definitely lead to a freezing state. Only a non-standard k (less than n in this ordering) will be chosen, and in fact that least such k in the natural ordering of ω will be.

4.2   Lemmas

Lemma 1. *There is a program which, on input e, diverges if e computes (the characteristic function of) the truth set of a model of some computable theory T, and freezes otherwise.*

We assume here some standard coding of syntax into arithmetic. The model can be taken to be a structure on, say, the odd integers, so that the even integers can be used for the symbols of the language, and formulas with parameters can be considered. Of course, this program can easily be converted into one that halts instead of diverges: ask the oracle about this program, and if the answer comes back "divergent," then halt. It will be easy to see that in some instances it can be recognized that e does not compute such a set, and our program could return that instead of diverging; but if, say, $\langle e \rangle(0)$ freezes, then any such program as ours would have to freeze, and there seemed to be no benefit in a program that sometimes recognizes when e is not as desired and sometimes freezes.

Proof. It is feedback Turing computable to dovetail the generation of T, the computations of $\langle e \rangle(n)$ for all n, and the check that latter theory is complete, consistent, and contains T. If e computes such a model, this procedure will never end; if e finds some violation, the procedure can be taken to freeze. If some $\langle e \rangle(n)$ freezes, the procedure will necessarily freeze.

We will be using this to see if e codes a model of $V = L_\alpha$. We do not sharply distinguish between the Σ_1 truth set of some L_α and the full truth set, since this computational paradigm can easily shuttle between them.

Lemma 2. *There is a program such that, if e computes a partial order on a subset of ω, on input e it will return 0 if e's order is well-founded and 1 if ill-founded.*

Proof. This is a lot like the proof of the computability of $\mathcal{O}$.

For pre-processing, check whether the domain of e is finite. If so, you have your answer. Else, continue.

First we check for well-foundedness. Go through the natural numbers, and for each such n, if n has no e-predecessors (determined by an oracle call), halt, else run this same procedure, via the fixed-point or recursion theorem, on the same order restricted to those elements e-less than n. In the tree of sub-computations, the children of a node given by n are exactly the e-predecessors of n. So this tree is well-founded iff $<_e$ is well-founded. So this procedure diverges iff $<_e$ is well-founded, else it freezes.

To check for ill-foundedness, run in parallel the following procedure on each $n \in \omega$. If n has no predecessor, freeze. Else, by the fixed point theorem, run this same procedure on the same order restricted to those elements e-less than n. In the tree of runs, the children of a node given by n are exactly the e-predecessors of n. So this tree is well-founded iff $<_e$ is well-founded. Since the terminal nodes all freeze, the only possible non-freezing semantics is an infinite descending path, which exists exactly when $<_e$ is ill-founded.

Now run both of those checks in parallel. Whichever one does not freeze is what tells you whether $<_e$ is well- or ill-founded.

As usual, it is easy to see that what can be computed is exactly some initial segment of L. We will shortly see just what this initial segment is. Before that, we will prove some lemmas which handle some simpler cases, partly to get the reader (and author!) used to the kind of arguments employed, and partly so in the main theorem we can ignore some of the cases of weaker, messier ordinals, and focus on just the more strongly closed ones.

Lemma 3. *The supremum α of the computable ordinals is admissible.*

Proof. Suppose not. Let $f : \omega \to \alpha$ witness α's inadmissibility. For each n, using the previous lemmas, one can check whether $\langle e \rangle$ codes a model of "$V = L_\gamma$ is the least admissible set in which $f(n)$ is defined," and if so whether the model so coded is well-founded. On many inputs this will freeze, but since by hypothesis α is the least non-computable ordinal, there is at least one e_n on which this halts (possibly more, allowing for some flexibility in the coding). By making a parallel call of all natural numbers, one can produce such an e_n.

To see whether a Σ_1 formulas ϕ is in the Σ_1 truth set for L_α, consider the procedure which runs through each n, finds a truth set for $f(n)$ as above, and stops whenever ϕ shows up as true in one of those sets. Now ask the oracle whether that procedure halts. If so ϕ is true in L_α, else not.

Lemma 4. *α is greater than the least recursively inaccessible.*

Proof. The following procedure will generate the Σ_1 truth set of the first recursively inaccessible.

Start with (a code for) the truth set of $L_{\omega_1^{CK}}$. We will describe a procedure which pieces larger and larger initial segments of L together, which diverges (continues indefinitely) as long as it's still working on the first inaccessible, and which freezes whenever it finds a contradiction in what it has done so far.

At any stage along the way, there will be a well-founded model of $V = L_\gamma$, as well as a finite set of Π_1 sentences the procedure is committed to making true. As soon as the model at hand falsifies one of those sentences, then the procedure freezes, because it sees that the jig is up.

Dovetail consideration of all countably many Σ_1 formulas $\phi(x, y, z)$ and all countably many sets A and tuples b that show up in the models produced in this construction. At stage n we are considering a certain ϕ, A, and b, and will decide whether we think $\forall a \in A \; \exists y \; \phi(a, y, b)$ is true or false in the first recursively inaccessible. In parallel, choose either true or false. Moreover, if you choose true, then you must provide a well-founded model of $V = L_\gamma$ extending the previously chosen model by at least one admissible, in which the chosen formula with parameters is true, and which also models there is no recursively inaccessible. If you choose false, then you must also choose a specific $a \in A$, and include in the set of sentences "$\forall y \ \neg\phi(a, y, \boldsymbol{b})$".

Since this construction has no halting condition, the only way it can not freeze is if it diverges. It cannot diverge by always making the chosen formula false, if for no other reason than there are infinitely many total Σ_1 functions in the starting model, and they cannot consistently be made partial. So infinitely often the model under consideration will be extended by at least one admissible. Hence the limit model will be an initial segment of L which is a limit of admissibles. Let ϕ be Σ_1 and $A, \boldsymbol{b}$ be in the limit model. Suppose it's true in this model that $\forall a \in A \ \exists y \ \phi(a, y, \boldsymbol{b})$. When that formula came under consideration, it could not have been deemed false, because then we would have committed ourselves to a specific counter-example, and that counter-example would have been seen to be invalid at some point, leading to a freezing computation. So the formula was deemed to be true. Hence a model was picked in which the induced relation was total, thereby providing a bound on the range. Hence the limit model is admissible. Since it's a limit of models of "there is no recursively inaccessible," it is itself the least recursively inaccessible.

We have just argued that any divergent run of this program produces the least recursively inaccessible. Furthermore, there are divergent runs, by always choosing whatever is in fact true of that ordinal.

4.3   Main Theorems

Definition 6. *Let Γ be a collection of formulas, X a class of ordinals, and ν^{+X} the least member of X greater than ν. We say that α is Γ-reflecting on X if, for all $\phi \in \Gamma$, if $L_{\alpha+x} \models \phi(\alpha)$, then for some $\beta < \alpha$, $L_{\beta+x} \models \phi(\beta)$.*

We are interested in the case $\Gamma = \Pi_1$ and $X =$ the collection of admissible ordinals. For this choice of X, we abbreviate ν^{+X} by ν^+, which is standard notation for the next admissible anyway. This is called Π_1 **gap-reflection on admissibles**. Let γ be the least such ordinal.

It may seem like a strange notion. But this is not the first time it has come up. Extending work in [5], it was shown in [3] that such ordinals are exactly the Σ_1^1 reflecting ordinals. (In this context, the superscript 1 refers not to reals but to subsets of the structure over which the formula is being evaluated). The reason this topic came up in the latter paper is that a particular case of its main theorem is that γ is the closure point of Σ_2-definable sets of integers in the μ-calculus. (The μ-calculus is first-order logic augmented with least and greatest fixed-point operators. In this context, Σ_2 refers to the complexity of the fixed points in the formula, namely, in normal form, a least fixed point in front, followed by a greatest fixed point, followed by a fixed-point-free matrix). In [5] it was also shown that the least Σ_1^1 reflecting ordinal is also the closure point of

Σ_1^1 monotone inductive definitions. (Here the superscript does refer to reals). Furthermore, that is the same least ordinal over which winning strategies for all Σ_2^0 games are definable (Solovay, see [4] 7C.10 or [10]). As though that weren't enough, [9] shows the equivalence of closure under Σ_1^1 monotone inductive definitions with the Σ_1^1 Ramsey property. (For all Σ_1^1 partitions P of ω there is an infinite set $H \subseteq \omega$ such that the infinite subsets of H are either all in P or all not in P). With all of these applications, this definition counts as natural.

Theorem 2. *The ordinals so computable are exactly those less than γ.*

So there is an intimate connection between parallel feedback computability and all of the other notions listed above. This was not expected. In the simpler case of feedback Turing computability [1], it was really no surprise that it turned out to be the same as hyperarithmeticity, as both are essentially adjoining well-foundedness to computation. But we have no intuition, even after the fact, in support of the current result.

Proof. For one direction, we will argue that no computation $\langle e \rangle(n)$ can be witnessed to converge or diverge from stage γ onwards. Notice that for any $\gamma' > \gamma$, if $T_{\gamma'}^{(e,n)}$ is different from $T_{\gamma}^{(e,n)}$, that can only be because some other computation $\langle e' \rangle(n')$ was seen to converge or diverge at some stage at least γ and less than γ'. Tracing back the computation of $\langle e' \rangle(n')$, we are eventually led to a computation that was seen to converge or diverge at exactly stage γ. Since γ is a limit of admissibles, there are no new terminal nodes on any tree of runs at stage γ. Hence there is some computation $\langle e \rangle(n)$ such that $T_{\gamma}^{(e,n)}$ is ill-founded, but $T_{\beta}^{(e,n)}$ is well-founded for any $\beta < \gamma$. How could the ill-foundedness of $T_{\gamma}^{(e,n)}$ be most economically expressed? Since γ is the γth admissible ordinal, $T_{\gamma}^{(e,n)}$ is definable over L_γ. It is a basic result of admissibility theory that a tree in an admissible set is well-founded iff there is a rank function from the tree to the ordinals in that very same admissible set. So the ill-foundedness of such a tree is witnessed by the non-existence of such a function in any admissible set containing the tree. In the case at hand, that is a Π_1 statement in L_{γ^+} with parameter γ. By the choice of γ, this reflects down to some smaller β. So $T_{\beta}^{(e,n)}$, for some smaller β, was already seen to be ill-founded. So there can be no new computation values at stage γ, and hence not beyond either.

For the converse, let β be strictly less than γ; by lemmas 2 and 3, we can assume that β is a limit of admissibles. Assume inductively that for each $\alpha < \beta$ there is an e such that $\langle e \rangle(\cdot)$ is the characteristic function of the Σ_1 truth set of L_α. Let ϕ witness that β is not Π_1 gap-reflecting on admissibles: so ϕ is Π_1, and $L_{\beta^+} \models \phi(\beta)$, but if $\alpha < \beta$ then $L_{\alpha^+} \not\models \phi(\alpha)$. We must show that (the characteristic function of) the Σ_1 truth set of L_β is computable.

As in lemma 4, start with (a code for the Σ_1 truth set of) $L_{\omega_1^{CK}}$. At any stage along the way, there will be a well-founded model of $V = L_\alpha$, as well as two finite sets (both empty at the beginning) of sentences. The intent of this construction is that, if it continues for ω-many steps, the union of the L_α's so chosen will be L_β, all of the sentences in the first set will be true in L_β, and the second set will provide a term model of $V = L_{\beta+}$.

The action at any stage is much as in the previous lemma. First, check for the consistency of a theory, to be described below. If an inconsistency is found, freeze. Else we are going to continue building the ultimate model. This involves interleaving steps to make sure that the union of the chosen L_α's, L_δ, is admissible (and $\delta \le \beta$), with steps to insure that $L_{\delta+} \models \phi(\delta)$ (guaranteeing $\delta \ge \beta$). We assume a dovetailing, fixed at the beginning, of all (countably many) formulas ψ with parameters. For the formulas in the first set, the parameters are the sets in the L_α's chosen along the way. For the formulas in the second set, the parameters include, in addition to the members of the L_α's, also constants c_i for the term model, as well as a dedicated constant we will ambiguously call δ, since the ordinal δ is its intended interpretation.

At any even stage $2n$, consider the nth formula of the form $\forall a \in A\ \exists y\ \psi(a, y, \boldsymbol{b})$, where ψ is Σ_1 and the parameters are from the L_α at hand. In parallel, choose it to be either true or false. Moreover, you must provide a well-founded model of $V = L_\alpha$, extending the previously chosen model by at least one admissible. Furthermore, if you had deemed the formula to be true, then it must hold in the chosen L_α; if false, then you must also choose a specific $a \in A$, and include in the first set of sentences "$\forall y\ \neg\psi(a, y, \boldsymbol{b})$". Notice that this step includes as a degenerate case those instances in which ψ does not depend on a, thereby forcing us to decide all Σ_1 and Π_1 formulas. Finally, it must be the case that $\alpha < \beta$, which can be verified computably, since it needs only a well founded model of $V = L_{\alpha+}$ (which exists by the inductive hypothesis and the choice of β) which also satisfies "$\neg\phi(\alpha) \wedge \forall \nu < \alpha\ L_{\nu+} \not\models \phi(\nu)$".

At an odd stage $2n + 1$, consider similar to the above the nth formula of the form $\forall a \in A\ \exists y\ \psi(a, y, \boldsymbol{b})$, where ψ is Σ_1, only this time the parameters are for the second set (that means the parameters are from an already chosen L_α and the c_i's and δ). Include in the second set either "$\forall a \in A\ \exists y \in \tau\ \psi(a, y, \boldsymbol{b})$", for some term τ, or "$\tau \in A \wedge \forall y\ \neg\psi(\tau, y, \boldsymbol{b})$," for some term τ. Of course, this step is meant to include all possible degenerate cases, such as Σ_1 assertions, even quantifier-free sentences. Also, if "$\tau < \delta$" for some term τ is ever included in the second set, then, extending L_α if need be, for some $\epsilon < \alpha$ the sentence "$\tau = \epsilon$" is included in the second set.

With regard to the theory referenced above but there left unspecified, at any stage along the way it will be "$V = L_{\delta+}$ is admissible, and δ is admissible, and $\alpha < \delta$ (where L_α is the model we have at this stage), and everything in the first set is true in L_δ, and everything in the second set is true in V".

For this computation, the tree of runs has neither halting nor divergence nodes (since, whenever it does not freeze, it makes another oracle call). It is ill-founded, since there is a run of the computation which does not halt, namely one using the truth about L_β and $L_{\beta+}$ to make decisions along the way. We would like to show that along any infinite path in the tree of runs, the induced δ equals β.

Consider the term model induced by the second set. There is an isomorphism between the term δ and the union of the α's chosen along the way: on the one hand, the assertion "$\alpha < \delta$" was included in the theory along the way, and on the other, anything ever deemed less than δ was forced to be less than some α. So we can consider the term model as including some (standard) ordinal δ. Also, this δ is at most β, since each α is less than β. The next observation is that this term model satisfies "$V = L_{\delta^+}$ is admissible," by the Henkinization (choice of explicit witnesses) performed on the second set. Of course, the term model might well be ill-founded. But its well-founded part has ordinal height the real δ^+. By the downward persistence of Π_1 sentences, since $\phi(\delta)$ holds in the term model, it holds in the actual L_{δ^+}. By the choice of ϕ, δ is at least as big as β.

We must turn this procedure into a way of getting the characteristic function for the truth set of L_β. For any Σ_1 sentence χ, run the procedure as above, with χ and $\neg\chi$ each separately, in parallel, included in the first set. The false option is inconsistent and so any such computation will freeze, so the answer you will get is the true option, along with the information that the procedure diverges.

Corollary 1. *For $\beta < \gamma$, the order-types of the $\Sigma_1(L_\beta)$-definable well-orderings of ω are the ordinals less than β^+.*

This is a generalization of the earlier result that the order-types of the Π_1^1 well-orderings are cofinal in ω_2^{CK}. Sacks [7], giving this special case as an exercise (p. 51, 7.10), attributes it to Richard Platek, who never published a proof. Although Platek may have been the first to notice this (Sacks in personal correspondence dates it from the '60s), Tanaka [8] seems to have discovered it independently.

The corollary as stated is not the optimal result, since the conclusion holds for any β which is Σ_1 projectible, by arguments similar to Tanaka's. It's just that this more general result is no longer a corollary to the theorem.

Proof. For simplicity, assume that β is a limit of admissibles. The construction of the theorem is of an ill-founded tree T_β, Σ_1 definable over L_β, such that any infinite path yields a term model of $V = L$ with ordinal standard part β^+. If the well-founded nodes all had rank less than some $\beta' < \beta^+$, then they could all be distinguished from the non-well-founded nodes definably over $L_{\beta'}$. So an infinite path, and hence such a term model, is also definable over $L_{\beta'}$. It is then easy (which we can here take to mean "definable over $L_{\beta'}$") to read off all the reals in this model. This includes reals with L-rank cofinal in β^+. This is a contradiction. Hence, for any $\beta' < \beta$, there is a node in T_β with that rank. The nodes of T_β are labeled with pairs (e, n). They also have associated with them two finite sets of formulas. The formulas are just finite pieces of syntax, except for the parameters from L_β's. But L_β is the Σ_1 Skolem hull of ω, which provides an integer name for each of its members (for instance, a Σ_1 formula that it uniquely satisfies). So each formula can be coded by a natural number. All told, each node can be represented by a natural number. This produces an ordering of a subset of ω with rank β'. To get this to be a well-ordering, it suffices to take the Kleene-Brouwer ordering of that tree.

Happily, the work done also enables us to determine at least an upper bound for the non-deterministic computations.

Theorem 3. *Any relation computable via a non-deterministic parallel feedback Turing machine, as in the previous section, is $\Sigma_1(L_\gamma)$.*

Proof. By much the same argument as before. The only possible values come from halting nodes, divergent nodes, and the ill-foundedness of trees. A node is seen to halt at a successor stage, and γ is not a successor ordinal. A node is seen to diverge at a stage of the form $\alpha + \omega$, and γ is not of that form. As for the last possibility, the tree $T_\gamma^{(e,n)}$ is Δ_1 definable in L_{γ^+} with parameter γ. If it's not well-founded, that fact is Π_1 expressible in L_{γ^+}. By the choice of γ, a smaller $T_\alpha^{(e,n)}$ was already ill-founded, so divergence was already a value for $\langle e \rangle(n)$. Hence there are no new possible values for any computation at or after stage γ.

References

1. Ackerman, N., Freer, C., Lubarsky, R.: Feedback turing computability, and turing computability as feedback. In: Proceedings of LICS 2015, Kyoto, Japan (2015). Accessed https://math.fau.edu/lubarsky/pubs.html
2. Lubarsky, R.: ITTMs with feedback. In: Schindler, R. (ed.) Ways of Proof Theory, pp. 341–354. Ontos (2010). Accessed http://math.fau.edu/lubarsky/pubs.html
3. Lubarsky, R.: μ-definable set of integers. J. Symbolic Log. **58**(1), 291–313 (1993)
4. Moschovakis, Y.: Descriptive Set Theory. First edition North Holland (1987); second edition AMS (2009)
5. Richter, W., Aczel, P.: Inductive definitions and reflecting properties of admissible ordinals. In: Fenstad, H. (eds.) Generalized Recursion Theory, pp. 301–381. North-Holland (1974)
6. Rogers, H.: Theory of Recursive Functions and Effective Computability. McGraw-Hill, New York (1967)
7. Sacks, G.: Higher Recursion Theory. Springer, New York (1990)
8. Tanaka, H.: On analytic well-orderings. J. Symbolic Log. **35**(2), 198–204 (1970)
9. Tanaka, K.: The Galvin-Prikry theorem and set existence axioms. Ann. Pure Appl. Log. **42**(1), 81–104 (1989)
10. Tanaka, K.: Weak axioms of determinacy and subsystems of analysis II (Σ_2^0 Games). Ann. Pure Appl. Log. **52**(1–2), 181–193 (1991)

Compactness in the Theory of Continuous Automata

Scott Messick$^{(\boxtimes)}$

Department of Mathematics, Cornell University, Ithaca, NY 14850, USA
sbm73@cornell.edu

Abstract. We develop a topological theory of continuous-time automata which replaces finiteness assumptions in the classical theory of finite automata by compactness assumptions. The theory is designed to be as mathematically simple as possible while still being relevant to the question of physical feasibility. We include a discussion of which behaviors are and are not permitted by the framework, and the physical significance of these questions. To illustrate the mathematical tractability of the theory, we give basic existence results and a Myhill-Nerode theorem. A major attraction of the theory is that it covers finite automata and continuous automata in the same abstract framework.

Keywords: Continuous automata · Continuous-time computation · Analog computation · Topological monoids · Myhill-Nerode theorem · Young measures

1 Introduction

Continuous-time computations present a number of novel theoretical challenges. Our main purpose here is to suggest a simple set of effectiveness criteria which a continuous-time process, however it is mathematically defined, ought to satisfy in order that it may be feasibly implemented, at least approximately, by a physical device.

The main strategy we mean to employ is to reason by analogy with the theory of finite automata. The basic idea is to replace the finite set of states and finite alphabet each by compact topological spaces and to insist that all maps be continuous. In carrying out this idea, we will have to make precise what we mean by a continuous-time process. This clarification will take some work—in fact about half the paper, including a detour through the subject of topologies on spaces of functions.

Much previous work has also touched on notions of effectiveness for continuous-time computation. For example, [7] considers automata over continuous-time but which nonetheless have finite (discrete) sets of states. Note that for us, since all maps are continuous including state evolution, the state cannot move between different connected components of the state space. So we are dealing more or less strictly with a continuous-time, continuous-space model of computation.

© Springer International Publishing Switzerland 2016
S. Artemov and A. Nerode (Eds.): LFCS 2016, LNCS 9537, pp. 251–266, 2016.
DOI: 10.1007/978-3-319-27683-0_18

Jeandel [5] goes the other way and considers discrete-time, continuous-space computation. The concept of hybrid automata [4,6] includes a direct axiomatization of the possibility of both discrete and continuous state evolution, with various possible models considered for the continuous part. Hybrid automata theory has been very successful in practical applications to real-world hybrid systems (systems which include interacting analog and digital components). See [3] for a survey on all these and related ideas. Here we isolate a purely continuous notion of computation for detailed theoretical study.

Our motivation for requiring all maps to be continuous stems from the insight of constructive analysis that for any totally defined function to be computable in any practical sense, it has to be continuous. Roughly, the idea is that we have no hope of exactly specifying a point in an infinite space (at least in general), so we are going to have to make do with approximations. A topology on a space specifies precisely what it means to approximate a point in the space, at least in the limit. A metric goes further and actually gives an absolute notion of quality of approximation. Either way, for a map from an input space to an output space to be computable, it is necessary that an approximation of the output can be determined from a sufficiently good approximation of the input.

The above considerations about continuity and effectiveness have been given rigorous treatment from a number of different perspectives, including constructive analysis, such as in Bishop's classic purely constructive text [2], and in computable analysis—a good modern reference is [8]. Brouwer controversially claimed, as part of his intuitionistic philosophy of mathematics, that *all* functions must be continuous. Here we do not attempt to be fully constructive and do not adopt any of these frameworks. We are content to take them as the motivation for the requirement of continuity which we view as a necessary, if not quite sufficient, condition for effectiveness.

1.1 A Sample Menu of Continuous Devices

Before proceeding further, let us establish context by giving examples of behaviors which we might, in our naïveté, expect from a continuous device. To be clear, we are *not* saying that all of these examples are actually effectively implementable. In fact some of them will be definitively ruled out by our criteria given later. They are examples to illustrate what is at stake.

Throughout this paper we assume a deterministic, automatic style of computation.[1] By "automatic", we mean that the computation updates its state instanteously in response to ongoing input. There is no external memory and no waiting for the machine to halt. For the time being we treat the input as a function of time which takes values in some alphabet Σ. To emphasize that input is changing over time, we may use the phrase *input signal*.

---

[1] While we are interested in continuous computation generally, the theory of deterministic automata provides a nice starting point because of its simplicity. In addition, the necessity of constant, instantaneous updates is a feature of many real-world continuous-time problems.

Within these restrictions, the computation style of the examples still varies over a couple of dimensions:

Alphabet Type. Discrete or continuous? Basic respective examples are $\Sigma = \{0, 1\}$ and $\Sigma = [0, 1]$.

Output Style. Recognizer or transducer? Traditionally, an automaton would fit in one of these two categories. A recognizer has a set of accept states and there by computes some language (set of inputs). A transducer would actively put out an output signal of a similar kind as the input, perhaps over a different alphabet. Here we also allow another style which we call a *deducer*. A deducer is like a recognizer but with more than two, usually infinitely many, outcomes. It computes a function from all possible input signals into some static space of possible outcomes. In the discrete world all deducers could be reduced to a finite collection of recognizers working together; but there is no obvious analogue in the continuous case.

Input Restrictions. What sorts of functions are we going to allow? A priori there are many possibilities. Continuous? Lipschitz? Smooth? Piecewise versions of one of these? Piecewise constant? We have deliberately not answered this question yet because it turns out to be very important and will get much attention later.

Here are the examples.

Time Counter. For any alphabet Σ, a deducer which tracks the amount of time the signal spends inside a fixed subset $A \subseteq \Sigma$.

Delay. A transducer which, for any alphabet, outputs a signal on the same alphabet delayed in time by a fixed value τ. The output for the first τ time-units is a fixed constant.

Integrator. $\Sigma = [0, 1]$ or more generally, Σ may be a subset of a topological vector space. An integrator would simply integrate the signal over time, providing its value as an output signal. The value of the integral up to the current time is the only state information needed, with some arbitrary initial state.

Differentiator. A transducer which outputs the derivative of the input signal. Some sort of state information is needed about the infinitesimal past.

Alternation Counter. A deducer for a discrete alphabet, say $\Sigma = \{0, 1\}$, which counts the number of changes in the input value.

Switching Controller. The alphabet is discrete, say $\Sigma = \{0, 1\}$. We imagine an object which can be switched between two different physical behaviors, each represented by a vector field on the state space. At any given time, the input value determines which behavior is in effect, the state moving along the flow curve for the corresponding vector field. This example may be generalized for a continuous alphabet by considering a continuously parametrized family of vector fields.

1.2 Continuity and Compactness Principles

Now we examine in more detail how to state effectiveness criteria in terms of continuity and compactness. To begin, what do we mean when exactly when we say *all* maps should be continuous? Consider the following particular maps associated to a continuous process. Note that whenever we say a map should be continuous, we are also implicitly asserting that its domain and codomain should have a specified topology.

Dynamics of the Computation. The update rule which determines a new state given an old state and the intervening input should be continuous in both arguments. This map should be continuous.

Outcome Map for a Deducer. The map which associates states to outcomes. This map should be continuous. Note that this criterion almost makes traditional recognizers impossible: the input space is likely connected, so any continuous recognizer to two outcomes would be trivial.

Overall Input-output Map. For a deducer with a given start state, the continuity of this map follows from the above by composition. For a transducer, the map from input to output signals (for any given time interval and start state) should be continuous. We will define deducers and transducers precisely in Sect. 3.2.

Input as a Function of Time. We do *not* require this function to be continuous. It is not part of the computation, but rather is given to us. Also, it would not be an advantage for elegance, because it would mean that there is implicit state information not attached to the state space: if the input takes on a certain value at a certain time, future input would be required to have a matching left one-sided limit at that time.

Similarly, let us consider exactly which spaces we are asserting should be compact. Since points in spaces have to be specified by approximation, to say a space is compact is to say that for any given degree of approximation, specifying a point requires only finitely much information.

The Alphabet Σ. The alphabet should be compact.

The Fixed-Interval Input Space. By this we mean the space of all possible input signals over a given time interval. This space should be compact. Note that in the discrete analogue comes for free in finite automata theory: if Σ is finite, then so is Σ^n for any n.

The Unrestricted Input Space. By this we mean the space of all possible input functions over any time interval. We do *not* require this space to be compact. It is analogous to Σ^*.

The State Space. The state space should be compact.

In discussing the examples, we sometimes will gloss over relatively innocuous non-compact state spaces such as the non-negative reals. The reason is convenience; in real-world problems real variables come with bounds and the example is easily patched by restricting the variable not to leave these bounds; or to cycle through. We are *not* saying that the restriction is unimportant theoretically; without it the theory would become degenerate because we could use a copy of the unrestricted input space as the state space.

1.3   Moving Toward Formalization

The discussion so far leads us to the following formal definition.

Definition 1. *A continuous automaton is a topological space S, the state space, together with a topological monoid M, the input monoid, and a continuous right action of M on S, the update rule. A continuous right action of M on S is a continuous map*

$$S \times M \to S$$

typically written as $s \cdot m$ for $s \in S$, $m \in M$, satisfying the action law *or* law of causality:

$$(s \cdot m_1) \cdot m_2 = s \cdot m_1 m_2. \tag{1}$$

Note that a version of this definition appeared in [5] (which does not consider continuous time), and that special cases include discrete-time continuous-space automata, as in that paper, and the classical theory of finite automata, when all sets are given the discrete topology and $M = \Sigma^*$.

The reason we have included the continuity principles in the definition but not the compactness principles is one of mathematical convenience. We will sometimes have need to consider non-compact automata, but never discontinuous automata.

The definition leaves us some flexibility in how the input signals are defined. A prototypical input monoid may be defined as follows. An element $u \in M$ is a function $u : [0, \ell(u)] \to \Sigma$ where Σ is a fixed compact alphabet. For concreteness, we may require the function to be Lebesgue measurable. The monoid operation is concatenation:

$$uv : [0, \ell(u) + \ell(v)] \to \Sigma$$

$$(uv)(t) = \begin{cases} u(t) & 0 \le t < \ell(u) \\ v(t - \ell(u)) & \ell(u) \le t \le \ell(u) + \ell(v). \end{cases}$$

Here we have just shifted v over and joined it with u. This monoid together with Definition 1 formalize what we so far have called a "continuous-time process". This monoid is intended to be analogous to the free monoid Σ^* under concatenation. However, it does not satisfy all the compactness principles.

1.4   Compactness of M_ℓ

Let M_ℓ be the set of input signals of some fixed length ℓ. Our second compactness principle should now say that M_ℓ is a compact subspace of M. Unfortunately, almost no function space is even locally compact, no matter how compact the domain and codomain might be. In particular, neither Lebesgue measurable functions nor any of the other classes of functions mentioned earlier yields a locally compact space when endowed with any standard topology, including uniform convergence, L^p-like metrics, or convergence in measure.

We mentioned that the discrete analogue of this principle, finiteness of Σ^n, came for free, and now it appears we have no obvious way to achieve it at all.

Nonetheless, we really would like to have this compactness and will devote special effort to finding a suitable monoid which satisfies this property. We insist for two reasons. First, theoretically, a glance at standard automata theory shows that almost all interesting results ultimately depend on the fact that Σ^n is finite, not just that Σ is finite. Second, it makes physical sense to require compactness. We are finitary beings–even if a physical device somehow did correctly process what is essentially an infinite amount of information in a finite time, we would have no way to verify that it did so correctly.

One way to get M_ℓ to be compact is to really restrict the class of functions, say to lipschitz functions with lipschitz constant less than some fixed bound K. This set of functions is equicontinuous and thus the Arzelà-Ascoli theorem makes it compact. However, we earlier mentioned drawbacks to assuming continuous input functions, and it seems to us that in real-world problems it is rarely reasonable assume a uniform lipschitz bound.

In the next section we will show a method to make M_ℓ compact by using a topology different from all those mentioned above. Accompanying the change in topology will be a a completion, which is necessary to actually achieve compactness. Points in the completion are technically not functions but equivalence classes, similar to the situation with L^p, which is why we have so far been coy about the exact nature of the "input signals".

To motivate the compactification construction, consider that the reason why function spaces are not compact boils down to oscillations of unboundedly high frequency. For example, if $f_n : [0, 1] \to \{0, 1\}$ is the function which alternates 2^n times between the values 0 and 1 on intervals of equal width, then (f_n) has no convergent subsequence. Problems with high frequency oscillations are not limited to the realm of pure mathematics; for example, electronic devices which are sensitive to the frequency of their input break at some point if the frequency is too high. Our mathematical solution will be correspond to requiring devices to tolerate high frequency noise by somehow averaging it out.

So what happens to the examples when we consider the possibility of high-frequency inputs? The counter is fine. It will give an approximately correct value even if the input is shifted around or averaged out. The integrator and delay[2] are also fine.

The differentiator and alternation counter do *not* work as described. They are sensitive to oscillations in the input. For example, the derivative of a function can change dramatically and become not well-defined even if the function is given only a very small uniform bump, if that bump happens to have high-frequency noise in it. Similarly, there is no way the alternation counter can be interpreted as needing only a finite amount of information from any interval of input, unless we somehow know that the input literally cannot oscillate at more than a certain frequency.

---

[2] The delay is interesting here because it connects the compactness principles for input and state space. Since the delay needs to store the last τ part of the input in its state, it has compact state space only if the space of τ-length inputs is compact.

Later we will see (Corollary 11) that the switching controller does satisfy our criteria.

2   Young Measures as Continuous-Time Words

The goal of this section is to give a complete, rigorous development of the compactification of M_ℓ to which we referred in Sect. 1.4. This space was actually discovered by L.C. Young [9] for the purpose of solving non-convex optimization problems in the calculus of variations and control theory. We will give a different definition, conceptualized in terms of *words*, analogous to elements of Σ^* in the discrete world, and defer proof of their equivalence to a future publication.

Let Σ be a compact metric space. As a simple nontrivial example, we may imagine $\Sigma = \{0, 1\}$. We are going to build a notion of continuous-time words of length 1 by starting with a small class $\mathcal{F}_1$ of functions $[0, 1] \to \Sigma$, defining a metric which we call the *Young metric*, and taking the completion of that metric. The completion will be M_1. (M_ℓ will be a scaling of M_1, so the following development applies to words of any length. We consider $\ell = 1$ for simplicity of exposition). Recall that a complete metric space is compact if and only if it is totally bounded, which means that for any $\varepsilon > 0$, there is a finite set whose ε-neighborhood is the entire space. So, after defining the Young metric, all we have to do is show that $\mathcal{F}_1$ is totally bounded. Then M_1 will also be totally bounded and therefore compact. As a bonus, we will also be able to give a concrete description of M_1 in terms of measures.

Let $\mathcal{F}_1$ be the set of piecewise constant functions $[0, 1] \to \Sigma$. The construction is not too sensitive to this exact choice. We could also choose piecewise continuous functions or Lebesgue measurable functions and achieve the same outcome—all of these sets of functions will be dense subsets of M_1.

To motivate the definition of the Young metric, imagine comparing two words u and v which are constant on all the tiny intervals $[k/N, (k+1)/N)$ for some fixed large N. These are just discrete words of length N, scaled down to have "letter width" $1/N$ and thereby fit in the interval $[0, 1]$ as continuous-time words. How different are these words?

$$01111100101110001110101111011$$
$$01101100101110011111010111011$$

They differ in two places, so their Hamming distance is 2. Say we scale that down appropriately for the continuous-time words in M_1, and their distance would be $2/N$. (In this example $N = 32$). If we defined our metric using a scaled Hamming distance like this, we would end up with essentially an L^1 function space, a well-known topology that does not solve our problems. In fact, we need a weaker topology, which is to say that at least some words need to be closer together than they were before. Consider these:

$$0101010101010101010101010101010101$$
$$1010101010101010101010101010101010$$

These words have the maximum possible Hamming distance. But we want to say they're actually close together. The reason is this: we could easily have some noise in our measurements of time that would make these words totally indistinguishable. Or to put it another way, if we look at the values of these words on any subinterval of time, we are going to get about the same number of 0 s and 1 s regardless of which word we look at. To be precise, the difference in occurrence of any letter on any subinterval is at most 1, so let's say these words have distance $1/N$. A less obvious example:

001001010100001010010010001010010101000

001010001100100001001001010101000

These words have a scaled Hamming distance of $13/N$, but the new proposal gives them distance just $2/N$. Note that on any subinterval we get about two-thirds 0 s and one-third 1 s.

We are almost ready to define the Young metric formally. We need to be precise about the "difference in occurrence" of values on a subinterval, but unfortunately this issue is more complicated in the case where Σ is not discrete. For example, if $\Sigma = [0,1]$ and we have the constant words $u(t) = 0.5$ and $u(t) = 0.501$, these words should also be close to each other. The solution is to use the Kantorovich-Wasserstein metric on measures.

When comparing two words u and v on an interval $[s,t]$, we only mean to look at their values on that interval, not where the values occur. So naturally we look at the pushforward measures $u_*(\lambda_{[s,t]})$ and $v_*(\lambda_{[s,t]})$ where $\lambda_{[s,t]}$ is Lebesgue measure on $[s,t]$. If u and v are piecewise constant, these measures are a finite sum of point-masses at the values of the functions u and v. The mass at each value is the total width of intervals on which the word took that value. In symbols,

$$u_*(\lambda_{[s,t]})(\{\sigma\}) = \lambda(\{r \in [s,t] : u(r) = \sigma\}). \tag{2}$$

The Kantorovich-Wasserstein metric is the infimal cost of transforming one measure into another by moving measure around: to move M measure over a distance D costs $M \cdot D$. If the measures are finite sums of point masses, then only finitely many discrete moves are needed. The formal definition follows.

Definition 2. *Let μ and ρ be positive Borel measures on a compact metric space X. A coupling γ of μ and ρ is a measure on $X \times X$ such that $\pi_{1,*}(\gamma) = \mu$ and $\pi_{2,*}(\gamma = \rho)$. Then the Kantorovich-Wasserstein metric $d_{KW}(\mu, \rho)$ is defined as*

$$d_{KW}(\mu, \rho) = \inf_{\gamma} \int_{X \times X} d_X d\gamma$$

where d_X is the metric on X and the infimum is taken over all couplings γ.

For more information about this metric, see [1], Chaps. 6–7.

To keep the formalism as simple as possible, note that a simple triangle-inequality argument shows that we can restrict our attention to initial subintervals $[0,s]$: if we want to consider $[s,t]$ we can compare on both $[0,s]$ and $[0,t]$.

Definition 3. *For a measurable function* $u : [0, 1] \to \Sigma$, *the* accumulant *of* u, *denoted* $\overline{u}$, *is the measure-valued function defined by*

$$\overline{u}(s) = u_*(\lambda_{[0,s]}) \tag{3}$$

or equivalently

$$\overline{u}(s)(X) = \lambda \left(\{ t \in [0, s] : u(t) \in X \} \right). \tag{4}$$

So the accumulant $\overline{u}$ is a measure-valued function of time. Roughly speaking, it tells us, up to this time, how much time the function u has spent at each value so far.

Definition 4. *Let* $u, v : [0, 1] \to \Sigma$ *be measurable functions. Then the* Young *metric* $d(u, v)$ *is*

$$d(u, v) = \sup_{s \in [0,1]} d_{KW}(\overline{u}(s), \overline{v}(s)). \tag{5}$$

Note that the Young metric is technically only defined on equivalence classes of functions up to almost-everywhere equality. We are quite happy with this, however, because it naturally wipes away concerns about endpoint conventions and when we take the completion, we are going to have to give up on input signals being functions anyway.

Proposition 5. *The set* $\mathcal{F}_1$ *of piecewise constant functions* $[0, 1] \to \Sigma$, *modulo almost-everywhere equality, is totally bounded in the Young metric.*

We omit the proof of Proposition 5 due to space limitations.[3]
    Then M_1 may be defined as the completion of $\mathcal{F}_1$. As mentioned, this construction works equally well to define any M_ℓ. However, to define M, we have to resolve two more technical issues. First, the metric needs to be extended so words of different length may be compared. Second, we need to check that the concatenation operation is well-defined on the completion, for which it suffices to show that it is uniformly continuous in the Young metric.

Definition 6. *Let* $\mathcal{F}$ *be the set of piecewise constant functions* $[0, \ell] \to \Sigma$ *for any* $\ell \geq 0$. *The Young metric is extended to* $\mathcal{F}$ *as follows. Let* $u, v \in \mathcal{F}$ *and assume without loss of generality that* $\ell(u) \leq \ell(v)$. *Then*

$$d(u, v) = d(u, v \restriction [0, \ell(u)]) + (\ell(v) - \ell(u)).$$

Proposition 7. *Concatenation gives a uniformly continuous map* $\mathcal{F} \times \mathcal{F} \to \mathcal{F}$.

Briefly, the proof of Proposition 7 is to first consider the (easy) case where the two first words have the same length, and then to handle the general case by noting that shifting a word by ε moves its accumulant by at most ε in the Young metric. This argument actually shows that concatenation is 1-lipschitz in each argument.

---

[3] A full account of all of these results will appear in my forthcoming dissertation.

Definition 8. *Let Σ be a compact metric space. The Young monoid over Σ, here denoted M, is the metric monoid defined as the completion of $\mathcal{F}$ (the piecewise constant functions) under the Young metric. Multiplication in M is the extension by uniform continuity of concatenation in $\mathcal{F}$. Elements of M are referred to as* measurable words *or* continuous-time words.

Note that the length $\ell(u)$ is also uniformly continuous in $u \in M$, so we may safely regard M_ℓ as a subspace of M. A few more observations, with more or less routine proofs:

- $M_{\leq \ell}$, the space of words of length *at most* ℓ, is compact.
- The new points in the completion do not necessarily correspond to functions of any kind. To see why, note simply that for any fixed time t, the map $u \mapsto u(t)$ is *not* uniformly continuous.
- Though $\mathcal{F}$ was defined as piecewise constant functions, we could just as well have used measurable functions. Each measurable function $[0, \ell] \to \Sigma$ defines an element of M. Consider sequences of piecewise constant functions which converge in the L^1 sense—they will also be Cauchy in the weaker Young metric. However, elements of this kind do not exhaust the completion.
- Anything with a natural definition in terms of accumulants, such as the integral of a word $u : [0, 1] \to [0, 1]$, is well-defined by virtue of uniform continuity.

Although not strictly necessary, it may be intuitively helpful to have a self-contained description of the Young monoid. The following definition provides such a description in terms of generalized accumulants, making use of the full generality of the Kantorovich-Wasserstein metric for atomless measures.

Definition 9. *Let Σ be a compact metric space. The Young monoid M of measurable continuous-time words over Σ may alternately be defined as follows. M_ℓ is the set of functions $\overline{u}$ on $[0, \ell]$ whose values are positive Borel measures on Σ with the following two properties:*

- *$\overline{u}(t)$ has total measure t, i.e., $\overline{u}(t)(\Sigma) = t$, and*
- *for each $A \subseteq \Sigma$, the function $t \mapsto \overline{u}(t)(A)$ is increasing.*

(Note that these properties are shared by all accumulants of measurable functions). The metric on M_ℓ is defined by

$$d(\overline{u}, \overline{v}) = \sup_{s \in [0,1]} d_{KW}(\overline{u}(s), \overline{v}(s)) \tag{6}$$

and extended to M as above. Multiplication in M is defined as concatenation:

$$(\overline{uv})(t) = \begin{cases} \overline{u}(t) & t \leq \ell(\overline{u}) \\ \overline{u}(\ell(\overline{u})) + \overline{v}(t - \ell(\overline{u})) & \ell(\overline{u}) \leq t \leq \ell(\overline{u}) + \ell(\overline{v}). \end{cases} \tag{7}$$

Now we may view the new points in the completion as words which accumulate values simultaneously at more than one element of Σ. For example, the limit of piecewise constant functions which alternate ever more rapidly between 0 and 1, spending equal time at each value, is a word (generalized accumulant) which constantly accumulates value at both 0 and 1, each at a rate of $1/2$.

3   Existence of Continuous Automata

3.1   Bottom-Up: Continuous Automata from Diffential Equations

In light of how we constructed the Young monoid M, we now have a natural strategy for formally proving the existence of a continuous automaton M with some desired behavior. We first define the behavior on piecewise constant inputs, and then show that it is uniformly continuous with respect to the Young metric. It then automatically extends to all of M. We can go further: by the law of causality (1), behavior on piecewise constant inputs is already determined by behavior on constant inputs, and, for that matter, constant inputs of arbitrarily small length. So we really only have to specify the behavior on constant inputs, which amounts to specifying a topological dynamical system for each $\sigma \in \Sigma$. Furthermore, this method should be adequate to produce any continuous automaton whatsoever, because any continuous automaton can be restricted to its behavior on constant inputs of small length, and uniform continuity follows by compactness.

We are left with the question: what conditions on a family of dynamical systems, parametrized by Σ, will produce a behavior that is uniformly continuous in the Young metric? The following theorem, which we do not prove here, gives a partial answer to this question.

Theorem 10. *Let S be a metric space and let $\varphi : \Sigma \times \mathbb{R}_{\geq 0} \times S \to S$ be a continuously Σ-parametrized family of flows in S, denoted as $(\sigma, t, x) \mapsto \varphi_{\sigma,t}(x)$. Assume furthermore that*

– *There are constants A, B, and K, such that*

$$d_S(\varphi_{\sigma,s}(x), \varphi_{\tau,t}(y)) \leq e^{Kt} d_S(x,y) + B d_\Sigma(\sigma,\tau) K^{-1}(e^{Kt} - 1) + A\,|s - t|\ (8)$$

for any $s, t \geq 0$, any $\sigma, \tau \in \Sigma$, and any $x, y \in S$.
– *There is a constant C such that*

$$d_S\left(\varphi_{\sigma,t}(\varphi_{\tau,t}(x)), \varphi_{\tau,t}(\varphi_{\sigma,t}(x))\right) \leq Ct^2 \tag{9}$$

for any $t \geq 0$, any $x \in S$, and any $\sigma, \tau \in \Sigma$.

Then the action of the constant words defined by φ_σ for $\sigma \in \Sigma$ extends uniquely to a continuous automaton.

Informally speaking, the inequalities amount to the following:

– Each flow has uniformly bounded speed. (Third term of (8)).
– For each flow, running the flow for any fixed amount of time t results in a lipschitz map $\varphi_{\sigma,t} : S \to S$. Furthermore note that for a larger time such as $2t$, $\varphi_{\sigma,2t}$ would also be forced to be lipschitz; the exponential in the inequality says that the lipschitz constants are all compatible in this way. They are generated infinitesimally, so to speak. (First term of (8)).

- The second term of (8) says that flows parametrized by different letters may differ locally according to the distance between those letters in Σ. The exponential factor is present because over time the discrepancy also feeds on itself in accordance with the previous remark.
- Equation (9) says that any two of the flows commute with one another at second-order. It is a kind of mutual smoothness condition that is tailored to ensure uniform continuity in the Young metric.

While (8) is true for almost any natural family of flows that one might want to write down, (9) is more restrictive. Still, it is true provided all of the flows are in fact smooth flows on a manifold. So we have the following corollary, which is the main use of the theorem.

Corollary 11. *Suppose Σ is a compact metric space and $S \subseteq \mathbb{R}^n$ is compact. Consider a parametrized system of differential equations*

$$\frac{d}{dt}\boldsymbol{y} = f(\sigma, \boldsymbol{y}) \tag{10}$$

where $f : \Sigma \times S \to \mathbb{R}^n$ is lipschitz and everywhere tangent to S. Then there is a unique continuous automaton over the Young monoid for Σ such that for each $\sigma \in \Sigma$, constant input of value σ causes the state to evolve according to (10).

The proof of Corollary 11 proceeds by first using the Picard-Lindelöf Theorem to produce unique global solutions for each fixed σ. Then using standard estimates, the hypotheses of Theorem 10 are satisfied, which gives us a unique automaton.

3.2   Top-Down: A Myhill-Nerode Theorem

For a Myhill-Nerode Theorem, we need to consider the full input-output semantics. Recall the notions of deducer and transducer from the introduction. There will be a version of the theorem for each of these. Let us start by formally defining a deducer.

Definition 12. *A deducer is a continuous automaton (S, M) together with a distinguished start state s_0 and a continuous map $o : S \to \mathcal{O}$, where $\mathcal{O}$ is a topological space, the outcome space.*

A deducer computes a unique map $c : M \to \mathcal{O}$, defined by $c(m) = o(s_0 \cdot m)$. The Myhill-Nerode Theorem gives a reverse statement: for any continuous map $c : M \to \mathcal{O}$, there is a canonical deducer $(\mathscr{S}, M)$ which computes c. $\mathscr{S}$ is minimum in the following sense: there is a natural morphism from $\mathscr{S}$ to any other automaton computing c. A morphism is a continuous map on states which respects both the dynamics and the input-output semantics. It follows that $\mathscr{S}$ is unique up to isomorphism (homeomorphism of state spaces which respects input-output semantics).

The elements of $\mathscr{S}$ are interpreted as *intrinsic states*. (Notationally, $\mathscr{S}$ is reserved for this purpose, whereas S may refer to the state space of an arbitrary automaton). To motivate the definition, notice that if we already have a deducer, each state $s \in S$ is associated naturally to a continuous function $M \to \mathcal{O}$, namely $m \mapsto o(s \cdot m)$.

Definition 13. *Suppose* $c : M \to \mathcal{O}$ *is given.*

- *An* intrinsic state *is a continuous function* $f : M \to \mathcal{O}$.
- *The* universal update rule *is* $(f \cdot m)(n) = f(mn)$.
- *The* initial intrinsic state *is* $f_0(m) = c(m)$.
- *By* $\mathscr{S}_c$ *or simply* $\mathscr{S}$ *we mean the space of all intrinsic states reachable from the initial intrinsic state via the universal update rule, endowed with the compact-open topology. Abusively, we may also mean the deducer with* $\mathscr{S}$ *as the state space, the universal update rule as the dynamics, the initial intrinic state as the start state, and the outcome map* $f \mapsto f(1_M)$.

The compact-open topology on a function space F of continuous functions $X \to Y$, briefly, is the topology of uniform convergence on compact sets. It is a standard topology on function spaces used in homotopy theory and many other areas. It has two key properties, provided that X and Y are sufficiently well-behaved (locally compact Hausdorff). One asserts that certain functions defined on F are continuous, and the other asserts that certain functions into F are continuous:

- The application map $F \times X \to Y$ is continuous.
- If P is a topological space and $g : P \times X \to Y$ is a continuous map such that for each fixed $p \in P$, g restricts to a map $X \to Y$ which is an element of F, then the curried map $h : P \to F$ defined by $h(p)(x) = g(p, x)$ is continuous. Here we think of P as continuously parametrizing a family of functions in F.

The proof of the theorem amounts to using these properties to check that all the relevant maps are indeed continuous.

Theorem 14. *Assume* X, M *and* $\mathcal{O}$ *are locally compact Hausdorff. Given a continuous function* $c : X \to \mathcal{O}$, *the corresponding deducer* $\mathscr{S}_c$ *defined above is indeed a continuous automaton. Furthermore, there is a canonical morphism from* $\mathscr{S}_c$ *to any other deducer (with the same* M *and* $\mathcal{O}$) *computing* c. *If* M *and* $\mathcal{O}$ *are both metrizable, then so is* $\mathscr{S}_c$.

Corollary 15. *If* c *can be computed by any deducer with a compact state space, then in particular* $\mathscr{S}_c$ *must be compact.*

The corresponding theorem for transducers is slightly more complicated.

Definition 16. *A* transducer *is a continuous automaton* (S, M) *together with a distinguished start state* s_0 *and a continuous map* $T : S \times M \to N$, *written* $(s, m) \mapsto T_s(m)$, *where* N *is another topological monoid, which satisfies the transduction law:*

$$T_s(m_1 m_2) = T_s(m_1) T_{s \cdot m_1}(m_2). \tag{11}$$

The transduction law (11) is in the same spirit as the law of causality (1), and it ensures that the transducer does in fact compute a well-defined function $C : M \to N$, namely $C = T_{s_0}$. C is also a causal function in the sense defined below.

Remark 17. Given any measurable word as input, the state of an automaton has to be a continuous function of time. If the transduction output depended only on state, it would itself be a continuous function of time, rather than a more general measurable word. The definition just given avoids this restriction. For example, it allows an identity transducer which copies input to output (requiring only one state).

In defining the canonical automaton for a continuous function $C : M \to N$, there is an additional difficulty that C has no hope of being computed by any automaton at all unless it is *causal*, meaning that $C(x \cdot m)$ can always be written as $C(x) \cdot n$ for some n. To obtain a canonical minimum automaton, we will need that this n is uniquely defined and obtained continuously from x and m.

Convention 18. For the remainder of this section, we make the following assumptions about N.

- N is left-cancellative: $ab = ac \implies b = c$. Consequently, we can define $a \backslash b$ as the unique n such that $an = b$, provided it exists. Then $a(a \backslash b) = b$.
- The map $(a, b) \mapsto a \backslash b$ is continuous on its domain.

Note that these conditions do hold in case N is a Young monoid.

Definition 19. *A map* $f : M \to N$ *is causal if for every* $x \in M$ *and* $m \in M$, *there exists* $n \in N$ *such that* $f(x \cdot m) = f(x) \cdot n$.

Definition 20. *Suppose* $C : M \to N$ *is given.*

- *An* intrinsic state *is a causal continuous function* $f : M \to N$ *such that* $f(1_M) = 1_N$.
- *The* universal update rule *is* $(f \cdot m_1)(m_2) = f(m_1) \backslash f(m_1 m_2)$.
- *The* initial intrinsic state *is* $f_0(m) = C(m)$.
- *By* $\mathscr{S}_C$ *or simply* $\mathscr{S}$ *we mean the space of all intrinsic states reachable from the initial intrinsic state via the universal update rule, endowed with the compact-open topology. Abusively, we may also mean the transducer with* $\mathscr{S}$ *as the state space, the universal update rule as the dynamics, the initial intrinic state as the start state, and the transduction output* $\mathscr{T}_f(m) = f(m)$.

Theorem 21. *Assume* M *and* N *are locally compact Hausdorff and also satisfy the assumptions of Convention 18. Given a causal continuous function* $C : M \to N$, $\mathscr{S}_C$ *is in fact a well-defined transducer. (The maps are continuous and the transduction law is satisfied). Furthermore, there is a canonical morphism from* $\mathscr{S}_C$ *to any other transducer (with the same* M *and* N) *computing* C. *If* M *and* N *are both metrizable, then so is* $\mathscr{S}_C$.

4   A Fixed Alternation-Counter

Earlier we noted that the differentiator and alternation-counter do not satisfy our effectiveness criteria. To be precise, they do not correspond to any continuous automaton over the measurable word monoid. Nonetheless, there are real devices referred to as "differentiators", for example. Here we suggest a way to explain such behavior by defining a sort of time-scale dependent approximation to the behavior of the alternation-counter. The case of the differentiator is more complicated, but should be amenable to a similar treatment. Note that the work in this section is of a more preliminary nature.

To make the behavior continuous in the Young metric, we are going to have the device measure alternations only relative to some specified time-scale constant $\varepsilon > 0$, by waiting for the input to accumulate value at a new alphabet letter. To keep the state space compact, we will loop back to 0 when we reach some large number L.

- Recall that $\Sigma = \{0, 1\}$.
- The state space is $S = M_\varepsilon \times [0, L] / \{0 \sim L\}$.
- If the state is written (v, a), then v continuously updates to record the last ε time units of input, and a updates as

$$\frac{da}{dt} = \frac{2\overline{v}(\varepsilon)(1 - u(t))}{\varepsilon}$$

where $u(t)$ is the letter currently being read.

If the input contains alternations which are separated by at least ε, then the a part of the state simply counts their number, assuming it started at zero. (The initial state of v also matters for whether the first input read is considered an alternation). The one exception is when an alternation happened more recently than ε time units ago, in which case it has not yet been fully counted, allowing a to change continuously. If ε is very small, we might realistically never see a taking a non-integer value.

5   Conclusions and Future Work

To summarize our main findings:

- The abstract framework of continuous monoid actions neatly covers finite automata theory as well as continuous-time automata.
- The construction of an appropriate topology allows development of a continuous-time theory along the same lines as the finite theory.
- Since most of the literature on continuous processes uses differential equations whose solutions are guaranteed by the Picard-Lindelöf conditions, those models are covered, at least in principle, by this one. (This point does not apply to any discontinuous behavior).

- Furthermore, the Myhill-Nerode theorems suggest that quite a bit more behavior can be considered in the same framework. For example, the delay was easy for us to understand here, but cannot be described using differential equations.
- This theory is defined using a more abstract description level than differential equations and is correspondingly simpler, at least if we are willing to accept the definition of the Young metric as simple.

My hope, at which I have hinted throughout the paper, is that this work will help bolster the theoretical understanding of many issues of continuous processes which have previously been examined mainly from a more practical angle. A prime example is the question of whether and how we might be able to translate between continuous, discrete, and hybrid computational and control systems. Having a simple unified framework should make it easier to state and examine such questions. Much work remains to be done.

Acknowledgements. I am grateful to my friends, instructors, and others who have helped me develop this research into its current form. There are a few I want to thank by name. Iian Smythe gave me an invaluable mathematical pointer early on. Bob Constable taught me the foundations of constructive thought, which subtly permeate the ideas presented here. Anonymous reviewers of earlier versions of this paper provided extremely useful feedback on wide-ranging matters, including among many other things the writing style and how to explain what the Young metric actually does. Lastly, I thank my advisor Anil Nerode deeply for his support and insight.

References

1. Ambrosio, L., Gigli, N., Savaré, G.: Gradient Flows: in Metric Spaces and in the Space of Probability Measures. Springer, Heidelberg (2006)
2. Bishop, M.: Foundations of Constructive Analysis. McGraw-Hill Book Company, New York (1967)
3. Bournez, O., Campagnolo, M.L.: A survey on continuous time computations. In: Cooper, S.B., Löwe, B., Sorbi, A. (eds.) New Computational Paradigms, pp. 383–423. Springer, New York (2008)
4. Henzinger, T.A.: The theory of hybrid automata. In: Inan, M.K., Kurshan, R.P. (eds.) Verification of Digital and Hybrid Systems. NATO ASI Series, vol. 170, pp. 265–292. Springer, Heidelberg (2000)
5. Jeandel, E.: Topological automata. Theor. Comput. Syst. **40**(4), 397–407 (2007)
6. Lynch, N., Segala, R., Vaandrager, F.: Hybrid i/o automata. Inf. Comput. **185**(1), 105–157 (2003)
7. Rabinovich, A.: Automata over continuous time. Theor. Comput. Sci. **300**(1), 331–363 (2003)
8. Weihrauch, K.: Computable Analysis: An Introduction. Springer Science & Business Media, Heidelberg (2012)
9. Young, L.C.: Lectures on the Calculus of Variations and Optimal Control Theory, vol. 304. American Mathematical Society, London (1980)

Measure Quantifier in Monadic Second Order Logic

Henryk Michalewski[1] and Matteo Mio[2]([✉])

[1] University of Warsaw, Warsaw, Poland
[2] CNRS/ENS-Lyon, Lyon, France
matteo.mio@ens-lyon.fr

Abstract. We study the extension of Monadic Second Order logic with the "for almost all" quantifier $\forall^{=1}$ whose meaning is, informally, that $\forall^{=1} X.\phi(X)$ holds if $\phi(X)$ holds almost surely for a randomly chosen X. We prove that the theory of MSO $+ \forall^{=1}$ is undecidable both when interpreted on $(\omega, <)$ and the full binary tree. We then identify a fragment of MSO $+ \forall^{=1}$, denoted by MSO $+ \forall_\pi^{=1}$, and reduce some interesting problems in computer science and mathematical logic to the decision problem of MSO $+ \forall_\pi^{=1}$. The question of whether MSO $+ \forall_\pi^{=1}$ is decidable is left open.

Keywords: Monadic second order logic · Lebesgue measure

1 Introduction

Monadic Second Order logic (MSO) is the extension of first order logic with quantification over subsets of the domain. For example, when interpreted over the relational structure $(\omega, <)$ of natural numbers with the standard order, the formula $\exists A.\forall n.\exists m.(n < m \wedge m \in A)$ expresses the existence of set A of natural numbers which is infinite (see Sect. 2 for definitions).

One of the first results about MSO was proved by Robinson [14] in 1958. He showed, answering a question of Tarski, that the theory $\mathrm{MSO}(\omega, +, <)$ is undecidable. In 1962 Büchi [5] proved that the weaker theory $\mathrm{MSO}(\omega, <)$ is decidable and in 1969 Rabin [13] extended this positive result to the MSO theory of the full binary tree (see Sect. 3 for definitions). Büchi and Rabin's theorems are widely regarded among the deepest decidability results in theoretical computer science. Their importance stems from the fact that many problems in the field of formal verification of programs can be reduced to these logics.

A long standing open problem in the field of verification of *probabilistic* programs is the decidability of the SAT(isfability) problem of probabilistic temporal

H. Michalewski—Author supported by Poland's National Science Centre grant no. 2012/07/D/ST6/02443
M. Mio—Author supported by grant "Projet Émergent PMSO" of the École Normale Supérieure de Lyon and Poland's National Science Centre grant no. 2014-13/B/ST6/03595.

© Springer International Publishing Switzerland 2016
S. Artemov and A. Nerode (Eds.): LFCS 2016, LNCS 9537, pp. 267–282, 2016.
DOI: 10.1007/978-3-319-27683-0_19

logics such as pCTL* and its extensions (see, e.g., [2,4]). In the attempt of making some progress, it seems worthwhile to formulate some aspects of the SAT problem as questions expressed in the logical framework of MSO. Given the vast literature on MSO, this might facilitate the application of known results and would make the SAT problem of pCTL* simpler to access by a broader group of logicians.

As a first step in this direction, following the seminal work of Harvey Friedman, who introduced and investigated similar concepts in the context of First Order logic in unpublished manuscripts in 1978–79[1], we have recently considered in [12] the extension of MSO on the full binary tree with Friedman's "for almost all" quantifier ($\forall^*$) interpreted using the concept of Baire Category as:

$$\forall^* X.\phi(X) \text{ holds } \Leftrightarrow \text{ the set} \{A \mid \phi(A) \text{ holds}\} \text{ is topologically large}$$

where "topologically large" means *comeager* in the Cantor space topology of subsets of the full binary tree. We proved in [12] that the sets definable using the quantifier $\forall^*$ can actually be defined without it: $\text{MSO} = \text{MSO} + \forall^*$. This is a result of some independent interest but, most importantly, it fits into the research program outlined above since we successfully used it to prove [12] the decidability of the *finite-SAT* problem (a variant of the SAT problem mentioned above) for the qualitative fragment of pCTL* and similar logics.

In this paper we consider a natural variant of the above extension. We introduce the logic $\text{MSO} + \forall^{=1}$, interpreted both on $(\omega, <)$ and on the binary tree, obtained by extending MSO with Friedman's "for almost all" quantifier ($\forall^{=1}$) interpreted using the concept of Lebesgue measure as:

$$\forall^{=1} X.\phi(X) \text{ holds } \Leftrightarrow \text{ the set} \{A \mid \phi(A) \text{ holds}\} \text{ is of Lebesgue measure } 1.$$

Thus, informally, $\forall^{=1} X.\phi(X)$ holds if $\phi(A)$ is true for a random A. We prove, using results from [1] and [7], that unlike the case of $\text{MSO} + \forall^*$:

Theorem 1. *The theory of* $\text{MSO} + \forall^{=1}$ *on* $(\omega, <)$ *is undecidable.*

The proof of this result is presented in Sect. 5. As a consequence also the theory of $\text{MSO} + \forall^{=1}$ on the full binary tree is undecidable (Corollary 1).

Motivated by this negative result, we investigate the theory of a weaker fragment of $\text{MSO} + \forall^{=1}$ on trees which we denote by $\text{MSO} + \forall_\pi^{=1}$. Informally, $\forall_\pi^{=1} X.\phi(X)$ holds if $\phi(P)$ is true for a random *path* P in the full binary tree. We observe (Proposition 3) that $\text{MSO} + \forall_\pi^{=1}$ is strictly more expressive than MSO. However we have not been able to answer the following question[2]:

Problem 1. *Is the theory of* $\text{MSO} + \forall_\pi^{=1}$ *on the binary tree decidable?*

This problem, which we leave open, seems to deserve some attention. Indeed in Sect. 7 we show that the decidability of $\text{MSO} + \forall_\pi^{=1}$ would have some interesting applications. Most importantly, from the point of view of our research

---

[1] See [15] for an overview of Friedman's research.

[2] Further open problems regarding $\text{MSO} + \forall_\pi^{=1}$ are formulated in Sect. 8.

program, if the theory of MSO $+ \forall_\pi^{=1}$ is decidable then the SAT problem for the qualitative fragment of pCTL* is decidable (Theorem 5). Regarding applications in mathematical logic, we prove (Theorem 8) that the first order theory of the lattice of F_σ subsets of the Cantor space with the predicates $C(X) \Leftrightarrow$ "X is a closed set" and $N(X) \Leftrightarrow$ "X is a Lebesgue null set" is interpretable in MSO$+\forall_\pi^{=1}$. As another example, we show (Theorem 9) that the first order theory of the Lebesgue measure algebra with Scott's closure operator is interpretable in MSO $+ \forall_\pi^{=1}$. Hence if MSO $+ \forall_\pi^{=1}$ is decidable, these two theories are also decidable. Lastly, we also establish (Theorem 6) that the *qualitative languages* of trees, recently investigated in [6], are definable by MSO $+ \forall_\pi^{=1}$ formulas.

2   Measure and Probabilistic Automata

The set of natural numbers and their standard total order are denoted by the symbols ω and $<$, respectively. Given sets X and Y we denote with X^Y the space of functions $X \to Y$. We can view elements of X^Y as Y-indexed sequences $\{x_i\}_{i\in Y}$ of elements of X. We refer to Σ^ω as the collection of *ω-words over* Σ. The collection of *finite* sequences of elements in Σ is denoted by Σ^*. As usual we denote with ϵ the empty sequence and with ww' the concatenation of $w, w' \in \Sigma^*$.

The set $\{0,1\}^\omega$ of ω-words over $\{0,1\}$, endowed with the product topology (where $\{0,1\}$ is given the discrete topology) is called the *Cantor space*. Given a finite set Σ, the spaces Σ^ω and $\{0,1\}^{\Sigma^*}$ are homeomorphic to the Cantor space. The Cantor space is *zero-dimensional*, i.e., it has a basis of *clopen* (both open and closed) sets. A subset of $\{0,1\}^\omega$ is a F_σ set if it is expressible as a countable union of closed sets. For a detailed exposition of these topological notions see introductory chapters of [9]. We summarize below the basic concepts related to Borel measures. For more details see, e.g., Chap. 17 of [9]. The smallest σ-algebra of subsets of $\{0,1\}^\omega$ containing all open sets is denoted by $\mathcal{B}$ and its elements are called *Borel sets*. Given a $A \in \mathcal{B}$ we denote its complement by $\neg B$. A *Borel probability measure* on $\{0,1\}^\omega$ is a function $\mu : \mathcal{B} \to [0,1]$ such that: $\mu(\emptyset) = 0$, $\mu(\{0,1\}^\omega) = 1$ and, if $\{B_n\}_{n\in\omega}$ is a sequence of disjoint Borel sets, $\mu(\bigcup_n B_n) = \sum_n \mu(B_n)$. Every Borel measure μ on the Cantor space is *regular*: for every Borel set B there exists a F_σ set $A \subseteq B$ such that $\mu(A) = \mu(B)$. We will be mostly interested in one specific Borel measure on the Cantor space which we refer to as the *Lebesgue measure*. This is the unique Borel measure satisfying the equality $\mu(B_{n=0}) = \mu(B_{n=1}) = \frac{1}{2}$, where $B_{n=0} = \{(b_i)_{i\in\omega} \mid b_n = 0\}$ and $B_{n=1} = \{(b_i)_{i\in\omega} \mid b_n = 1\}$, respectively. Intuitively, the Lebesgue measure on $\{0,1\}^\omega$ generates an infinite sequence $(b_0, b_1, \dots)$ by deciding to fix $b_n = 0$ or $b_n = 1$ by tossing a fair coin, for every $n \in \omega$.

2.1   Probabilistic Büchi Automata

In this section we define the class of *probabilistic Büchi automata* introduced in [1] and state the undecidability of their *emptiness problem under the probable semantics* [1, Theorem 7.2]. This is the key technical result used in our proof of undecidability of MSO $+\forall^{=1}$ in Sect. 5.

Definition 1 (Probabilistic Büchi Automaton). *A probabilistic Büchi automaton is a tuple* $\mathcal{A} = \langle \Sigma, Q, q_I, F, \Delta \rangle$ *where:* Σ *is a finite nonempty input alphabet,* Q *is a finite nonempty set of* states, $q_I \in Q$ *is the initial state,* $F \subseteq Q$ *is the set of* accepting states *and* $\Delta : Q \to (\Sigma \to \mathcal{D}(Q))$ *is the* transition function, *where* $\mathcal{D}(Q)$ *denotes the collection of probability distributions on* Q.

To illustrate the above definition consider the probabilistic Büchi automaton (from [1, Lemma 4.3]) $\mathcal{A} = \langle \{a, b\}, Q, q_1 F, \Delta \rangle$ where $Q = \{q_1, q_2, \bot\}$, $F = \{q_1\}$ and Δ is defined as in Fig. 1.

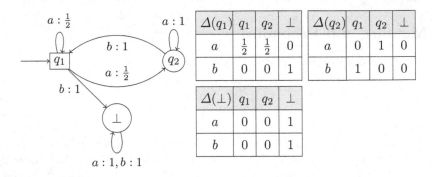

Fig. 1. A probabilistic Büchi automaton with three states. Boxes denote accepting states and circles denote not accepting states.

We now describe the intended interpretation of probabilistic Büchi automata. As for ordinary Büchi Automata (see [17] for a detailed introduction to this classical concept) a probabilistic Büchi automaton "reads" ω-words over the finite alphabet Σ. However, unlike ordinary Büchi automata, a probabilistic Büchi automaton "accepts" an input ω-word w with some *probability* $\mathbb{P}_w^{\mathcal{A}}$. We now describe this notion.

A probabilistic Büchi automaton starts reading a ω-word $w = (a_0, a_1, \dots) \in \Sigma^\omega$ from the state $q_0 = q_I$. After reading the first letter a_0, the automaton moves to state $q \in Q$ with probability $\Delta(q_0, a_0, q)$. If the state q is reached, after the second letter a_1 is read, the automaton reaches the state q' with probability $\Delta(q, a_1, q')$. More generally, if at stage n the automaton is in state q, after reading the letter a_n of w, the automaton reaches the state q' with probability $\Delta(q, a_n, q')$. Hence, a ω-word w induces a random walk on the set of states Q of the automaton $\mathcal{A}$. One can naturally formalize this random walk as a Borel probability measure $\mu_w^{\mathcal{A}}$ on the space Q^ω (see [1, §3.1] for detailed definitions).

Considering the example in Fig. 1 and the ω-word $a^\omega = (a, a, a \dots)$, the probability measure $\mu_w^{\mathcal{A}}$ assigns probability $\frac{1}{4}$ to the set of sequences $q_1 q_1 q_1 Q^\omega$ starting with three consecutive q_1's.

A sequence $(q_0 q_1 \dots q_n \dots) \in Q^\omega$ of states of $\mathcal{A}$ is *accepting* if for infinitely many $i \in \omega$, the state q_i belongs to the set F of accepting states. We denote with $Acc \subseteq Q^\omega$ the set of accepting sequences of states. Clearly Acc is a Borel set.

We say that $\mathcal{A}$ accepts the ω-word $w \in \Sigma^\omega$ with probability $\mathbb{P}_w^{\mathcal{A}} = \mu_w^{\mathcal{A}}(Acc)$. We are now ready to state a fundamental result about probabilistic Büchi automata.

Theorem 2 (Theorem 7.2 in [1]). *It is undecidable if for a given probabilistic Büchi automaton $\mathcal{A}$ there exists $w \in \Sigma^\omega$ such that $\mathbb{P}_w^{\mathcal{A}} > 0$.*

An inspection of the proof of Theorem 2 from [1] reveals that the problem remains undecidable if we restrict to the class of probabilistic Büchi automaton $\mathcal{A}$ such that, for some $k \in \omega$, all probabilities appearing in (the matrices of) Δ of $\mathcal{A}$ belong[3] to the set $\{0, \frac{1}{2^k}, \ldots, \frac{i}{2^k}, \ldots, 1\}$. We can further restrict attention to the class of *simple* probabilistic Büchi automata defined below.

Definition 2. *A probabilistic Büchi automaton $\mathcal{A}$ is simple if, for some $k \in \omega$ all probabilities appearing in (the matrices of) Δ are either 0 or $\frac{1}{2^k}$.*

Proposition 1. *It is undecidable if for a given simple probabilistic Büchi automaton $\mathcal{A}$ there exists $w \in \Sigma^\omega$ such that $\mathbb{P}_w^{\mathcal{A}} > 0$.*

Proof. We can transform an automaton $\mathcal{A}$ with probabilities in $\{0, \frac{1}{2^k} \ldots \frac{i}{2^k}, \ldots, 1\}$ to an equivalent one having only probabilities in $\{0, \frac{1}{2^k}\}$ by "splitting probabilities" introducing new copies of the states.   □

3   Syntax and Semantics of Monadic Second Order Logic

In this section we define the syntax and the semantics of the MSO logic interpreted over the linear order of natural numbers ("MSO on ω-words") and over the full binary tree ("MSO on trees"). This material is standard and a more detailed exposition can be found in [17].

MSO on ω-words. We first define the syntax and the semantics of MSO on $(\omega, <)$. We follow the standard presentation of MSO on $(\omega, <)$ where only second order variables are considered. We refer to Sect. 2.3 of [17] for more details.

Definition 3 (Syntax). *The set of formulas of the logic MSO is generated by the following grammar: $\phi ::= \mathrm{Sing}(X) \mid X < Y \mid X \subseteq Y \mid \neg\phi \mid \phi_1 \vee \phi_2 \mid \forall X.\phi$, where X, Y range over a countable set of variables. We write $\phi(X_1, \ldots, X_n)$ to indicate that $\overrightarrow{X} = (X_1, \ldots, X_n)$ is the list of free variables in ϕ.*

This presentation is convenient since MSO formulas can be regarded as first-order formulas over the signature $\mathcal{S}$ consisting of the unary symbol $Sing$ and the two binary symbols $<$ and $\subseteq$. MSO formulas are interpreted over the collection of subsets of ω (i.e., the collection $\{0,1\}^\omega$ of ω-words over $\{0,1\}$) with the following interpretations of the symbols in $\mathcal{S}$:

---

[3] As observed in [1, Remark 7.3], a proof of Theorem 2 can be derived from the decidability of a similar problem for finite probabilistic automata obtained by Gimbert and Oualhadj in [7, Theorem 4]. In [7, Proposition 2] the authors notice that the problem remains undecidable even if all probabilities appearing in the automaton belongs to $\{0, \frac{1}{4}, \frac{2}{4}, \frac{3}{4}, 1\}$.

- $Sing^I(X) \Leftrightarrow X = \{n\}$, for some $n \in \omega$, i.e., $X \subseteq \omega$ is a singleton.
- $<^I (X, Y) \Leftrightarrow X = \{n\}$, $Y = \{m\}$ and $n < m$.
- $\subseteq^I (X, Y) \Leftrightarrow X \subseteq Y$, i.e., X is a subset of Y.

Definition 4 (Semantics). *Let $\mathbb{W}$ be the structure for the signature $\mathcal{S}$ defined as $\mathbb{W} = (\{0,1\}^\omega, Sing^I, <^I, \subseteq^I)$. The truth of MSO formulas ϕ is given by the relation $\mathbb{W} \models \phi$, where $\models$ is the standard first-order satisfaction relation. Given parameters $A_1, \ldots, A_n \in \{0,1\}^\omega$, we write $\overrightarrow{A} \in \phi(\overrightarrow{X})$ to indicate that $\mathbb{W} \models \phi(\overrightarrow{A})$, i.e., that $\mathbb{W}$ satisfies the formula ϕ with parameters $\overrightarrow{A}$.*

Thus a formula $\phi(X_1, \ldots, X_n)$ defines a subset of $(\{0,1\}^\omega)^n$ or, equivalently, a subset of $(\{0,1\}^n)^\omega$ that is a set of ω-words over $\Sigma = \{0,1\}^n$. The subsets of Σ^ω definable by a MSO formula ϕ are called *regular*.

Remark 1. The presentation of $MSO(\omega, <)$ as the first order theory of $\mathbb{W}$ is technically convenient. Yet it is often useful to express concisely formulas such as $\forall x.(x \in Y \to \phi(x, Z))$ where the lowercase letter x ranges over natural numbers and the relation symbol $\in$ is interpreted as membership, as expected. Formulas of this kind can always be rephrased in the language of the signature $\{Sing, <, \subseteq\}$. For example the formula above can be expressed as: $\forall X.(Sing(X) \to (X \subseteq Y \to \phi(X, Z))$. We refer to [17] for a detailed exposition.

MSO on Trees. We now introduce, following a similar approach, the syntax and the semantics of MSO on trees.

Definition 5 (Full Binary Tree). *The collection $\{L, R\}^*$ of finite words over the alphabet $\{L, R\}$ can be seen as the set of vertices of the infinite binary tree. We refer to $\{L, R\}^*$ as the* full binary tree. *We use the letters v and w to range over elements of the full binary tree.*

Definition 6 (Syntax). *The set of formulas of the logic MSO on the full binary tree is generated by the following grammar:*

$$\phi ::= Sing(X) \mid Succ_L(X, Y) \mid Succ_R(X, Y) \mid X \subseteq Y \mid \neg\phi \mid \phi_1 \vee \phi_2 \mid \forall X.\phi$$

where X, Y range over a countable set of variables.

Hence MSO formulas are conventional first-order formulas over the signature $\mathcal{S}$ consisting of one unary symbol $Sing$ and three binary symbols $Succ_L, Succ_R, \subseteq$. We interpret MSO formulas over the collection $\{0,1\}^* \to \{0,1\}$ of subsets of the full binary. To improve the notation, given a set Σ we write $\mathcal{T}_\Sigma$ to denote the set $\{0,1\}^* \to \Sigma$. Thus MSO formulas are interpreted over the universe $\mathcal{T}_{\{0,1\}}$ with the following interpretations of the symbols in $\mathcal{S}$:

- $Sing^I(X) \Leftrightarrow X = \{v\}$, for some $v \in \{L, R\}^*$, i.e., if $X \in \mathcal{T}_{\{0,1\}}$ is a singleton.
- $Succ_L^I(X, Y) \Leftrightarrow$ "$X = \{v\}, Y = \{w\}$ and $w = vL$.
- $Succ_R^I(X, Y) \Leftrightarrow$ "$X = \{v\}, Y = \{w\}$ and $w = vR$.
- $\subseteq^I (X, Y) \Leftrightarrow X \subseteq Y$, i.e., if X is a subset of Y.

Definition 7 (Semantics). *Let* $\mathbb{T}$ *be the structure for the signature* S *defined as* $\langle \mathcal{T}_{\{0,1\}}, Sing^I, \mathbf{Succ}_L^I, \mathbf{Succ}_R^I, \subseteq^I \rangle$. *The truth of a MSO formula* ϕ *is given by the relation* $\mathbb{T} \models \phi$. *Given parameters* $\overrightarrow{A} \in \mathcal{T}_{\{0,1\}}$, *we write* $\overrightarrow{A} \in \phi(\overrightarrow{X})$ *to indicate that* $\mathbb{T} \models \phi(A_1, \ldots, A_n)$, *i.e., that* $\mathbb{T}$ *satisfies the formula* ϕ *with parameters* $\overrightarrow{A}$.

Thus a formula $\phi(X_1, \ldots, X_n)$ defines a subset of $(\mathcal{T}_{\{0,1\}})^n$ or, equivalently, a subset of $\mathcal{T}_\Sigma$ with $\Sigma = \{0,1\}^n$.

4   MSO with Measure Quantifier: MSO $+\forall^{=1}$

In this section we introduce the logic MSO $+ \forall^{=1}$, interpreted both on ω-words and on trees, obtained by extending ordinary MSO with Friedman's "for almost all" quantifier interpreted using the concept of Lebesgue measure.

4.1   MSO $+\forall^{=1}$ on ω-words

Definition 8. *The syntax of* MSO $+ \forall^{=1}$ *on* ω*-words is obtained by extending that of MSO (Definition 3) with the quantifier* $\forall^{=1}X.\phi$ *as follows:*

$$\phi ::= Sing(X) \mid X < Y \mid X \subseteq Y \mid \neg\phi \mid \phi_1 \vee \phi_2 \mid \forall X.\phi \mid \forall^{=1}X.\phi$$

The following definition specifies the semantics MSO $+ \forall^{=1}$ on ω-words.

Definition 9. *Each formula* $\phi(X_1, \ldots, X_n)$ *of* MSO $+ \forall^{=1}$ *is interpreted as a subset of* $(\{0,1\}^\omega)^n$ *by extending Definition 4 with the following clause:*

$$(A_1, \ldots, A_n) \in \forall^{=1}X.\phi(X, Y_1, \ldots, Y_n)$$
$$\Leftrightarrow$$
$$\mu_{\{0,1\}^\omega}\left(\{B \mid (B, A_1, \ldots, A_n) \in \phi(X, Y_1, \ldots, Y_n)\}\right) = 1$$

where A_i, B *range over* $\{0,1\}^\omega$ *and* $\mu_{\{0,1\}^\omega}$ *is the Lebesgue measure on* $\{0,1\}^\omega$. *For a given formula* ϕ *we define* $\exists^{>0}X.\phi$ *as a shorthand for* $\neg\forall^{=1}X.\neg\phi$.

The set denoted by $\forall^{=1}X.\phi(X, \overrightarrow{Y})$ can be illustrated as in Fig. 2, as the collection of tuples $\overrightarrow{A}$ having a *large* section $\phi(X, \overrightarrow{A})$, that is a section having Lebesgue measure 1. Informally, $(A_1, \ldots, A_n)$ satisfies $\forall^{=1}X.\phi(X, \overrightarrow{Y})$ if "for almost all" $B \in \{0,1\}^\omega$, the tuple $(B, A_1, \ldots, A_n)$ satisfies ϕ. Similarly, $\overrightarrow{A} \in \exists^{>0}X.\phi(X, \overrightarrow{Y})$ iff the section $\phi(X, \overrightarrow{A})$ has positive measure.

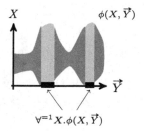

Fig. 2. The large sections selected by the quantifier $\forall^{=1}$ are marked in grey.

Remark 2. Every relation on $\{0,1\}^\omega$ definable by a MSO + $\forall^{=1}$ formula clearly belongs to a finite level of the projective hierarchy. However, since the family of MSO + $\forall^{=1}$ definable relations is closed under Boolean operations and projections, it is not clear if every MSO + $\forall^{=1}$ definable relation is Lebesgue measurable. We formulate this as Problem 2 is Sect. 8. In the rest of the paper we assume sufficiently strong set-theoretical assumptions (e.g., Projective Determinacy, see [9, Sect. 38.C]) to guarantee that Definition 9 is well specified, i.e., that all considered sets are measurable.

On Fubini's Theorem. The Fubini theorem is a classical result in analysis which states that the measure of a set $A \subseteq X \times Y$ can be expressed by iterated integration over the X and Y axis. In terms of MSO + $\forall^{=1}$, the Fubini theorem corresponds (see [9, Sect. 17.A]) to the fact that

$$\forall^{=1}X.\forall^{=1}Y.\phi(X, Y, \overrightarrow{Z}) = \forall^{=1}X.\forall^{=1}Y.\phi(X, Y, \overrightarrow{Z})$$

and, importantly for the proof of Theorem 3, that:

$$(A_1, \ldots, A_n) \in \forall^{=1}X.\forall^{=1}Y.\phi(X, Y, \overrightarrow{Z})$$
$$\Leftrightarrow$$
$$\mu_{(\{0,1\}^2)^\omega}\left(\{(B, C) \mid (B, C, A_1, \ldots, A_n) \in \phi(X, Y, Z_1 \ldots, Z_n)\}\right) = 1$$

where $\mu_{(\{0,1\}^2)^\omega}$ is the Lebesgue measure on the product space $(\{0,1\}^2)^\omega = \{0,1\}^\omega \times \{0,1\}^\omega$ defined as the product measure $\mu_{\{0,1\}^\omega} \otimes \mu_{\{0,1\}^\omega}$.

4.2   MSO + $\forall^{=1}$ on Trees

The definition of MSO + $\forall^{=1}$ on trees is similar to that of MSO + $\forall^{=1}$ on words and extends the syntax of MSO on trees (Definition 6) with the new quantifier $\forall^{=1}X.\phi$. The semantics of MSO + $\forall^{=1}$ on trees is obtained by extending Definition 7 by the following interpretation of $\forall^{=1}$:

$$(A_1, \ldots, A_n) \in (T_{\{0,1\}})^n \in \forall^{=1}X.\phi(X, Y_1, \ldots, Y_n)$$
$$\Leftrightarrow$$
$$\mu_{T_{\{0,1\}}}\left(\{B \mid (B, A_1, \ldots, A_n) \in \phi(X, Y_1, \ldots, Y_n)\}\right) = 1$$

where $\mu_{T_{\{0,1\}}}$ is the Lebesgue measure on $\mu_{T_{\{0,1\}}}$.

The Lebesgue measure $\mu_{T_{\{0,1\}}}$ can be seen as the random process of generation of a tree $A \in T_{\{0,1\}}$ by fixing the label (either 0 or 1) of each vertex $v \in \{L, R\}^*$ of the binary tree by tossing a fair coin. Hence, intuitively, the formula $\forall^{=1}X.\phi(X)$ holds true if $\phi(A)$ holds for a random tree $A \in T_{\{0,1\}}$.

5   Undecidability of MSO + $\forall^{=1}$

In this section we prove that the theory of MSO+$\forall^{=1}$ on ω-words is undecidable. This is done by reducing the (undecidable by Proposition 1) emptiness problem of simple probabilistic Büchi automaton $\mathcal{A}$ to the decision problem of MSO+$\forall^{=1}$.

The reduction closely resembles the standard translation between ordinary Büchi automata and MSO on ω-words (see, e.g., [17, Sect. 3.1]).

In what follows, let us fix an arbitrary k-simple probabilistic Büchi automaton $\mathcal{A} = \langle\{a,b\}, q_I, Q, F, \Delta\rangle$ with $Q = \{q_1, q_2, \ldots, q_n\}$. We write $q_i < q_j$ if $i < j$. Without loss of generality, let us assume that $|\Sigma| = 2^m$, for some number $m \in \omega$, so that we can identify Σ with $\{0,1\}^m$. Hence a ω-word $w \in \Sigma^\omega$ can be uniquely identified with a tuple $\overrightarrow{X} = (X_1, \ldots, X_m)$ with $X_i \in \{0,1\}^\omega$.

Since $\mathcal{A}$ is k-simple (see Definition 2), for each state $q \in Q$ and letter $a \in \Sigma$, there are exactly 2^k possible transitions in Δ, each one having probability $\frac{1}{2^k}$. Therefore, for each state q and letter a, we can identify the available transitions by numbers in $\{0, \ldots, 2^k - 1\} = \{0,1\}^k$ as follows: the number i denotes the transition to the i-th (with respect to the total order $<$ on Q) reachable (with probability $\frac{1}{2^k}$) state. We can identify an infinite sequence of transitions with an infinite sequence $\overrightarrow{Y} = (\{0,1\}^k)^\omega$, i.e., by a tuple $(Y_1, \ldots, Y_k)$ with $Y_j \in \{0,1\}^\omega$. The existence of a ω-word $w \in \Sigma^\omega$ such that $\mathbb{P}_w^\mathcal{A} > 0$ is expressed in MSO+$\forall^{=1}$ by

$$\phi_\mathcal{A} = \exists\overrightarrow{X}.\exists^{>0}\overrightarrow{Y}.\psi_\mathcal{A}(\overrightarrow{X}, \overrightarrow{Y})$$

where $\exists\overrightarrow{X}$ and $\exists^{>0}\overrightarrow{Y}$ stand for $\exists X_1.\exists X_2. \ldots .\exists X_m$ and $\exists^{>0}Y_1.\exists^{>0}Y_2. \ldots .\exists^{>0}Y_k$, respectively. The formula $\psi_\mathcal{A}$, which we define below, expresses that when interpreting $\overrightarrow{X}$ as a ω-word $w \in \Sigma^\omega$ and $\overrightarrow{Y}$ as an infinite sequence of transitions, the infinite sequence (q_n) of states visited in $\mathcal{A}$, which is uniquely determined by $\overrightarrow{X}$ and $\overrightarrow{Y}$, contains infinitely many accepting states, that is, $(q_n) \in Acc$.

Due to the Fubini Theorem (see Sect. 4) and the fact that an ω-word $A \in (\{0,1\}^k)^\omega$ randomly generated with the Lebesgue measure on $\{0,1\}^k$ assumes at a given position $n \in \omega$ a value in $\{0,1\}^k$ with uniform probability $\frac{1}{2^k}$, the formula $\phi_\mathcal{A}$ indeed expresses that there exists $w \in \Sigma^\omega$ such that $\mathbb{P}_w^\mathcal{A} > 0$. The formula $\psi_\mathcal{A}(\overrightarrow{X}, \overrightarrow{Y})$ is defined using standard ideas (see, e.g., [17, Sect. 3.1]):

$$\exists Q_1, \ldots, Q_n. \Big((a) \text{ for all } i \in \omega \text{ there is a unique } j \in \{1, \ldots, n\} \text{ such that } i \in Q_j$$

$$\text{and (b) } \forall i \in \omega, \text{ if } \overrightarrow{X}(i) = a \text{ and } \overrightarrow{Y}(i) = t \text{ then } i+1 \in Q_{(q,a,t)}$$

$$\text{and (c) } \exists_{j \in F} \text{ for infinitely many } i \in \omega, i \in Q_j\Big)$$

The formula expresses that: (a) there exists an assignment of states to positions $i \in \omega$ such that each position is assigned a unique state; that (b) if position i is labeled by state q, $(X_1^i, \ldots, X_m^i)$ represents the letter $a \in \Sigma$ and $(X_1^i, \ldots, X_k^i)$ represent the transition $0 \leq t < 2^k$, then $i + 1$ belongs to the state (denoted in the formula by (q, a, t)) which is the t-th reachable state from q on letter a; (c) the sequence contains infinitely many accepting states. Hence we get a more detailed version of Theorem 1 stated in the Introduction:

Theorem 3. *For each simple probabilistic Büchi automaton $\mathcal{A}$, the MSO $+\forall^{=1}$ sentence $\phi_\mathcal{A}$ is true if and only there exists $w \in \Sigma^\omega$ such that $\mathbb{P}_w^\mathcal{A} > 0$. Hence the theory of MSO $+ \forall^{=1}$ is undecidable.*

Undecidability of MSO $+ \forall^{=1}$ *on Trees.* The theory of MSO $+\forall^{=1}$ on $(\omega, <)$ can be interpreted within the theory of MSO $+ \forall^{=1}$ on the full binary tree by the standard interpretation of $(\omega, <)$ as the set of vertices of the leftmost branch in the full binary tree and it is not difficult to see that this interpretation preserves the meaning of the $\forall^{=1}$ measure quantifier. We only state the following result. A detailed proof will be published elsewhere.

Corollary 1. *The theory of* MSO $+ \forall^{=1}$ *on the full binary tree is undecidable.*

6   The Logic MSO+$\forall_\pi^{=1}$ on Trees

In this section we identify a variant of MSO $+ \forall^{=1}$ on trees which we denote by MSO $+ \forall_\pi^{=1}$. This logic is obtained by extending ordinary MSO with the quantifier $\forall_\pi^{=1}$. Intuitively, the quantifier $\forall_\pi^{=1}$ is defined by restricting the range of the measure quantifier $\forall^{=1}$ to the collection of paths in the full binary tree (see definition 10 below) so that the formula $\forall_\pi^{=1} X . \phi(X)$ holds if a randomly chosen path X satisfies the property ϕ with probability 1. More precisely,

Definition 10. *A subset* $X \subseteq \{L, R\}^*$ *(equivalently,* $X \in \mathcal{T}_{\{0,1\}}$*) is called a path if it satisfies the following conditions: (1) X is closed downward: for all* $x, y \in \{L, R\}^*$*, if* $x \in X$ *and* y *is a prefix of* x *then* $y \in X$*; (2) X is not empty:* $\epsilon \in X$*; and (3) X branches uniquely: for every* $x \in \{L, R\}^*$*, if* $x \in X$ *then either* $xR \in X$ *or* $xL \in X$ *but not both. Let* $\mathcal{P} \subseteq \mathcal{T}_{\{0,1\}}$ *be the collection of all paths.*

In other words, $X \in \mathcal{P}$ if the set of vertices in X describe an infinite branch in the full binary tree. Clearly $\mathcal{P}$ is homeomorphic as a subspace of $\mathcal{T}_{\{0,1\}}$ to the set $\{L, R\}^\omega$ of ω-words over the alphabet $\{L, R\}$ and it is simple to verify that:

Proposition 2. *The equality* $\mu_{\mathcal{T}_{\{0,1\}}}(\mathcal{P}) = 0$ *holds.*

However the space $\mathcal{P}$ carries the natural Lebesgue measure $\mu_{\{L,R\}^\omega}$ which we use below to define the semantics of MSO $+ \forall_\pi^{=1}$.

Definition 11 (Syntax of MSO $+ \forall_\pi^{=1}$**).** *The syntax of* MSO $+ \forall_\pi^{=1}$ *formulas* ϕ *is generated by the following grammar:*

$$\phi ::= \mathrm{Sing(X)} \mid \mathit{Succ}_L(X, Y) \mid \mathit{Succ}_R(X, Y) \mid X \subseteq Y \mid \neg \phi \mid \phi_1 \vee \phi_2 \mid \forall X . \phi \mid \forall_\pi^{=1} X . \phi$$

Definition 12 (Semantics of MSO $+ \forall_\pi^{=1}$**).** *The semantics of* MSO $+ \forall_\pi^{=1}$ *is defined by extending the semantics of* MSO *on trees (Definition 7) as follows:*

$$(A_1, \ldots, A_n) \in \forall_\pi^{=1} X . \phi(X, Y_1, \ldots, Y_n)$$
$$\Leftrightarrow$$
$$\mu_{\{L,R\}^\omega}\left(\{B \in \mathcal{P} \mid (B, A_1, \ldots, A_n) \in \phi(X, Y_1, \ldots, Y_n)\}\right) = 1$$

Hence, informally, $(A_1, \ldots, A_n) \in \forall_\pi^{=1} X.\phi(X, Y_1, \ldots, Y_n)$ if, for a randomly chosen $B \in \mathcal{P}$, the formula $\phi(B, A_1, \ldots, A_n)$ holds almost surely.

It is not immediately clear from the previous definition if the quantifier $\forall_\pi^{=1}$ can be expressed in $MSO + \forall^{=1}$ on trees. Indeed note that, since $\mu_{\mathcal{T}_{\{0,1\}}}(\mathcal{P}) = 0$, the naive definition $\forall_\pi^{=1} X.\phi(X) = \forall^{=1} X.(\text{``X is a path''} \wedge \phi(X))$, where the predicate "$X$ is a path" is easily expressible in MSO, but does not work. Indeed the $MSO + \forall^{=1}$ expression on the right always defines the empty set because the collection of $X \in \mathcal{T}_{\{0,1\}}$ satisfying the conjunction is a subset of $\mathcal{P}$ and therefore has $\mu_{\mathcal{T}_{\{0,1\}}}$ measure 0.

Nevertheless the quantifier $\forall_\pi^{=1}$ can be expressed in $MSO + \forall^{=1}$ on trees with a more elaborate encoding presented below. The main ingredient of the encoding is a MSO definable continuous function f which maps a tree $X \in \mathcal{T}_{\{0,1\}}$ to a path $f(X) \in \mathcal{P}$ preserving measure in the sense stated in Lemma 1.2 below.

Definition 13. *Define the binary relation $f(X, Y)$ on $\mathcal{T}_{\{0,1\}}$ by the following MSO formula:* "Y is a path" *and* $\forall y \in Y. \exists z. (Succ_L(y, z)$ *and* $(z \in Y \Leftrightarrow y \in X))$.

Lemma 1. *For every $X \in \mathcal{T}_{\{0,1\}}$ there exists exactly one $Y \in \mathcal{P} \subseteq \mathcal{T}_{\{0,1\}}$ such that $f(X, Y)$. Hence the relation f is a function $f : \mathcal{T}_{\{0,1\}} \to \mathcal{P}$. Furthermore f satisfies the following properties:*

1. *f is a continuous, open and surjective function,*
2. *Assume $B \subseteq \mathcal{P}$ is $\mu_{\{L,R\}^\omega}$ measurable. Then $\mu_{\{L,R\}^\omega}(B) = \mu_{\mathcal{T}_{\{0,1\}}}(f^{-1}(B))$.*

A proof of Lemma 1 will be published elsewhere.

We can now present the correct $MSO + \forall^{=1}$ encoding of the quantifier $\forall_\pi^{=1}$.

Theorem 4. *For every $MSO + \forall_\pi^{=1}$ formula $\psi(\overrightarrow{Z})$ there exists a $MSO + \forall^{=1}$ formula $\psi'(\overrightarrow{Z})$ such that ψ and ψ' denote the same set.*

Proof. The proof goes by induction on the complexity of ψ with the interesting case being $\phi(\overrightarrow{Z}) = \forall_\pi^{=1} Y.\psi(Y, \overrightarrow{Z})$. By induction hypothesis, there exists a $MSO + \forall^{=1}$ formula ψ' defining the same set as ψ. Then the $MSO + \forall^{=1}$ formula ϕ' corresponding to ϕ is: $\phi'(\overrightarrow{Z}) = \forall^{=1} X.(\exists Y.(f(X, Y) \wedge \psi'(Y, \overrightarrow{Z})))$. We now show that ϕ and ϕ' indeed define the same set. The following are equivalent:

1. $\overrightarrow{C} \in \forall_\pi^{=1} Y.\psi(Y, \overrightarrow{Z})$,
2. (by Definition of $\forall_\pi^{=1} X$) The set $A = \{Y \in \mathcal{P} \mid \psi(Y, \overrightarrow{C})\}$ is such that $\mu_{\{L,R\}^\omega}(A) = 1$,
3. (by Lemma 1.(2)) The set $B \subseteq \mathcal{T}_{\{0,1\}}$, defined as $B = f^{-1}(A)$, i.e., as $B = \{X \in \mathcal{T}_{\{0,1\}} \mid \exists Y.(f(X, Y) \wedge \psi(Y, \overrightarrow{C}))\}$. is such that $\mu_{\mathcal{T}_{\{0,1\}}}(B) = 1$.
4. (by definition of $\forall^{=1}$ and using $\psi = \psi'$) $\overrightarrow{C} \in \forall^{=1} X.(\exists Y.(f(X, Y) \wedge \phi'(Y, \overrightarrow{Z})))$.

$\square$

7 On the Expressive Power of MSO+$\forall_\pi^{=1}$

In Sect. 5 we proved that the theory of MSO + $\forall^{=1}$ on ω-words and trees is undecidable. Motivated by this negative result, in Sect. 6 we introduced the logic MSO + $\forall_\pi^{=1}$ on trees and in Theorem 4 we proved that it can be regarded as a syntactical fragment of MSO + $\forall^{=1}$ on trees.

We have not been able to establish if the logic MSO + $\forall_\pi^{=1}$ on trees is decidable or not (Problem 1 in the Introduction). On the one hand we observe (Proposition 3 below), by applying a result of [6], that MSO + $\forall_\pi^{=1}$ can define non-regular sets. On the other hand, it does not seem possible to apply the methods utilized in this paper to prove its undecidability.

In the rest of this section we investigate the expressive power of MSO+$\forall_\pi^{=1}$. We show that the the decidability of MSO + $\forall_\pi^{=1}$ implies the decidability of the SAT problem for the qualitative fragment of the probabilistic logic pCTL*. We establish a connection between MSO + $\forall_\pi^{=1}$ and automata theory by showing that the class of *qualitative languages* of trees of [6] can be expressed by MSO + $\forall_\pi^{=1}$ formulas (Theorem 6). We prove that the first order theory of the lattice of F_σ subsets of the Cantor space with the predicates $C(X) \Leftrightarrow$ "X is a closed set" and $N(X) \Leftrightarrow$ "X is a Lebesgue null set" is interpretable in MSO + $\forall_\pi^{=1}$ (Theorem 9). Lastly, we show that the first order theory of the Lebesgue measure algebra equipped with Scott's closure operator is interpretable in MSO + $\forall_\pi^{=1}$.

7.1 SAT Problem of Probabilistic Temporal Logics

In this subsection we sketch the essential arguments that allow to reduce the SAT problem of the qualitative fragment of pCTL* and similar logics to the decision problem of MSO + $\forall_\pi^{=1}$. We assume the reader is familiar with the logic pCTL*. We refer to the textbook [2] for a detailed introduction.

The logic pCTL* and its variants are designed to express properties of Markov chains. The following is a long standing open problem (see, e.g., [4]).

SAT Problem. *Given a pCTL* state-formula ϕ, is there a Markov chain M and a vertex $v \in M$ such that v satisfies ϕ?*

Without loss of generality (see, e.g., Sect. 5 of [12] for details), we can restrict the statement of the SAT problem to range over Markov chains M whose underlying directed graph has the structure of the full binary tree, where each edge (connecting a vertex to one of its two children) has probability $\frac{1}{2}$. This is a convenient restriction that allows to interpret pCTL* formulas $\phi(P_1, \ldots, P_n)$ with n propositional variables as denoting sets $[\![\phi]\!] \subseteq \mathcal{T}_\Sigma$ for $\Sigma = \{0, 1\}^n$.

It is well known that there exists pCTL* formulas such that $[\![\phi]\!] \neq \emptyset$ but $[\![\phi]\!]$ does not contain any regular tree. This means the logic pCTL* can define non-regular sets of trees. We show now that every pCTL* definable set $[\![\phi]\!]$ is MSO + $\forall_\pi^{=1}$ definable. The argument is similar[4] to the one used in [16] to prove that sets

---

[4] In fact, following the work of [16], the logic pCTL* is also definable in a weaker logic such as Thomas' *chain logic* extended with the quantifier $\forall_\pi^{=1}$.

of trees defined in the logic CTL* can be defined in MSO. Each pCTL* state formula $\phi(P_1, \ldots, P_n)$ is translated to a MSO $+ \forall_\pi^{=1}$ formula $F_\phi(X_1, \ldots, X_n, y)$ and each pCTL* path formula $\psi(P_1, \ldots, P_n)$ is translated to a MSO $+ \forall_\pi^{=1}$ formula $F_\psi(X_1, \ldots, X_n, Y)$ such that:

- a vertex v of the full binary tree satisfies the pCTL* state-formula $\phi(P_1, \ldots, P_n)$ if and only if $F_\phi(P_1, \ldots, P_n, v)$ is a valid MSO $+ \forall_\pi^{=1}$ formula with parameters,
- a path $A \in \mathcal{P}$ in the the full binary tree satisfies the pCTL* path-formula $\psi(P_1, \ldots, P_n)$ if and only if $F_\psi(P_1, \ldots, P_n, A)$ is a valid MSO $+ \forall_\pi^{=1}$ formula with parameters.

The only case different from [16] is for a pCTL state-formula of the form $\phi = \mathbb{P}_{>0}\psi(P_1, \ldots, P_n)$ which holds at a vertex v if the collection of paths starting from v and satisfying ψ has positive measure; ϕ is is translated to MSO $+ \forall_\pi^{=1}$ as follows:

$$F_\phi(X_1, \ldots, X_n, y) = \exists_\pi^{>0} Y. \big(Y \text{ is a path containing } x, \text{ and}$$
$$F_\psi(X_1 \ldots, X_n, Z) \text{ holds where } Z \text{ is the set of descendants of } x \text{ in } Y\big)$$

We state the correctness of this translation as the following

Theorem 5. *The decidability of the SAT problem for the qualitative fragment of pCTL* is reducible to the decidability of* MSO $+ \forall_\pi^{=1}$.

7.2 On the Qualitative Languages of Carayol, Haddad and Serre

In a recent paper [6] Carayol, Haddad and Serre have considered a probablistic interpretation of standard nondeterministic tree automata. Below we briefly discuss this interpretation referring to [6] for more details. The standard interpretation of a nondeterministic tree automaton $\mathcal{A}$ over the alphabet Σ is the set $\mathcal{L}(A) \subseteq \mathcal{T}_\Sigma$ of trees $X \in \mathcal{T}_\Sigma$ such that there exists a *run* ρ of X on $\mathcal{A}$ such that for all paths π in ρ, the path π is accepting. The probabilistic interpretation in [6] associates to each nondeterministic tree automaton the language $\mathcal{L}^{=1}(\mathcal{A}) \subseteq \mathcal{T}_\Sigma$ of trees $X \in \mathcal{T}_\Sigma$ such that there exists a *run* ρ of X on $\mathcal{A}$ such that for *almost all* paths π in ρ, the path π is accepting, where "almost all" means having Lebesgue measure 1. Using the language of MSO $+\forall_\pi^{=1}$ the language $\mathcal{L}^{=1}(\mathcal{A})$ can be naturally expressed by the following formula $\psi_A(\overrightarrow{X})$:

$$\psi_A(\overrightarrow{X}) = \exists \overrightarrow{Y}.\left(\text{``}\overrightarrow{Y} \text{ is a run of } \overrightarrow{X} \text{ on } \mathcal{A}\text{''} \wedge \forall_\pi^{=1} Z.(\text{``}Z \text{ is an accepting path of } \overrightarrow{Y}\text{''})\right)$$

Theorem 6. *Let $L \subseteq \mathcal{T}_\Sigma$ be a set of trees definable by a nondeterministic tree automaton with probabilistic interpretation. Then L is definable in* MSO $+ \forall_\pi^{=1}$.

Let $L \subseteq \mathcal{T}_{\{0,1\}}$ consists of $A \in \mathcal{T}_{\{0,1\}}$ such that the set of branches having infinitely many vertices labeled by 1 has measure 1. In [6, Example 7] it is proved that L is not regular and definable by a nondeterministic tree automata with probabilistic interpretation. Therefore:

Proposition 3. MSO $+ \forall_\pi^{=1}$ *is a proper extensions of* MSO *on trees.*

7.3   An Extension of Rabin's Theory of the Lattice of F_σ Sets

Rabin in [13] proved the decidability of MSO on the full binary tree and as corollaries obtained several decidability results. One of them ([13, Theorem 2.8]) states that the first order theory of the lattice of F_σ subsets of the Cantor space $\{0,1\}^\omega$, with the predicate $C(X) \Leftrightarrow$ "X is a closed set", is decidable. Formally this result can be stated as follows.

Theorem 7 (Rabin). *The FO theory of the structure $\langle F_\sigma, \cup, \cap, C \rangle$ is decidable.*

Rabin proved this theorem by means of a reduction to the MSO theory of the full binary tree. He observed that the Cantor space $\{0,1\}^\omega$ is a homeomorphic copy of the set of paths $\mathcal{P}$ in the full binary tree (see Definition 10). He then noted that an arbitrary set of vertices $X \in T_{\{0,1\}}$ can be viewed as a set $\langle X \rangle \subseteq \mathcal{P}$ of paths by the MSO expressible definition $\langle X \rangle = \{Y \in \mathcal{P} \mid Y \cap X \text{ isfinite}\}$. He showed that a set of paths $A \subseteq \mathcal{P}$ is F_σ if and only if there exists some $X \in T_{\{0,1\}}$ such that $A = \langle X \rangle$ and that it is possible to express in MSO that $\langle X \rangle$ is closed. For details we refer to [13, §2].

We now consider an extension of the structure $\langle F_\sigma, \cup, \cap, C \rangle$ by a new predicate $N(X) \Leftrightarrow$ "X is a Lebesgue null set".

Theorem 8. *The first order theory of the structure $\langle F_\sigma, \cup, \cap, C, N \rangle$ is interpretable in $MSO + \forall_\pi$.*

Proof. It is straightforward to extend Rabin's interpretation by an appropriate $MSO + \forall_\pi$ interpretation of the predicate N. Let $\phi(X)$ be the formula with one free-variable defined as: $\forall_\pi^{=1} Y.(Y \in \langle X \rangle)$ where, in accordance with Rabin's interpretation, the predicate $Y \in \langle X \rangle$ is defined as "$Y \cap X$ is a finite set", which is easily expressible in MSO. Then one has $\langle X \rangle \in N$ if and only if $\phi(X)$ holds, and this completes the proof.   □

Hence if the theory of $MSO + \forall_\pi^{=1}$ is decidable then the first order theory of $\langle F_\sigma, \cup, \cap, C, N \rangle$ is also decidable.

7.4   On the Measurable Algebra with Scott's Closure Operation

In the classic paper "*The algebra of Topology*" [11] McKinsey and Tarski defined *closure algebras* as pairs $\langle B, \Diamond \rangle$ where B is a Boolean algebra and $\Diamond : B \to B$ is unary operation satisfying the axioms: $\Diamond \Diamond x = \Diamond x$, $x \leq \Diamond x$, $\Diamond(x \vee y) = \Diamond x \vee \Diamond y$ and $\Diamond \top = \top$.

Let $\mathcal{B}$ denote the collection of Borel subsets of the Cantor space $\{0,1\}^\omega$. Define the equivalence relation $\sim$ on $\mathcal{B}$ as $X \sim Y$ if $\mu_{\{0,1\}^\omega}(X \triangle Y) = 0$, where $X \triangle Y = (X \setminus Y) \cup (Y \setminus X)$. The quotient $\mathcal{B}/\sim$ is a complete Boolean algebra with operations defined as $[X]_\sim \vee [Y]_\sim = [X \cup Y]_\sim$ and $\neg[X]_\sim = [\{0,1\}^\omega \setminus X]_\sim$. It is called the (Lebesgue) *measure algebra* (see, e.g., [9, 17.A]) and denoted by $\mathcal{M}$.

Recently Dana Scott has observed[5] that the *(Lebesgue) measure algebra* $\mathcal{M}$ naturally carries the structure of a closure algebra.

---

[5] Result announced by Scott during a seminar entitled "Mixing Modality and Probability" given in Edinburgh, June 2010.

Definition 14. *An element $[X]_\sim \in \mathcal{M}$ is called* closed *if it contains a closed set, i.e., if there exists a closed set Y such that $Y \in [X]_\sim$. Let $\Diamond_\mathcal{M} : \mathcal{M} \to \mathcal{M}$ be defined as follows: $\Diamond([X]_\sim) = \bigwedge \{[Y]_\sim \mid [X]_\sim \le [Y]_\sim$ and $[Y]_\sim$ is closed$\}$.* Note that the infimum exists because $\mathcal{M}$ is complete.

Proposition 4 (Scott). *The pair $\mathcal{S} = \langle \mathcal{M}, \Diamond_\mathcal{M} \rangle$ is a closure algebra.*

Interestingly, it was proved in [10, Theorem 6.3] that $\mathcal{S}$ is universal among the class of all closure algebras: an equation holds in $\mathcal{S}$ if and only if it holds in all closure algebras. We make the following observation.

Theorem 9. *The first order theory of $\mathcal{S}$ is interpretable in $\mathrm{MSO} + \forall_\pi$.*

Proof. By Theorem 8 it is sufficient to observe that the theory of $\mathcal{S}$ can be interpreted within the theory of $\langle F_\sigma, \cup, \cap, C, N \rangle$. This is possible as, by regularity of Borel measures, any element $[X]_\sim$ contains an F_σ sets. A detailed proof will be published elsewhere. □

Hence if $\mathrm{MSO} + \forall_\pi^{=1}$ is decidable then the first order theory of $\mathcal{S}$ is also decidable.

8 Open Problems

In the Introduction we formulated Problem 1 regarding the decidability of the theory of $\mathrm{MSO} + \forall_\pi^{=1}$. In light of Theorems 8 and 9, the decidability of the theories of $\langle F_\sigma, \cup, \cap, C, N \rangle$ and $\langle \mathcal{M}, \Diamond_\mathcal{M} \rangle$ is a closely related problem. In particular, if one of these two theories is undecidable, then also $\mathrm{MSO} + \forall_\pi^{=1}$ is undecidable.

In Sect. 4 in Remark 2 we noticed that the definition of the semantics of $\mathrm{MSO} + \forall^{=1}$ involves potentially non-measurable sets. One encounters the same problem in the definition of $\mathrm{MSO} + \forall_\pi^{=1}$. Hence:

Problem 2. *Are relations defined by $\mathrm{MSO} + \forall_\pi^{=1}$ formulas Lebesgue measurable?*

In previous work [8] we proved that the all regular sets of trees are $\mathcal{R}$-sets and, as a consequence, Lebesgue measurable. Therefore a variant of Problem 2 above asks whether all $\mathrm{MSO} + \forall_\pi^{=1}$ definable sets are $\mathcal{R}$-sets. In the other direction, $\mathcal{R}$-sets belong to the $\mathbf{\Delta}_2^1$ class of the projective hierarchy. So we can ask:

Problem 3. *Is the class of sets definable by $\mathrm{MSO} + \forall_\pi^{=1}$ formulas contained in a certain fixed level of the projective hierarchy?*

A negative answer would likely lead to undecidability of $\mathrm{MSO} + \forall_\pi^{=1}$ (see [3]).

References

1. Baier, C., Grösser, M., Bertrand, N.: Probabilistic ω-automata. J. ACM **59**(1), 1 (2012)
2. Baier, C., Katoen, J.P.: Principles of Model Checking. The MIT Press, Cambridge (2008)

3. Bojańczyk, M., Gogacz, T., Michalewski, H., Skrzypczak, M.: On the decidability of MSO+U on infinite trees. In: Esparza, J., Fraigniaud, P., Husfeldt, T., Koutsoupias, E. (eds.) ICALP 2014, Part II. LNCS, vol. 8573, pp. 50–61. Springer, Heidelberg (2014)

4. Brázdil, T., Forejt, V., Kretínský, J., Kucera, A.: The satisfiability problem for probabilistic CTL. In: Proceedings of LICS, pp. 391–402 (2008)

5. Büchi, J.R.: On a decision method in restricted second order arithmetic. In: Logic, Methodology and Philosophy of Science, Proceedings, pp. 1–11. American Mathematical Society (1962)

6. Carayol, A., Haddad, A., Serre, O.: Randomization in automata on infinite trees. ACM Trans. Comput. Logic 15(3), 24 (2014)

7. Gimbert, H., Oualhadj, Y.: Probabilistic automata on finite words: decidable and undecidable problems. In: Abramsky, S., Gavoille, C., Kirchner, C., Meyer auf der Heide, F., Spirakis, P.G. (eds.) ICALP 2010. LNCS, vol. 6199, pp. 527–538. Springer, Heidelberg (2010)

8. Gogacz, T., Michalewski, H., Mio, M., Skrzypczak, M.: Measure properties of game tree languages. In: Csuhaj-Varjú, E., Dietzfelbinger, M., Ésik, Z. (eds.) MFCS 2014, Part I. LNCS, vol. 8634, pp. 303–314. Springer, Heidelberg (2014)

9. Kechris, A.S.: Classical Descriptive Set Theory. Springer Verlag, New York (1994)

10. Lando, T.A.: Completeness of S4 for the lebesgue measure algebra. J. Philos. Logic (2010)

11. McKinsey, J.C.C., Tarski, A.: The algebra of topology. Ann. Math. (1944)

12. Michalewski, H., Mio, M.: Baire category quantifier in monadic second order logic. In: Halldórsson, M.M., Iwama, K., Kobayashi, N., Speckmann, B. (eds.) ICALP 2015. LNCS, vol. 9135, pp. 362–374. Springer, Heidelberg (2015)

13. Rabin, M.O.: Decidability of second-order theories and automata on infinite trees. Trans. Am. Math. Soc. 141, 1–35 (1969)

14. Robinson, R.M.: Restricted set-theoretical definitions in arithmetic. Proc. Amer. Math. Soc. 9, 238–242 (1958)

15. Steinhorn, C.I.: Borel Structures and Measure and Category Logics. In: Model-theoretic logics. Perspectives in Mathematical Logic, vol. 8, Chap. XVI. Springer-Verlag, New York (1985). http://projecteuclid.org/euclid.pl/1235417282

16. Thomas, W.: On chain logic, path logic, and first-order logic over infinite trees. In: Proceedings of LICS, pp. 245–256 (1987)

17. Thomas, W.: Languages, automata, and logic. In: Rozenberg, G., Salomaa, A. (eds.) Handbook of Formal Languages, pp. 389–455. Springer, Berlin (1996)

A Cut-Free Labelled Sequent Calculus for Dynamic Epistemic Logic

Shoshin Nomura[✉], Hiroakira Ono, and Katsuhiko Sano

Japan Advanced Institute of Science and Technology, Ishikawa, Japan
{nomura,ono,v-sano}@jaist.ac.jp

Abstract. Dynamic Epistemic Logic is a logic that is aimed at formally expressing how a person's knowledge changes. We provide a cut-free labelled sequent calculus (**GDEL**) on the background of existing studies of Hilbert-style axiomatization **HDEL** by Baltag et al. (1989) and labelled calculi for Public Announcement Logic by Maffezioli et al. (2011) and Nomura et al. (2015). We first show that the *cut* rule is admissible in **GDEL**. Then we show **GDEL** is sound and complete for Kripke semantics. Lastly, we touch briefly on our on-going work of an automated theorem prover of **GDEL**.

Keywords: Dynamic Epistemic Logic · Action models · Labelled sequent calculus · Admissibility of cut · Validity of sequents

1 Introduction

The purpose of this paper is to provide a cut-free labelled sequent calculus[1] for Dynamic Epistemic Logic (DEL for short). The logic DEL, one of whose original idea was proposed by Baltag et al. [4] and elaborated in several papers (e.g., [4–6,10]), introduces a basic ideas of how to express formally how knowledge states change. A simple version of DEL is called Public Announcement Logic (PAL) by Plaza [17], and it introduced a basic idea of knowledge change by using Kripke semantics. As the name PAL shows, it only deals mainly with 'public announcement,' by which every agent receives the same information; however, the state of knowledge may be changed not only by public announcements but also announcements to a specific group in a community. A typical example is 'private announcements,' in which someone informs something to only a single person (e.g., a personal letter). On the other hand, DEL, an expansion of PAL, is a logic which can express not only public announcement, but more delicate and complicated flows of information such as private announcement, and such a factor that causes change of knowledge state is called an *event* (or *action*) as a whole.

---

[1] Labelled sequent calculus is one of the most prevailing methods of sequent calculus for modal logic (cf. Negri et al. [12]).

© Springer International Publishing Switzerland 2016
S. Artemov and A. Nerode (Eds.): LFCS 2016, LNCS 9537, pp. 283–298, 2016.
DOI: 10.1007/978-3-319-27683-0_20

For the last decade, several studies of semantical developments of DEL have emerged for the sake of capturing characteristics regarding knowledge; nevertheless, compared to the development of semantical variants of DEL, proof-theoretical aspects of DEL seems undeveloped. However, recently, a labelled sequent calculus for PAL was proposed by Maffezioli et al. [11], and its revised version by Balbiani et al. [3] and Nomura et al. [15]; a labelled sequent calculus for intutionistic PAL was also given by Nomura et al. [14]. We note that there are some proof-theoretic studies of DEL such as a tableaux calculus for DEL by Aucher et al. [1,2], a display calculus for it by Frittella et al. [9] and a nested calculus for it by Dyckhoff et al. [7]. We, in this paper, construct a labelled calculus of DEL on the recent background of studies of a labelled system of PAL, especially on [15].

The outline of the paper is as follows. In Sect. 2, we introduce the syntax and Kripke semantics of DEL, and look at a specific example of knowledge change with DEL. In Sect. 3, we see Hilbert-style axiomatization (**HDEL**) of DEL, and then we give our labelled sequent calculus for DEL (**GDEL**) based on the study of a labelled system of PAL. In Sect. 4, we establish admissibility of the *cut* rule in **GDEL**. In Sect. 5, we prove the soundness theorem and then give a proof of the completeness theorem of **GDEL** as a corollary. In the last section, we give a brief explanation of our automated theorem prover of **GDEL**.

2   Dynamic Epistemic Logic

First of all, we define the syntax and Kripke semantics of DEL. In this paper, we mainly follow the definition of DEL as given in van Ditmarsch et al. [6].[2] Let $\mathsf{Agt} = \{a, b, c, \ldots\}$ be a finite set of *agents* and $\mathsf{Prop} = \{p, q, r, \ldots\}$ a countably infinite set of *propositional atoms*. An (S5) *event frame* is a pair $(\mathsf{S}, (\sim_a)_{a \in \mathsf{Agt}})$ where S is a non-empty *finite* set of events and $\sim_a$ is an equivalence relation on S ($a \in \mathsf{Agt}$), which represents agent a's uncertainty. In what follows, we use an element of a countable set $\mathsf{Evt} = \{\mathsf{a}, \mathsf{b}, \mathsf{c}, \mathsf{s}, \mathsf{t}, \ldots\}$ as a *meta-variable* to refer to an event.

Definition 1. *We define the set* $\mathcal{L}_{DEL} = \{A, B, \ldots\}$ *of all formulas of DEL and the set of all (S5) event models* $\mathsf{M} = (\mathsf{S}, (\sim_a)_{a \in \mathsf{Agt}}, \mathsf{pre})$ *by simultaneous induction as follows:*

$$A ::= p \mid \neg A \mid (A \wedge A) \mid \Box_a A \mid [\mathsf{a}^\mathsf{M}]A, \quad (p \in \mathsf{Prop}, a \in \mathsf{Agt}, and\, \mathsf{a} \in \mathsf{S})$$

where $(\mathsf{S}, (\sim_a)_{a \in \mathsf{Agt}})$ *is an event frame,* pre *is a function which assigns an* $\mathcal{L}_{DEL}$*-formula* $\mathsf{pre}(\mathsf{b})$ *to each event* $\mathsf{b} \in \mathsf{S}$*, and an expression* a^M *is an abbreviation of a pointed event model* (M, a)*. We read* $[\mathsf{a}^\mathsf{M}]A$ *as 'after an event* a^M *occurs, A holds'.[3] Boolean connectives such as* $\rightarrow, \vee$ *are defined as usual.*

---

[2] DEL is called Logic of Epistemic Action in [4] and Action Model Logic in [6].

[3] Ditmarsch et al. [6] includes union of events $\mathsf{a}^\mathsf{M} \cup \mathsf{a}'^{\mathsf{M}'}$ in the language, but we do not include it since $[\mathsf{a}^\mathsf{M} \cup \mathsf{a}'^{\mathsf{M}'}]A$ can be handled as a defined connective by $[\mathsf{a}^\mathsf{M}]A \wedge [\mathsf{a}'^{\mathsf{M}'}]A$.

For any event model $M = (S, (\sim_a)_{a \in \mathsf{Agt}}, \mathsf{pre})$, we use M as a superscript of S, $\sim_a$ and pre such as S^M, $\sim_a^M$ and pre^M to emphasize that they belong to the event model M. PEvt is used to denote the set $\{a^M, b^N, \ldots\}$ of all pointed event models.

Definition 2 (Composition of Events). *Given any two event models* M *and* M′, *the* composition *of the events* (M; M′) *is the event model such that:*

$$S^{M;M'} = S^M \times S^{M'},$$
$$(a, a') \sim_a^{M;M'} (b, b') \quad iff \quad a \sim_a^M b \ and \ a' \sim_a^{M'} b',$$
$$\mathsf{pre}^{M;M'}((a, a')) = \mathsf{pre}^M(a) \wedge [a^M]\mathsf{pre}^{M'}(a').$$

Given any pointed event models (a^M) *and* $(a'^{M'})$, *the composition of the pointed event models,* $(a^M); (a'^{M'})$, *is the pointed event model such that* $(a, a')^{M;M'}$ *with* (M; M′).

Note that the above event model $(a, a')^{M;M'}$ *is a pointed event model (but only with a complex name of event (a, a')) by the definition above, and so it is included in PEvt.

Let us move onto the Kripke semantics of DEL. A *Kripke model* $\mathfrak{M}$ is a triple $(W, (R_a)_{a \in \mathsf{Agt}}, V)$ such that W is a non-empty set of *worlds* (W of $\mathfrak{M}$ is also written as $\mathcal{D}(\mathfrak{M})$), $(R_a)_{a \in \mathsf{Agt}}$ is an Agt-indexed family of binary relations on W (a ranges over Agt) and $V : \mathsf{Prop} \to \mathcal{P}(W)$ is a valuation function. We note that epistemic logics are usually based on modal logic S5, but here we define DEL based on modal logic K as a starting point for constructing its sequent calculus system; therefore, we do not assume any frame property on R_a in this paper.

Given a Kripke model $\mathfrak{M}$ and a world $w \in \mathcal{D}(\mathfrak{M})$, the satisfaction relation $\mathfrak{M}, w \models A$ for a formula A is inductively defined as follows:

$$\mathfrak{M}, w \models p \quad iff \ w \in V(p),$$
$$\mathfrak{M}, w \models \neg A \quad iff \ \mathfrak{M}, w \not\models A,$$
$$\mathfrak{M}, w \models A \wedge B \ iff \ \mathfrak{M}, w \models A \ and \ \mathfrak{M}, w \models B,$$
$$\mathfrak{M}, w \models \Box_a A \quad iff \ for \ all \ w' \in W : wR_a w' \ implies \ \mathfrak{M}, w' \models A,$$
$$\mathfrak{M}, w \models [a^M]A \ iff \ \mathfrak{M}, w \models \mathsf{pre}(a) \ implies \ \mathfrak{M}^{\otimes M}, (w, a) \models A,$$

where $\mathfrak{M}^{\otimes M} = (W^{\otimes M}, (R_a^{\otimes M})_{a \in \mathsf{Agt}}, V^{\otimes M})$ is the *updated Kripke model* of $\mathfrak{M}$ by an event model M and it is defined as:

$$W^{\otimes M} = \{(w, a) \in W \times S^M \mid \mathfrak{M}, w \models \mathsf{pre}^M(a)\},$$
$$(w, a)R_a^{\otimes M}(w', a') \ iff \ wR_a w' \ and \ a \sim_a^M a',$$
$$(w, a) \in V^{\otimes M}(p) \ iff \ w \in V(p),$$

where $a \in \mathsf{Agt}$ and $p \in \mathsf{Prop}$. A formula A is *valid* if $\mathfrak{M}, w \models A$ holds in any Kripke model $\mathfrak{M}$ and any world $w \in \mathcal{D}(\mathfrak{M})$. Intuitively, $\mathfrak{M}^{\otimes M}$ means $\mathfrak{M}$ updated by event M. We briefly give an example which will show a way how DEL expresses a changing knowledge state, by taking an example of an event model Read (the simplest example of 'private announcement') in [6, p. 166].

Example 1. Suppose there are two agents a and b, and neither of them knows whether p. Then only a reads a letter where p is written. As a consequence, a's knowledge changes and she knows p, but b does not. Let $\mathsf{Agt} = \{a, b\}$. Then a Kripke model $\mathfrak{M}$ and an event model M are defined as follows:

$\mathfrak{M} = (W, R_a, R_b, V)$
  where $W = \{w_1, w_2\}$, $R_a = R_b = W^2$ and $V(p) = \{w_1\}$,
$\mathsf{Read} = (\mathsf{S}, \sim_a, \sim_b, \mathsf{pre})$
  where $\mathsf{S} = \{\mathsf{p}, \mathsf{np}\}$, $\sim_a = \{(\mathsf{p}, \mathsf{p}), (\mathsf{np}, \mathsf{np})\}$, $\sim_b = \mathsf{S}^2$ and $\mathsf{pre}(\mathsf{p}) = p$, $\mathsf{pre}(\mathsf{np}) = \neg p$.

This situation of the agent a and b can be semantically formalized by a pointed Kripke model $(\mathfrak{M}, w_1)$, an event $(\mathsf{Read}, \mathsf{p})$ and the pointed updated model $(\mathfrak{M}^{\otimes \mathsf{Read}}, (w_1, \mathsf{p}))$. The models are depicted as follows (assuming reflexivity).

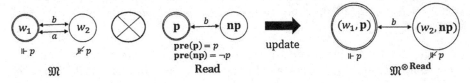

Each double circle indicates the given or resulting point (world or event) which stands for the actual world or actual event. Intuitively, for any agent x, a bidirectional arrow of x between two worlds (or events) stands for that x cannot distinguish between two, and he/she is ignorant of the reality (or actually what occurred) if one side of the arrow is the actual world (or event). $(\mathfrak{M}, w_1)$ stands for the initial knowledge state of both a and b where both are ignorant of p (the actual world w_1). $(\mathsf{Read}, \mathsf{p})$ stands for an event such that only a reads the letter containing information p, and that is because a does not have her bidirectional arrow between the two worlds. The updated model $(\mathfrak{M}^{\otimes \mathsf{Read}}, (w_1, \mathsf{p}))$ stands for the knowledge state of both a and b where a knows p but b is still ignorant of p.

Additionally, *multiple updates* $(\cdots (\mathfrak{M}^{\otimes \mathsf{M}_1})^{\otimes \cdots})^{\otimes \mathsf{M}_n}$ on $\mathfrak{M}$ are also possible, which we write by $\mathfrak{M}^{\otimes \mathsf{M}_1 \otimes \cdots \otimes \mathsf{M}_n}$ for simplicity. Each of Greek letters $\alpha, \beta, \ldots$ indicates a finite list $(\mathsf{a}_1^{\mathsf{M}_1}), \ldots, (\mathsf{a}_n^{\mathsf{M}_n})$ of pointed event models, and use ϵ for the empty list. Moreover, if α is a list $(\mathsf{a}_1^{\mathsf{M}_1}), \ldots, (\mathsf{a}_n^{\mathsf{M}_n})$ of pointed event models, then we define $\alpha_{evt} := (\mathsf{a}_1, \ldots, \mathsf{a}_n)$ and $\alpha_{mdl} := (\mathsf{M}_1, \ldots, \mathsf{M}_n)$, and $\alpha_{evt} := \epsilon$ and $\alpha_{mdl} := \epsilon$ if α is ϵ. The symbol $\mathfrak{M}^{\otimes \alpha_{mdl}}$ indicates $\mathfrak{M}^{\otimes \mathsf{M}_1 \otimes \mathsf{M}_2 \otimes \cdots \otimes \mathsf{M}_n}$ when $\alpha_{mdl} = (\mathsf{M}_1, \mathsf{M}_2, \ldots, \mathsf{M}_n)$, and $\mathfrak{M}^{\otimes \alpha_{mdl}}$ indicates $\mathfrak{M}$ when $\alpha = \epsilon$.

3   Labelled Sequent Calculus for DEL

Hilbert-style axiomatization **HDEL** of DEL was introduced by Baltag et al. [4] and the completeness was shown. The axiomatization is defined in Table 1 where the axioms for event operators are added to the axiomatization of modal logic K. These additional axioms (from $(RA1)$ to $(RA5)$) are often called *recursion axioms*, as they express a way of reducing each formula of **HDEL** equivalently into a formula of K. The completeness theorem of **HDEL** can be shown by an argument in [4, Proposition 4.5].

Table 1. Hilbert-style axiomatization of DEL: **HDEL**

Modal Axioms	all instantiations of propositional tautologies
	(K) $\Box_a(A \to B) \to (\Box_a A \to \Box_a B)$
Recursion Axioms	$(RA1)$ $[\mathsf{a}^M]p \leftrightarrow (\mathsf{pre(a)} \to p)$
	$(RA2)$ $[\mathsf{a}^M]\neg A \leftrightarrow (\mathsf{pre(a)} \to \neg[\mathsf{a}^M]A)$
	$(RA3)$ $[\mathsf{a}^M](A \wedge B) \leftrightarrow [\mathsf{a}^M]A \wedge [\mathsf{a}^M]B$
	$(RA4)$ $[\mathsf{a}^M]\Box_a A \leftrightarrow (\mathsf{pre(a)} \to \bigwedge_{\mathsf{a}\sim_a^M \mathsf{x}} \Box_a[\mathsf{x}^M]A)$
	$(RA5)$ $[\mathsf{a}^M][\mathsf{a'}^{M'}]A \leftrightarrow [(\mathsf{a}^M); (\mathsf{a'}^{M'})]A$
Inference Rules	(MP) From A and $A \to B$, infer B
	$(Nec\Box_a)$ From A, infer $\Box_a A$

Proposition 1 (Completeness of HDEL). *For any formula $A \in \mathcal{L}_{DEL}$, A is valid in all Kripke models iff A is provable in* **HDEL**.

Now we define our labelled calculus **GDEL** for DEL. For any $i \in \mathbb{N}$ and any finite list α of pointed event models, a pair $\langle i, \alpha \rangle$ is called a *label*, and we *abuse i* for an abbreviation of $\langle i, \epsilon \rangle$, when it does not cause any confusion. For any formula A and label $\langle i, \alpha \rangle$, we call $\langle i, \alpha \rangle{:}A$ a *labelled formula*. Similarly, for any agent $a \in \mathsf{Agt}$ and any lists α, α' of events, an expression $\langle i, \alpha \rangle R_a \langle j, \alpha' \rangle$ is defined to be a *relational atom* if (1) $\alpha_{mdl} = \alpha'_{mdl}$ and (2) $\mathsf{a}_i \sim_a^{M_i} \mathsf{b}_i$ holds for any $1 \leq i \leq n$, where $\alpha = (\mathsf{a}_1^{M_1}, \ldots, \mathsf{a}_n^{M_n})$ and $\alpha' = (\mathsf{b}_1^{M_1}, \ldots, \mathsf{b}_n^{M_n})$. A *labelled expression* $\mathfrak{A}$ is either a labelled formula or a relational atom.

Definition 3. *The length of a labelled expression $\ell(\mathfrak{A})$, a formula $\ell(A)$ and an event model $\ell(M)$ is defined as follows:*

$$\ell(\langle i, \alpha \rangle{:}A) = \ell(\alpha) + \ell(A), \ell(\langle i, \alpha \rangle R_a \langle j, \alpha' \rangle) = \ell(\alpha),$$

$$\ell(p) = 1, \quad \ell(A \wedge B) = \ell(A) + \ell(B) + 1, \ell(\neg A) = \ell(\Box_a A) = \ell(A) + 1,$$

$$\ell([\mathsf{a}^M]A) = \ell(M) + \ell(A) + 1,$$

$$\ell(M) = \max\{\ell(\mathsf{pre}^M(\mathsf{x})) \mid \mathsf{x} \in \mathsf{S}^M\}, \quad \ell(\mathsf{a}_1^{M_1}, \ldots, \mathsf{a}_n^{M_n}) = \ell(M_1) + \cdots + \ell(M_n).$$

A *sequent* $\Gamma \Rightarrow \Delta$ is a pair of multi-sets Γ, Δ of labelled expressions The existence of the pointed event models a^M in our syntax of DEL forces us to handle many *branches* in a naturally constructed sequent calculus. For example, we may consider a set

$$\{1{:}\mathsf{pre(x)} \Rightarrow 2{:}\mathsf{pre(y)} \mid \mathsf{p} \sim_b \mathsf{x} \text{ and } \mathsf{x} \sim_a \mathsf{y}\}$$

of sequents in the setting of Example 1. In order to handle such several branches simultaneously in a sequent calculus, we introduce the notation

$$1{:}\mathsf{pre(x)} \Rightarrow 2{:}\mathsf{pre(y)} \parallel \mathsf{p} \sim_b \mathsf{x}, \mathsf{x} \sim_a \mathsf{y}$$

Table 2. Labelled sequent calculus for DEL : **GDEL**

(Initial c-sequent)
$$\mathfrak{A} \Rightarrow \mathfrak{A} \parallel \Sigma$$

(Structural Rules)

$$\frac{\Gamma \Rightarrow \Delta \parallel \Sigma}{\mathfrak{A}, \Gamma \Rightarrow \Delta \parallel \Sigma} \ (Lw) \qquad \frac{\Gamma \Rightarrow \Delta \parallel \Sigma}{\Gamma \Rightarrow \Delta, \mathfrak{A} \parallel \Sigma} \ (Rw) \qquad \frac{\mathfrak{A}, \mathfrak{A}, \Gamma \Rightarrow \Delta \parallel \Sigma}{\mathfrak{A}, \Gamma \Rightarrow \Delta \parallel \Sigma} \ (Lc) \qquad \frac{\Gamma \Rightarrow \Delta, \mathfrak{A}, \mathfrak{A} \parallel \Sigma}{\Gamma \Rightarrow \Delta, \mathfrak{A} \parallel \Sigma} \ (Rc)$$

(Rules for the propositional connectives)

$$\frac{\Gamma \Rightarrow \Delta, \langle i, \alpha \rangle{:}A \parallel \Sigma}{\langle i, \alpha \rangle{:}\neg A, \Gamma \Rightarrow \Delta \parallel \Sigma} \ (L\neg) \qquad \frac{\langle i, \alpha \rangle{:}A, \Gamma \Rightarrow \Delta \parallel \Sigma}{\Gamma \Rightarrow \Delta, \langle i, \alpha \rangle{:}\neg A \parallel \Sigma} \ (R\neg)$$

$$\frac{\langle i, \alpha \rangle{:}A, \Gamma \Rightarrow \Delta \parallel \Sigma}{\langle i, \alpha \rangle{:}A \wedge B, \Gamma \Rightarrow \Delta \parallel \Sigma} \ (L\wedge 1) \qquad \frac{\langle i, \alpha \rangle{:}B, \Gamma \Rightarrow \Delta \parallel \Sigma}{\langle i, \alpha \rangle{:}A \wedge B, \Gamma \Rightarrow \Delta \parallel \Sigma} \ (L\wedge 2)$$

$$\frac{\Gamma \Rightarrow \Delta, \langle i, \alpha \rangle{:}A \parallel \Sigma \quad \Gamma \Rightarrow \Delta, \langle i, \alpha \rangle{:}B \parallel \Sigma}{\Gamma \Rightarrow \Delta, \langle i, \alpha \rangle{:}A \wedge B \parallel \Sigma} \ (R\wedge)$$

(Rules for the knowledge operator)

$$\frac{\Gamma \Rightarrow \Delta, \langle i, \epsilon \rangle R_a \langle j, \epsilon \rangle \parallel \Sigma \quad \langle j, \epsilon \rangle{:}A, \Gamma \Rightarrow \Delta \parallel \Sigma}{\langle i, \epsilon \rangle{:}\Box_a A, \Gamma \Rightarrow \Delta \parallel \Sigma} \ (L\Box_a 1) \qquad \frac{\langle i, \epsilon \rangle R_a \langle j, \epsilon \rangle, \Gamma \Rightarrow \Delta, \langle j, \epsilon \rangle{:}A \parallel \Sigma}{\Gamma \Rightarrow \Delta, \langle i, \epsilon \rangle{:}\Box_a A \parallel \Sigma} \ (R\Box_a 1) *_2$$

$$\frac{\Gamma \Rightarrow \Delta, \langle i, a_1^{M_1}, ..., a_k^{M_k} \rangle R_a \langle j, b_1^{M_1}, ..., b_k^{M_k} \rangle \parallel \Sigma \quad \langle j, b_1^{M_1}, ..., b_k^{M_k} \rangle{:}A, \Gamma \Rightarrow \Delta \parallel \Sigma}{\langle i, a_1^{M_1}, ..., a_k^{M_k} \rangle{:}\Box_a A, \Gamma \Rightarrow \Delta \parallel \Sigma} \ (L\Box_a 2) *_1$$

$$\frac{\langle i, a_1^{M_1}, ..., a_k^{M_k} \rangle R_a \langle j, v_1^{M_1}, ..., v_k^{M_k} \rangle, \Gamma \Rightarrow \Delta, \langle j, v_1^{M_1}, ..., v_k^{M_k} \rangle{:}A \parallel \Sigma, a_1 {\sim}_a^{M_1} v_1, ..., a_k {\sim}_a^{M_k} v_k}{\Gamma \Rightarrow \Delta, \langle i, a_1^{M_1}, ..., a_k^{M_k} \rangle{:}\Box_a A \parallel \Sigma} \ (R\Box_a 2) *_2$$

(Rules for the event operators)

$$\frac{\langle i, \alpha \rangle{:}p, \Gamma \Rightarrow \Delta \parallel \Sigma}{\langle i, \alpha, a^M \rangle{:}p, \Gamma \Rightarrow \Delta \parallel \Sigma} \ (Lat) \qquad \frac{\Gamma \Rightarrow \Delta, \langle i, \alpha \rangle{:}p \parallel \Sigma}{\Gamma \Rightarrow \Delta, \langle i, \alpha, a^M \rangle{:}p \parallel \Sigma} \ (Rat)$$

$$\frac{\Gamma \Rightarrow \Delta, \langle i, \alpha \rangle{:}pre^M(a) \parallel \Sigma \quad \langle i, \alpha, a^M \rangle{:}A, \Gamma \Rightarrow \Delta \parallel \Sigma}{\langle i, \alpha \rangle{:}[a^M]A, \Gamma \Rightarrow \Delta \parallel \Sigma} \ (L[.]) \qquad \frac{\langle i, \alpha \rangle{:}pre^M(a), \Gamma \Rightarrow \Delta, \langle i, \alpha, a^M \rangle{:}A \parallel \Sigma}{\Gamma \Rightarrow \Delta, \langle i, \alpha \rangle{:}[a^M]A \parallel \Sigma} \ (R[.])$$

(Rules for the relational atoms)

$$\frac{\langle i, \alpha \rangle R_a \langle j, \alpha' \rangle, \Gamma \Rightarrow \Delta \parallel \Sigma}{\langle i, \alpha, a^M \rangle R_a \langle j, \alpha', b^M \rangle, \Gamma \Rightarrow \Delta \parallel \Sigma} \ (Lrel1)$$

$$\frac{\langle i, \alpha \rangle{:}pre^M(a), \Gamma \Rightarrow \Delta \parallel \Sigma}{\langle i, \alpha, a^M \rangle R_a \langle j, \alpha', b^M \rangle, \Gamma \Rightarrow \Delta \parallel \Sigma} \ (Lrel2) \qquad \frac{\langle j, \alpha' \rangle{:}pre^M(b), \Gamma \Rightarrow \Delta \parallel \Sigma}{\langle i, \alpha, a^M \rangle R_a \langle j, \alpha', b^M \rangle, \Gamma \Rightarrow \Delta \parallel \Sigma} \ (Lrel3)$$

$$\frac{\Gamma \Rightarrow \Delta, \langle i, \alpha \rangle R_a \langle j, \alpha' \rangle \parallel \Sigma \quad \Gamma \Rightarrow \Delta, \langle i, \alpha \rangle{:}pre^M(a) \parallel \Sigma \quad \Gamma \Rightarrow \Delta, \langle j, \alpha' \rangle{:}pre^M(b) \parallel \Sigma}{\Gamma \Rightarrow \Delta, \langle i, \alpha, a^M \rangle R_a \langle j, \alpha', b^M \rangle \parallel \Sigma} \ (Rrel)$$

$*_1$ Each b_h is an event such that $a {\sim}_a^{M_h} b_h$.
$*_2$ $j \in \mathbb{N}$ and each $v_h \in \mathsf{CVar}$ do not appear in lower c-sequent.

for representing the above set. In general, we keep a countable proper subset $\mathsf{CVar} = \{x, y, z, \ldots\}$ of Evt for comprehension variables and define that a *collective sequent* (simply a *c-sequent*) is an expression:

$$\Gamma \Rightarrow \Delta \parallel \Sigma$$

where $\Gamma \Rightarrow \Delta$ is a sequent and $\Sigma = \{s_1 \sim_{a_1} t_1, \ldots, s_n \sim_{a_n} t_n\}$, s_i or t_i from Evt is assumed to be an element of CVar in $s_i \sim_{a_i} t_i$ and all the variables from CVar occuring in $\Gamma \Rightarrow \Delta$ are bounded in Σ, i.e., they are a subset of all the variables from CVar occuring in Σ. The c-sequent above represents the following set of sequents:

$$\{\Gamma \Rightarrow \Delta \mid s_1 \sim_{a_1} t_1 \text{ and } \cdots \text{ and } s_n \sim_{a_n} t_n\}.$$

Throughout the paper we use Greek letter Σ for a finite set $\{s_1 \sim_{a_1} t_1, \ldots, s_n \sim_{a_n} t_n\}$ of event relations.

We now introduce the set of rules of **GDEL** which is presented in Table 2. We call labelled expression $\mathfrak{A}$ in the lower c-sequent at each inference rule *principal* if $\mathfrak{A}$ is not in either Γ or Δ. Moreover, the following usual inference rules for the defined logical connectives are all derivable in **GDEL**:

$$\frac{\Gamma \Rightarrow \Delta, \langle i, \alpha\rangle{:}A, \langle i, \alpha\rangle{:}B \parallel \Sigma}{\Gamma \Rightarrow \Delta, \langle i, \alpha\rangle{:}A \vee B, \parallel \Sigma} \ (R\vee) \qquad \frac{\langle i, \alpha\rangle{:}A, \Gamma \Rightarrow \Delta \parallel \Sigma \quad \langle i, \alpha\rangle{:}B, \Gamma \Rightarrow \Delta \parallel \Sigma}{\langle i, \alpha\rangle{:}A \vee B, \Gamma \Rightarrow \Delta \parallel \Sigma} \ (L\vee)$$

$$\frac{\langle i, \alpha\rangle{:}A, \Gamma \Rightarrow \Delta, \langle i, \alpha\rangle{:}B \parallel \Sigma}{\Gamma \Rightarrow \Delta, \langle i, \alpha\rangle{:}A \to B, \parallel \Sigma} \ (R\to) \qquad \frac{\Gamma \Rightarrow \Delta, \langle i, \alpha\rangle{:}A \parallel \Sigma \quad \langle i, \alpha\rangle{:}B, \Gamma \Rightarrow \Delta \parallel \Sigma}{\langle i, \alpha\rangle{:}A \to B, \Gamma \Rightarrow \Delta \parallel \Sigma} \ (L\to).$$

A *derivation* of c-sequent $\Gamma \Rightarrow \Delta \parallel \Sigma$ in **GDEL** is a tree of c-sequents satisfying the following conditions:

1. The uppermost c-sequent of the tree is an initial sequent.
2. Every c-sequent in the tree except the lowest c-sequent is an upper c-sequent of an inference rule.
3. The lowest c-sequent is $\Gamma \Rightarrow \Delta \parallel \Sigma$.

Given a c-sequent $\Gamma \Rightarrow \Delta \parallel \Sigma$, it is *provable in* **GDEL** and we write $\vdash_{\mathbf{GDEL}} \Gamma \Rightarrow \Delta \parallel \Sigma$ if there is a derivation of the c-sequent; and especially if there exists a derivation of the c-sequent which is restricted to event models $M_1, \ldots, M_n$ which appear in the derivation, we say, for emphasizing the fact, it is *provable in* **GDEL** *under event models* $M_1, \ldots, M_n$ and write $M_1, \ldots, M_n \vdash_{\mathbf{GDEL}} \Gamma \Rightarrow \Delta \parallel \Sigma$. (In the case that c-sequent $\Gamma \Rightarrow \Delta \parallel \Sigma$ and a derivation of it do not include any event model, we write $\varepsilon \vdash_{\mathbf{GDEL}} \Gamma \Rightarrow \Delta \parallel \Sigma$ to emphasize the case. We remark that this should be distinguished from $\vdash_{\mathbf{GDEL}} \Gamma \Rightarrow \Delta \parallel \Sigma$ defined above.) Let us look at a specific derivation of **GDEL** to help capture the essence of **GDEL**. Using the event model Read in Example 1, and we show one of the exercises in [6, p.166].

Example 2. In the setting of Example 1, we can easily show that $\Rightarrow 0{:}[\mathsf{p}^{\mathsf{Read}}]\Box_b(\Box_a p \vee \Box_a \neg p)$ is provable in **GDEL** as follows (intuitively, this formula means that after the agent a reads a letter containing p, the agent b knows that a knows whether p):

$$
\dfrac{
\dfrac{
\dfrac{
\dfrac{
\dfrac{
\dfrac{
\dfrac{
\dfrac{
\dfrac{
\dfrac{
\dfrac{
\dfrac{
\begin{array}{c}
\dfrac{\textit{Initial Seq.}}{2{:}p,3{:}p \Rightarrow 2{:}p}\ (Lw) \\[4pt]
\vdots\ \{x\mid p{\sim}_a x\}=\{p\} \\
2{:}\mathrm{pre}(y),3{:}\mathrm{pre}(z),3{:}p \Rightarrow 2{:}p \parallel p{\sim}_a y, p{\sim}_a z
\end{array}
\qquad
\dfrac{
\dfrac{\textit{Initial Seq.}}{2{:}\neg p,3{:}p \Rightarrow 2{:}p,3{:}p}\ (Lw)/(Rw)
}{
\begin{array}{c}
2{:}\neg p,3{:}\neg p,3{:}p \Rightarrow 2{:}p\ (L\neg) \\[4pt]
\vdots\ \{x\mid np{\sim}_a x\}=\{np\} \\
2{:}\mathrm{pre}(y),3{:}\mathrm{pre}(z),3{:}p \Rightarrow 2{:}p \parallel np{\sim}_a y, np{\sim}_a z
\end{array}
}
}{
\begin{array}{c}
\vdots\ \{x\mid p{\sim}_b x\}=\{p,np\} \\
2{:}\mathrm{pre}(y),3{:}\mathrm{pre}(z),3{:}p \Rightarrow 2{:}p \parallel p{\sim}_b x, x{\sim}_a y, x{\sim}_a z
\end{array}
}{2{:}\mathrm{pre}(y),3{:}\mathrm{pre}(z),\langle 3,z^{\mathrm{Read}}\rangle{:}p \Rightarrow 2{:}p \parallel p{\sim}_b x, x{\sim}_a y, x{\sim}_a z}\ (Lat)
}{\langle 1,x^{\mathrm{Read}}\rangle R_a\langle 2,y^{\mathrm{Read}}\rangle,\langle 1,x^{\mathrm{Read}}\rangle R_a\langle 3,z^{\mathrm{Read}}\rangle,\langle 3,z^{\mathrm{Read}}\rangle{:}p \Rightarrow 2{:}p \parallel p{\sim}_b x, x{\sim}_a y, x{\sim}_a z}\ (Lrel3)
}{\langle 1,x^{\mathrm{Read}}\rangle R_a\langle 2,y^{\mathrm{Read}}\rangle,\langle 1,x^{\mathrm{Read}}\rangle R_a\langle 3,z^{\mathrm{Read}}\rangle \Rightarrow 2{:}p,\langle 3,z^{\mathrm{Read}}\rangle{:}\neg p \parallel p{\sim}_b x, x{\sim}_a y, x{\sim}_a z}\ (R\neg)
}{\langle 1,x^{\mathrm{Read}}\rangle R_a\langle 2,y^{\mathrm{Read}}\rangle \Rightarrow 2{:}p,\langle 1,x^{\mathrm{Read}}\rangle{:}\Box_a\neg p \parallel p{\sim}_b x, x{\sim}_a y}\ (R\Box_a 2)
}{\langle 1,x^{\mathrm{Read}}\rangle R_a\langle 2,y^{\mathrm{Read}}\rangle \Rightarrow \langle 2,y^{\mathrm{Read}}\rangle{:}p,\langle 1,x^{\mathrm{Read}}\rangle{:}\Box_a\neg p \parallel p{\sim}_b x, x{\sim}_a y}\ (Rat)
}{\Rightarrow \langle 1,x^{\mathrm{Read}}\rangle{:}\Box_a p,\langle 1,x^{\mathrm{Read}}\rangle{:}\Box_a\neg p \parallel p{\sim}_b x}\ (R\Box_a 2)
}{\Rightarrow \langle 1,x^{\mathrm{Read}}\rangle{:}\Box_a p \vee \Box_a\neg p \parallel p{\sim}_b x}\ (R\vee)
}{0{:}\mathrm{pre}(p),\langle 0,p^{\mathrm{Read}}\rangle R_b\langle 1,x^{\mathrm{Read}}\rangle \Rightarrow \langle 1,x^{\mathrm{Read}}\rangle{:}\Box_a p \vee \Box_a\neg p \parallel p{\sim}_b x}\ (Lw)
}{0{:}\mathrm{pre}(p) \Rightarrow \langle 0,p^{\mathrm{Read}}\rangle{:}\Box_b(\Box_a p \vee \Box_a\neg p)}\ (R\Box_b 2)
}{\Rightarrow 0{:}[p^{\mathrm{Read}}]\Box_b(\Box_a p \vee \Box_a\neg p)}\ (R[.])
$$

The derivation above is restricted to the event model Read, and so

$$
\mathrm{Read} \vdash_{\mathbf{GDEL}} \Rightarrow 0{:}[p^{\mathrm{Read}}]\Box_b(\Box_a p \vee \Box_a\neg p).
$$

4   Admissibility of Cut in GDEL

In this section, we provide a proof of the admissibility of the following rule (Cut) in **GDEL**:

$$
\dfrac{\Gamma \Rightarrow \Delta, \mathfrak{A} \parallel \Sigma \quad \mathfrak{A}, \Gamma' \Rightarrow \Delta' \parallel \Sigma}{\Gamma, \Gamma' \Rightarrow \Delta, \Delta' \parallel \Sigma}\ (Cut).
$$

For preparations for the proof of the theorem, we show the substitution lemma. The result of substitution $\mathfrak{A}[^n_m]$ (m is substituted by n in $\mathfrak{A}$) is defined as follows:

Definition 4. *Let n, m be any elements in $\mathbb{N}$.*

$$
i[^n_m] := i \ (\textit{if } n \neq i), \qquad i[^n_m] := n \ (\textit{if } n = i),
$$
$$
(\langle i,\alpha\rangle{:}A)[^n_m] := \langle i[^n_m],\alpha\rangle{:}A, \ (\langle i,\alpha\rangle R_a\langle j,\alpha'\rangle)[^n_m] := \langle i[^n_m],\alpha\rangle R_a\langle j[^n_m],\alpha'\rangle.
$$
For a multi-set Γ of labelled expressions, $\Gamma[^n_m]$ denotes the set $\{\mathfrak{A}[^n_m] \mid \mathfrak{A} \in \Gamma\}$.

Lemma 1 (Substitution Lemma). *If $\vdash_{\mathbf{GDEL}} \Gamma \Rightarrow \Delta \parallel \Sigma$, then $\vdash_{\mathbf{GDEL}}$ $\Gamma[^n_m] \Rightarrow \Delta[^n_m] \parallel \Sigma$ with the same derivation height, for any $n, m \in \mathbb{N}$.*

Proof. This proof is done in a similar manner to the proof in Negri et al. [13, p. 194]. □

Theorem 1 (Admissibility of (Cut)). *For any c-sequent $\Gamma \Rightarrow \Delta \parallel \Sigma$, if $\vdash_{\textbf{GDEL}} \Gamma \Rightarrow \Delta \parallel \Sigma$ with (Cut), then $\vdash_{\textbf{GDEL}} \Gamma \Rightarrow \Delta \parallel \Sigma$ without (Cut). In particular, if $\mathsf{M}_1, ..., \mathsf{M}_k \vdash_{\textbf{GDEL}} \Gamma \Rightarrow \Delta \parallel \Sigma$ with (Cut), then $\mathsf{M}_1, ..., \mathsf{M}_k \vdash_{\textbf{GDEL}} \Gamma \Rightarrow \Delta \parallel \Sigma$ without (Cut).*

Proof. The proof is carried out with Ono et al.'s method of $(Ecut)$ [16]. $(Ecut)$ is given as follows:

$$\frac{\Gamma \Rightarrow \Delta, \mathfrak{A}^m \parallel \Sigma \quad \mathfrak{A}^n, \Gamma' \Rightarrow \Delta' \parallel \Sigma}{\Gamma, \Gamma' \Rightarrow \Delta, \Delta' \parallel \Sigma} \ (Ecut)$$

where $n, m \geq 0$ and $\mathfrak{A}$ is called a *cut expression*. The theorem is proven by double induction on the height of the derivation and the length of the cut expression $\ell(\mathfrak{A})$ of $(Ecut)$. The proof is divided into four cases:

1. At least one of the upper c-sequents of $(Ecut)$ is an initial c-sequent,
2. The last inference rule of either upper c-sequent of $(Ecut)$ is a structural rule,
3. The last inference rule of either upper c-sequent of $(Ecut)$ is a non-structural rule, and the principal expression introduced by the rule is not a cut expression, and
4. The last inference rules of two upper c-sequents of $(Ecut)$ are both non-structural rules, and the principal expressions introduced by the rules used on the upper c-sequents of $(Ecut)$ are both cut expressions.

Here we consider only the following critical subcases.

Subcase of 3: the last inference rule of left upper c-sequents of $(Ecut)$ is $(R\square_a 2)$ which is not cut expression.

In this case, we obtain the following derivation:

$$\frac{\dfrac{\begin{array}{c} \vdots \ \mathcal{D}_1 \\ \langle i, a_1, ..., a_k\rangle R_a \langle j, v_1, ..., v_k\rangle, \Gamma \Rightarrow \Delta, \langle j, v_1, ..., v_k\rangle{:}A, \mathfrak{A}^m \parallel \Sigma, a_1{\sim}_a^{M_1} v_1, ..., a_k{\sim}_a^{M_k} v_k \end{array}}{\Gamma \Rightarrow \Delta, \langle i, a_1, ..., a_k\rangle{:}\square_a A, \mathfrak{A}^m \parallel \Sigma} \ (R\square_a 2) \qquad \begin{array}{c} \vdots \ \mathcal{D}_2 \\ \mathfrak{A}^n, \Gamma' \Rightarrow \Delta' \parallel \Sigma \end{array}}{\Gamma, \Gamma' \Rightarrow \Delta, \Delta' \parallel \Sigma} \ (Ecut)$$

Since each $v_h \in \mathsf{CVar}$ does not appear in the lower c-sequents, it does not also appear in $\mathfrak{A}, \Gamma'$ and Δ'. Therefore, even if $a_1{\sim}_a^{M_1} v_1, \ldots, a_k{\sim}_a^{M_k} v_k$ are added to Σ, its provability does not obviously change with the same height of the derivation, and we obtain $\vdash_{\textbf{GDEL}} \mathfrak{A}^n, \Gamma' \Rightarrow \Delta' \parallel \Sigma, a_1{\sim}_a^{M_1} v_1, \ldots, a_k{\sim}_a^{M_k} v_k$. Then we may transform the derivation into the following:

$$\frac{\dfrac{\begin{array}{c} \vdots \ \mathcal{D}_1 \\ \langle i, a_1, ..., a_k\rangle R_a \langle j, v_1, ..., v_k\rangle, \Gamma \Rightarrow \Delta, \langle j, v_1, ..., v_k\rangle{:}A, \mathfrak{A}^m \parallel \Sigma, a_1{\sim}_a^{M_1} v_1, ..., a_k{\sim}_a^{M_k} v_k \end{array} \quad \begin{array}{c} \vdots \ \mathcal{D}_2' \\ \mathfrak{A}^n, \Gamma' \Rightarrow \Delta' \parallel \Sigma, a_1{\sim}_a^{M_1} v_1, ..., a_k{\sim}_a^{M_k} v_k \end{array}}{\langle i, a_1, ..., a_k\rangle R_a \langle j, v_1, ..., v_k\rangle, \Gamma, \Gamma' \Rightarrow \Delta, \Delta', \langle j, v_1, ..., v_k\rangle{:}A \parallel \Sigma, a_1{\sim}_a^{M_1} v_1, ..., a_k{\sim}_a^{M_k} v_k}}{\Gamma, \Gamma' \Rightarrow \Delta, \Delta' \parallel \Sigma} \ (R\square_a 2)$$

with $(Ecut)$ applied above.

Subcase of 4: both sides of $\mathfrak{A}$ in $(Ecut)$ are $\langle i, a_1^{M_1}, \ldots, a_k^{M_k} \rangle : \Box_a A$ and principal expressions.

Let us consider the case where $k = 1$ for simplicity, and $\mathfrak{A} = \langle i, a^M \rangle : \Box_a A$. In this case, we obtain the following derivation.

$$
\cfrac{
\cfrac{\vdots \; \mathcal{D}_1}{\langle i, a^M \rangle R_a \langle v, x^M \rangle, \Gamma \Rightarrow \Delta, \langle v, x^M \rangle : A, \mathfrak{A}^{m-1} \parallel \Sigma, a \sim_a^M x} {\Gamma \Rightarrow \Delta, \mathfrak{A}^m \parallel \Sigma} (R\Box_a 2)
\qquad
\cfrac{\cfrac{\vdots \; \mathcal{D}_2}{\mathfrak{A}^{n-1}, \Gamma' \Rightarrow \Delta', \langle i, a^M \rangle R_a \langle j, b^M \rangle \parallel \Sigma} \quad \cfrac{\vdots \; \mathcal{D}_3}{\langle j, b^M \rangle : A, \mathfrak{A}^{n-1}, \Gamma' \Rightarrow \Delta' \parallel \Sigma}}{\mathfrak{A}^n, \Gamma' \Rightarrow \Delta' \parallel \Sigma}(L\Box_a 2)
}{\Gamma, \Gamma' \Rightarrow \Delta, \Delta' \parallel \Sigma}(Ecut)
$$

First, replace v with j in the left upper c-sequent by Lemma 1. Next, since we know that $a \sim_a^M b$ by $\langle i, a^M \rangle R_a \langle j, b^M \rangle$ in the middle upper c-sequent and the condition of an event relation, we have the following.

$$
\vdash_{\textbf{GDEL}} \langle i, a^M \rangle R_a \langle j, b^M \rangle, \Gamma \Rightarrow \Delta, \langle j, b^M \rangle : A, \mathfrak{A}^{m-1} \parallel \Sigma
$$

Then the derivation above can be transformed into the following:

$$
\cfrac{
\cfrac{\cfrac{\vdots \; \mathcal{D}_1^+}{\Gamma \Rightarrow \Delta, \mathfrak{A}^m \parallel \Sigma} \quad \cfrac{\vdots \; \mathcal{D}_2}{p\mathfrak{A}^{n-1}, \Gamma' \Rightarrow \Delta', \langle i, a^M \rangle R_a \langle j, b^M \rangle \parallel \Sigma}}{\Gamma, \Gamma' \Rightarrow \Delta, \Delta', \langle i, a^M \rangle R_a \langle j, b^M \rangle \parallel \Sigma}(Ecut)_2
\quad
\cfrac{\cfrac{\vdots \; \mathcal{D}_1 \text{ and Lemma 1}}{\langle i, a^M \rangle R_a \langle j, b^M \rangle, \Gamma \Rightarrow \Delta, \langle j, b^M \rangle : A, \mathfrak{A}^{m-1} \parallel \Sigma} \quad \cfrac{\vdots \; \mathcal{D}_{23}^+}{\mathfrak{A}^n, \Gamma' \Rightarrow \Delta' \parallel \Sigma}}{\langle i, a^M \rangle R_a \langle j, b^M \rangle, \Gamma', \Gamma \Rightarrow \Delta, \Delta', \langle j, b^M \rangle : A \parallel \Sigma}(Ecut)_1
}{\Gamma, \Gamma, \Gamma', \Gamma' \Rightarrow \Delta, \Delta, \Delta', \Delta', \langle j, b^M \rangle : A \parallel \Sigma}(Ecut)_4
$$

$$
\vdots
$$

$$
\cfrac{
\Gamma, \Gamma, \Gamma', \Gamma' \Rightarrow \Delta, \Delta, \Delta', \Delta', \langle j, b^M \rangle : A \parallel \Sigma
\qquad
\cfrac{\cfrac{\vdots \; \mathcal{D}_1^+}{\Gamma \Rightarrow \Delta, \mathfrak{A}^m \parallel \Sigma} \quad \cfrac{\vdots \; \mathcal{D}_3}{\langle j, b^M \rangle : A, \mathfrak{A}^{n-1}, \Gamma' \Rightarrow \Delta' \parallel \Sigma}}{\langle j, b^M \rangle : A, \Gamma, \Gamma' \Rightarrow \Delta, \Delta' \parallel \Sigma}(Ecut)_3
}{\cfrac{\Gamma, \Gamma, \Gamma, \Gamma', \Gamma', \Gamma' \Rightarrow \Delta, \Delta, \Delta, \Delta', \Delta', \Delta' \parallel \Sigma}{\Gamma, \Gamma' \Rightarrow \Delta, \Delta' \parallel \Sigma}(Lc)/(Rc)}(Ecut)_5
$$

where $(Ecut)_{1,2,3}$ are applicable by induction hypothesis, since the derivation height of $(Ecut)$ is reduced by comparison with the original derivation. Besides, the application of $(Ecut)_{4,5}$ is also allowed by induction hypothesis, where $\ell(\mathfrak{A})$ is reduced as follows: $\ell(\langle i, a^M \rangle : \Box_a A) > \ell(\langle i, b^M \rangle : A)$ and $\ell(\langle i, a^M \rangle : \Box_a A) > \ell(\langle i, a^M \rangle R_a \langle i, b^M \rangle)$. □

5   Soundness and Completeness of GDEL

Our task in this sections is to establish that our sequent system **GDEL** is sound and complete for Kripke semantics through Proposition 1 (Completeness of **HDEL**). We first show the theorem that every provable formula in **HDEL** is also provable in **GDEL**. To show the theorem requires the following trivial derivable rules in **GDEL** (for the case of (RA4)) and one lemma (for (RA5)).

$$
\cfrac{\Gamma \Rightarrow \Delta, \langle i, \alpha \rangle : A(\mathsf{v}) \parallel \Sigma, a \sim_a^M \mathsf{v}}{\Gamma \Rightarrow \Delta, \langle i, \alpha \rangle : \bigwedge_{a \sim_a^M \mathsf{x}} A \parallel \Sigma}(R\wedge\wedge)\dagger
\qquad
\cfrac{\langle i, \alpha \rangle : A(\mathsf{b}), \Gamma \Rightarrow \Delta \parallel \Sigma}{\langle i, \alpha \rangle : \bigwedge_{a \sim_a^M \mathsf{x}} A, \Gamma \Rightarrow \Delta \parallel \Sigma}(L\wedge\wedge)\ddagger
$$

† $v \in \mathsf{CVar}$ does not appear in the lower c-sequent.
‡ b is in $\sim_a(a)$.

where $A(b)$ (or $A(v)$) means that b (or v) possibly appears in formula A. These rules are only generalizations of $(L\wedge)$ and $(R\wedge)$.

Lemma 2. *For any finite lists α, β of events, any formula A, any finite set Σ of relational atoms, the following hold:*

(i) $\vdash_{\mathbf{GDEL}}$ $\langle i, \alpha, (\mathsf{a}, \mathsf{a}')^{\mathsf{M};\mathsf{M}'}, \beta \rangle R_a \langle j, \alpha', (\mathsf{b}, \mathsf{b}')^{\mathsf{M};\mathsf{M}'}, \beta' \rangle \Rightarrow \langle i, \alpha, \mathsf{a}^{\mathsf{M}}, \mathsf{a}'^{\mathsf{M}'}, \beta \rangle$
$R_a \langle j, \alpha', \mathsf{b}^{\mathsf{M}}, \mathsf{b}'^{\mathsf{M}'}, \beta \rangle \parallel \Sigma$

(ii) $\vdash_{\mathbf{GDEL}}$ $\langle i, \alpha, \mathsf{a}^{\mathsf{M}}, \mathsf{a}'^{\mathsf{M}'}, \beta \rangle R_a \langle j, \alpha', \mathsf{b}^{\mathsf{M}}, \mathsf{b}'^{\mathsf{M}'}, \beta \rangle \Rightarrow \langle i, \alpha, (\mathsf{a}, \mathsf{a}')^{\mathsf{M};\mathsf{M}'}, \beta \rangle$
$R_a \langle j, \alpha', (\mathsf{b}, \mathsf{b}')^{\mathsf{M};\mathsf{M}'}, \beta' \rangle \parallel \Sigma$

(iii) $\vdash_{\mathbf{GDEL}} \langle i, \alpha, (\mathsf{a}, \mathsf{a}')^{\mathsf{M};\mathsf{M}'}, \beta \rangle{:}A \Rightarrow \langle i, \alpha, \mathsf{a}^{\mathsf{M}}, \mathsf{a}'^{\mathsf{M}'}, \beta \rangle{:}A \parallel \Sigma$

(iv) $\vdash_{\mathbf{GDEL}} \langle i, \alpha, \mathsf{a}^{\mathsf{M}}, \mathsf{a}'^{\mathsf{M}'}, \beta \rangle{:}A \Rightarrow \langle i, \alpha, (\mathsf{a}, \mathsf{a}')^{\mathsf{M};\mathsf{M}'}, \beta \rangle{:}A \parallel \Sigma$

Proof. The proofs of (i), (ii), (iii) and (iv) are simultaneously conducted by double induction on $\ell(\mathfrak{A})$ and the length of β $(= \beta')$. We only look at the proof of (i) (other cases can be shown similarly).

Base case: (i) where $\beta = \epsilon$.

We show the following, and it is straightforward to construct a derivation with $(Rrel)/(Lreli)$ and $(R\wedge)/(L \wedge i)$. Note that $(\mathsf{a}, \mathsf{a}')^{\mathsf{M};\mathsf{M}'}$ is included in PEvt by Definition 2.

$\vdash_{\mathbf{GDEL}} \langle i, \alpha, (\mathsf{a}, \mathsf{a}')^{\mathsf{M};\mathsf{M}'} \rangle R_a \langle j, \alpha', (\mathsf{b}, \mathsf{b}')^{\mathsf{M};\mathsf{M}'} \rangle \Rightarrow \langle i, \alpha, \mathsf{a}^{\mathsf{M}}, \mathsf{a}'^{\mathsf{M}'} \rangle R_a \langle j, \alpha', \mathsf{b}^{\mathsf{M}}, \mathsf{b}'^{\mathsf{M}'} \rangle \parallel \Sigma$

Induction step of (i) where $\beta = (\gamma, \mathsf{c}^{\mathsf{M}''})$ and $\beta' = (\gamma', \mathsf{d}^{\mathsf{M}''})$

$$\cfrac{\cfrac{\overset{\text{\footnotesize Induction hypothesis of (i)}}{\langle i, \alpha, (\mathsf{a},\mathsf{a}')^{\mathsf{M};\mathsf{M}'}, \gamma \rangle R_a \langle j, \alpha', (\mathsf{b},\mathsf{b}')^{\mathsf{M};\mathsf{M}'}, \gamma' \rangle \Rightarrow \langle i, \alpha, \mathsf{a}^{\mathsf{M}}, \mathsf{a}'^{\mathsf{M}'}, \gamma \rangle R_a \langle j, \alpha', \mathsf{u}^{\mathsf{M}}, \mathsf{b}'^{\mathsf{M}'}, \gamma' \rangle \parallel \Sigma}}{\langle i, \alpha, (\mathsf{a},\mathsf{a}')^{\mathsf{M};\mathsf{M}'}, \gamma, \mathsf{c}^{\mathsf{M}''} \rangle R_a \langle j, \alpha', (\mathsf{b},\mathsf{b}')^{\mathsf{M};\mathsf{M}'}, \gamma', \mathsf{d}^{\mathsf{M}''} \rangle \Rightarrow \langle i, \alpha, \mathsf{a}^{\mathsf{M}}, \mathsf{a}'^{\mathsf{M}'}, \gamma \rangle R_a \langle j, \alpha', \mathsf{b}^{\mathsf{M}}, \mathsf{b}'^{\mathsf{M}'}, \gamma' \rangle \parallel \Sigma}(Lrel) \quad \vdots\ \mathcal{D}_1 \quad \vdots\ \mathcal{D}_2}{\langle i, \alpha, (\mathsf{a},\mathsf{a}')^{\mathsf{M};\mathsf{M}'}, \gamma, \mathsf{c}^{\mathsf{M}''} \rangle R_a \langle j, \alpha', (\mathsf{b},\mathsf{b}')^{\mathsf{M};\mathsf{M}'}, \gamma', \mathsf{d}^{\mathsf{M}''} \rangle \Rightarrow \langle i, \alpha, \mathsf{a}^{\mathsf{M}}, \mathsf{a}'^{\mathsf{M}'}, \gamma, \mathsf{c}^{\mathsf{M}''} \rangle R_a \langle j, \alpha', \mathsf{b}^{\mathsf{M}}, \mathsf{b}'^{\mathsf{M}'}, \gamma', \mathsf{d}^{\mathsf{M}''} \rangle \parallel \Sigma}(Rrel)$$

$\mathcal{D}_1$ (and similarly $\mathcal{D}_2$) is immediately given by $(Lrel2)$ and induction hypthesis of (iii). $\square$

Theorem 2. *For any formula A, if $\vdash_{\mathbf{HDEL}} A$, then $\vdash_{\mathbf{GDEL+}} \Rightarrow \langle i, \epsilon \rangle{:}A$ for any $i \in \mathbb{N}$.*

Proof. Suppose $\vdash_{\mathbf{HDEL}} A$, and fix any $i \in \mathbb{N}$ (let i be 0). The proof is conducted by induction on the height of derivation of **HDEL**. We pick up some significant base cases (the derivation height of **HDEL** is equal to 0).
(RA4: Right to Left)

$$\dfrac{\dfrac{\dfrac{\text{Initial Seq.}}{0R_a1 \Rightarrow \langle 1, y^M\rangle{:}A, 0R_a1 \parallel a{\sim}_a^M y}\ (Rw)}{\mathfrak{A} \Rightarrow \langle 1, y^M\rangle{:}A, 0R_a1 \parallel a{\sim}_a^M y}\ (Lrel1) \qquad \dfrac{\dfrac{\text{Initial Seq.}}{1{:}\text{pre}(y) \Rightarrow \langle 1, y^M\rangle{:}A, 1{:}\text{pre}(y) \parallel a{\sim}_a^M y}\ (Rw)}{\dfrac{\mathfrak{A} \Rightarrow \langle 1, y^M\rangle{:}A, 1{:}\text{pre}(y) \parallel a{\sim}_a^M y}{1{:}[y^M]A, \mathfrak{A} \Rightarrow \langle 1, y^M\rangle{:}A \parallel a{\sim}_a^M y}\ (L[.])}\ (Lrel3) \qquad \dfrac{\dfrac{\text{Initial Seq.}}{\langle 1, y^M\rangle{:}A, \mathfrak{A} \Rightarrow \langle 1, y^M\rangle{:}A \parallel a{\sim}_a^M y}\ (Lw)}{}}{\qquad} (L\square_a 2)$$

$$\mathbf{Q} \qquad \dfrac{\dfrac{\dfrac{\dfrac{0{:}\square_a[y^M]A, \mathfrak{A} \Rightarrow \langle 1, y^M\rangle{:}A \parallel a{\sim}_a^M y}{0{:}\bigwedge_{a{\sim}_a^M x}\square_a[x^M]A, \mathfrak{A} \Rightarrow \langle 1, y^M\rangle{:}A \parallel a{\sim}_a^M y}\ (L\wedge\wedge)}{0{:}\bigwedge_{a{\sim}_a^M x}\square_a[x^M]A \Rightarrow \langle 0, a^M\rangle{:}\square_a A}\ (R\square_a)}{}}{}$$

$$\dfrac{\dfrac{\text{Initial Seq.}}{0{:}\text{pre}(a) \Rightarrow \langle 0, a^M\rangle{:}\square_a A, 0{:}\text{pre}(a)}\ (Rw) \qquad 0{:}\text{pre}(a), 0{:}\bigwedge_{a{\sim}_a^M x}\square_a[x^M]A \Rightarrow \langle 0, a^M\rangle{:}\square_a A}{\dfrac{0{:}\text{pre}(a), 0{:}\text{pre}(a) \to \bigwedge_{a{\sim}_a^M x}\square_a[x^M]A \Rightarrow \langle 0, a^M\rangle{:}\square_a A}{\dfrac{0{:}\text{pre}(a) \to \bigwedge_{a{\sim}_a^M x}\square_a[x^M]A \Rightarrow 0{:}[a^M]\square_a A}{\Rightarrow 0{:}(\text{pre}(a) \to \bigwedge_{a{\sim}_a^M x}\square_a[x^M]A) \to [a^M]\square_a A}\ (R \to)}\ (R[.])}\ (L \to)}$$

where $\mathfrak{A} = \langle 0, a^M\rangle R_a \langle 1, y^M\rangle$.
(RA5: Right to Left)

$$\dfrac{\dfrac{\dfrac{\dfrac{\text{Initial Seq.}}{\mathfrak{A}, 0{:}\text{pre}(a) \Rightarrow 0{:}\text{pre}(a)}\ (Lw) \qquad \dfrac{\dfrac{\text{Initial Seq.}}{\mathfrak{A}, 0{:}\text{pre}(a) \Rightarrow \mathfrak{A}}\ (Lw)}{\dfrac{\mathfrak{A}, 0{:}\text{pre}(a) \Rightarrow 0{:}[a^M]\text{pre}'(a')}\ (R[.])}\ (R\wedge)}{\dfrac{\mathfrak{A}, 0{:}\text{pre}(a) \Rightarrow 0{:}\text{pre}^{M{:}M'}((a, a'))}{\mathfrak{A}, 0{:}\text{pre}(a) \Rightarrow \langle 0, a^M, a'^{M'}\rangle{:}A, 0{:}\text{pre}^{M{:}M'}((a, a'))}\ (Lw)} \qquad \dfrac{\dfrac{\text{Lemma 2}}{\langle 0, (a, a')^{M{:}M'}\rangle{:}A \Rightarrow \langle 0, a^M, a'^{M'}\rangle{:}A}{\langle 0, (a, a')^{M{:}M'}\rangle{:}A, \mathfrak{A}, 0{:}\text{pre}(a) \Rightarrow \langle 0, a^M, a'^{M'}\rangle{:}A}\ (Lw)}{}}{\dfrac{\mathfrak{A}, 0{:}\text{pre}(a), 0{:}[(a^M);(a'^{M'})]A \Rightarrow \langle 0, a^M, a'^{M'}\rangle{:}A}{\dfrac{0{:}\text{pre}(a), 0{:}[(a^M);(a'^{M'})]A \Rightarrow \langle 0, a^M\rangle{:}[a'^{M'}]A}{\dfrac{0{:}[(a^M);(a'^{M'})]A \Rightarrow 0{:}[a^M][a'^{M'}]A}{\Rightarrow 0{:}[(a^M);(a'^{M'})]A \to [a^M][a'^{M'}]A}\ (R \to)}\ (R[.])}\ (R[.])}\ (L[.])}\ (L[.])$$

where $\mathfrak{A} = \langle 0, a^M\rangle{:}\text{pre}'(a')$. Other base cases can also be shown easily. In induction step, we show the admissibility of the inference rules **HDEL**, such as (MP) and $(Nec\square_a)$. The case of (MP) is shown with (Cut). The case of $(Nec\square_a)$ is shown with Lemma 1, (Lw) and $(R \to)$ as follows:

$$\dfrac{\dfrac{\begin{array}{c}\textit{Assumption}\\ \hline \Rightarrow \langle i, \epsilon\rangle{:}A\\ \vdots\\ \textit{Lemma 1}\\ \hline \Rightarrow \langle j, \epsilon\rangle{:}A\end{array}}{\dfrac{\langle i, \epsilon\rangle R_a\langle j, \epsilon\rangle \Rightarrow \langle j, \epsilon\rangle{:}A}{\Rightarrow \langle i, \epsilon\rangle{:}\square_a A}\ (R\square_a)}\ (Lw)}{}$$

$\square$

Let us move on to a proof of the soundness theorem of **GDEL**. For the soundness theorem, we expand the definition of the satisfaction relation to the labelled expression and the c-sequent. Hereinafter we denote $(w, (a_1, a_2, \ldots, a_n))$ for $(\cdots ((w, a_1), a_2), \ldots, a_n)$.

Definition 5. *Let $\mathfrak{M}$ be a Kripke model and f be an assignment function $f : \mathbb{N} \to \mathcal{D}(\mathfrak{M})$, α be any finite list of events. $\mathfrak{M}, f \models \mathfrak{A}$ is defined as follows:*

$$\mathfrak{M}, f \models \langle i, \alpha \rangle{:}A \qquad \text{iff} \qquad \mathfrak{M}^{\otimes \alpha_{mdl}}, (f(i), \alpha_{evt}) \models A,$$
$$\text{and} \quad (f(i), \alpha_{evt}) \in \mathcal{D}(\mathfrak{M}^{\otimes \alpha_{mdl}}),$$

$$\mathfrak{M}, f \models \langle i, \epsilon \rangle \mathsf{R}_a \langle j, \epsilon \rangle \qquad \text{iff} \qquad (f(i), f(j)) \in R_a,$$
$$\mathfrak{M}, f \models \langle i, \alpha, \mathsf{a}^{\mathsf{M}} \rangle \mathsf{R}_a \langle j, \alpha', \mathsf{b}^{\mathsf{M}} \rangle \quad \text{iff} \qquad \mathfrak{M}, f \models \langle i, \alpha \rangle \mathsf{R}_a \langle j, \alpha' \rangle$$
$$\text{and} \ \mathfrak{M}^{\otimes \alpha_{mdl}}, (f(i), \alpha_{evt}) \models \mathsf{pre}^{\mathsf{M}}(\mathsf{a})$$
$$\text{and} \ \mathfrak{M}^{\otimes \alpha'_{mdl}}, (f(j), \alpha'_{evt}) \models \mathsf{pre}^{\mathsf{M}}(\mathsf{b}).$$

Nomura et al. [15] gave light on the notion of *surviveness* of a world in the definition of satisfaction of the labelled expressions in PAL. The notion should be also considered in DEL, otherwise the soundness does not hold like in the case of PAL shown in [15]. Specifically, note that at the satisfaction of the labelled formula $\langle i, \alpha \rangle{:}A$, not only the labelled formula is true by the valuation, but also a corresponding world $(f(i), \alpha_{evt})$ must exist or survive in the updated domain $\mathcal{D}(\mathfrak{M}^{\otimes \alpha_{mdl}})$. Otherwise $\mathfrak{M}^{\otimes \alpha_{mdl}}, (f(i), \alpha_{evt}) \models A$ is ill-defined. Following the idea of [15], it is sufficient to pay attention to the negated form of the labelled expression $\overline{\mathfrak{A}}$ taking into the condition of surviveness of a world which must also survive in the updated domain. With the notion of surviveness, $\mathfrak{M}, f \models \overline{\mathfrak{A}}$ is defined as follows:

Definition 6. *Let f be an assignment function $f : \mathbb{N} \to \mathcal{D}(\mathfrak{M})$ (for any $\mathfrak{M}$), α be any finite list of events. $\mathfrak{M}, f \models \overline{\mathfrak{A}}$ is defined as follows:*

$$\mathfrak{M}, f \models \overline{\langle i, \alpha \rangle{:}A} \qquad \text{iff} \ \mathfrak{M}^{\otimes \alpha_{mdl}}, (f(i), \alpha_{evt}) \not\models A$$
$$\text{and} \quad (f(i), \alpha_{evt}) \in \mathcal{D}(\mathfrak{M}^{\otimes \alpha_{mdl}}),$$

$$\mathfrak{M}, f \models \overline{\langle i, \epsilon \rangle \mathsf{R}_a \langle j, \epsilon \rangle} \qquad \text{iff} \ (f(i), f(j)) \notin R_a,$$
$$\mathfrak{M}, f \models \overline{\langle i, \alpha, \mathsf{a}^{\mathsf{M}} \rangle \mathsf{R}_a \langle j, \alpha', \mathsf{b}^{\mathsf{M}} \rangle} \ \text{iff} \ \mathfrak{M}, f \models \langle i, \alpha \rangle \mathsf{R}_a \langle j, \alpha' \rangle$$
$$\text{or} \ \mathfrak{M}^{\otimes \alpha_{mdl}}, (f(i), \alpha_{evt}) \not\models \mathsf{pre}^{\mathsf{M}}(\mathsf{a})$$
$$\text{or} \ \mathfrak{M}^{\otimes \alpha'_{mdl}}, (f(j), \alpha'_{evt}) \not\models \mathsf{pre}^{\mathsf{M}}(\mathsf{b}).$$

Additionally, it should be clarified that these semantic definitions for relational atoms are connected with an accessibility relation as follows:

Lemma 3. *The following equivalent relations hold.*

(1) $\mathfrak{M}, f \models \langle i, \alpha \rangle \mathsf{R}_a \langle j, \alpha' \rangle$ *iff* $((f(i), \alpha_{evt}), (f(j), \alpha'_{evt})) \in R_a^{\otimes \alpha_{mdl}}$
(2) $\mathfrak{M}, f \models \overline{\langle i, \alpha \rangle \mathsf{R}_a \langle j, \alpha' \rangle}$ *iff* $((f(i), \alpha_{evt}), (f(j), \alpha'_{evt})) \notin R_a^{\otimes \alpha_{mdl}}$

Proof. Both can be straightforwardly shown by induction on the length of $\alpha \ (= \alpha')$. $\qquad \square$

The validity of c-sequents is defined as follows:

Definition 7 (Validity of a c-sequent). *We say that sequent $\Gamma \Rightarrow \Delta$ is t-valid in $\mathfrak{M}$ if there is no assignment $f : \mathbb{N} \to \mathcal{D}(\mathfrak{M})$ such that $\mathfrak{M}, f \models \mathfrak{A}$ for all $\mathfrak{A} \in \Gamma$, and $\mathfrak{M}, f \models \overline{\mathfrak{B}}$ for all $\mathfrak{B} \in \Delta$. Furthermore, c-sequent $\Gamma \Rightarrow \Delta \parallel \Sigma$ is t-valid if every sequent in $\{ \Gamma \Rightarrow \Delta \mid \Sigma \}$ is valid.*

The reader might think that the validity of the sequent looks strange, since it is usually defined as follows, which we call here *s-validity*.

$\Gamma \Rightarrow \Delta$ *is s−valid in* $\mathfrak{M}$ *if for all assignment* $f : \mathbb{N} \rightarrow \mathcal{D}(\mathfrak{M})$ *such that* $\mathfrak{M}, f \models$ $\mathfrak{A}$ *for all* $\mathfrak{A} \in \Gamma$ *implies* $\mathfrak{M}, f \models \mathfrak{B}$ *for some* $\mathfrak{B} \in \Delta$.

However, the soundness theorem with respect to s-validity fails, as the following proposition shows.[4]

Proposition 2. *There is a Kripke model* $\mathfrak{M}$ *such that* $(R\neg)$ *of* **GDEL** *does not preserve s-validity in* $\mathfrak{M}$.

Proof. Let $\mathsf{Agt} = \{a\}$, $\mathsf{M} = (\mathsf{S}, \sim_a, \mathsf{pre})$ with $\mathsf{S} = \{\mathsf{n}\}$, $\sim_a = \{(\mathsf{n},\mathsf{n})\}$, and $\{(\mathsf{n}, \neg p)\})$; moreover, $\mathfrak{M} = (W, R_a, V)$ with $W = \{w_1, w_2\}$, $R_a = W^2$ and $V(p) = \{w_1\}$. Therefore, $\mathfrak{M}^{\otimes \mathsf{M}} = (W^{\otimes \mathsf{M}}, R_a^{\otimes \mathsf{M}}, V^{\otimes \mathsf{M}})$ with $W^{\otimes \mathsf{M}} = \{(w_2, \mathsf{n})\}$, $R_a^{\otimes \mathsf{M}} = \{(w_2, \mathsf{n})\}$ and $V^{\otimes \mathsf{M}}(p) = \varnothing$. A particular instance of the application of $(R\neg)$ is as follows:

$$\frac{\langle 1, \mathsf{n}^{\mathsf{M}} \rangle : p \Rightarrow \| \varnothing}{\Rightarrow \langle 1, \mathsf{n}^{\mathsf{M}} \rangle : \neg p \| \varnothing} \ (R\neg)$$

We can easily show that the upper c-sequent is *s*-valid in $\mathfrak{M}^{\otimes \mathsf{M}}$ but the lower c-sequent is not. $\qquad\square$

Theorem 3 (Soundness of GDEL). *For any c-sequent* $\Gamma \Rightarrow \Delta \parallel \Sigma$, *if* $\vdash_{\mathbf{GDEL}} \Gamma \Rightarrow \Delta \parallel \Sigma$, $\Gamma \Rightarrow \Delta \parallel \Sigma$ *is t-valid in every Kripke model* $\mathfrak{M}$.

Proof. The proof is carried out by induction on the height of the derivation of $\Gamma \Rightarrow \Delta \parallel \Sigma$ in **GDEL**. We only confirm the case where the last applied rule is $(L\Box_a 2)$.

Fix any event variables $\mathsf{x}_1, \ldots, \mathsf{x}_n$ in Σ. Then we show the contraposition such that if the lower c-sequent of the rule $(L\Box_a 2)$ is not *t*-valid, then the upper c-sequent of it is also not *t*-valid. Suppose that the lower c-sequent of $(L\Box_a 2)$ is not t-valid, and by Definition 7, there is some $f : \mathbb{N} \rightarrow W$ such that $\mathfrak{M}, f \models \mathfrak{A}$ for all $\mathfrak{A} \in \Gamma$ and $\mathfrak{M}, f \models \langle i, \mathsf{a}_1^{\mathsf{M}_1}, \ldots, \mathsf{a}_n^{\mathsf{M}_n} \rangle : \Box_a A$ and $\mathfrak{M}, f \models \overline{\mathfrak{B}}$ for all $\mathfrak{B} \in \Delta$. Fix such f. Then it suffices to show $\mathfrak{M}, f \models$ $\overline{\langle i, \mathsf{a}_1^{\mathsf{M}_1}, \ldots, \mathsf{a}_n^{\mathsf{M}_n} \rangle R_a \langle j, \mathsf{b}_1^{\mathsf{M}_1}, \ldots, \mathsf{b}_n^{\mathsf{M}_n} \rangle}$ or $\mathfrak{M}, f \models \langle j, \mathsf{b}_1^{\mathsf{M}_1}, \ldots, \mathsf{b}_n^{\mathsf{M}_n} \rangle : A$. From the supposition, i.e., $\mathfrak{M}^{\otimes \mathsf{M}_1 \otimes \cdots \otimes \mathsf{M}_k}, (f(i), \mathsf{a}_1, \ldots, \mathsf{a}_n) \models \Box_a A$ and $(f(i), \mathsf{a}_1, \ldots, \mathsf{a}_n) \in$ $\mathcal{D}(\mathfrak{M}^{\otimes \mathsf{M}_1 \otimes \cdots \otimes \mathsf{M}_k})$, we obtain for all $v \in \mathcal{D}(\mathfrak{M}^{\otimes \mathsf{M}_1 \otimes \cdots \otimes \mathsf{M}_k})$, $((f(i), \mathsf{a}_1, \ldots, \mathsf{a}_k), v) \notin$ $R_a^{\otimes \mathsf{M}_1 \otimes \cdots \otimes \mathsf{M}_k}$ or $\mathfrak{M}^{\otimes \mathsf{M}_1 \otimes \cdots \otimes \mathsf{M}_k}, b \models A$. Take v as $(f(j), \mathsf{b}_1, \ldots, \mathsf{b}_n)$. Then by Lemma 3 and Definition 5, we obtain what we desired. $\square$

Combining Theorems 2 and 3 with Proposition 1, we have the following.

Corollary 1. (Completeness of GDEL). *Given any formula A, the following are equivalent:* (i) *A is valid on all Kripke models.* (ii) $\vdash_{\mathbf{HDEL}} A$. (iii) $\vdash_{\mathbf{GDEL}} \Rightarrow \langle i, \epsilon \rangle : A$ *for any* $i \in \mathbb{N}$.

---

[4] This counter-example of the soundness theorem with s-valid is pointed out already in [15, Proposition4], and the proposition of PAL is also applicable to DEL.

6   Future Works

As we have seen above, we introduced a cut-free labelled sequent calculus **GDEL** which is sound and complete for Kripke semantics. Based on our cut-free sequent calculus **GDEL**, we are now developing an automated theorem prover in the programming language Haskell.[5] Our prover gives a proof of a given c-sequent when it is valid. Since a formula of DEL which stands for a knowledge state is likely to be quite complicated, an automated prover for DEL, which will enable us to quickly and accurately calculate the provability (and validity via completeness) of a formula, would be quite helpful. The below screen-shot is an example of proving one direction of $(RA1) : [\mathsf{a}^\mathsf{M}]p \to (\mathsf{pre}(\mathsf{a}) \to p)$ by our prover. But, at the present moment, we have not completed yet showing that our prover will eventually terminate and give the answer "FAIL" when a c-sequent is not valid.

Moreover, we may consider some other tasks from our labelled calculus such as expansion of the basis of the calculus from modal logic K to other modal logics (especially S5), adding 'common knowledge' in the language of DEL and developing **GDEL** to a contraction-free calculus (G3-system). These will be left to our future works.

Acknowledgement. We would like to thank anonymous reviewers for their constructive comments to our manuscript. The first author would like to thank his superviser, Prof. Satoshi Tojo, for his helpful comments to a draft. Finally, We thank Sean Arn for his proofreading of the final version of the paper. The first author is supported by Grant-in-Aid for JSPS Fellows in pursuing the present research, the third author was supported by JSPS KAKENHI, Grant-in-Aid for Young Scientists 15K21025. The first and third authors were also supported by JSPS Core-to-Core Program (A. Advanced Research Networks).

References

1. Aucher, G., Maubert, B., Schwarzentruber, F.: Generalized DEL-sequents. In: del Cerro, L.F., Herzig, A., Mengin, J. (eds.) JELIA 2012. LNCS, vol. 7519, pp. 54–66. Springer, Heidelberg (2012)
2. Aucher, G., Schwarzentruber, F.: On the complexity of dynamic epistemic logic. In: Schipper, B.C. (ed.) Proceedings of TARK, pp. 19–28 (2013)
3. Balbiani, P., Demange, V., Galmiche, D.: A sequent calculus with labels for PAL. Presented in Advances in Modal Logic (2014)
4. Baltag, A., Moss, L., Solecki, S.: The logic of public announcements, common knowledge and private suspicions. In: Proceedings of TARK, pp. 43–56. Morgan Kaufmann Publishers (1989)
5. van Benthem, J.: Dynamic logic for belief revision. J. App. Non-Classical Logics **14**(2), 129–155 (2004)
6. van Ditmarsch, H., Hoek, W., Kooi, B.: Dynamic Epistemic Logic. Springer, The Netherlands (2008)

---

[5] As relational works, there exist automated semantic tools of DEL such as DEMO [8] and Aximo [18].

7. Dyckhoff, R., Sadrzadeh, M.: A cut-free sequent calculus for algebraic dynamic epistemic logic. Technical report RR-10-11, OUCL, June 2010
8. van Eijck, J.: DEMO - a demo of epistemic modelling. In: Interactive Logic - Proceedings of the 7th Augustus de Morgan Workshop, pp. 305–363 (2007)
9. Frittella, S., Greco, G., Kurz, A., Palmigiano, A., Sikimić, V.: Multi-type display calculus for dynamic epistemic logic. J. Logic Comput. (2014). doi:10.1093/logcom/exu068. (First published online: December 5)
10. Gerbrandy, J., Groeneveld, W.: Reasoning about information change. J. Logic Lang. Inf. **6**(2), 147–169 (1997)
11. Maffezioli, P., Negri, S.: A gentzen-style analysis of public announcement Logic. In: Proceedings of the International Workshop on Logic and Philosophy of Knowledge, Communication and Action, pp. 293–313 (2010)
12. Negri, S., von Plato, J.: Structural Proof Theory. Cambridge University Press, Cambridge (2001)
13. Negri, S., von Plato, J.: Proof Analysis. Cambridge University Press, Cambridge (2011)
14. Nomura, S., Sano, K., Tojo, S.: A labelled sequent calculus for intuitionistic public announcement logic. In: The Proceedings of 20th International Conferences on Logic for Programming, Artificial Intelligence and Reasoning (LPAR-20) (forthcoming)
15. Nomura, S., Sano, K., Tojo, S.: Revising a sequent calculus for public announcement logic. In: Structural Analysis of Non-classical Logics-The Proceedings of the Second Taiwan Philosophical Logic Colloquium (TPLC-2014) (forthcoming)
16. Ono, H., Komori, Y.: Logics without contraction Rule. J. Symbolic Logic **50**(1), 169–201 (1985)
17. Plaza, J.: Logic of public communications. In: Proceedings of the 4th International Symposium on Methodologies for Intellingent Systems: Poster Session Program, pp. 201–216 (1989)
18. Richards, S., Sadrzadeh, M.: Aximo: automated axiomatic reasoning for information update. Electron. Notes Theor. Comput. Sci. **231**, 211–225 (2009)

The Urysohn Extension Theorem for Bishop Spaces

Iosif Petrakis[(⊠)]

University of Munich, Theresienstr. 39, 80333 Munich, Germany
petrakis@math.lmu.de

Abstract. Bishop's notion of function space, here called Bishop space, is a function-theoretic analogue to the classical set-theoretic notion of topological space. Bishop introduced this concept in 1967, without exploring it, and Bridges revived the subject in 2012. The theory of Bishop spaces can be seen as a constructive version of the theory of the ring of continuous functions. In this paper we define various notions of embeddings of one Bishop space to another and develop their basic theory in parallel to the classical theory of embeddings of rings of continuous functions. Our main result is the translation within the theory of Bishop spaces of the Urysohn extension theorem, which we show that it is constructively provable. We work within Bishop's informal system of constructive mathematics BISH, inductive definitions with countably many premises included.

Keywords: Constructive topology · Bishop spaces · Embeddings · Urysohn extension theorem

1 Introduction

The theory of Bishop spaces (TBS) is a constructive approach to general topology based on the notion of function space, here called Bishop space, that it was introduced by Bishop in [1], p. 71, but it was not really studied until Bridges's paper [7], that was followed by Ishihara'a paper [16], and our development of TBS in [22–24]. The main characteristics of TBS are the following:

1. Points are accepted from the beginning, hence it is not a point-free approach to topology.
2. Most of its notions are function-theoretic. Set-theoretic notions are avoided or play a secondary role to its development.
3. It is constructive. We work within Bishop's informal system of constructive mathematics BISH (see [4,5]), inductive definitions with rules of countably many premises included, a system connected to Martin-Löf's constructivism [17] and type theory [18]. The underlying logic of BISH is intuitionistic, while Myhill's system CST* of constructive Set Theory with inductive definitions, or Martin-Löf's extensional type theory, can be considered as formalizations of its underlying set theory.

© Springer International Publishing Switzerland 2016
S. Artemov and A. Nerode (Eds.): LFCS 2016, LNCS 9537, pp. 299–316, 2016.
DOI: 10.1007/978-3-319-27683-0_21

4. It has simple foundation and it follows the style of standard mathematics.

In other words, TBS is an approach to constructive point-function topology. The main motivation behind the introduction of Bishop spaces is that function-based concepts suit better to constructive study rather than set-based ones. Instead of having space-structures on a set X and $\mathbb{R}$, that determine a posteriori which functions of type $X \to \mathbb{R}$ are continuous with respect to them, we start from a given class of "continuous" functions of type $X \to \mathbb{R}$ that determines a posteriori a topological space-structure on X. "Continuity" in TBS is a primitive notion, a starting point similar to Spanier's theory of quasi-topological spaces in [27], or to the theory of limit spaces of Fréchet in [13].

TBS permits a "communication" with the classical theory of the rings of continuous functions, since many concepts, questions and results from the classical theory of $C(X)$, where X is a topological space, can be translated into TBS. Although this communication does not imply a direct translation from the theory of $C(X)$ to TBS, since the logic of TBS is intuitionistic, it is one of the features of TBS which makes it, in our view, so special as an approach to constructive topology. One could see TBS as an abstract, constructive version of the classical theory of $C(X)$, which we hope to be of interest to a classical mathematician too.

In this paper we develop the constructive basic theory of embeddings of Bishop spaces in parallel to the classical basic theory of embeddings of rings of continuous functions which is found in the book [11] of Gillman and Jerison. Our main result is the incorporation of the fundamental Urysohn extension theorem within the theory of embeddings of Bishop spaces.

2   Basic Definitions and Facts

In order to be self-contained we include in this section some basic definitions and facts necessary to the rest of the paper, that are partly found in [23]. For all proofs not included in this paper we refer to [24].

If X, Y are sets and $\mathbb{R}$ is the set of the constructive reals, we denote by $\mathbb{F}(X, Y)$ the functions of type $X \to Y$, by $\mathbb{F}(X)$ the functions of type $X \to \mathbb{R}$, by $\mathbb{F}_b(X)$ the bounded elements of $\mathbb{F}(X)$, and by $\mathrm{Const}(X)$ the subset of $\mathbb{F}(X)$ of all constant functions $\overline{a}$, where $a \in \mathbb{R}$. A function $\phi : \mathbb{R} \to \mathbb{R}$ is called *Bishop-continuous*, if ϕ is uniformly continuous on every bounded subset of $\mathbb{R}$, and we denote their set by $\mathrm{Bic}(\mathbb{R})$. If $f, g \in \mathbb{F}(X)$, $\epsilon > 0$, and $\Phi \subseteq \mathbb{F}(X)$, we define $U(g, f, \epsilon)$ and $U(\Phi, f)$ by

$$U(g, f, \epsilon) := \forall_{x \in X}(|g(x) - f(x)| \leq \epsilon),$$

$$U(\Phi, f) := \forall_{\epsilon > 0} \exists_{g \in \Phi}(U(g, f, \epsilon)).$$

Definition 1. *A Bishop space is a pair $\mathcal{F} = (X, F)$, where X is an inhabited set and $F \subseteq \mathbb{F}(X)$, a Bishop topology on X, or simply a topology on X, satisfies the following conditions:*

(BS_1) $a \in \mathbb{R} \to \overline{a} \in F$.
(BS_2) $f \in F \to g \in F \to f + g \in F$.
(BS_3) $f \in F \to \phi \in \mathrm{Bic}(\mathbb{R}) \to \phi \circ f \in F$,

$$X \xrightarrow{\ f\ } \mathbb{R}$$

$$F \ni \phi \circ f \qquad\qquad \phi \in \mathrm{Bic}(\mathbb{R})$$

$$\mathbb{R}.$$

(BS_4) $f \in \mathbb{F}(X) \to U(F, f) \to f \in F$.

Bishop used the term *function space* for $\mathcal{F}$ and *topology* for F. Since the former is used in many different contexts, we prefer the term Bishop space for $\mathcal{F}$, while we use the latter, as the *topology of functions* F on X corresponds nicely to the standard *topology of opens* $\mathcal{T}$ on X. Using BS_2 and BS_3 we get that if F is a topology on X, then fg, λf, $-f$, $\max\{f, g\} = f \vee g$, $\min\{f, g\} = f \wedge g$ and $|f| \in F$, for every $f, g \in F$ and $\lambda \in \mathbb{R}$. By BS_4 F is closed under uniform limits, where $f_n \xrightarrow{u} f$ denotes that f is the uniform limit of $(f_n)_{n \in \mathbb{N}}$. Moreover, $\mathrm{Const}(X) \subseteq F \subseteq \mathbb{F}(X)$, where $\mathrm{Const}(X)$ is the *trivial* topology on X and $\mathbb{F}(X)$ is the *discrete* topology on X. If F is a topology on X, the set F_b of all bounded elements of F is also a topology on X that corresponds to the ring $C^*(X)$ of the bounded elements of $C(X)$, for some topological space X. It is easy to see that $\mathrm{Bic}(\mathbb{R})$ is a topology on $\mathbb{R}$, and the structure $\mathcal{R} = (\mathbb{R}, \mathrm{Bic}(\mathbb{R}))$ is the *Bishop space of reals*.

The importance of the notion of a Bishop topology lies on Bishop's inductive concept of the least topology including a given subbase F_0, found in [1], p. 72, and in [4], p. 78, where the definitional clauses of a Bishop topology are turned into inductive rules.

Definition 2. *The least topology* $\mathcal{F}(F_0)$ *generated by a set* $F_0 \subseteq \mathbb{F}(X)$, *called a subbase of* $\mathcal{F}(F_0)$, *is defined by the following inductive rules:*

$$\frac{f_0 \in F_0}{f_0 \in \mathcal{F}(F_0)}, \qquad \frac{a \in \mathbb{R}}{\overline{a} \in \mathcal{F}(F_0)}, \qquad \frac{f, g \in \mathcal{F}(F_0)}{f + g \in \mathcal{F}(F_0)},$$

$$\frac{f \in \mathcal{F}(F_0), \ \phi \in \mathrm{Bic}(\mathbb{R})}{\phi \circ f \in \mathcal{F}(F_0)}, \qquad \frac{(g \in \mathcal{F}(F_0), \ U(g, f, \epsilon))_{\epsilon > 0}}{f \in \mathcal{F}(F_0)}.$$

If F_0 is inhabited, then the rule of the inclusion of the constant functions is redundant to the rule of closure under composition with $\mathrm{Bic}(\mathbb{R})$. The most complex inductive rule above can be replaced by the rule

$$\frac{g_1 \in \mathcal{F}(F_0) \ \wedge \ U(g_1, f, \frac{1}{2}), \ g_2 \in \mathcal{F}(F_0) \ \wedge \ U(g_2, f, \frac{1}{2^2}), \dots}{f \in \mathcal{F}(F_0)},$$

which has the "structure" of Brouwer's F-inference with countably many conditions in its premiss (see e.g., [19]). The above rules induce the following induction principle $\mathrm{Ind}_{\mathcal{F}}$ on $\mathcal{F}(F_0)$:

$$\forall_{f_0 \in F_0}(P(f_0)) \rightarrow$$
$$\forall_{a \in \mathbb{R}}(P(\overline{a})) \rightarrow$$
$$\forall_{f,g \in \mathcal{F}(F_0)}(P(f) \rightarrow P(g) \rightarrow P(f+g)) \rightarrow$$
$$\forall_{f \in \mathcal{F}(F_0)} \forall_{\phi \in \mathrm{Bic}(\mathbb{R})}(P(f) \rightarrow P(\phi \circ f)) \rightarrow$$
$$\forall_{f \in \mathcal{F}(F_0)}(\forall_{\epsilon > 0} \exists_{g \in \mathcal{F}(F_0)}(P(g) \ \wedge \ U(g, f, \epsilon)) \rightarrow P(f)) \rightarrow$$
$$\forall_{f \in \mathcal{F}(F_0)}(P(f)),$$

where P is any property on $\mathbb{F}(X)$. Hence, starting with a constructively acceptable subbase F_0 the generated least topology $\mathcal{F}(F_0)$ is a constructively graspable set of functions exactly because of the corresponding principle $\mathrm{Ind}_{\mathcal{F}}$. Despite the seemingly set-theoretic character of the notion of a Bishop space the core of TBS is the study of the inductively generated Bishop spaces. For example, since $\mathrm{id}_{\mathbb{R}} \in \mathrm{Bic}(\mathbb{R})$, where $\mathrm{id}_{\mathbb{R}}$ is the identity on $\mathbb{R}$, we get by the closure of $\mathcal{F}(\mathrm{id}_{\mathbb{R}})$ under BS_3 that $\mathrm{Bic}(\mathbb{R}) = \mathcal{F}(\mathrm{id}_{\mathbb{R}})$. Moreover, most of the new Bishop spaces generated from old ones are defined through the concept of the least topology. A property P on $\mathbb{F}(X)$ is *lifted* from a subbase F_0 to the generated topology $\mathcal{F}(F_0)$, if

$$\forall_{f_0 \in F_0}(P(f_0)) \rightarrow \forall_{f \in \mathcal{F}(F_0)}(P(f)).$$

It is easy to see inductively that boundedness is a lifted property. If (X, d) is a metric space and the elements of F_0 are bounded and uniformly continuous functions, then uniform continuity is also a lifted property.

Since Bishop did not pursue a constructive reconstruction of topology in [1], he didn't mention $\mathrm{Ind}_{\mathcal{F}}$, or some related lifted property. Apart from the notion of a Bishop space, Bishop introduced in [1], p. 68, the inductive notion of the least algebra $\mathcal{B}(B_{0,F})$ of Borel sets generated by a given set $B_{0,F}$ of F-complemented subsets, where F is an arbitrary subset of $\mathbb{F}(X)$. Since this notion was central to the development of constructive measure theory in [1], Bishop explicitly mentioned there the corresponding induction principle $\mathrm{Ind}_{\mathcal{B}}$ on $\mathcal{B}(B_{0,F})$ and studied specific lifted properties in that setting. Brouwer's inductive definition of the countable ordinals in [8] and Bishop's inductive notion of a Borel set were the main non-elementary inductively defined classes of mathematical objects used in constructive mathematics and motivated the formal study of inductive definitions in the 60s and the 70s (see [9]). Since then the use of inductive definitions in constructive mathematics and theoretical computer science became a common practice. In [3] Bishop and Cheng developed though, a reconstruction of constructive measure theory independently from the inductive definition of Borel sets, that replaced the old theory in [4]. In [2] Bishop, influenced by

Gödel's Dialectica interpretation, discussed a formal system Σ that would "efficiently express" his informal system of constructive mathematics. Since the new measure theory was already conceived and the theory of Bishop spaces was not elaborated at all, Bishop found no reason to extend Σ to subsume inductive definitions. In [20] Myhill proposed instead the formal theory CST of sets and functions to codify [1]. He also took Bishop's inductive definitions at face value and showed that the existence and disjunction properties of CST persist in the extended with inductive definitions system CST*.

Definition 3. *If $\mathcal{F} = (X, F)$ and $\mathcal{G} = (Y, G)$ are Bishop spaces, a Bishop morphism, or simply a morphism, from $\mathcal{F}$ to $\mathcal{G}$ is a function $h : X \to Y$ such that $\forall_{g \in G}(g \circ h \in F)$*

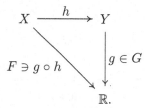

We denote by $\mathrm{Mor}(\mathcal{F}, \mathcal{G})$ the set of morphisms from $\mathcal{F}$ to $\mathcal{G}$, which are the arrows in the category of Bishop spaces **Bis**. It is easy to see that if $\mathcal{F} = (X, F)$ is a Bishop space, then $F = \mathrm{Mor}(\mathcal{F}, \mathcal{R})$. If $\mathcal{F} = (X, F)$ and $\mathcal{G}_0 = (Y, \mathcal{F}(G_0))$ are Bishop spaces, a function $h : X \to Y \in \mathrm{Mor}(\mathcal{F}, \mathcal{G}_0)$ if and only if $\forall_{g_0 \in G_0}(g_0 \circ h \in F)$

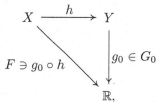

a very useful property that it is proved inductively and we call the *lifting of morphisms*. If $h \in \mathrm{Mor}(\mathcal{F}, \mathcal{G})$ is onto Y, then h is called a *set-epimorphism*, and we denote their set by $\mathrm{setEpi}(\mathcal{F}, \mathcal{G})$. We call some $h \in \mathrm{Mor}(\mathcal{F}, \mathcal{G})$ *open*, if $\forall_{f \in F} \exists_{g \in G}(f = g \circ h)$. Clearly, if $h \in \mathrm{Mor}(\mathcal{F}, \mathcal{G})$ such that h is 1-1 and onto Y, then $h^{-1} \in \mathrm{Mor}(\mathcal{G}, \mathcal{F})$ if and only if h is open. In this case h is called an *isomorphism* between $\mathcal{F}$ and $\mathcal{G}$. In [23] we showed that in the case of a set-epimorphism h, openness of h is also a lifted property.

Definition 4. *If $\mathcal{F} = (X, F)$ is a Bishop space and $A \subseteq X$ is inhabited, the relative Bishop space of $\mathcal{F}$ on A is the structure $\mathcal{F}_{|A} = (A, F_{|A})$, where $F_{|A} := \mathcal{F}(\{f_{|A} \mid f \in F\})$. We also call $\mathcal{F}_{|A}$ a subspace of $\mathcal{F}$. If $\mathcal{F} = (X, F)$ and $\mathcal{G} = (Y, G)$ are given Bishop spaces, their product is the structure $\mathcal{F} \times \mathcal{G} = (X \times Y, F \times G)$, where $F \times G := \mathcal{F}(\{f \circ \pi_1 \mid f \in F\} \cup \{g \circ \pi_2 \mid g \in G\})$, and π_1, π_2 are the projections of $X \times Y$ to X and Y, respectively.*

If F_0 is a subbase of F, we get inductively that $F_{|A} = \mathcal{F}(\{f_{0|A} \mid f_0 \in F_0\})$. It is straightforward to see that $\mathcal{F} \times \mathcal{G}$ satisfies the universal property for products and that $F \times G$ is the least topology which turns the projections π_1, π_2 into morphisms. If F_0 is a subbase of F and G_0 is a subbase of G, then we get inductively that $\mathcal{F}(F_0) \times \mathcal{F}(G_0) = \mathcal{F}(\{f_0 \circ \pi_1 \mid f_0 \in F_0\} \cup \{g_0 \circ \pi_2 \mid g_0 \in G_0\})$. Consequently, $\mathrm{Bic}(\mathbb{R}) \times \mathrm{Bic}(\mathbb{R}) = \mathcal{F}(\{\mathrm{id}_{\mathbb{R}} \circ \pi_1\} \cup \{\mathrm{id}_{\mathbb{R}} \circ \pi_2\}) = \mathcal{F}(\pi_1, \pi_2)$. The arbitrary product $\prod_{i \in I} \mathcal{F}_i$ of a family $(\mathcal{F}_i)_{i \in I}$ of Bishop spaces indexed by some I is defined similarly. Using the lifting of morphisms it is easy to show the following proposition.

Proposition 1. *Suppose that* $\mathcal{F} = (X, F)$, $\mathcal{G} = (Y, G)$, $\mathcal{H} = (Z, H)$ *are Bishop spaces and* $A \subseteq X$, $B \subseteq Y$.

(i) $j \in \mathrm{Mor}(\mathcal{H}, \mathcal{F} \times \mathcal{G})$ *if and only if* $\pi_1 \circ j \in \mathrm{Mor}(\mathcal{H}, \mathcal{F})$ *and* $\pi_2 \circ j \in \mathrm{Mor}(\mathcal{H}, \mathcal{G})$.
(ii) *If* $e : X \to B$, *then* $e \in \mathrm{Mor}(\mathcal{F}, \mathcal{G}) \leftrightarrow e \in \mathrm{Mor}(\mathcal{F}, \mathcal{G}_{|B})$.
(iii) $(F \times G)_{|A \times B} = F_{|A} \times G_{|B}$.

Note that Proposition 1(i) and (iii) hold for arbitrary products too. If $\mathcal{F}_i = (X_i, F_i)$ is a family of Bishop spaces indexed by some inhabited set I and $x = (x_i)_{i \in I} \in \prod_{i \in I} X_i$, then the *slice* $S(x; j)$ through x parallel to x_j, where $j \in I$, is the set $S(x; j) := X_j \times \prod_{i \neq j} \{x_i\} \subseteq \prod_{i \in I} X_i$ of all I-tuples where all components other the j-component are the ones of x, while the j-component ranges over X_j. The next fact is used in the proof of the Proposition 11 and it is a direct consequence of the Proposition 1.

Proposition 2. *If* $\mathcal{F}_i = (X_i, F_i)$ *is a family of Bishop spaces indexed by some inhabited set* I *and* $x = (x_i)_{i \in I} \in \prod_{i \in I} X_i$, *then the function* $s_j : X_j \to S(x; j)$, *defined by* $x_j \mapsto x_j \times \prod_{i \neq j} \{x_i\}$, *where* $S(x; j)$ *is the slice through* x *parallel to* x_j, *is an isomorphism between* $\mathcal{F}_j$ *and* $\mathcal{S}(x; j) = (S(x; j), F(x; j))$, *where* $F(x; j) = (\prod_{i \in I} F_i)_{|S(x;j)}$.

Definition 5. *If* $\mathcal{G} = (Y, G)$ *is a Bishop space,* X *is an inhabited set and* $\theta : X \to Y$, *the weak topology* $F(\theta)$ *on* X *induced by* θ *is defined as* $F(\theta) := \mathcal{F}(\{g \circ \theta \mid g \in G\})$. *The space* $\mathcal{F}(\theta) = (X, F(\theta))$ *is called the weak Bishop space on* X *induced by* θ. *If* $\mathcal{F} = (X, F)$ *is a Bishop space,* Y *is an inhabited set and* $e : X \to Y$ *is onto* Y, *the set of functions* $G_e := \{g \in \mathbb{F}(Y) \mid g \circ e \in F\}$ *is a topology on* Y. *We call* $\mathcal{G}_e = (Y, G_e)$ *the quotient Bishop space, and* G_e *the quotient topology on* Y, *with respect to* e.

The weak topology $F(\theta)$ is the least topology on X which makes θ a morphism. If θ is onto Y, then $\theta \in \mathrm{setEpi}(\mathcal{F}(\theta), \mathcal{G})$, and by the lifting of openness we get that $F(\theta) = \{g \circ \theta \mid g \in G\}$, a fact that we use in the proof of the Proposition 6. In analogy to classical topology, the quotient topology G_e is the largest topology on Y which makes e a morphism.

In [4], pp. 91–92, it is shown[1] that if $D \subseteq X$ is a dense subset of the metric space X, Y is a complete metric space, and $f : D \to Y$ is uniformly continuous

---

[1] The uniqueness property is included, for example, in [21], p. 238.

with modulus of continuity ω, then there exists a unique uniform continuous extension $g : X \to Y$ of f with modulus of continuity $\frac{1}{2}\omega$. The next lemma is a useful generalization of it[2] that we proved in [24] and we use it here in the proof of the Proposition 3(vi).

Lemma 1. *Suppose that X is an inhabited metric space, $D \subseteq X$ is dense in X and Y is a complete metric space. If $f : D \to Y$ is uniformly continuous on every bounded subset of D, then there exists a unique extension $g : X \to Y$ of f which is uniformly continuous on every bounded subset of X with modulus of continuity $\omega_{g,B}(\epsilon) = \frac{1}{2}\omega_{f,B\cap D}(\epsilon)$, for every inhabited, bounded and metric-open subset B of X. Moreover, if f is bounded by some $M > 0$, then g is also bounded by M.*

Within BISH a *compact* metric space is defined as a complete and totally bounded space. A *locally compact* metric space X is a space in which every bounded subset of X is included in a compact one. If X is locally compact, the set $\mathrm{Bic}(X)$, defined like $\mathrm{Bic}(\mathbb{R})$, is a topology on X. Using the definition of a continuous function on a locally compact metric space, given in [4], p. 110, Bishop's formulation of the Tietze theorem for metric spaces becomes as follows.

Theorem 1. *Let Y be a locally compact subset of a metric space X and $I \subset \mathbb{R}$ an inhabited compact interval. Let $f : Y \to I$ be uniformly continuous on the bounded subsets of Y. Then there exists a function $g : X \to I$ which is uniformly continuous on the bounded subsets of X, and which satisfies $g(y) = f(y)$, for every $y \in Y$.*

Corollary 1. *If Y is a locally compact subset of $\mathbb{R}$ and $g : Y \to I \in \mathrm{Bic}(Y)$, where $I \subset \mathbb{R}$ is an inhabited compact interval, then there exists a function $\phi : \mathbb{R} \to I \in \mathrm{Bic}(\mathbb{R})$ which satisfies $\phi(y) = g(y)$, for every $y \in Y$.*

We use the Corollary 1 in the proof of the Propositions 3(v) and 8, while in [24] we used it to show the following fundamental fact, which is used here in the proof of the Proposition 9.

Theorem 2. *Suppose that (X, F) is a Bishop space and $f \in F$ such that $f \geq \overline{c}$, for some $c > 0$. Then, $\frac{1}{f} \in F$.*

---

[2] According to Bishop and Bridges [4], p. 85, if $B \subseteq X$, where (X, d) is an inhabited metric space, B is a *bounded subset* of X, if there is some $x_0 \in X$ such that $B \cup \{x_0\}$ with the induced metric is a bounded metric space. If we suppose that the inclusion map of a subset is the identity (see [4], p. 68), the induced metric on $B \cup \{x_0\}$ is reduced to the relative metric on $B \cup \{x_0\}$. We may also denote a bounded subset B of an inhabited metric space X by (B, x_0, M), where $M > 0$ is a bound for $B \cup \{x_0\}$. If (B, x_0, M) is a bounded subset of X then $B \subseteq \mathcal{B}(x_0, M)$, and $(\mathcal{B}(x_0, M), x_0, 2M)$ is also a bounded subset of X. I.e., a bounded subset of X is included in an inhabited bounded subset of X which is also metric-open i.e., it includes an open ball of every element of it.

Definition 6. *If (X, F) is a Bishop space, the relations defined by*

$$x_1 \bowtie_F x_2 :\leftrightarrow \exists_{f \in F}(f(x_1) \bowtie_{\mathbb{R}} f(x_2)),$$

$$A \bowtie_F B :\leftrightarrow \exists_{f \in F} \forall_{a \in A} \forall_{b \in B}(f(a) = 0 \wedge f(b) = 1)$$

where $x_1, x_2 \in X$, $a \bowtie_{\mathbb{R}} b :\leftrightarrow a > b \vee a < b \leftrightarrow |a - b| > 0$, for every $a, b \in \mathbb{R}$, and $A, B \subseteq X$, are the canonical point-point and set-set apartness relations on X. If $\bowtie$ is a point-point apartness relation on X^3, F is called $\bowtie$-Hausdorff, if $\bowtie \subseteq \bowtie_F$. The F-zero sets $Z(F)$ of (X, F) are the subsets of X of the form $\zeta(f) = \{x \in X \mid f(x) = 0\}$, where $f \in F$.

In [24] we showed within BISH that $Z(F)$ is closed under countably infinite intersections, and the sets $[f \leq \bar{a}] = \{x \in X \mid f(x) \leq a\}, [f \geq \bar{a}] = \{x \in X \mid f(x) \geq a\}$, where $a \in \mathbb{R}$, are in $Z(F)$. We also used the Theorem 2 to show the Urysohn lemma for the zero sets of a Bishop space. According to the classical Urysohn lemma for $C(X)$-zero sets, the disjoint zero sets of any topological space X are separated by some $f \in C(X)$ (see [11], p. 17). Constructively, we need to replace the negative notion of disjointness of two zero sets by a positive notion.

Theorem 3 (Urysohn Lemma for F-Zero Sets). *If (X, F) is a Bishop space and $A, B \subseteq X$, then $A \bowtie_F B \leftrightarrow \exists_{f,g \in F} \exists_{c>0}(A \subseteq \zeta(f) \wedge B \subseteq \zeta(g) \wedge |f| + |g| \geq \bar{c})$.*

3   Embeddings of Bishop Spaces

If $\mathcal{G}, \mathcal{F}$ are Bishop spaces, the notions "$\mathcal{G}$ is embedded in $\mathcal{F}$" and "$\mathcal{G}$ is bounded-embedded in $\mathcal{F}$" translate into TBS the notions "Y is C-embedded in X" and "Y is C^*-embedded in X", for some $Y \subseteq X$ and a given topology of opens $\mathcal{T}$ on X (see [11], p. 17). If F is a topology on X, $f \in F$ and $a, b \in \mathbb{R}$ such that $a \leq b$, we say that a, b *bound* f, if $\forall_{x \in X}(a \prec f(x) \prec b)$, where $\prec \in \{<, \leq\}$.

Definition 7. *If $\mathcal{F} = (X, F)$, $\mathcal{G} = (Y, G)$ are Bishop spaces and $Y \subseteq X$, then*

(i) $\mathcal{G}$ *is embedded in* $\mathcal{F}$, *if* $\forall_{g \in G} \exists_{f \in F}(f_{|Y} = g)$.

(ii) $\mathcal{G}$ *is bounded-embedded in* $\mathcal{F}$, *if* $\mathcal{G}_b$ *is embedded in* $\mathcal{F}_b$.

(iii) $\mathcal{G}$ *is full bounded-embedded in* $\mathcal{F}$, *if* $\mathcal{G}$ *is bounded-embedded in* $\mathcal{F}$, *and for every* $g \in G_b$, *if* a, b *bound* g, *then* a, b *bound some extension* f *of* g *in* F_b.

(iv) $\mathcal{G}$ *is dense-embedded in* $\mathcal{F}$, *if* $\forall_{g \in G} \exists!_{f \in F}(f_{|Y} = g)$.

(v) $\mathcal{G}$ *is dense-bounded-embedded in* $\mathcal{F}$ *and* $\mathcal{G}$ *is dense-full bounded-embedded in* $\mathcal{F}$ *are defined similarly to (iv).*

(vi) $\mathcal{F}$ *extends* $\mathcal{G}$, *if* $\forall_{f \in F}(f_{|Y} \in G)$.

---

[3] See definition 2.1 in [4], p. 72. It is also easy to see that $a \bowtie_{\mathbb{R}} b \leftrightarrow a \bowtie_{\mathrm{Bic}(\mathbb{R})} b$, for every $a, b \in \mathbb{R}$.

Clearly, (X, G) is embedded in (X, F) if and only if $G \subseteq F$. The Definition 7(vi) is necessary, since a topology F on some X does not necessarily behave like $C(X)$, where every $f \in C(X)$ restricted to Y belongs to $C(Y)$. By the definition of the relative Bishop space we get immediately that $\mathcal{F}$ extends $\mathcal{F}_{|Y}$. If $\mathcal{G}$ is embedded in $\mathcal{F}$, then $\mathcal{G}'$ is embedded in $\mathcal{F}$, where $\mathcal{G}' = (Y, G')$ and $G' \subseteq G$. If (X, F) is a Bishop space and $Y \subseteq X$, a *retraction* of X onto Y is a function $r : X \to Y$ such that $r(y) = y$, for every $y \in Y$, and $r \in \mathrm{Mor}(\mathcal{F}, \mathcal{F}_{|Y})$. In this case Y is called a *retract* of X. For example, the Cantor space with the product topology on $(2, \mathbb{F}(2))$ is a retract of the Baire space with the product topology on $(\mathbb{N}, \mathbb{F}(\mathbb{N}))$.

Proposition 3. *Suppose that $Y \subseteq X$ and $\bowtie$ is a point-point apartness relation on X.*

(i) $(Y, \mathrm{Const}(Y))$ is embedded in every Bishop space (X, F).

(ii) If $\forall_{x \in X}(x \in Y \lor x \notin Y)$, then $(Y, \mathbb{F}(Y))$ is embedded in $(X, \mathbb{F}(X))$.

(iii) If $Y = \{x_1, \ldots, x_n\}$, where $x_i \bowtie x_j$, for every $i \neq j \in \{1, \ldots, n\}$, and F is a topology on X which is $\bowtie$-Hausdorff, then $(Y, \mathbb{F}(Y))$ is full bounded-embedded in (X, F).

(iv) $(\mathbb{N}, \mathbb{F}(\mathbb{N}))$ is full bounded-embedded in $(\mathbb{Q}, \mathrm{Bic}(\mathbb{Q}))$.

(v) If $X = \mathbb{R}$ and Y is locally compact, then $(Y, \mathrm{Bic}(Y))$ is bounded-embedded in $\mathcal{R}$.

(vi) If X is a locally compact metric space and Y is dense in X, then $(Y, \mathrm{Bic}(Y))$ is dense-embedded and dense-bounded-embedded in $(X, \mathrm{Bic}(X))$.

(vii) If F is a topology on X and Y is a retract of X, then $\mathcal{F}_{|Y}$ is embedded in $\mathcal{F}$.

Proof. (i) and (ii) are trivial. To show (iii) we fix some $g \in \mathbb{F}(Y)$ and let $g(x_i) = a_i$, for every i. If we consider the $(n - 1) + (n - 2) + \ldots + 1$ functions $f_{ij} \in F$ such that $f_{ij}(x_i) \bowtie_{\mathbb{R}} f_{ij}(x_j)$, for every $i < j$, then the function f on X, defined by $f(x) := \sum_{i=1}^{n} a_i A_i(x)$, where

$$A_i(x) := \prod_{k=i+1}^{n} \frac{f_{ik}(x) - f_{ik}(x_k)}{f_{ik}(x_i) - f_{ik}(x_k)} \prod_{k=1}^{i-1} \frac{f_{ki}(x_k) - f_{ki}(x)}{f_{ki}(x_k) - f_{ki}(x_i)},$$

is in F and $A_i(x_j) = 1$, if $j = i$, $A_i(x_j) = 0$, if $j \neq i$. Hence, f extends g, and clearly $(Y, \mathbb{F}(Y))$ is full-bounded embedded in (X, F). We need the $\bowtie$-Hausdorff condition on F so that $(f_{ij}(x_i) - f_{ij}(x_j)) \bowtie_{\mathbb{R}} 0$ and then $(f_{ij}(x_i) - f_{ij}(x_j))^{-1}$ is well-defined, for every $i < j$.

(iv) If q is a rational such that $q \geq 0$, there is a unique $n \in \mathbb{N}$ such that $q \in [n, n+1)$. If $g : \mathbb{N} \to \mathbb{R}$, we define $\phi^*(q) = \gamma_n(q)$, where $\gamma_n : \mathbb{Q} \cap [n, n+1) \to \mathbb{R}$ is defined by $\gamma_n(q) = (g(n+1) - g(n))q + (n+1)g(n) - g(n+1)n$ i.e., $\gamma_n(\mathbb{Q} \cap [n, n+1))$ is the set of the rational values in the linear segment between $g(n)$ and $g(n+1)$. Of course, $\phi^*(n) = g(n)$. Next we define $\phi^*(q) = g(0)$, for every $q < 0$. To show that $\phi^* \in \mathrm{Bic}(\mathbb{Q})$, and since ϕ^* is constant on $\mathbb{Q}_-$, it suffices to show that $\phi^* \in \mathrm{Bic}(\mathbb{Q}_+)$. For that we fix a bounded subset (B, q_0, M) of $\mathbb{Q}_+$, where without loss of generality $M \in \mathbb{N}$. Since $B \subseteq \mathcal{B}(q_0, M)$, we have that $B \subseteq [n, N]$, where $n, N \in \mathbb{N}$, $n < N$, $q_0 - M \in [n, n+1)$ and $q_0 + M \in [N, N+1)$. Each γ_i is uniformly continuous on $[i, i+1) \cap \mathbb{Q}$ with modulus of continuity $\omega_i(\epsilon) = \frac{\epsilon}{|g(i+1)-g(i)|+1}$, for every $\epsilon > 0$. Hence, ϕ^* is uniformly continuous on B with modulus of continuity $\omega_{\phi^*,B}(\epsilon) = \min\{\omega_i(\epsilon) \mid n \leq i \leq N\}$, for every $\epsilon > 0$. If g is bounded, then by its definition ϕ^* is also bounded and if a, b bound g, then a, b bound ϕ^*.

(v) If $M > 0$ such that $f(Y) \subseteq [-M, M]$, then we use the Corollary 1.

(vi) Since $\mathbb{R}$ is a complete metric space, we use the Lemma 1.

(vii) We show first that r is a quotient map i.e., $F_{|Y} = G_r = \{g : Y \to \mathbb{R} \mid g \circ r \in F\}$. By the definition of $r \in \mathrm{Mor}(\mathcal{F}, \mathcal{F}_{|Y})$, we have that $\forall_{g \in F_{|Y}}(g \circ r \in F)$ i.e., $F_{|Y} \subseteq G_r$. For that we can also use our remark in Sect. 2 that the quotient topology G_r is the largest topology such that r is a morphism. If $g \in G_r$, then $(g \circ r)_{|Y} = g \in F_{|Y}$ i.e., $F_{|Y} \supseteq G_r$. Hence, if $g \in F_{|Y} = G_r$, the function $g \circ r \in F$ extends g.

Proposition 4. *Suppose that* $\mathcal{F} = (X, F)$, $\mathcal{G} = (Y, G)$ *are Bishop spaces and* $Y \subseteq X$. *If* $\mathcal{G}$ *is embedded in* $\mathcal{F}$, *then* $\mathcal{G}$ *is bounded-embedded in* $\mathcal{F}$.

Proof. We show that if $g \in G_b$ and $\exists_{f \in F}(f_{|Y} = g)$, then $\exists_{f \in F_b}(f_{|Y} = g)$; if f extends g and $|g| \leq M$, then $h = (-\overline{M} \vee f) \wedge \overline{M} \in F_b$ and $h_{|Y} = g$. I.e., $\mathcal{G}$ is bounded-embedded in $\mathcal{F}$, if $\forall_{g \in G_b} \exists_{f \in F}(f_{|Y} = g)$. Since $G_b \subseteq G$ and G is embedded in $\mathcal{F}$, $\mathcal{G}$ is bounded-embedded in $\mathcal{F}$.

There are trivial counterexamples to the converse of the previous proposition; if Y is an unbounded locally compact subset of $\mathbb{R}$, then by the Proposition 3(v) $(Y, \mathrm{Bic}(Y))$ is full bounded-embedded in $\mathcal{R}_b$, while $(Y, \mathrm{Bic}(Y))$ is not embedded in $\mathcal{R}_b$, since $\mathrm{id}_Y \in \mathrm{Bic}(Y)$ and any extension of id_Y is an unbounded function.

Proposition 5. *If* $Z \subseteq Y \subseteq X$, $\mathcal{H} = (Z, H)$, $\mathcal{G} = (Y, G)$, $\mathcal{F} = (X, F)$ *are Bishop spaces,* $\mathcal{F}$ *extends* $\mathcal{G}$ *and* $\mathcal{G}$ *is embedded in* $\mathcal{F}$, *then* $\mathcal{H}$ *is embedded in* $\mathcal{F}$ *if and only if* $\mathcal{H}$ *is embedded in* $\mathcal{G}$.

Proof. If $\forall_{h \in H} \exists_{f \in F}(f_{|Z} = h)$, we show that $\forall_{h \in H} \exists_{g \in G}(g_{|Z} = h)$. If $h \in H$ and we restrict some $f \in F$ which extends h to Y, we get an extension of h in G. For the converse if $h \in H$, we extend it to some $g \in G$, and g is extended to some $f \in F$, since $\mathcal{G}$ is embedded in $\mathcal{F}$.

The next three propositions show how the embedding of $\mathcal{G}$ in $\mathcal{F}$ generates new embeddings under the presence of certain morphisms.

Proposition 6. *Suppose that* $\mathcal{F} = (X, F)$, $\mathcal{G} = (Y, G)$ *and* $\mathcal{H} = (B, H)$ *are Bishop spaces, where* $B \subseteq Y$. *If* $\mathcal{H}$ *is embedded in* $\mathcal{G}$ *and* $e \in \mathrm{setEpi}(\mathcal{F}, \mathcal{G})$, *then the weak Bishop space* $\mathcal{F}(e_{|A})$ *on* $A = e^{-1}(B)$ *induced by* $e_{|A}$ *is embedded in* $\mathcal{F}$.

Proof. Since $e : X \to Y$ is onto Y, we have that $e_{|A} : A \to B$ is onto B and $e_{|A} \in \mathrm{setEpi}(\mathcal{F}(e_{|A}), \mathcal{H})$, where by a remark following the definition of weak topology in Sect. 2 we have that $F(e_{|A}) = \{h \circ e_{|A} \mid h \in H\}$. If we fix some $h \circ e_{|A} \in F(e_{|A})$, where $h \in H$, then, since $\mathcal{H}$ is embedded in $\mathcal{G}$, there is some $g \in G$ such that $g_{|B} = h$. Since $e \in \mathrm{setEpi}(\mathcal{F}, \mathcal{G}) \subseteq \mathrm{Mor}(\mathcal{F}, \mathcal{G})$, we get that $g \circ e \in F$. If $a \in A$, then $(g \circ e)(a) = g(b) = h(b)$, where $b = e(a)$. Since $(h \circ e_{|A})(a) = h(e(a)) = h(b)$, we get that $(g \circ e)_{|A} = h \circ e_{|A}$ i.e., $\mathcal{F}(e_{|A})$ is embedded in $\mathcal{F}$.

Proposition 7. *If $\mathcal{F} = (X, F), \mathcal{G} = (Y, G)\ \mathcal{H} = (Z, H)$ are Bishop spaces, $Y \subseteq X$, $\mathcal{G}$ is embedded in $\mathcal{F}$ and $e \in \mathrm{Mor}(\mathcal{F}, \mathcal{H})$ is open, then the quotient Bishop space $\mathcal{G}_{e_{|Y}} = (e(Y), G_{e_{|Y}})$ is embedded in $\mathcal{H}$.*

Proof. Let $g' : e(Y) \to \mathbb{R} \in G_{e_{|Y}}$ i.e., $g' \circ e_{|Y} \in G$. Since $\mathcal{G}$ is embedded in $\mathcal{F}$, there exists some $f \in F$ such that $f_{|Y} = g' \circ e_{|Y}$. Since e is open, there exists some $h \in H$ such that $f = h \circ e$. We show that $h_{|e(Y)} = g'$; if $b = e(y) \in e(Y)$, for some $y \in Y$, then $h(b) = h(e(y)) = f(y) = (g' \circ e_{|Y})(y) = g'(e(y)) = g'(b)$.

Next we translate to TBS the classical fact that if an element of $C(X)$ carries a subset of X homeomorphically onto a closed set S in $\mathbb{R}$, then S is C-embedded in X (see [11], p. 20).

Proposition 8. *Suppose that A is a locally compact subset of $\mathbb{R}$, $\mathcal{F} = (X, F)$ is a Bishop space, $Y \subseteq X$ and $f \in F$ such that $f_{|Y} : Y \to A$ is an isomorphism between $\mathcal{F}_{|Y}$ and $(A, \mathrm{Bic}(A)_b)$. Then $\mathcal{F}_{|Y}$ is embedded in $\mathcal{F}$.*

Proof. Since $f_{|Y}$ is an isomorphism between $\mathcal{F}_{|Y}$ and $(A, \mathrm{Bic}(A)_b)$, its inverse θ is an isomorphism between $(A, \mathrm{Bic}(A)_b)$ and $\mathcal{F}_{|Y}$. We fix some $g \in F_{|Y}$. Since $\theta \in \mathrm{Mor}((A, \mathrm{Bic}(A)_b), \mathcal{F}_{|Y})$, we have that $g \circ \theta \in \mathrm{Bic}(A)_b$. By the Corollary 1 there exists some $\phi \in \mathrm{Bic}(\mathbb{R})$ which extends $g \circ \theta$. By BS$_3$ we have that $\phi \circ f \in F$ and for every $y \in Y$ we have that $(\phi \circ f)(y) = ((g \circ \theta) \circ f)(y) = (g \circ (\theta \circ f))(y) = (g \circ (\theta \circ f_{|Y}))(y) = (g \circ \mathrm{id}_{|Y})(y) = g(y)$.

If $(X, \mathcal{T})$ is a topological space and $Y \subseteq X$ is C^*-embedded in X, then if Y is also C-embedded in X, it is (completely) separated in $C(X)$ from every $C(X)$-zero set disjoint from it (see [11], pp. 19–20). If we add within TBS a positive notion of disjointness between Y and $\zeta(f)$ though, we avoid the corresponding hypothesis of $\mathcal{G}$ being embedded in $\mathcal{F}$.

Definition 8. *If F is a topology on X, $f \in F$ and $Y \subseteq X$, we say that Y and $\zeta(f)$ are separated, $\mathrm{Sep}(Y, \zeta(f))$, if $\forall_{y \in Y}(|f(y)| > 0)$, and Y and $\zeta(f)$ are uniformly separated, $\mathrm{Usep}(Y, \zeta(f))$, if there is some $c > 0$ such that $\forall_{y \in Y}(|f(y)| \geq c)$.*

Of course, $\mathrm{Usep}(Y, \zeta(f)) \to \mathrm{Sep}(Y, \zeta(f))$. If $f, g \in F$ such that $|f| + |g| \geq \bar{c}$ (see the formulation of the Theorem 3), then we get $\mathrm{Usep}(\zeta(g), \zeta(f))$ and $\mathrm{Usep}(\zeta(f), \zeta(g))$. Since the sets $U(f) = \{x \in X \mid f(x) > 0\}$, where $f \in F$, are basic open sets in the induced neighborhood structure on X by F (see [4], p. 77), we call Y a *uniform G_δ-set*, if there exists a sequence $(f_n)_n$ in F such that $Y = \bigcap_{n \in \mathbb{N}} U(f(n))$ and $\mathrm{Usep}(Y, \zeta(f_n))$, for every $n \in \mathbb{N}$.

Proposition 9. *If $\mathcal{F} = (X, F)$, $\mathcal{G} = (Y, G)$ are Bishop spaces, $Y \subseteq X$, $\mathcal{F}$ extends $\mathcal{G}$, $\mathcal{G}$ is bounded-embedded in $\mathcal{F}$, and $f \in F$, then $\mathrm{Usep}(Y, \zeta(f)) \to Y \bowtie_F \zeta(f)$.*

Proof. Since $|f| \in F$ and $\mathcal{F}$ extends $\mathcal{G}$, we have that $|f|_{|Y} \in G$, and $|f|_{|Y} \geq \bar{c}$. By Theorem 2 we get that $\frac{1}{|f|_{|Y}} \in G$. Since $\bar{0} < \frac{1}{|f|_{|Y}} \leq \frac{\bar{1}}{c}$, we actually have that $\frac{1}{|f|_{|Y}} \in G_b$. Since $\mathcal{G}$ is bounded-embedded in $\mathcal{F}$, there exists $h \in F$ such that

$h_{|Y} = \frac{1}{|f|_{|Y}}$. Since $|h| \in F$ satisfies $|h|_{|Y} = \frac{1}{|f|_{|Y}}$ too, we suppose without loss of generality that $h \geq \overline{0}$. If we define $g := h|f|$, then $g \in F$, $g(y) = h(y)|f(y)| = \frac{1}{|f(y)|}|f(y)| = 1$, for every $y \in Y$, and $g(x) = h(x)|f(x)| = h(x)0 = 0$, for every $x \in \zeta(f)$.

Corollary 2. *Suppose that $\mathcal{F} = (X, F)$, $\mathcal{G} = (Y, G)$ are Bishop spaces, $Y \subseteq X$, and $\mathcal{G}$ is full bounded-embedded in $\mathcal{F}$. If Y is a uniform G_δ-set, then Y is an F-zero set.*

Proof. Suppose that $Y = \bigcap_{n \in \mathbb{N}} U(f_n)$ and $\forall_{y \in Y}(|f_n(y)| \geq c_n)$, where $c_n > 0$, for every $n \in \mathbb{N}$. Since $U(f) = U(f \vee \overline{0})$ and $\mathrm{Usep}(Y, \zeta(f)) \to \mathrm{Usep}(Y, \zeta(f \vee \overline{0}))$, we assume without loss of generality that $f_n \geq \overline{0}$, for every $n \in \mathbb{N}$. By the proof the Proposition 9 we have that there is a function $h_n \in F$ such that $h_n \geq \overline{0}$, $(h_n f_n)(Y) = 1$ and $(h_n f_n)(\zeta(f_n)) = 0$, for every $n \in \mathbb{N}$. Therefore, $Y \subseteq \zeta(g_n)$, where $g_n = (h_n f_n - \frac{2}{3}) \wedge \overline{0}$, for every $n \in \mathbb{N}$. Next we show that $\zeta(g_n) \subseteq U(f_n)$, for every $n \in \mathbb{N}$. Since $\mathcal{G}$ is full bounded-embedded in $\mathcal{F}$ and according to the proof the Proposition 9, $\overline{0} < \frac{1}{f_{n|Y}} \leq \frac{1}{c_n}$, we get that $\overline{0} < h_n \leq \frac{1}{c_n}$. If $z \in X$ such that $g_n(z) = 0$, then $h_n(z) f_n(z) \geq \frac{2}{3}$, and since $h_n(z) > 0$, we conclude that $f_n(z) \geq \frac{2}{3 h_n(z)} > 0$. Thus, $Y \subseteq \bigcap_{n \in \mathbb{N}} \zeta(g_n) \subseteq \bigcap_{n \in \mathbb{N}} U(f_n) = Y$, which implies that $Y = \bigcap_{n \in \mathbb{N}} \zeta(g_n) = \zeta(g)$, for some $g \in F$, since $Z(F)$ is closed under countably infinite intersections.

Without the condition of $\mathcal{G}$ being full bounded-embedded in $\mathcal{F}$ in the previous proposition we can show only that $\neg(f_n(z) = 0)$. Although $f_n(z) \geq 0$, we cannot infer within BISH that $f_n(z) > 0$; the property of the reals $\forall_{x,y \in \mathbb{R}}(\neg(x \geq y) \to x < y)$ is equivalent to Markov's principle (MP) (see [5], p. 14), and it is easy to see that this property is equivalent to $\forall_{x \in \mathbb{R}}(x \geq 0 \to \neg(x = 0) \to x > 0)$. Next we translate to TBS the classical result that if Y is C^*-embedded in X such that Y is (completely) separated from every $C(X)$-zero set disjoint from it, then Y is C-embedded in X. Constructively it is not clear, as it is in the classical case, how to show that the expected positive formulation of the previous condition provides an inverse to Proposition 4. The reason is that if (X, F) is an arbitrary Bishop space, it is not certain that $\tan \circ f \in F$, for some $f : X \to (-\frac{\pi}{2}, \frac{\pi}{2}) \in F$ (note that $\tan^{-1} = \arctan \in \mathrm{Bic}(\mathbb{R})$). If $\Phi_1, \Phi_2 \subseteq \mathbb{F}(X)$, we denote by $\Phi_1 \vee \Phi_2$ the least topology including them. The proof of the interesting case of the next theorem is in BISH + MP.

Theorem 4. *Suppose that $\mathcal{F} = (X, F)$, $\mathcal{G} = (Y, G)$ are Bishop spaces, $Y \subseteq X$, $a > 0$, $e : (-a, a) \to \mathbb{R}$ such that $e^{-1} : \mathbb{R} \to (-a, a) \in \mathrm{Bic}(\mathbb{R})$, and $\mathcal{F}(a) = (X, F(a))$, where*

$$F(a) = F \vee \{e \circ f \mid f \in F \text{ and } f(X) \subseteq (-a, a)\}.$$

(i) If $\mathcal{G}$ is full bounded-embedded in $\mathcal{F}$, then $\mathcal{G}$ is embedded in $\mathcal{F}(a)$.

(ii) (MP) If $\forall_{f \in F}(\mathrm{Sep}(Y, \zeta(f)) \to Y \bowtie_F \zeta(f))$ and $\mathcal{G}$ is bounded-embedded in $\mathcal{F}$, then $\mathcal{G}$ is embedded in $\mathcal{F}(a)$.

Proof. We fix some $g \in G$. Since $e^{-1} \in \text{Bic}(\mathbb{R})$, by the condition BS$_3$ we have that $e^{-1} \circ g : Y \to (-a, a) \in G_b$. Since $\mathcal{G}$ is bounded-embedded in $\mathcal{F}$, there is some $f \in F_b$ such that $f_{|Y} = e^{-1} \circ g$.

(i) If $\mathcal{G}$ is full bounded-embedded in $\mathcal{F}$, then we have that $f : X \to (-a, a)$. Hence, $e \circ f \in F(a)$, and $(e \circ f)_{|Y} = e \circ f_{|Y} = e \circ (e^{-1} \circ g) = g$.

(ii) In [24] we showed within BISH that $[|f| \geq \bar{a}] = \{x \in X \mid |f|(x) \geq a\} = \zeta(f^*)$, where $f^* = (|f| - \bar{a}) \wedge \bar{0} \in F$. If $y \in Y$, then $|f^*(y)| = |(|f(y)| - a) \wedge 0| = |(|(e^{-1} \circ g)(y)| - a) \wedge 0| = ||(e^{-1} \circ g)(y)| - a| = a - |(e^{-1} \circ g)(y)| > 0$, since $|(e^{-1} \circ g)(y)| \in [0, a)$ (if $-a < x < a$, then $|x| < a$). Since $\text{Sep}(Y, \zeta(f^*))$, by our hypothesis there exists some $h \in F$ such that $0 \leq h \leq 1$, $h(Y) = 1$ and $h(\zeta(f^*)) = 0$. There is no loss of generality if we assume that $0 \leq h \leq 1$, since if $h \in F$ separates Y and $\zeta(f^*)$, then $|h| \wedge \bar{1} \in F$ separates them too. We define $J := f \cdot h \in F$. If $y \in Y$, we have that $J(y) = f(y)h(y) = f(y)$. Next we show that $\forall_{x \in X}(\neg(|J(x)| \geq a))$. If $x \in X$ such that $|J(x)| \geq a$, then $|f(x)| \geq |f(x)||h(x)| = |j(x)| \geq a$, therefore $x \in \zeta(f^*)$. Consequently, $h(x) = 0$, and $0 = |J(x)| \geq a > 0$, which leads to a contradiction. Because of MP we get that $\forall_{x \in X}(|J(x)| < a)$, in other words, $J : X \to (-a, a)$. Hence $e \circ J \in F(a)$, and $(e \circ J)_{|Y} = e \circ J_{|Y} = e \circ f = e \circ (e^{-1} \circ g) = g$.

4   The Urysohn Extension Theorem

In this section we show the Urysohn extension theorem within TBS, an adaptation of Urysohn's theorem that any closed set in a normal topological space is C^*-embedded (see [11], p. 266). As Gillman and Jerison note in [11], p. 18, it is "the basic result about C^*-embedding". According to it, a subspace Y of a topological space X is C^*-embedded in X if and only if any two (completely) separated sets in Y are (completely) separated in X. Here we call Urysohn extension theorem the appropriate translation to TBS of the non-trivial sufficient condition. Next follows the translation to TBS of the trivial necessity condition. The hypothesis "$\mathcal{F}$ extends $\mathcal{G}$" of the Theorem 5 is not necessary to its proof.

Proposition 10. *Suppose that $\mathcal{F} = (X, F)$, $\mathcal{G} = (Y, G)$ are Bishop spaces and $Y \subseteq X$. If $\mathcal{G}$ is bounded-embedded in $\mathcal{F}$, then $\forall_{A, B \subseteq Y}(A \bowtie_{G_b} B \to A \bowtie_{F_b} B)$.*

Proof. If $A, B \subseteq Y$ such that A, B are separated by some $g \in G_b$, then, since $\mathcal{G}$ is bounded-embedded in $\mathcal{F}$, there is some $f \in F_b$ which extends g, hence f separates A and B.

Next we show that the proof of the classical Urysohn extension theorem can be carried out within BISH. Recall that if $x \in \mathbb{R}$, then $x = (x_n)_{n \in \mathbb{N}}$, where $x_n \in \mathbb{Q}$, for every $n \in \mathbb{N}$, such that $\forall_{n, m \in \mathbb{N}+}(|x_m - x_n| \leq m^{-1} + n^{-1})$. Moreover, $x > 0 :\leftrightarrow \exists_{n \in \mathbb{N}}(x_n > \frac{1}{n})$, and $x \geq 0 :\leftrightarrow \forall_{n \in \mathbb{N}}(x_n \geq -\frac{1}{n})$ (see [4], pp. 18–22). If $q \in \mathbb{Q}$, then $q = (q_n)_{n \in \mathbb{N}} \in \mathbb{R}$, where $q_n = q$, for every $n \in \mathbb{N}$. Using MP one shows immediately that $\neg(x \leq -q) \to \neg(x \geq q) \to |x| < q$, where $x \in \mathbb{R}$ and $q \in \mathbb{Q}$. Without MP and completely within BISH, we show that under the same hypotheses one gets that $|x| \leq q$, which is what we need in order to get a constructive proof of the Urysohn extension theorem.

Lemma 2. $\forall_{q \in \mathbb{Q}} \forall_{x \in \mathbb{R}} (\neg(x \leq -q) \to \neg(x \geq q) \to |x| \leq q)$.

Proof. We fix some $q \in \mathbb{Q}$, $x = (x_n)_n \in \mathbb{R}$ and we suppose that $\neg(x \leq -q)$ and $\neg(x \geq q)$. Since $|x| = (\max\{x_n, -x_n\})_{n \in \mathbb{N}}$, we show that $q \geq |x| \leftrightarrow q - |x| \geq 0 \leftrightarrow \forall_n(q - \max\{x_n, -x_n\} \geq -\frac{1}{n})$. If we fix some $n \in \mathbb{N}$, and since $x_n \in \mathbb{Q}$, we consider the following case distinction.

(i) $x_n \geq 0$: Then $q - \max\{x_n, -x_n\} = q - x_n$ and we get that $q - x_n < -\frac{1}{n} \to x_n - q > \frac{1}{n} \to x > q \to x \geq q \to \perp$, by our second hypothesis. Hence, $q - x_n \geq -\frac{1}{n}$.

(ii) $x_n \leq 0$: Then $q - \max\{x_n, -x_n\} = q + x_n$ and we get that $q + x_n < -\frac{1}{n} \to -q - x_n > \frac{1}{n} \to -q > x \to -q \geq x \to \perp$, by our first hypothesis. Hence, $q + x_n \geq -\frac{1}{n}$.

Theorem 5 (Urysohn Extension Theorem for Bishop Spaces). *Suppose that $\mathcal{F} = (X, F)$, $\mathcal{G} = (Y, G)$ are Bishop spaces, $Y \subseteq X$ and $\mathcal{F}$ extends $\mathcal{G}$. If $\forall_{A,B \subseteq Y}(A \bowtie_{G_b} B \to A \bowtie_{F_b} B)$, then $\mathcal{G}$ is bounded-embedded in $\mathcal{F}$.*

Proof. We fix some $g \in G_b$, and let $|g| \leq \overline{M}$, for some natural $M > 0$. In order to find an extension of g in F_b we define a sequence $(g_n)_{n \in \mathbb{N}^+}$, such that $g_n \in G_b$ and

$$|g_n| \leq \overline{3r_n}, \qquad r_n := \frac{M}{2}\left(\frac{2}{3}\right)^n,$$

for every $n \in \mathbb{N}^+$. For $n = 1$ we define $g_1 = g$, and we have that $|g_1| \leq \overline{M} = \overline{3r_1}$. Suppose next that we have defined some $g_n \in G_b$ such that $|g_n| \leq \overline{3r_n}$. We consider the sets

$$A_n = [g_n \leq \overline{-r_n}] = \{y \in Y \mid g_n(y) \leq -r_n\},$$

$$B_n = [g_n \geq \overline{r_n}] = \{y \in Y \mid g_n(y) \geq r_n\}.$$

Clearly, $g_n^*(A_n) = -r_n$ and $g_n^*(B_n) = r_n$, where $g_n^* = (\overline{-r_n} \vee g_n) \wedge \overline{r_n} \in G_b$. Since $g_n^*(A_n) \bowtie_{\mathbb{R}} g_n^*(B_n)$, we get that $A_n \bowtie_{G_b} B_n$, therefore there exists some $f \in F_b$ such that $A_n \bowtie_f B_n$. Without loss of generality we assume that $f_n(A_n) = -r_n$, $f_n(B_n) = r_n$ and $|f_n| \leq \overline{r_n}$. Next we define

$$g_{n+1} := g_n - f_{n|Y} \in G_b,$$

since $\mathcal{F}$ extends $\mathcal{G}$. If $y \in A_n$ we have that

$$|g_{n+1}(y)| = |(g_n - f_{n|Y})(y)| = |g_n(y) - (-r_n)| = |g_n(y) + r_n| \leq 2r_n,$$

since $-3r_n \leq g_n(y) \leq -r_n \to -2r_n \leq g_n(y) + r_n \leq 0$. If $y \in B_n$ we have that

$$|g_{n+1}(y)| = |(g_n - f_{n|Y})(y)| = |g_n(y) - r_n| = g_n(y) - r_n \leq 2r_n,$$

since $r_n \leq g_n(y) \leq 3r_n \to 0 \leq g_n(y) - r_n \leq 2r_n$. Next we show that

$$\forall_{y \in Y}(|g_{n+1}(y)| \leq 2r_n).$$

We fix some $y \in Y$ and we suppose that $|g_{n+1}(y)| > 2r_n$. This implies that $y \notin A_n \cup B_n$, since if $y \in A_n \cup B_n$, then by the previous calculations we get that $|g_{n+1}(y)| \leq 2r_n$, which contradicts our hypothesis. Hence we have that $\neg(g_n(y) \leq -r_n)$ and $\neg(g_n(y) \geq r_n)$. By the Lemma 2 we get that $|g_n(y)| \leq r_n$, therefore $|g_{n+1}(y)| \leq |g_n(y)| + |f_n(y)| \leq r_n + r_n = 2r_n$, which contradicts our assumption $|g_{n+1}(y)| > 2r_n$. Thus we get that $|g_{n+1}(y)| \leq 2r_n$, and since y is arbitrary we get

$$|g_{n+1}| \leq \overline{2r_n} = \overline{3r_{n+1}}.$$

By the condition BS_4 the function $f := \sum_{n=1}^{\infty} f_n$ belongs to F, since the partial sums converge uniformly to f. Note that the infinite sum is well-defined by the Weierstrass comparison test (see [4], p. 32). Note also that

$$(f_1 + \ldots + f_n)_{|Y} = (g_1 - g_2) + (g_2 - g_3) + \ldots + (g_n - g_{n+1}) = g_1 - g_{n+1}.$$

Since $r_n \xrightarrow{n} 0$, we get $g_{n+1} \xrightarrow{n} 0$, hence $f_{|Y} = g_1 = g$. Note that f is also bounded by M:

$$|f| = |\sum_{n=1}^{\infty} f_n| \leq \sum_{n=1}^{\infty} |f_n| \leq \sum_{n=1}^{\infty} \frac{M}{2}(\frac{2}{3})^n = \frac{M}{2} \sum_{n=1}^{\infty} (\frac{2}{3})^n = \frac{M}{2} 2 = M.$$

The main hypothesis of the Urysohn extension theorem

$$\forall_{A,B \subseteq Y} (A \bowtie_{G_b} B \to A \bowtie_{F_b} B)$$

requires quantification over the power set of Y, therefore it is against the practice of predicative constructive mathematics. It is clear though by the above proof that we do not need to quantify over all the subsets of Y, but only over the ones which have the form of A_n and B_n. If we replace the initial main hypothesis by the following

$$\forall_{g,g' \in G_b} \forall_{a,b \in \mathbb{R}} ([g \leq \overline{a}] \bowtie_{G_b} [g' \geq \overline{b}] \to [g \leq \overline{a}] \bowtie_{F_b} [g' \geq \overline{b}]),$$

we get a stronger form of the Urysohn extension theorem, since this is the least condition in order the above proof to work. Actually, this stronger formulation of the Urysohn extension theorem applies to the classical setting too. A slight variation of the previous new main hypothesis, which is probably better to use, is

$$\forall_{g,g' \in G_b} (\zeta(g) \bowtie_{G_b} \zeta(g') \to \zeta(g) \bowtie_{F_b} \zeta(g')),$$

since the sets of the form A_n and B_n are G_b-zero sets.

Definition 9. *If (X, F) is a Bishop space and $Y \subseteq X$ is inhabited, we say that Y is a Urysohn subset of X, if $\forall_{g,g' \in (F_{|Y})_b} (\zeta(g) \bowtie_{(F_{|Y})_b} \zeta(g') \to \zeta(g) \bowtie_{F_b} \zeta(g'))$.*

Next follows a direct corollary of the Theorem 5 and the previous remark.

Corollary 3. *Suppose that $\mathcal{F} = (X, F)$ is a Bishop space, $Y \subseteq X$ is a Urysohn subset of X and $g : Y \to \mathbb{R}$ is in $(F_{|Y})_b$. Then there exists $f : X \to \mathbb{R}$ in F_b which extends g.*

An absolute retract for normal topological spaces is a space that can be substituted for $\mathbb{R}$ in the formulation of the Tietze theorem, according to which a continuous real-valued function on a closed subset of a normal topological space has a continuous extension (see [10], p. 151).

Definition 10. *If Q is a property on sets, a Bishop space $\mathcal{H} = (Z, H)$ is called an absolute retract with respect to Q, or $\mathcal{H}$ is $AR(Q)$, if for every Bishop space $\mathcal{F} = (X, F)$ and $Y \subseteq X$ we have that*

$$Q(Y) \to \forall_{e \in \mathrm{Mor}(\mathcal{F}_{|Y}, \mathcal{H})} \exists_{e^* \in \mathrm{Mor}(\mathcal{F}, \mathcal{H})} (e^*_{|Y} = e).$$

Clearly, the Corollary 3 says that $\mathcal{R}$ is AR(Urysohn). The next proposition shows that there exist many absolute retracts. In particular, the products $\mathcal{R}^n, \mathcal{R}^\infty$ are AR(Urysohn).

Proposition 11. *Suppose that $\mathcal{H}_i = (Z_i, H_i)$ is a Bishop space, for every $i \in I$. Then $\prod_{i \in I} \mathcal{H}_i$ is $AR(Q)$ if and only if $\mathcal{H}_i$ is $AR(Q)$, for every $i \in I$.*

Proof. ($\leftarrow$) If $Y \subseteq X$ such that $Q(Y)$ and if $\mathcal{H}_i$ is $AR(Q)$, for every $i \in I$, then by the Proposition 1(i) we have that

$$e : Y \to \prod_{i \in I} Z_i \in \mathrm{Mor}(\mathcal{F}_{|Y}, \prod_{i \in I} \mathcal{H}_{i \in I}) \leftrightarrow \forall_{i \in I} (\pi_i \circ e \in \mathrm{Mor}(\mathcal{F}_{|Y}, \mathcal{H}_i))$$

$$\to \forall_{i \in I} (\exists_{e^*_i \in \mathrm{Mor}(\mathcal{F}, \mathcal{H}_i)} (e^*_{i|Y} = \pi_i \circ e)).$$

We define $e^* : X \to \prod_{i \in I} Z_i$ by $x \mapsto (e^*_i(x))_{i \in I}$. Clearly, $e^*(y) = e^*_i(y))_{i \in I} = ((\pi_i \circ e)(y))_{i \in I} = e(y)$ and $e^* \in \mathrm{Mor}(\mathcal{F}, \prod_{i \in I} \mathcal{H}_{i \in I})$, by the Proposition 1(i) and the fact that $e^*_i = \pi_i \circ e^* \in \mathrm{Mor}(\mathcal{F}, \mathcal{H}_i)$, for every $i \in I$.

($\to$) Suppose that $\prod_{i \in I} \mathcal{H}_i$ is $AR(Q)$ and $e_i : Y \to Z_i \in \mathrm{Mor}(\mathcal{F}_{|Y}, \mathcal{H}_i)$. If we fix $z = (z_i)_{i \in I} \in \prod_{i \in I} Z_i$, then by the Proposition 2 the function

$$s_i : Z_i \to S(z; i) = Z_i \times \prod_{j \neq i} \{z_j\} \subseteq \prod_{i \in I} Z_i$$

$$z_i \mapsto z_i \times \prod_{j \neq i} \{z_j\}$$

is an isomorphism between $\mathcal{H}_i$ and the slice space $\mathcal{S}(z; i) = (S(z; i), H(z; i))$, where $H(z; i) = (\prod_{i \in I} H_i)_{|S(z;i)}$. Hence, the mapping $s_i \circ e_i : Y \to \prod_{i \in I} Z_i \in \mathrm{Mor}(\mathcal{F}_{|Y}, \prod_{i \in I} \mathcal{H}_{i \in I})$. By our hypothesis there exists some $e^* : X \to \prod_{i \in I} Z_i \in \mathrm{Mor}(\mathcal{F}_{|Y}, \prod_{i \in I} \mathcal{H}_{i \in I})$ which extends $s_i \circ e_i$. Thus, $\pi_i \circ e^* : X \to Z_i \in \mathrm{Mor}(\mathcal{F}, \mathcal{H}_i)$, for every $i \in I$. But $\pi_i \circ e^* = e_i$, since for every $y \in Y$ we have that $(\pi_i \circ e^*)(y) = \pi(e^*(y)) = \pi_i((s_i \circ e_i)(y)) = \pi_i(e_i(y) \times \prod_{j \neq i} \{z_j\}) = e_i(y)$.

5   Concluding Comments

In this paper we presented the basic theory of embeddings of Bishop spaces and we showed that the classical proof of the Urysohn extension theorem for topological spaces generates a constructive proof of the Urysohn extension theorem for Bishop spaces. Our results form only the very beginning of a theory of embeddings of Bishop spaces. If we look at the classical theory of embeddings of rings of continuous functions, we will see too many topics that at first sight it seems difficult, to say the least, to develop constructively. The Stone-Čech compactification and Hewitt's realcompactification depend on the existence of non-trivial ultrafilters, while many facts in the characterizations of the maximal ideals of $C(X)$ or $C^*(X)$ depend on non-constructive formulations of compactness.

Nevertheless, we find encouraging that quite "soon" one can start developing a theory of embeddings within TBS, and also rewarding that non-trivial theorems, like the Urysohn extension theorem, belong to it. Behind these partial "successes" lies, in our view, the function-theoretic character of TBS which offers the direct "communication" between TBS and the theory of $C(X)$ that we mentioned in the Introduction. Maybe, this is the main advantage of TBS with respect to other approaches to constructive topology.

The apartness relations mentioned already here show the connection of TBS with the theory of apartness spaces of Bridges and Vîţă in [6]. Both these theories start from a notion of space that differs from a topological space treated intuitionistically, as in [28] or [12], or from a constructive variation of the notion of a base of a topological space, the starting point of the point-free formal topology of Martin-Löf and Sambin (see [25,26]) and Bishop's theory of neighborhood spaces, as it is developed mainly by Ishihara in [14,15]. In our opinion, if the notion of space in constructive topology "mimics" that of topological space, then it is more difficult to constructivise topology than starting from a notion of space which by its definition is more suitable to constructive study. The function-theoretic character of the notion of Bishop space and of Bishop morphism, in contrast to the set-theoretic character of an apartness space and of a strongly continuous function, seems to facilitate a constructive reconstruction of topology and a possible future translation of TBS to type theory.

References

1. Bishop, E.: Foundations of Constructive Analysis. McGraw-Hill, New York (1967)
2. Bishop, E.: Mathematics as a numerical language. In: Kino, A., Myhill, J., Vesley, R.E. (eds.) Intuitionism and Proof Theory, pp. 53–71. North-Holland, Amsterdam (1970)
3. Bishop, E., Cheng, H.: Constructive measure theory. Mem. Amer. Math. Soc. **116** (1972)
4. Bishop, E., Bridges, D.: Constructive Analysis, Grundlehren der Math. Wissenschaften 279. Springer, Heidelberg (1985)
5. Bridges, D.S., Richman, F.: Varieties of Constructive Mathematics. Cambridge University Press, Cambridge (1987)

6. Bridges, D., Vîţă, L.S.: Apartness and Uniformity: A Constructive Development. CiE Series Theory and Applications of Computability. Springer, Heidelberg (2011)
7. Bridges, D.: Reflections on function spaces. Ann. Pure Appl. Logic **163**, 101–110 (2012)
8. Brouwer, L.E.J.: Zur Begründung der intuitionistische Mathematik III. Math. Ann. **96**, 451–488 (1926)
9. Buchholz, W., Feferman, S., Pohlers, W., Sieg, W.: Iterated Inductive Definitions and Subsystems of Analysis: Recent Proof-Theoretic Studies. LNM, vol. 897. Springer, Heidelberg (1981)
10. Dugundji, J.: Topology. Wm. C. Brown Publishers, Dubuque (1989)
11. Gillman, L., Jerison, M.: Rings of Continuous Functions. Van Nostrand, Princeton (1960)
12. Grayson, R.J.: Concepts of general topology in constructive mathematics and in sheaves. Ann. Math. Logic **20**, 1–41 (1981)
13. Fréchet, M.: Sur quelques points du calcul fonctionnel. Rend. Circ. Mat. di Palermo **22**, 1–74 (1906)
14. Ishihara, H., Mines, R., Schuster, P., Vîţă, L.S.: Quasi-apartness and neighborhood spaces. Ann. Pure Appl. Logic **141**, 296–306 (2006)
15. Ishihara, H.: Two subcategories of apartness spaces. Ann. Pure Appl. Logic **163**, 132–139 (2013)
16. Ishihara, H.: Relating Bishop's function spaces to neighborhood spaces. Ann. Pure Appl. Logic **164**, 482–490 (2013)
17. Martin-Löf, P.: Notes on Constructive Mathematics. Almqvist and Wiksell, Stockholm (1968)
18. Martin-Löf, P.: Intuitionistic Type Theory: Notes by Giovanni Sambin on a Series of Lectures Given in Padua, June 1980. Bibliopolis, Napoli (1984)
19. Martino, E., Giaretta, P.: Brouwer Dummett, and the bar theorem. In: Bernini, S. (ed.) Atti del Congresso Nazionale di Logica, Montecatini Terme, 1–5 Ottobre 1979, Napoli, pp. 541–558 (1981)
20. Myhill, J.: Constructive set theory. J. Symbolic Logic **40**, 347–382 (1975)
21. Palmgren, E.: From intuitionistic to point-free topology: on the foundations of homotopy theory. In: Lindström, S., et al. (eds.) Logicism, Intuitionism, and Formalism. Synthese Library, vol. 341, pp. 237–253. Springer, Heidelberg (2009)
22. Petrakis, I.: Bishop spaces: constructive point-function topology, In: Mathematisches Forschungsinstitut Oberwolfach Report No. 52/2014, Mathematical Logic: Proof Theory, Constructive Mathematics, pp. 26–27 (2014)
23. Petrakis, I.: Completely regular bishop spaces. In: Beckmann, A., Mitrana, V., Soskova, M. (eds.) CiE 2015. LNCS, vol. 9136, pp. 302–312. Springer, Heidelberg (2015)
24. Petrakis, I.: Constructive topology of Bishop spaces. Ph.D. thesis, LMU Munich (2015)
25. Sambin, G.: Intuitionistic formal spaces - a first communication. In: Skordev, D. (ed.) Mathematical Logic and its Applications, pp. 187–204. Plenum Press, New York (1987)
26. Sambin, G.: The Basic Picture: Structures for Constructive Topology. Oxford University Press (in press)
27. Spanier, E.: Quasi-topologies. Duke Math. J. **30**(1), 1–14 (1963)
28. Troelstra, A.S.: Intuitionistic general topology. Ph.D. thesis, Amsterdam (1966)

An Arithmetical Interpretation of Verification and Intuitionistic Knowledge

Tudor Protopopescu[(✉)]

The Graduate Center, City University of New York, New York, USA
`tprotopopescu@gradcenter.cuny.edu`

Abstract. Intuitionistic epistemic logic introduces an epistemic operator, which reflects the intended BHK semantics of intuitionism, to intuitionistic logic. The fundamental assumption concerning intuitionistic knowledge and belief is that it is the product of verification. The BHK interpretation of intuitionistic logic has a precise formulation in the Logic of Proofs and its arithmetical semantics. We show here that this interpretation can be extended to the notion of verification upon which intuitionistic knowledge is based, thereby providing the systems of intuitionistic epistemic logic extended by an epistemic operator based on verification with an arithmetical semantics too.

Keywords: Intuitionistic epistemic logic · Logic of proofs · Arithmetic interpretation · Intuitionistic knowledge · BHK semantics · Verification

1 Introduction

The intended semantics for intuitionistic logic is the Brouwer-Heyting-Kolmogorov (BHK) interpretation, which holds that a proposition is true if proved. The systems of intuitionistic epistemic logic, the IEL family introduced in [5], extend intuitionistic logic with an epistemic operator and interpret it in a manner reflecting the BHK semantics. The fundamental assumption concerning knowledge interpreted intuitionistically is that knowledge is the product of verification, where a verification is understood to be a justification sufficient to warrant a claim to knowledge which is not necessarily a strict proof.

In [5] the notion of verification was treated intuitively. Here we show that verification can also be given an arithmetical interpretation, thereby showing that the notion of verification assumed in an intuitionistic interpretation of knowledge has an exact model.

Following Gödel [11] it is well known that intuitionistic logic can be embedded into the classical modal logic S4 regarded as a provability logic. Artemov [2] formulated the Logic of Proofs, LP, and showed that S4 in turn can be interpreted in LP, and that LP has an arithmetical interpretation as a calculus of

---

T. Protopopescu—Many thanks to Sergei Artemov for helpful suggestions and inspiring discussions.

explicit proofs in Peano Arithmetic PA.[1] Accordingly this makes precise the BHK semantics for intuitionistic logic. Intuitionistic logic, then, can be regarded as an implicit logic of proofs, and its extension with an epistemic/verification operator in the systems IEL⁻ and IEL (given in Sect. 2) can be regarded as logics of implicit proofs, verification and their interaction.

This is of interest for a number of reasons. It shows that the notion of verification on which intuitionistic epistemic logic is based is coherent and can be made concrete, and does so in a manner consonant with the intended BHK interpretation of the epistemic operator. Further, given intuitionistic logic's importance in computer science as well as the need for a constructive theory of knowledge, finding a precise provability model for verification and intuitionistic epistemic logic (see Sect. 5) is well-motivated.

2   Intuitionistic Epistemic Logic

According to the BHK semantics a proposition, A, is true if there is a proof of it and false if the assumption that there is a proof of A yields a contradiction. This is extended to complex propositions by the following clauses:

- a proof of $A \wedge B$ consists in a proof of A and a proof of B;
- a proof of $A \vee B$ consists in giving either a proof of A or a proof of B;
- a proof of $A \rightarrow B$ consists in a construction which given a proof of A returns a proof of B;
- $\neg A$ is an abbreviation for $A \rightarrow \bot$, and $\bot$ is a proposition that has no proof.

The salient property of verification-based justification, in the context of the BHK semantics, is that it follows from intuitionistic truth, hence

$$A \rightarrow \mathbf{K}A \qquad \qquad \text{(Co-Reflection)}$$

is valid on a BHK reading. Since any proof is a verification, the intuitionistic truth of a proposition yields that the proposition is verified.

By similar reasoning the converse principle

$$\mathbf{K}A \rightarrow A \qquad \qquad \text{(Reflection)}$$

is not valid on a BHK reading. A verification need not be, or yield a method for obtaining, a proof, hence does not guarantee the intuitionistic truth of a proposition. Reflection expresses the factivity of knowledge in a classical language, intuitionistically factivity is expressed by

$$\mathbf{K}A \rightarrow \neg\neg A. \qquad \qquad \text{(Intuitionistic Factivity)}$$

The basic system of intuitionistic epistemic logic, incorporating minimal assumptions about the nature of verification, is the system IEL⁻. IEL⁻ can be seen as the system formalising intuitionistic belief.

---

[1] As opposed to provability in PA, the calculus of which is the modal logic GL, see [6].

Definition 1 (IEL⁻). *The list of axioms and rules of* IEL⁻ *consists of:*

IE0. Axioms of propositional intuitionistic logic.
IE1. $\mathbf{K}(A \to B) \to (\mathbf{K}A \to \mathbf{K}B)$
IE2. $A \to \mathbf{K}A$

Modus Ponens.

It is consistent with IEL⁻ that false propositions can be verified. It is desirable, however, that false propositions not be verifiable; to be a logic of knowledge the logic should reflect the truth condition on knowledge, i.e. factivity – that it is not possible to know falsehoods. The system IEL incorporates the truth condition and hence can be viewed as an intuitionistic logic of knowledge.

Definition 2 (IEL). *The list of axioms and rules for* IEL *are those for* IEL⁻ *with the additional axiom:*

IE3. $\mathbf{K}A \to \neg\neg A.$

Given Axiom IE2 the idea that it is not possible to know a falsehood can be equivalently expressed by $\neg\mathbf{K}\bot$.[2] For the following we will use this form of the truth condition in place of Axiom IE3.

Kripke models were defined for both systems, and soundness and completeness shown with respect to them, see [5].

3 Embedding Intuitionistic Epistemic Logic into Classical Modal Logic of Verification

The well known Gödel translation yields a faithful embedding of the intuitionistic propositional calculus, IPC, into the classical modal logic S4.[3] By extending S4 with a verification modality $\mathbf{V}$, the embedding can be extended to IEL⁻ and IEL, and shown to remain faithful, see [14].

S4V⁻ is the basic logic of provability and verification.

Definition 3 (S4V⁻ **Axioms**). *The list of axioms and rules of* S4V⁻ *consists of:*

A0. The axioms of S4 *for* $\Box$.
A1. $\mathbf{V}(A \to B) \to (\mathbf{V}A \to \mathbf{V}B)$
A2. $\Box A \to \mathbf{V}A$

---

[2] Or indeed, $\neg(\mathbf{K}A \wedge \neg A)$, $\neg A \to \neg\mathbf{K}A$ or $\neg\neg(\mathbf{K} \to A)$, all are equivalent to Axiom IE3 given Axiom IE2, see [5].

[3] The soundness of the translation was proved by Gödel [11] while the faithfulness was proved by McKinsey and Tarski [12]. See [8] for a semantic, and [15] for a syntactic proof.

R1. Modus Ponens

R2. □-Necessitation $\frac{\vdash A}{\vdash \Box A}$

As with IEL we add the further condition that verifications should be consistent.

Definition 4 (S4V). S4V *is* S4V$^-$ *with the additional axiom:*[4]

A3. $\neg\Box\mathbf{V}\bot$.

Kripke models for each system were outlined in [14] and the systems shown to be sound and complete with respect to them.

For IEL$^-$ and IEL their embedding into S4V$^-$ and S4V respectively, is faithful. For an IEL$^-$ or IEL formula F, $tr(F)$ is the translation of F according to the rule

box every sub-formula

into the language of S4V$^-$ or S4V respectively.

Theorem 1 (Embedding). *The Gödel translation faithfully embeds* IEL$^-$ *and* IEL *into* S4V$^-$ *and* S4V, *respectively:*

$$\text{IEL}^-, \text{IEL} \vdash F \Leftrightarrow \text{S4V}^-, \text{S4V} \vdash tr(F).$$

Proof. See [14]. □

4   Logics of Explicit Proofs and Verification

Gödel [11] suggested that the modal logic S4 be considered as a provability calculus. This was given a precise interpretation by Artemov, see [2,4], who showed that explicit proofs in Peano Arithmetic, PA, was the model of provability which S4 described. The explicit counter-part of S4 is the Logic of Proofs LP in which each □ in S4 is replaced by a term denoting an explicit proof. Since intuitionistic logic embeds into S4 the intended BHK semantics for IPC as an implicit calculus of proofs is given an explicit formulation in LP, and hence an arithmetical semantics. Here we show that this arithmetical interpretation can be further extended to the Logic of Proofs augmented with a verification modality, providing S4V$^-$ and S4V, and therefore IEL$^-$ and IEL with an arithmetical semantics. Similarly to the foundational picture regarding the relation between IPC, S4 and LP (see [2]) we have that[5]

$$\text{IEL} \hookrightarrow \text{S4V} \hookrightarrow \text{LPV}$$

The basic system of explicit proofs and verifications LPV$^-$ is defined thus:

---

[4] [14] presented a stronger version of S4V with $\neg\mathbf{V}\bot$ instead of $\neg\Box\mathbf{V}\bot$. The weaker axiom presented here is sufficient for the embedding; one can readily check that the Gödel translation of $\neg\mathbf{K}\bot$, $\Box\neg\Box\mathbf{V}\Box\bot$, is derivable in S4V as formulated here. The weaker axiom allows for a uniform arithmetical interpretation of verification.

[5] Similar embeddings hold for IEL$^-$, S4V$^-$, and LPV$^-$.

Definition 5 (Explicit Language). *The language of* LPV⁻ *consists of:*

1. *The language of classical propositional logic;*
2. *A verification operator* **V**;
3. Proof variables, *denoted by* $x, y, x_1, x_2 \ldots$;
4. Proof constants, *denoted by* $a, b, c, c_1, c_2 \ldots$;
5. Operations on proof terms, *building complex proof terms from simpler ones of three types:*
    (a) *Binary operation* · *called* application;
    (b) *Binary operation* + *called* plus;
    (c) *Unary operation* ! *called* proof checker;
6. Proof terms: *any proof variable or constant is a proof term; if t and s are proof terms so are* $t \cdot s$, $t + s$ *and* !t.
7. Formulas: *A propositional letter p is a formula; if A and B are formulas then so are* ¬A, $A \wedge B$, $A \vee B$, $A \to B$, **V**A, t:A.

Formulas of the type $t{:}A$ are read as "*t is a proof A*".

Definition 6 (LPV⁻). *The list of axioms and rules of* LPV⁻ *consists of:*

E0. *Axioms of propositional classical logic.*
E1. $t{:}(A \to B) \to (s{:}A \to (t \cdot s){:}B)$
E2. $t{:}A \to A$
E3. $t{:}A \to !t{:}t{:}A$
E4. $t{:}A \to (s + t){:}A$, $t{:}A \to (t + s){:}A$

E5. $\mathbf{V}(A \to B) \to (\mathbf{V}A \to \mathbf{V}B)$
E6. $t{:}A \to \mathbf{V}A$

R1. *Modus Ponens*
R2. *Axiom Necessitation:* $\frac{\vdash A}{\vdash c{:}A}$ *where A is any of Axioms E0 to E6 and c is some proof constant.*

Definition 7 (LPV). *The system* LPV *is* LPV⁻ *with the additional axiom:*

E7. $\neg t{:}\mathbf{V}\bot$

A *constant specification*, $\mathcal{CS}$, is a set $\{c_1{:}A_1, c_2{:}A_2 \ldots \}$ of formulas such that each A_i is an axiom from the list E0 to E6 above, and each c_i is a proof constant. This set is generated by each use of the constant necessitation rule in an LPV⁻ proof. The axiom necessitation rule can be replaced with a 'ready made' constant specification which is added to LPV⁻ as a set of extra axioms. For such a $\mathcal{CS}$ let LPV⁻-$\mathcal{CS}$ mean LPV⁻ minus the axiom necessitation rule plus the members of $\mathcal{CS}$ as additional axioms.

A proof term, t, is called a *ground term* if it contains no proof variables, but is built only from proof constants and operations on those constants.

LPV⁻ and LPV are able to internalise their own proofs, that is if

$$A_1 \ldots A_n, y_1{:}B_1 \ldots y_n{:}B_n \vdash F$$

then for some term $p(x_1 \ldots x_n, y_1 \ldots y_n)$

$$x_1{:}A_1 \ldots x_n{:}A_n, y_1{:}B_1 \ldots y_n{:}B_n \ \vdash p(x_1 \ldots x_n, y_1 \ldots y_n){:}F,$$

see [2]. As a consequence LPV$^-$ and LPV have the constructive necessitation rule: for some ground proof term t,

$$\frac{\vdash F}{\vdash t{:}F.}$$

This yields in turn:

Lemma 1 (V Necessitation). **V**-*Necessitation* $\frac{\vdash A}{\vdash \mathbf{V}A}$ *is derivable in* LPV$^-$ *and* LPV.

Proof. Assume $\vdash A$, then by constructive necessitation $\vdash t{:}A$ for some ground proof term t, hence by Axiom E6 $\vdash \mathbf{V}A$.

Note that the Deduction Theorem holds for both LPV$^-$ and LPV.

5   Arithmetical Interpretation of LPV$^-$ and LPV

We give an arithmetical interpretation of LPV$^-$ and LPV by specifying a translation of the formulas of LPV$^-$ and LPV into the language of Peano Arithmetic, PA. We assume that a coding of the syntax of PA is given. n denotes a natural number and $\overline{n}$ the corresponding numeral. $\ulcorner F \urcorner$ denotes the numeral of the Gödel number of a formula F. For readability we suppress the overline for numerals and corner quotes for the Gödel number of formulas, and trust that the appropriate number or numeral, as context requires, can be recovered.[6]

Definition 8 (Normal Proof Predicate). *A normal proof predicate is a provably Δ formula* Prf *(x, y) such that for every arithmetical sentence F the following holds:*

1. *PA $\vdash F \Leftrightarrow$ for some $n \in \omega$, $\mathsf{Prf}(n, F)$*
2. *A proof proves only a finite number of things; i.e. for every k the set $T(k) = \{l \,|\, \mathsf{Prf}(k, l)\}$ is finite.[7]*
3. *Proofs can be joined into longer proofs; i.e. for any k and l there is an n s.t. $T(k) \cup T(l) \subseteq T(n)$.*

Example 1. An example of a numerical relation that satisfies the definition of $\mathsf{Prf}(x, y)$ is the standard proof predicate $\mathsf{Proof}(x, y)$ the meaning of which is

"x is the Gödel number of a derivation of a formula with the Gödel number y".

---

[6] E.g. by techniques found in [6,9].
[7] I.e. $T(k)$ is the set of theorems proved by the proof k.

Theorem 2. *For every normal proof predicate* $\mathsf{Prf}(x, y)$ *there exist recursive functions* $m(x, y)$, $a(x, y)$ *and* $c(x)$ *such that for any arithmetical formulas F and G and all natural numbers k and n the following formulas hold:*

1. $(\mathsf{Prf}(k, F \to G) \wedge \mathsf{Prf}(n, F)) \to \mathsf{Prf}(m(k, n), G)$
2. $\mathsf{Prf}(k, F) \to \mathsf{Prf}(a(k, n), F), \quad \mathsf{Prf}(n, F) \to \mathsf{Prf}(a(k, n), F)$
3. $\mathsf{Prf}(k, F) \to \mathsf{Prf}(c(k), \mathsf{Prf}(k, F))$.

Proof. See [2]. $\qquad$

Definition 9 (Verification Predicate for LPV⁻). *A* verification predicate *is a provably Σ formula* $\mathsf{Ver}(x)$ *satisfying the following properties, for arithmetical formulas F and G:*

1. $\mathsf{PA} \vdash \mathsf{Ver}(F \to G) \to (\mathsf{Ver}(F) \to \mathsf{Ver}(G))$
2. *For each* n, $\mathsf{PA} \vdash \mathsf{Prf}(n, F) \to \mathsf{Ver}(F)$.

These are properties which a natural notion of verification satisfies.

Let $\mathsf{Bew}(x)$ be the standard provability predicate, and $\mathsf{Con(PA)}$ be the statement which expresses that PA is consistent, i.e. $\neg\mathsf{Bew}(\bot)$. $\neg\mathsf{Con(PA)}$ correspondingly is $\mathsf{Bew}(\bot)$.

Example 2. The following are examples of a verification predicate $\mathsf{Ver}(x)$:

1. "Provability in PA", i.e. $\mathsf{Ver}(x) = \mathsf{Bew}(x)$; for a formula F $\mathsf{Ver}(F)$ is $\exists x$ $\mathsf{Prf}(x, F)$.
2. "Provability in PA + Con(PA)" i.e. $\mathsf{Ver}(x) = \mathsf{Bew}(\mathsf{Con(PA)} \to x)$; one example of a formula for which $\mathsf{Ver}(x)$ holds in this sense is just the formula $\mathsf{Con(PA)}$. Such verification is capable of verifying propositions not provable in PA.
3. "Provability in PA + ¬Con(PA)" i.e. $\mathsf{Ver}(x) = \mathsf{Bew}(\neg\mathsf{Con(PA)} \to x)$; an example of a verifiable formula which is not provable in PA, is the formula $\neg\mathsf{Con(PA)}$. Such verification is capable of verifying false propositions.
4. $\top$, i.e. $\mathsf{Ver}(x) = \top$; that is for any formula F $\mathsf{Ver}(F) = \top$, hence any F is verified.

Lemma 2. $\mathsf{PA} \vdash F \;\Rightarrow\; \mathsf{PA} \vdash \mathsf{Ver}(F)$.

Proof. Assume $\mathsf{PA} \vdash F$, then by Definition 8 there is an n such that $\mathsf{Prf}(n, F)$ is true, hence $\mathsf{PA} \vdash \mathsf{Prf}(n, F)$, and by Definition 9 part 2 $\mathsf{PA} \vdash \mathsf{Ver}(F)$. $\qquad$

We now define an interpretation of the language of LPV⁻ into the language of Peano Arithmetic. An arithmetical interpretation takes a formula of LPV⁻ and returns a formula of Peano Arithmetic; we show the soundness of such an interpretation, if F is valid in LPV⁻ then for any arithmetical interpretation $*$ F^* is valid in PA.[8]

---

[8] A corresponding completeness theorem is left for future work, as is the development of a system with explicit verification terms, in addition to proof terms, realising the verification modality of S4V⁻ or S4V.

Definition 10 (Arithmetical Interpretation for LPV⁻). *An* arithmetical interpretation *for* LPV⁻ *has the following items:*

- *A normal proof predicate,* Prf, *with the functions* $m(x,y)$, $a(x,y)$ *and* $c(x)$ *as in Definition 8 and Theorem 2;*
- *A verification predicate,* Ver, *satisfying the conditions in Definition 9;*
- *An evaluation of propositional letters by sentences of* PA;
- *An evaluation of proof variables and constants by natural numbers.*

An arithmetical interpretation is given inductively by the following clauses:

$$(p)^* = p \text{ an atomic sentence of PA} \qquad (t \cdot s)^* = m(t^*, s^*)$$
$$\bot^* = \bot \qquad\qquad\qquad (t + s)^* = a(t^*, s^*)$$
$$(A \wedge B)^* = A^* \wedge B^* \qquad\qquad (!t)^* = c(t^*)$$
$$(A \vee B)^* = A^* \vee B^* \qquad\qquad (t{:}F)^* = \mathsf{Prf}(t^*, F^*)$$
$$(A \to B)^* = A^* \to B^* \qquad\qquad (\mathbf{V}F)^* = \mathsf{Ver}(F^*)$$

Let X be a set of LPV⁻ formulas, then X^* is the set of all F^*'s such that $F \in X$. For a constant specification, $\mathcal{CS}$, a *$\mathcal{CS}$-interpretation* is an interpretation * such that all formulas from $\mathcal{CS}^*$ are true. An LPV⁻ formula is *valid* if F^* is true under all interpretations *. F is *provably valid* if PA $\vdash F^*$ under all interpretations *. Similarly, F is *valid under constant specification* $\mathcal{CS}$ if F^* is true under all $\mathcal{CS}$-interpretations, and F is *provably valid under constant specification* $\mathcal{CS}$ if PA $\vdash F^*$ under any $\mathcal{CS}$-interpretation *.

Theorem 3 (Arithmetical Soundness of LPV⁻). *For any $\mathcal{CS}$-interpretation * with a verification predicate as in Definition 9 any LPV⁻-$\mathcal{CS}$ theorem, F, is provably valid under constant specification $\mathcal{CS}$:*

$$\text{LPV}^-\text{-}\mathcal{CS} \vdash F \;\Rightarrow\; \text{PA} \vdash F^*.$$

Proof. By induction on derivations in LPV⁻. The cases of the LP axioms are proved in [2].

Case 1. $(\mathbf{V}(A \to B) \to (\mathbf{V}A \to \mathbf{V}B))$.

$$[\mathbf{V}(A \to B) \to (\mathbf{V}A \to \mathbf{V}B)]^* \equiv \mathsf{Ver}(F \to G) \to (\mathsf{Ver}(F) \to \mathsf{Ver}(G)).$$

But PA $\vdash \mathsf{Ver}(F \to G) \to (\mathsf{Ver}(F) \to \mathsf{Ver}(G))$ by Definition 9.

Case 2. $(t{:}F \to \mathbf{V}F)$.

$$[t{:}F \to \mathbf{V}F]^* \equiv \mathsf{Prf}(t^*, F^*) \to \mathsf{Ver}(F^*).$$

Likewise PA $\vdash \mathsf{Prf}(t^*, F^*) \to \mathsf{Ver}(F^*)$ holds by Definition 9.

This arithmetical interpretation can be extended to LPV. Everything is as above except to Definition 9 we add the following item:

Definition 11 (Verification Predicate for LPV).

1. for any n, $\mathsf{PA} \vdash \neg\mathsf{Prf}(n, \mathsf{Ver}(\bot))$.

1–3 of Example 2 remain examples of a verification predicate which also satisfies the above consistency property. In each case respectively $\mathsf{Ver}(\bot)$ is

1. $\mathsf{Bew}(\bot)$
2. $\mathsf{Bew}(\neg\mathsf{Bew}(\bot) \to \bot)$, i.e. $\mathsf{Bew}(\neg\mathsf{Con}(\mathsf{PA}))$
3. $\mathsf{Bew}(\neg\neg\mathsf{Bew}(\bot) \to \bot)$, i.e. $\mathsf{Bew}(\mathsf{Con}(\mathsf{PA}))$.

All of these are false in the standard model of PA, and hence not provable in PA, hence for each n $\mathsf{PA} \vdash \neg\mathsf{Prf}(n, \mathsf{Ver}(\bot))$.

4. $\mathsf{Ver}(\bot) = \top$, is not an example of a verification predicate for LPV in the sense of Definition 11: $\mathsf{Ver}(\bot)$ would be provable in PA, and hence there would be an n for which $\mathsf{PA} \vdash \mathsf{Prf}(n, \mathsf{Ver}(\bot))$ holds, which contradicts Definition 11.

Theorem 4 (Arithmetical Soundness of LPV). *For any $\mathcal{CS}$-interpretation** *with a verification predicate as in Definition 11, if F is an LPV-$\mathcal{CS}$ theorem then it is provably valid under constant specification $\mathcal{CS}$:*

$$\mathsf{LPV}\text{-}\mathcal{CS} \vdash F \;\Rightarrow\; \mathsf{PA} \vdash F^*.$$

Proof. Add to the proof of Theorem 3 the following case:

Case 3. $(\neg t{:}\mathbf{V}\bot)$

$$[\neg t{:}\mathbf{V}\bot]^* \equiv \neg\mathsf{Prf}(n, \mathsf{Ver}(\bot)).$$

$\mathsf{PA} \vdash \neg\mathsf{Prf}(n, \mathsf{Ver}(\bot))$ *holds by Definition 11.*

6   Sequent Systems for $\mathsf{S4V}^-$ and $\mathsf{S4V}$

We give a sequent formulation of $\mathsf{S4V}^-$ and $\mathsf{S4V}$. We will denote the sequent formulations by $\mathsf{S4V}^-\mathsf{g}$, $\mathsf{S4Vg}$ respectively.

A sequent is a figure, $\Gamma \Rightarrow \Delta$, in which Γ, Δ are multi-sets of formulas.

Definition 12 ($\mathsf{S4V}^-\mathsf{g}$). *The axioms for the system $\mathsf{S4V}^-\mathsf{g}$ are:*

Axioms

$$P \Rightarrow P, P atomic \quad \bot \Rightarrow$$

*The structural and propositional rules are those of the system **G1c** from [15]. The modal rules are:*

$\Box$-*Rules*

$$\frac{\Gamma, X \Rightarrow \Delta}{\Gamma, \Box X \Rightarrow \Delta}(\Box \Rightarrow) \frac{\Box\Gamma \Rightarrow X}{\Box\Gamma \Rightarrow \Box X}(\Rightarrow \Box)$$

$\mathbf{V}$-*Rule*

$$\frac{\Gamma \Rightarrow X}{\mathbf{V}\Gamma \Rightarrow \mathbf{V}X}(\Rightarrow \mathbf{V})$$

Interaction-Rule

$$\frac{\Gamma, \mathbf{V}X \Rightarrow \Delta}{\Gamma, \Box X \Rightarrow \Delta}(\mathbf{V}/\Box \Rightarrow)$$

Definition 13 (S4Vg **Rules**)**.** *The system* S4Vg *is the system* S4V$^-$g *with the additional axiom:*
$\Box$**V**-*Axiom*

$$\Box \mathbf{V}\bot \Rightarrow$$

Soundness can be shown by induction on the rules of S4V$^-$g and S4Vg. Completeness and cut-elimination can be shown in a manner similar to that of [3].[9]

7   Realisation of S4V$^-$ and S4V

Here we show that each $\Box$ in an S4V$^-$ or S4V theorem can be replaced with a proof term so that the result is a theorem of LPV$^-$ or LPV, and hence that IEL$^-$ and IEL each have a proof interpretation. The converse, that for each LPV$^-$ or LPV theorem if all the proof terms are replaced with $\Box$'s the result is a theorem of S4V$^-$ or S4V also holds.

Definition 14 (Forgetful Projection). *The* forgetful projection, F^0 *of an* LPV$^-$ *or* LPV *formula is the result of replacing each proof term in F with a* $\Box$.

Theorem 5. LPV$^-$, LPV $\vdash F \Rightarrow$ S4V$^-$, S4V $\vdash F^0$ *respectively.*

Proof. By induction on S4V$^-$ derivations. The forgetful projections of Axioms E1 to E4 and E6 are $\Box(A \to B) \to (\Box A \to \Box B)$, $\Box A \to A$, $\Box A \to \Box\Box A$, $\Box A \to \Box A$ and $\Box A \to \mathbf{V}A$ respectively, which are all provable in S4V$^-$. The forgetful projection of $\neg t{:}\mathbf{V}\bot$ is $\neg\Box\mathbf{V}\bot$ which is provable in S4V. The rules are obvious.

Definition 15 (Realisation). *A* realisation, F^r, *of an* S4V$^-$ *or* S4V *formula F is the result of substituting a proof term for each* $\Box$ *in F, such that if* S4V$^-$, S4V $\vdash$ *F then* LPV$^-$, LPV $\vdash F^r$ *respectively.*

Definition 16 (Polarity of Formulas). *Occurrences of* $\Box$ *in F in $G \to F$,* $F \wedge G$, $G \wedge F$, $F \vee G$, $G \vee F$, $\Box G$ *and* $\Gamma \Rightarrow \Delta, F$ *have the same polarity as the occurrence of* $\Box$ *in F.*

Occurrences of $\Box$ *from $F \to G$, $\neg F$ and $F, \Gamma \Rightarrow \Delta$ have the polarity opposite to that of the occurrence of* $\Box$ *in F.*

Definition 17 (Normal Realisation). *A* realisation *r is called* normal *if all negative occurrences of* $\Box$ *are realised by proof variables.*

---

[9] See also [13] for another example of the method.

The informal reading of the S4 provability modality $\Box$ is existential, $\Box F$ means 'there is a proof of F' (as opposed to the Kripke semantic reading which is universal, i.e. 'F holds in all accessible states'), normal realisations are the ones which capture this existential meaning, see [2].

The realisation Theorem 6, shows that if a formula F is a theorem of S4V$^-$ then there is a substitution of proof terms for every $\Box$ occurring in F such that the result is a theorem of LPV$^-$. This means that every $\Box$ in S4V$^-$ can be thought of as standing for a (possibly complex) proof term in LPV$^-$, and hence, by Theorem 3, implicitly represents a specific proof in PA. The proof of the realisation theorem consists in a procedure by which such a proof term can be built, see [1, 2, 7, 10]. Given a (cut-free) proof in S4V$^-$g we show how to assign proof terms to each of the $\Box$'s occurring in the S4V$^-$g proof so that each sequent in the proof corresponds to a formula provable in LPV$^-$; this is done by constructing a Hilbert-style LPV$^-$ proof for the formula corresponding to each sequent, so as to yield the desired realisation.

Occurrences of $\Box$ in an S4V$^-$g derivation can be divided up into *families* of related occurrences. Occurrences of $\Box$ are related if they occur in related formulas of premises and conclusions of rules. A family of related occurrences is given by the transitive closure of such a relation. A family is called *essential* if it contains at least one occurrence of $\Box$ which is introduced by the ($\Rightarrow \Box$) rule. A family is called *positive (respectively negative)* if it consists of positive (respectively negative) occurrences of $\Box$. It is important to note that the rules of S4V$^-$g preserve the polarities of $\Box$. Any $\Box$ introduced by ($\Rightarrow \Box$) is positive, while $\Box$'s introduced by ($\Box \Rightarrow$), the interaction rule, and by the $\Box$**V** axiom in the case of S4V are negative.

Theorem 6 (S4V$^-$ Realisation). *If* S4V$^-$ $\vdash F$ *then* LPV$^-$ $\vdash F^r$ *for some normal realisation* r.

Proof. If S4V$^-$g $\vdash F$ then there exists a cut-free sequent proof, $\mathcal{S}$, of the sequent $\Rightarrow F$. The realisation procedure described below (following [1,2]) describes how to construct a normal realisation r for any sequent in $\mathcal{S}$.

Step 1. In every negative family and non-essential positive family replace each occurrence of $\Box B$ by $x{:}B$ for a fresh proof variable x.

Remark 1. Note that this means a $\Box$ introduced by means of the ($\Rightarrow$, **V**) rule (or additionally the axiom $\Box$**V**$\bot \Rightarrow$ in the case of LPV below) is realised by a proof variable.

Step 2. Pick an essential family, f, and enumerate all of the occurrences of the rule ($\Rightarrow \Box$) which introduce $\Box$'s in this family. Let n_f be the number of such introductions. Replace all $\Box$'s of family f by the proof term $v_1 + \ldots + v_{n_f}$ where v_i does not already appear as the result of a realisation. Each v_i is called a *provisional* variable which will later be replaced with a proof term.

After this step has been completed for all families of $\Box$ there are no $\Box$'s left in $\mathcal{S}$.

Step 3. This proceeds by induction on the depth of a node in S. For each sequent in S we show how to construct an LPV^- formula, F^r, corresponding to that sequent, such that $\mathsf{LPV}^- \vdash F^r$.

The realisation of a sequent $\mathcal{G} = \Gamma \Rightarrow \Delta$ is an LPV^- formula, $\mathcal{G}^r$, of the following form:

$$A_1^r \wedge \ldots \wedge A_n^r \to B_1^r \vee \ldots \vee B_m^r$$

The A^r's and B^r's denote realisations already performed. Let Γ^r stand for conjunctions of formulas and Δ^r for disjunctions of formulas; Γ^r prefixed with a $\mathbf{V}$ stands for conjunctions of $\mathbf{V}$'ed formulas, i.e. $\mathbf{V}\Gamma_n^r = \mathbf{V}A_1 \wedge \ldots \wedge \mathbf{V}A_n$.

The cases realising the rules involving the propositional connectives and $\square$ are shown in [2][10] (including how to replace provisional variables with terms). Let us check the rules involving $\mathbf{V}$.

Case 3. [Sequent $\mathcal{G}$ is the conclusion of a $(\Rightarrow \mathbf{V})$ rule: $\mathbf{V}\Gamma \Rightarrow \mathbf{V}X$]

$$\mathcal{G}^r = \mathbf{V}\Gamma_n^r \to \mathbf{V}X^r.$$

Now $\mathsf{LPV}^- \vdash (\Gamma_n^r \to X^r) \Rightarrow \mathsf{LPV}^- \vdash (\mathbf{V}\Gamma_n^r \to \mathbf{V}X^r)$, hence by the induction hypothesis the realisation of the premise of the rule, $\Gamma_n^r \to X^r$, is provable in LPV^-, and hence:

$$\mathsf{LPV}^- \vdash \mathbf{V}\Gamma_n^r \to \mathbf{V}X^r.$$

Case 4. [Sequent $\mathcal{G}$ is the conclusion of an interaction rule $(\mathbf{V}/\square \Rightarrow)$: $\Gamma, \square X \Rightarrow \Delta$]

$$\mathcal{G}^r = (\Gamma_n^r \wedge x{:}X^r) \to \Delta_m^r.$$

Since $x{:}A \to \mathbf{V}A$ is provable in LPV^- we have that

$$\mathsf{LPV}^- \vdash ((\Gamma_n^r \wedge \mathbf{V}X^r) \to \Delta_m^r) \to ((\Gamma_n^r \wedge x{:}X^r) \to \Delta_m^r).$$

By the induction hypothesis the realisation of the formula corresponding to the premise of the rule, $(\Gamma_n^r \wedge \mathbf{V}X^r) \to \Delta_m^r$, is provable, and hence:

$$\mathsf{LPV}^- \vdash (\Gamma_n^r \wedge x{:}X^r) \to \Delta_m^r.$$

Step 4. After applying the above three steps each $\mathcal{G} \in S$ has been translated into the language of LPV^-, and been shown to be derivable in LPV^-. Hence for the formula corresponding to the root sequent, $\Rightarrow F$, we have that

$$\mathsf{LPV}^- \vdash \top \to F^r.$$

Since $\mathsf{LPV}^- \vdash \top$

$$\mathsf{LPV}^- \vdash F^r.$$

Hence if $\mathsf{S4V}^- \vdash F$ there is a normal realisation r such that $\mathsf{LPV}^- \vdash F^r$.

---

[10] The procedure described in [2] gives an exponential increase in the size of the derivation of the desired F^r. [7] describes a modification of the procedure which gives only a polynomial increase.

Theorem 7 (S4V Realisation). *If* S4V $\vdash F$ *then* LPV $\vdash F^r$ *for some normal realisation* r.

Proof. We simply add the following case to Step 3 of Theorem 6. The rest is the same.

Case (Sequent $\mathcal{G}$ is the axiom: $\Box\mathbf{V}\bot \Rightarrow$).

$$\mathcal{G}^r = \neg x{:}\mathbf{V}\bot,$$

which is obviously derivable in LPV.

We are finally in a position to show that the systems of intuitionistic epistemic logic, IEL$^-$ and IEL, do indeed have an arithmetical interpretation.

Definition 18. *A formula of* IEL$^-$ *or* IEL *is called* proof realisable *if* $(tr(F))^r$ *is* LPV$^-$, *respectively* LPV, *valid under some normal realisation* r.

It follows that IEL$^-$ and IEL are sound with respect to proof realisability.

Theorem 8. *If* IEL$^-$, IEL $\vdash F$ *then* F *is proof realisable.*

Proof. By Theorem 1 if IEL$^-$, IEL $\vdash F$ then S4V$^-$, S4V $\vdash tr(F)$, respectively, and by Theorems 6 and 7 if S4V$^-$, S4V $\vdash tr(F)$ then LPV$^-$, LPV $\vdash (tr(F))^r$ respectively.

By Theorems 3 and 4 LPV$^-$ and LPV are sound with respect to their arithmetical interpretation, and hence by Theorem 8 so are IEL$^-$ and IEL.

8   Conclusion

Intuitionistic epistemic logic has an arithmetical interpretation, hence an interpretation in keeping with its intended BHK reading. Naturally verification in Peano Arithmetic, as outlined above, is not the only interpretation of verification for which the principles of intuitionistic epistemic logic are valid. IEL$^-$ and IEL may be interpreted as logics of the interaction between conclusive and non-conclusive evidence, e.g. mathematical proof vs. experimental confirmation, or observation vs. testimony. The question about exact interpretations for other intuitive readings of these logics is left for further investigation.

References

1. Artemov, S.: Operational modal logic. Technical report. MSI 95–29, Cornell University (1995)
2. Artemov, S.: Explicit provability and constructive semantics. Bull. Symbolic Logic **7**(1), 1–36 (2001)
3. Artemov, S.: Justified common knowledge. Theor. Comput. Sci. **357**, 4–22 (2006)
4. Artemov, S.: The logic of justification. Rev. Symbolic Logic **1**(4), 477–513 (2008)

5. Artemov, S., Protopopescu, T.: Intuitionistic epistemic logic. Technical report, December 2014. http://arxiv.org/abs/1406.1582
6. Boolos, G.: The Logic of Provability. Cambridge University Press, Cambridge (1993)
7. Brezhnev, V.N., Kuznets, R.: Making knowledge explicit: how hard it is. Theor. Comput. Sci. **357**(1), 23–34 (2006). http://dx.doi.org/10.1016/j.tcs.2006.03.010
8. Chagrov, A., Zakharyaschev, M.: Modal Logic. Clarendon Press, Oxford (1997)
9. Feferman, S.: Arithmetization of metamathematics in a general setting. Fundam. Math. **49**(1), 35–92 (1960)
10. Fitting, M.: The logic of proofs, semantically. Ann. Pure Appl. Logic **132**, 1–25 (2005)
11. Gödel, K.: An interpretation of the intuitionistic propositional calculus. In: Feferman, S., Dawson, J.W., Goldfarb, W., Parsons, C., Solovay, R.M. (eds.) Collected Works, vol. 1, pp. 301–303. Oxford University Press, Oxford (1933)
12. McKinsey, J.C.C., Tarski, A.: Some theorems about the sentential calculi of lewis and heyting. J. Symbolic Logic **13**(1), 1–15 (1948). http://www.jstor.org/stable/2268135
13. Mints, G.: A Short Introduction to Intuitionistic Logic. Springer, Berlin (2000)
14. Protopopescu, T.: Intuitionistic epistemology and modal logics of verification. In: van der Hoek, W., Holliday, W.H., Wang, W. (eds.) LORI 2015. LNCS, vol. 9394, pp. 295–307. Springer, Heidelberg (2015)
15. Troelstra, A., Schwichtenberg, H.: Basic Proof Theory. Cambridge University Press, Cambridge (2000)

Definability in First Order Theories of Graph Orderings

R. Ramanujam and R.S. Thinniyam$^{(\boxtimes)}$

The Institute of Mathematical Sciences, Chennai 600113, India
{jam,thinniyam}@imsc.res.in

Abstract. We study definability in the first order theory of graph order: that is, the set of all simple finite graphs ordered by either the minor, subgraph or induced subgraph relation. We show that natural graph families like cycles and trees are definable, as also notions like connectivity, maximum degree etc. This naturally comes with a price: bi-interpretability with arithmetic. We discuss implications for formalizing statements of graph theory in such theories of order.

Keywords: Graphs · Partial order · Logical theory · Definability

1 Introduction

Reasoning about graphs is a central occupation in computing science, since graphs are used to model many computational problems such as those in social networks, communication etc. In many cases, a single fixed graph is considered and some property has to be verified (e.g. bipartiteness) or some numerical parameter computed(e.g. independence number). However, as the complexity of the query increases, it can often be naturally recast as a question of relationships between graphs. For instance, asking if a graph is Hamiltonian is the same as looking for a cycle of the same order as the graph which occurs as a subgraph; asking for a k-colouring is the same as asking for a homomorphism of the graph to the k-clique. Studying the nature of relations on the set of all graphs has led to results such as the Graph Minor Theorem [18], whose algorithmic implications and influence on computer science cannot be overstated [1,2].

Consider the natural relations on graphs given by subgraph, induced subgraph and minor: these form partial orders over the set of all (simple, finite) graphs with interesting properties (see Fig. 1). Logical statements about these partial orders refer to graph families, and typically those given by some 'first order' closure condition, such as including/avoiding specific characteristics. Such statements are of immense interest to the theory of algorithms, motivating the logical study of graph order, and first order theories are the natural candidates for such a study.

Model theorists have taken up such studies. In a series of papers, Jezek and McKenzie [8–11] study the first order definability in substructure orderings on various finite ordered structures such as lattices, semilattices etc. Such a

© Springer International Publishing Switzerland 2016
S. Artemov and A. Nerode (Eds.): LFCS 2016, LNCS 9537, pp. 331–348, 2016.
DOI: 10.1007/978-3-319-27683-0_23

study is indeed foundational, and yet, it is of interest to study specific order structures on graphs to exploit their additional properties (if any). Indeed, the substructure order over graphs corresponds to the induced subgraph order, and this was investigated by Wires [19]. However, subgraph and minor orders are less amenable as substructure and hence deserve a closer look, which is the attempt initiated here. In the setting of directed graphs, the subdigraph order has been investigated by Kunos [16] recently. Work on word orders has been carried out by Kuske [17] as well as Kudinov et al. [14,15]. Other recent work on theories of classes of structures such as boolean algebras, linear orders and groups by Kach and Montalban [12] are different in spirit to ours, since they consider additive operations and the underlying structures may be infinite.

Our attempt here is not to study one graph order but rather to highlight the subtle differences in definability between different graph orders even while showing that they are all powerful enough to encode first order arithmetic. In fact, the subgraph and induced subgraph order are shown to be bi-interpretable with first order arithmetic. Many predicates which are interesting from a graph theoretic perspective such as connectivity, regularity, etc. are found to be first order definable, enabling us to articulate classical theorems of graph theory in such order theories.

We suggest that this paper as well as the related work mentioned are merely first steps of a larger programme of research, since we lack the tools as yet to address many related questions regarding indefinability, succinctness, algorithmic solutions, and so on.

The paper is organised as follows. After setting up the preliminaries, we study the subgraph order and show that certain numerical parameters such as order of a graph, commonly encountered graph families such as paths, cycles etc. and interesting graph predicates such as connectivity can be defined. We then show how such results can be lifted to the minor order. The machinery developed is used to show the bi-interpretability with arithmetic of the induced subgraph and subgraph orders and to interpret arithmetic in the minor order. Finally we display some interesting graph theoretical statements which can be stated using graph orders and discuss the research programme ahead.

2   Preliminaries

For the standard syntax and semantics of first order logic, we refer the reader to Enderton [4].

Definition 1 (Definability of Constants). *Fix a first order language $\mathcal{L}$. Let a be an element of the domain of an $\mathcal{L}$ structure $\mathcal{A}$. We say that a is definable in $\mathcal{A}$ if there exists an $\mathcal{L}$ formula $\phi_a(x)$ in one free variable such that $\mathcal{A}, a \vDash \phi_a(x)$ and for any $a' \neq a$ in the domain of $\mathcal{A}$, $\mathcal{A}, a' \nvDash \phi_a(x)$.*

We use a as a constant symbol representing the domain element a with the understanding that an equivalent formula can be written without the use of this constant symbol.

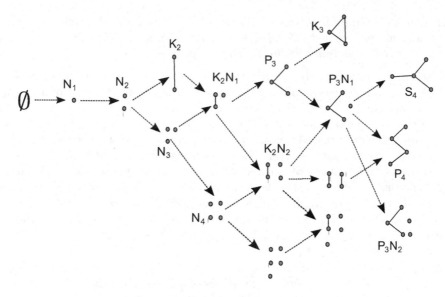

Fig. 1. The first few layers of the subgraph order.

Definition 2 (Covering Relation of a Poset). *Given an element x of a poset $(E, \leq)$, y is called an upper cover of x iff $x < y$ and there exists no element z of E such that $x < z < y$.*
Similarly y is called a lower cover of x iff $y < x$ and there exists no element z of E such that $y < z < x$.

Definition 3 (Graph Partial Orders). *Consider the following operations on graphs:*

- *A1. Deletion of a vertex (and all the edges incident on that vertex).*
- *A2. Deletion of an edge.*
- *A3. Contraction of an edge (given an edge $e = uv$, delete both u and v and introduce a new vertex w not in $V(g)$; connect all vertices which were adjacent to either u or v to w).*

For graphs g and g', g can be obtained from g' by any finite sequence of the operations:

1. *A1, A2 and A3 iff $g \leq_m g'$ (g is a minor of g').*
2. *A1 and A2 iff $g \leq_s g'$ (g is a subgraph of g').*
3. *A1 iff $g \leq_i g'$ (g is an induced subgraph of g').*

Let $\mathcal{G}$ denote the set of all simple graphs. We consider the base first order language $\mathcal{L}_0$ which has only the binary predicate symbol $\leq$ and an extension $\mathcal{L}_1$ that extends $\mathcal{L}_0$ with a constant symbol P_3 which stands for the path on three vertices. The latter is used in the case of the induced subgraph order in order to break the symmetry imposed by the automorphism which takes every graph to its complement.

Definition 4 (Graph Structures). *We denote the first order theories of the sub-graph and minor orders by* $\mathcal{L}_0$ *structures* $(\mathcal{G}, \leq_s)$ *and* $(\mathcal{G}, \leq_m)$ *respectively; and the induced subgraph order by the* $\mathcal{L}_1$ *structure* $(\mathcal{G}, \leq_i, P_3)$.

Notation: We use the letters x, y, z to denote variables representing graphs in formulas, u, v to represent nodes of a graph, e to represent the edge of a graph, g, h to represent graphs, $\mathcal{F}, \mathcal{G}$ to represent families of graphs. We write uv to denote the edge joining nodes u and v. We will denote by N_i, K_i, C_i, S_i, P_i the graph consisting of i isolated vertices, the i-clique, the cycle on i vertices, the star on i vertices and the path on i vertices respectively (Fig. 2); and by $\mathcal{N}, \mathcal{K}, \mathcal{C}, \mathcal{S}, \mathcal{P}$ the corresponding families of isolated vertices, cliques, cycles, stars and paths. $\mathcal{F}, \mathcal{T}$ represent forests and trees respectively. We will also on occasion, refer to certain fixed graphs or graph families by descriptive names (see Fig. 4).

k, l, m, n are used for natural numbers (also on occasion, members of the $\mathcal{N}$ family). All subscript or superscript variants such as x', x_i, etc. will be used to denote the same kind of object.

Given a graph g, $V(g)$ stands for the vertex set of g, $E(g)$ stands for the edge set of g, $|g|$ stands for the number of vertices of g (also called the order of g) and $|g|_{gr}$ stands for the graph consisting of only isolated vertices which has the same number of vertices as g. $||g||$ stands for the number of edges of g, also called the size of g. Given graphs g and h, $g \cup h$ stands for the disjoint union of g and h.

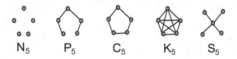

Fig. 2. Isolated points, path, cycle, clique and star of order 5 from left to right.

3 Definability in the Subgraph Order

We will take up definability in the subgraph order first. The defining formulae have been chosen such that most of them carry over in a straightforward way to the minor order. For a few predicates, significant modifications are required.

Constants, Covers and Cardinality

Lemma 1. *The upper and lower covering relations, the order of a graph, the family* $\mathcal{N}$ *and the graphs* N_1, K_2, K_3, S_4, P_4 *are definable in subgraph.*

The upper and lower covering relations for subgraph can immediately be defined:

$uc_s(x, y)$ iff x is an upper cover of y: $uc_s(x, y) := y <_s x \wedge \neg\exists z \, y <_s z <_s x$

$lc_s(x, y)$ iff x is a lower cover of y: $lc_s(x, y) := x <_s y \wedge \neg\exists z \, x <_s z <_s y$

Next we show that certain graphs in the first few layers of the subgraph order are definable. Refering to Fig. 1, the following formulae can easily be verified:

1. $\emptyset(x) := \forall y \; x \leq_s y$
2. $N_1(x) := uc_s(x, \emptyset); \quad N_2(x) := uc_s(x, N_1)$
3. $K_2(x) := uc_s(x, N_2) \wedge \exists y \; uc_s(y, x) \wedge \forall z \; uc_s(z, x) \supset z = y$
4. $N_3(x) := uc_s(x, N_2) \wedge x \neq K_2$
5. $K_2N_1(x) := uc_s(x, K_2); \quad K_2N_2(x) := uc_s(x, K_2N_1) \wedge uc_s(x, N_3)$
6. $P_3(x) := \exists! y \; uc_s(x, y) \wedge y = K_2N_1$ (where $\exists!$ is short for there exists unique)
7. $P_3N_1(x) := uc_s(x, P_3) \wedge uc_s(x, K_2N_2) \wedge \forall y \; uc_s(x, y) \supset (y = P_3 \vee y = K_2N_2)$
8. $S_4(x) := uc_s(x, P_3N_1) \wedge \forall y \; uc_s(x, y) \supset y = P_3N_1$
9. $K_3(x) := \exists! y \; lc_s(y, x) \wedge y = P_3$
10. $P_4(x) := uc_s(x, P_3N_1) \wedge x \neq S_4$

We note that if a family of totally ordered graphs is definable, then every member is definable as a constant by repeated use of the covering relation.

The family of isolated points is now easily seen to be definable via: $\mathcal{N}(x) := K_2 \not\leq_s x$. In addition, using the family $\mathcal{N}$ as a "yardstick", we can capture the cardinality (order) of a graph.

$order(n, x)$ iff $n \in \mathcal{N}$ and $|x| = |n|$:

$order(n, x) := \mathcal{N}(n) \wedge \forall m \; (\mathcal{N}(m) \wedge m \leq_s x) \supset m \leq_s n$.

For definable numerical predicates such as cardinality, we will simply use them as functions instead of predicates to simplify notation from here on i.e. $|x|_{gr}$ will denote the member of $\mathcal{N}$ whose order is the same as that of x.

Graph Families

Theorem 1. *The families* $\mathcal{K}, \mathcal{P}, \mathcal{C}, \mathcal{F}, \mathcal{T}, \mathcal{S}$ *are definable using subgraph.*

<u>Cliques</u>: Any graph to which an edge can be added contains at least two upper covers. The unique upper cover of a clique is formed by adding an isolated point to it. $\mathcal{K}(x) := \exists! y \; uc_s(y, x)$.

<u>Paths</u>

In order to define paths, we need to define a few additional families :

1. Disjoint unions of paths and cycles (denoted pac)
2. Disjoint unions of cycles i.e. sums of cycles (denoted soc)
3. Disjoint unions of paths i.e. forest of paths(denoted fop)

Assuming these, we can define paths :

$\mathcal{P}(x) := fop(x) \wedge \forall y \; |x|_{gr} = |y|_{gr} \wedge fop(y) \supset y \leq_s x$.

Out of all the fops of the same order n, the P_n forms the maximum element. Clearly by adding appropriate edges to a fop of the same order, one can form P_n. Adding any more edges to P_n gives a non-fop.

A graph is a disjoint union of paths and cycles iff it has maximum degree at most two: $pac(x) := S_4 \not\leq_s x$.

Assuming soc, fop can be defined: $fop(x) := pac(x) \wedge (\forall y \; soc(y) \supset y \not\leq_s x)$.

<u>if</u>: x is clearly a pac. Since x does not have any cycles as subgraph, it cannot have any soc as a subgraph.

only if: Let $x = c_1 \cup c_2 \cup ... \cup c_n$ where c_i is either a path or a cycle for all i. Suppose there is an i with c_i cycle. Then clearly $c_i \leq_s x$ but c_i is also a soc, which is a contradiction. Hence all components are paths and x is a fop.

It is only left to define disjoint unions (sums) of cycles i.e. soc (Fig. 3):

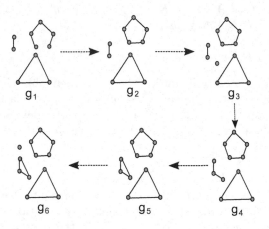

Fig. 3. $g_1, g_2, g_3, g_4, g_5, g_6$ all pac, only g_2, g_5, g_6 soc', only g_5 soc.

$$soc(x) = soc'(x) \wedge \forall y \, (uc_s(y, x) \wedge pac(y)) \supset soc'(y)$$
$$\text{where}$$
$$soc'(x) := x \neq \emptyset \wedge pac(x) \wedge \forall y \, (|y|_{gr} = |x|_{gr} \wedge pac(y)) \supset \neg x <_s y$$

Claim 1. $soc'(x)$ *iff every component of x is a cycle, N_1 or K_2 and x contains at most one copy of N_1 or one copy of K_2 but not both and x is not the empty graph.*

Proof. if: Clearly x is a pac. Suppose there exists a pac y of the same order as x and $x <_s y$. We can obtain y from x by addition of edges. But addition of any edge would introduce a degree three node, thus such a y cannot exist.

only if: Let $x = c_1 \cup c_2 \cup ... \cup c_n$ where c_i is either a cycle or a path. Suppose there is an i such that c_i is a path of order at least three. Let c_i' be the cycle formed by joining the ends of c_i. Now $x' = c_1 \cup ...c_{i-1} \cup c_i' \cup c_{i+1}... \cup c_n$ is also a pac, $|y| = |x|$ and x can obtained from y by deleting the newly added edge to get c_i from c_i'. Thus no path of length more than one can exist. Similarly, we can obtain a contradiction in the following cases by appropriately constructing x':

1. There are two copies of K_2 in x. Join the two copies end to end to form a path of length three, to get x'.

2. There are two copies of N_1 in x. Join the copies by an edge to get x'.
3. There is a K_2 and an N_1 as components in x. Join N_1 by an edge to K_2 to get a path of length two, to get x'.   □

Now we show the correctness of $soc(x)$.

if: Clearly x is a soc'. The only upper cover of x which is a pac is $x \cup N_1$ since adding any more edges would lead to a degree three node. $x \cup N_1$ is a soc'.

only if: Let $x = c_1 \cup c_2 \cup ... \cup c_n$ and x is a soc'. Suppose there is i such that c_i is K_2. Let $x' = x \cup N_1$. x' is an upper cover of x, is a pac but is not a soc' because it has an N_1 and a K_2 as components. Similarly we can rule out N_1 as a component of x.

Cycles, Forests, Trees, Stars

$$\mathcal{C}(x) :=pac(x) \wedge \exists y\, \mathcal{P}(y) \wedge |x|_{gr} = |y|_{gr} \wedge ucs(x, y)$$
$$forest(x) :=\forall y\, \mathcal{C}(y) \supset y \nleq_s x$$
$$\mathcal{T}(x) :=forest(x) \wedge \forall y\, (forest(y) \wedge |x|_{gr} = |y|_{gr}) \supset \neg x <_s y$$
$$\mathcal{S}(x) :=\mathcal{T}(x) \wedge P_4 \nleq_s x$$

It is clear that by deleting any edge from a cycle, we get a path which is a lower cover of the same order.

Conversely, consider any upper cover of a path with the same order. Adding an edge which joins the degree one vertices of the path gives a cycle, but adding an edge any where else creates a degree three vertex, which violates the condition that x is a pac. Thus only a cycle fulfills all the conditions.

A forest is a graph which contains no cycles. Of all forests with the same order, a tree is a maximal element since adding another edge gives a cycle. A non-tree forest can be made into a tree of same order by adding appropriate edges. A star is a tree which does not contain a path on four vertices as subgraph. Conversely, consider any tree with longest path on at most three vertices. Any other vertex must be connected to the midpoint of this longest path, thus it is a star.

Graph Predicates

Theorem 2. *Connectivity, maximum degree and maximum path length are definable in subgraph.*

Connectivity

$$conn(x) := \exists y\, \mathcal{T}(y) \wedge y \leq_s x \wedge |x|_{gr} = |y|_{gr}$$

A graph is connected iff it has a spanning tree.

Maximum path
$maxPath(n, x)$ iff $n \in \mathcal{N}$ and the largest path which is a subgraph of x is P_n.

$$maxPath(n, x) :=\mathcal{N}(n) \wedge \exists y\, \mathcal{P}(y) \wedge y \leq_s x \wedge |y|_{gr} = n\, \wedge$$
$$\forall z\, (\mathcal{P}(z) \wedge z \leq_s x) \supset z \leq_s y$$

Maximum degree
$maxDeg(n, x)$ iff $n \in \mathcal{N}$ and the maximum degree of x is $|n|$.

$$maxDeg(n, x) := \mathcal{N}(n) \wedge \exists y \, \mathcal{S}(y) \wedge y \leq_s x \wedge uc_s(|y|_{gr}, n) \wedge$$
$$\forall z \, (\mathcal{S}(z) \wedge z \leq_s x) \supset z \leq_s y$$

The maximum degree of x is one less than the order of the largest star which is a subgraph of x.

4   Definability in the Minor Order

Note that the minor order is identical to subgraph in an initial segment; the first additional relation which occurs in cycles and forests is shown in Fig. 4. This observation helps us reuse some of the machinery already developed for subgraph.

Observation 1. *The downclosure of S_5 and downclosure of K_3 are identical under subgraph and minor.*

Observation 2. *If $|x| = |y|$ then $x \leq_s y$ iff $x \leq_m y$ and $uc_s(x, y)$ iff $uc_m(x, y)$. Since the contraction operation reduces the number of vertices, restricting the orders to tuples of the same cardinality makes minor and subgraph equivalent.*

S_5 double3star K_3 C_4

Fig. 4. First difference between subgraph and minor.

We also have the following lemma on when the two orders can be taken to be equivalent.

Lemma 2. *Let x_n be an tree with at most one degree 3 node and no node of degree 4 or more. Then for any other graph x_0, $x_n \leq_m x_0$ iff $x_n \leq_s x_0$.*

Proof. It suffices to prove the only if direction since any subgraph is also a minor.
    We observe that there is a normal form for any sequence of minor operations. Let $x_n \leq_m x_0$ via a sequence of minor operations $o_1, o_2, ..., o_n$, then there exists a series of minor operations $o'_1, ..., o'_m$ on x_0 resulting in x_n such that no deletion operation occurs after a contraction operation and the number of contraction operations in the sequence $o'_1, ..., o'_m$ is at most the number of contractions in the original sequence $o_1, ..., o_n$.
    The result is proved by induction on the number of contraction operations in transforming x_0 to x_n. The details are given in the appendix.   □

The lemma and observations above help us transfer some results on definability from subgraph and minor order, simply by replacing the subgraph order by the minor order in the defining formulae.

Lemma 3. *The upper and lower covering relations, the order of a graph, the family $\mathcal{N}$ and the graphs N_1, K_2, K_3, S_4 are definable in minor order.*

Theorem 3. *The families $\mathcal{K}, \mathcal{P}, \mathcal{C}, \mathcal{F}, \mathcal{T}, \mathcal{S}$ are definable using minor.*

Firstly note that a graph contains a cycle as subgraph iff it contains K_3 as a minor (by contraction along the cycle). Hence forests are defined by: $\mathcal{F}(x) := K_3 \not\leq_m x$.

Observe that disjoint unions of paths and cycles (pac) can be defined by: $pac(x) := S_4 \not\leq_m x$. (By Lemma 2, we replace subgraph by minor in the defining formula). For forest of paths (fop), note that we can restrict a pac to be a forest, giving a fop: $fop(x) := \mathcal{F}(x) \wedge pac(x)$. Now, using Observation 2, we can immediately get paths, cliques, cycles and trees; and stars by Lemma 2:

$$\mathcal{P}(x) := fop(x) \wedge \forall y \; |x|_{gr} = |y|_{gr} \wedge fop(y) \supset y \leq_m x$$
$$\mathcal{K}(x) := \forall y \; |y|_{gr} = |x|_{gr} \supset y \leq_m x$$
$$\mathcal{C}(x) := pac(x) \wedge \exists y \; \mathcal{P}(y) \wedge |x|_{gr} = |y|_{gr} \wedge uc_m(x,y)$$
$$\mathcal{T}(x) := forest(x) \wedge \forall y \; (forest(y) \wedge |x|_{gr} = |y|_{gr}) \supset \neg x <_m y$$
$$\mathcal{S}(x) := \mathcal{T}(x) \wedge P_4 \not\leq_m x$$

Theorem 4. *Connectivity, maximum degree and maximum path length are definable in minor.*

$$conn(x) := \exists y \; \mathcal{T}(y) \wedge y \leq_m x \wedge |x|_{gr} = |y|_{gr}$$
$$maxPath(n,x) := \mathcal{N}(n) \wedge \exists y \; \mathcal{P}(y) \wedge y \leq_m x \wedge |y|_{gr} = n \wedge$$
$$\forall z \; (\mathcal{P}(z) \wedge z \leq_m x) \supset z \leq_m y$$

$maxDeg(n,x)$ *iff the maximum degree of x is $|n|$.*

Here we need to do some more work since the largest star which is a minor of x may be much larger than the maximum degree of the graph. The slightly involved construction is given in the appendix.

5 Arithmetic in Graph Orders

We define the ternary predicate version of arithmetic $(\mathbb{N}, plus, times)$ in the subgraph and minor orders. In order to do so, we need the following formulae: $N(g)$ iff g is a graph representing a number in our chosen representation. Let us denote by n_g the number denoted by g.
$plus(x,y,z)$ iff $N(x), N(y), N(z)$ hold and $n_x + n_y = n_z$ is true.
$times(x,y,z)$ iff $N(x), N(y), N(z)$ hold and $n_x \times n_y = n_z$ is true.

As can be gathered from the notation, our choice of (the unique) representation for natural number i is N_i, and from Lemmas 1 and 3, this family is definable in subgraph and minor.

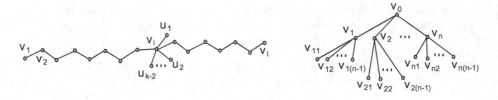

Fig. 5. Gadget graphs for addition and squaring.

To show the definability of the *plus* and *times* predicates, we will write out formulae using subgraph, but by Lemma 2 they can be transferred to minor.

Addition

$$plus(k,l,m) := \mathcal{N}(k) \wedge \mathcal{N}(l) \wedge \mathcal{N}(m) \wedge$$
$$(initial(k,l,m) \vee (N_3 \leq_s k \wedge N_3 \leq_s l \wedge$$
$$\exists x \; starTail(k,l,x) \wedge plus2(m,x))); \text{ where}$$
$$starTail(k,l,x) := starTail'(k,l,x) \wedge \forall x' \; starTail'(k,l,x') \supset |x|_{gr} \leq_s |x'|_{gr}$$
$$starTail'(k,l,x) := \mathcal{T}(x) \wedge maxDeg(x) = k \wedge maxPath(x) = l$$
$$plus2(m,x) := \exists m' \; uc_s(|m'|_{gr},x) \wedge uc_s(m,|m'|_{gr})$$
$$initial(k,l,m) := (k = \emptyset \wedge m = l) \vee (l = \emptyset \wedge m = k) \vee$$
$$(k = N_1 \wedge uc_s(m,l)) \vee (l = N_1 \wedge uc_s(m,k)) \vee$$
$$(k = N_2 \wedge \exists m' \; uc_s(|m'|_{gr},l) \wedge uc_s(m,|m'|_{gr})) \vee$$
$$(l = N_2 \wedge \exists m' \; uc_s(|m'|_{gr},k) \wedge uc_s(m,|m'|_{gr}))$$

When either k or l are strictly less than two, we hardcode the function value using *initial*.

When both are at least three, consider a tree with maximum degree k and maximum path l. A tree of least order with these properties is formed from a path by choosing some degree two vertex of the path v_i, adding $k-2$ new vertices $u_1, u_2, ..., u_{k-2}$ and adding the edges $u_1 v_i, u_2 v_i, ..., u_{k-2} v_i$ (see Fig. 5). The order of this tree is $k + l - 2$. This is captured in the formula *starTail* and in *plus2* we add two to its cardinality to get $k + l$.

Multiplication

We will show instead that squaring is definable, multiplication is easily obtained via the formula
$$(n_1 + n_2)^2 = n_1^2 + n_2^2 + 2 \times n_1 \times n_2$$
$square(n,m)$ iff $n, m \in \mathcal{N}$ and $|m| = |n|^2$

$$square(n,m) := \mathcal{N}(n) \wedge \mathcal{N}(m) \wedge \exists z \; ntree(n,z) \wedge uc_s(|z|_{gr},m) \wedge$$
$$\forall y \; ntree(n,y) \supset |y|_{gr} \leq_s |z|_{gr}; \text{ where}$$
$$ntree(n,z) := tree(z) \wedge maxDeg(z) = n \wedge P_6 \not\leq_s z$$

There exists a tree t of maximum order whose maximum degree is n and maximum path is P_5. To see that t has order $n^2 + 1$, observe that the tree has depth three (Fig. 5) and the total number of vertices is $1 + n + n \times (n - 1) = 1 + n^2$.

Thus by Lemmas 1, 3 and the definability of addition and multiplication shown above we have :

Theorem 5. *First order arithmetic is definable in the subgraph and minor orders.*

6   Encoding Graph Orders in Arithmetic

We will show that the structures $(\mathcal{G}, \leq_s)$ and $(\mathcal{G}, \leq_i, P_3)$ can be interpreted in first order arithmetic. In order to do this, we define the following formulae:

1. $isGraph(x)$ iff x is a number which represents a graph.
2. $sameGraph(x, y)$ iff x and y are numbers which represent the same graph.
3. $subGraph(x, y)$ iff the graph represented by x is a subgraph of the graph represented by y.
4. $inSubGraph(x, y)$ iff the graph represented by x is an induced subgraph of the graph represented by y.
5. $P_3(x)$ iff x represents the graph P_3.

Lemma 4 (Definable Arithmetical Predicates). *The following predicates are definable in first order arithmetic (defining formulae in appendix):*

1. *$bit(i, x)$ iff the ith bit of the binary representation of x is a 1.*
2. *$length(n, x)$ iff the length of the binary representation of x is n. We will denote this unique n by $|x|$.*
3. *$pow2(i, x)$ iff $x = 2^i$.*
4. *$rem(n, x, y)$ iff n is the remainder when x is divided by y; denoted $n = rem(x, y)$.*
5. *$div(n, x, y)$ iff n is the quotient when x is divided by y; denoted $n = x/y$.*
6. *$nchoose2(n, x)$ iff $x = \binom{n}{2}$ where $\binom{n}{2} = n \times (n - 1)/2$.*

6.1   Encoding Graphs

Any graph on n vertices has $\binom{n}{2}$ possible pairs of vertices. By fixing an appropriate order on these pairs, we may interpret any number whose binary representation has $\binom{n}{2} + 1$ bits as a graph on n vertices (we ignore the leading 1 since every binary representation has to start with a 1 except for the number 0). Let g be a graph on vertices $\{v_1, v_2, ..., v_n\}$. We define the ordering $\leq_e$ on tuples of vertices:

For $i < j, i' < j'$, $v_i v_j \leq_e v_{i'} v_{j'}$ iff $i < i'$, or $i = i'$ and $j \leq j'$.

Writing down the tuples in descending order, we get

$v_n v_{n-1}, v_n v_{n-2}, v_{n-1} v_{n-2}, v_n v_{n-3}, ..., v_3 v_1, v_2 v_1$. If we now replace the tuples by 0's for non-edges of g and 1's for edges and prefix a 1 to this string, we get a number m with bit length $\binom{n}{2} + 1$ which we say represents the (isomorphism class of) graph g. Note that as presented, there are multiple numbers which represent the same graph (upto isomorphism). We could choose the smallest of these to make the representation unique, instead we will stop at showing that there is a formula which identifies isomorphic graphs. We choose for the sake of completeness 0 as the unique number representing the empty graph. We denote the graph represented by x by g_x.

$$isGraph(x) := x = 0 \vee \exists n \; x = \left(1 + \binom{n}{2}\right)$$

Since we have arithmetical predicates available, we can define a formula for the order of a graph. Then to assert that there is an edge from v_i to v_j in x (where $i < j$), note that the tuple $\{v_i, v_j\}$ occurs at bit position $(2ni - 2n - i^2 - i + 2j)/2 = n(n-1)/2 - (n-i)(n-i+1)/2 + j - i$. We can further use arithmetical predicates to define formulas such as:

$perm(x, n)$ iff x represents a permutation on $[n]$.
$applyperm(x, i, j, n)$ iff x is a permutation on $[n]$ and sends i to j for $i, j \in [n]$.

We can then define the isomorphism of graphs:

$$sameGraph(x, y) := \exists n \; |x| = |y| = 1 + \binom{n}{2} \wedge \exists z \; perm(z, n) \wedge$$

$$\forall i \, \forall j \; 1 \leq i < j \leq n \supset (edgeExists(x, i, j) \iff (\exists i' \exists j' \; applyPerm(z, i, i', n)$$

$$\wedge \; applyPerm(z, j, j', n) \; \wedge edgeExists(y, i', j') \,))$$

The formula states that for x and y to represent the same graph, there must exist a permutation z such that for any tuple $\{v_i, v_j\}$ of vertices of x, $v_i v_j \in E(g_x)$ iff $v_{z(i)} v_{z(j)} \in E(g_y)$. Details are given in the appendix.

6.2   Subgraph and Induced Subgraph

$subGraph(x, y)$ iff x, y represent graphs and g_x is a subgraph of g_y.

$$subGraph(x, y) := isGraph(x) \wedge isGraph(y) \wedge |g_x| \leq |g_y| \wedge$$

$$\exists z \; sameGraph(y, z) \wedge \forall k \; 1 \leq k \leq |x| \supset (bit(k, x) \supset bit(k + |y| - |z|, z))$$

If x on vertices $u_1, u_2, ..., u_n$ is a subgraph of y on vertices $v_1, v_2, ..., v_m$ without regard for vertices, then there is a map $f : V(x) \to V(y)$ which witnesses it. Rename the vertices of y to get z by the map which sends $f(u_i)$ for any $u \in V(x)$ to v_{m-n+i} and fixes the other vertices. Then x is a subgraph of the graph on $v_m, v_{m-1} ..., v_{m-n+1}$ when considered with the labels.

Conversely, if the formula is true, the $sameGraph$ predicate gives us a witnessing permutation using which we can define the map witnessing that x is a subgraph of y.

We can define induced subgraph by a small modification in the subgraph formula as follows:

$$inSubGraph(x, y) := isGraph(x) \wedge isGraph(y) \wedge |g_x| \leq |g_y| \wedge$$

$$\exists z \; sameGraph(y, z) \wedge \forall k \; 1 \leq k \leq |x| \supset (bit(k, x) \iff bit(k, z))$$

Defining the constant P_3

The formula $P_3(x)$ can be easily defined as the disjunction over the formulae $x = c$ where c is a number representing P_3 since there are only finitely many of them. Thus we have:

Theorem 6. *The structures $(\mathcal{G}, \leq_s)$ and $(\mathcal{G}, \leq_i, P_3)$ are definable in $(\mathbb{N}, +, \times)$.*

Theorem 7 (Wires [19]). *Arithmetic is definable in $(\mathcal{G}, \leq_i, P_3)$.*

Combining the above result with Theorems 5 and 6 we have:

Theorem 8. *The structures $(\mathcal{G}, \leq_s)$ and $(\mathcal{G}, \leq_i, P_3)$ are bi-interpretable with first order arithmetic. The structure $(\mathcal{G}, \leq_m)$ can encode $(\mathbb{N}, plus, times)$.*

7 Discussion and Future Work

7.1 Decidability and Descriptive Complexity

An obvious corollary of our results is that the theories of the orders considered are undecidable, but it is natural to ask what the decidable fragments are. One may consider various restrictions: syntactic ones such as the $\forall^* \exists^*$ fragment, subclasses of graphs such as trees $(\mathcal{T}, \leq_s)$ or restrict the order e.g. theory of the covering relation $\mathrm{Th}(\mathcal{G}, uc_s)$. There is much work on general frameworks for graph theory, especially extremal graph theory, whose focus is on homomorphisms. In particular Hatami's paper [6] on the undecidability of inequalities over homomorphism densities underlines the difficulty of answering general questions about graphs. If our interest is in only obtaining undecidability results, ideas of recursive inseparability and other techniques (see [5]) may be more apt.

We also note that there is a large body of work on descriptive complexity [7], which takes the graph-as-a-model point of view. How definable families in our approach compare with the above is a matter of interest. In particular, every constant can be defined in the subgraph order using the methods of Wires [19], just as they can in descriptive complexity.

7.2 Extensions, Interdefinability and Graph Theory

We do not know if subgraph is definable using minor or vice versa. However, if we add the predicate $sameSize(x, y)$ which stands for x and y have the same number of edges, we can define subgraph using minor as shown below:

Suppose that x is a subgraph of y. Then we can think of y as being built from x in two steps. In the first step, we add to x a number of isolated points to give x' such that $|x'|_{gr} = |y|_{gr}$. In the second step, we only add extra edges to x' to get y.

We can formalize the two step construction as follows:

$$x \leq_s y := \exists x' \; vertdesc(x', x) \wedge edgedesc(y, x'); \text{ where}$$
$$edgedesc(x, y) := y \leq_m x \wedge |x|_{gr} = |y|_{gr}$$
$$vertdesc(x, y) := y \leq_m x \wedge samesize(x, y)$$

Such extensions also have serious implications for the kind of graph theoretic statements that can be made. This is because, though the structure of graph orders is rich, they have limited access to the "inner structure" of a graph. For instance, it is not clear how minimum degree of a graph can be defined using graph orders. We already know that we can do arithmetic over the order of a graph. By adding the predicate $sameSize(x, y)$, we can do arithmetic over the size. Consequently, concepts of minimum and average degree can be expressed and theorems about them written in the extended language. We can capture the size of a graph using $sameSize$:

$$||x||_{gr} = n := \mathcal{N}(n) \wedge \exists y \, \mathcal{C}(y) \, |y|_{gr} = n \wedge sameSize(x, y)$$

Minimum degree of a graph

$$minDeg(x, y) := \mathcal{N}(x) \wedge \exists z \, deleteNeighbours(y, z)$$
$$\wedge \, \forall z' \, deleteNeighbours(y, z') \supset ||z'||_{gr} \leq_s ||z||_{gr}$$
$$\wedge \, x = ||z||_{gr}; \text{ where}$$
$$deleteNeighbours(y, z) := z \leq_s y \wedge hasIso(z) \wedge |z|_{gr} = |y|_{gr}$$
$$hasIso(x) := \exists y \, uc_s(x, y) \wedge |y|_{gr} <_s |x|_{gr}$$

Average degree of a graph (integer ceiling)

$$\lceil AvgDeg(x, y) \rceil := \mathcal{N}(x) \wedge (||y||_{gr} \leq_s x \times |y|_{gr})$$
$$\wedge \, \forall z \, (\mathcal{N}(z) \wedge z <_s x) \supset (z \times |y|_{gr} <_s ||y||_{gr})$$

We can also define $\lfloor AvgDeg(x, y) \rfloor$ i.e. the floor instead of the ceiling defined above. Modifying the definition above by dividing by two gives us floor and ceiling versions of number of edges per vertex i.e. $\lfloor \epsilon(x, y) \rfloor$ and $\lceil \epsilon(x, y) \rceil$ respectively.

Theorem 9 (Diestel, Proposition 1.2.2 [3]). *Every graph g with at least one edge has a subgraph g' with $\delta(g') > \lfloor \epsilon(g') \rfloor \geq \lfloor \epsilon(g) \rfloor$ (where δ denotes minimum degree):*

$$\forall x \, \neg \mathcal{N}(x) \supset \exists y \, y \leq_s x \wedge \lfloor \epsilon(y) \rfloor <_s minDeg(y) \wedge \lfloor \epsilon(x) \rfloor \leq_s \lfloor \epsilon(y) \rfloor$$

7.3 Differences in Definability

From the work of Wires [19] it is known that all the graph families we defined in Lemmas 1 and 3 and many more are definable in $(\mathcal{G}, \leq_i, P_3)$. Thus they are definable in all three orders. But as we saw, while maximum degree was definable easily in subgraph, it takes more work in minor. Similarly, though cardinality is trivial in minor and subgraph, it seems to take much more work to define in induced subgraph. On the other hand, here is a predicate definable easily in induced minor which we do not know how to define in the other two :
$\alpha(n, x)$ iff $n \in N$ and $|n|$ is the independence number of x.

$$\alpha(n, x) := \mathcal{N}(n) \wedge n \leq_i x \wedge \forall y \, (\mathcal{N}(y) \wedge y \leq_i x) \supset y \leq_i n$$

Perhaps the most interesting direction of work lies in pinning down these differences, especially as pertains to the definability of predicates which are important from the point of view of graph theory and to determine what part of the "inner structure" of graphs can be determined by their relationships with other graphs. For this, we need to develop tools to prove indefinability, which is a challenging task.

Appendix

Proof of Lemma 2

Lemma 2: Let x_n be an tree with at most one degree 3 node and no node of degree 4 or more. Then for any other graph x_0, $x_n \leq_m x_0$ iff $x_n \leq_s x_0$.

We prove the result by induction on the number of contraction operations in transforming x_0 to x_n.

Base Case: There are no contraction operations, there is nothing to be done.

For the induction step there are two cases we consider:

Case 1 : x_n has no degree 3 node. Let x_n be a path $u_0, u_1, ..., u_m$. Let $o_1, .., o_n$ be the sequence of minor operations in normal form with x_i being obtained from x_{i-1} via operation o_i. o_n must be a contraction operation (else all operations are deletions and we are done). Therefore x_{n-1} is either a path of length $m + 1$ or a graph such that $V(x_{n-1}) = V(x_n) \cup \{u'\}$ and there exists an i with $E(x_{n-1}) = E(x_n) \cup \{u'u_i\}$ or $E(x_{n-1}) = E(x_n) \cup \{u'u_i, u'u_{i+1}\}$. In all cases, we can delete an endpoint of x_{n-1} or u' respectively in order to obtain x_n. Thus there is a sequence $o_1, .., o_{n-1}, o'_n$ of operations (o'_n is a deletion) to obtain x_n from x_0. Since this sequence has a smaller number of contractions, by induction hypothesis, x_n is a subgraph of x_0.

Case 2 : x_n has exactly one degree three node. Let x_n consist of a degree 3 node u with paths p_1, p_2, p_3. As before, consider the sequence of minor operations. In one case x_{n-1} is a graph with a degree 3 node attached to three paths exactly one of which has length one more than previously. We can delete the end point of the appropriate path to get x_n from x_{n-1}. Another possibility is that x_{n-1} is a graph with a vertex $u' \notin V(x_n)$ such that u' is attached to either one or two adjacent points of one of the paths p_1, p_2, p_3. As before, we can delete u' to get x_n from x_{n-1}. Then by induction hypothesis x_n is a subgraph of x_{n-1}.

Proof of Maximum Degree Definability in Theorem 4

Theorem 4: Connectivity, maximum degree and maximum path length are definable in minor.

In order to apply observation 2, we construct the following family:

$\mathcal{S} \cup \mathcal{N}(x)$ iff x is formed by addition of some arbitrary number of isolated vertices to a star.

$$\mathcal{S} \cup \mathcal{N}(x) := \mathcal{F}(x) \wedge \exists y \; hasStarComp(y, x) \wedge onlystar(x, y)$$

$$\text{where}$$

$$hasStarComp(y, x) := \mathcal{S}(y) \wedge y \leq_m x \wedge \forall z \; conn(z) \wedge z \leq_m x \supset z \leq_m y$$
$$onlyStar(x, y) := \forall x' \; \mathcal{F}(x') \wedge |x'|_{gr} = |x|_{gr} \wedge uc_m(x', x) \supset$$
$$\forall y' \; (conn(y') \wedge y' \leq_m x') \supset uc_m(|y'|_{gr}, |y|_{gr})$$

$onlyStarComp$ asserts that y is a star minor of x and in addition, every connected minor of x is also a minor of y. To fulfill this condition, x has to contain y as a connected component.

$onlyStar$ asserts that any forest x' which is formed by adding an edge to x (by observation 2) has the property that all its connected minors have order one more than the order of y.

Clearly, any graph formed by adding isolated vertices to a star has these properties.

$\mathcal{S} \cup \mathcal{N} subGraph$ states that there is a subgraph y of x which is a $\mathcal{S} \cup \mathcal{N}$ of same order as x. Note that for $S_n \cup N_m$ and $S_{n'} \cup N_{m'}$ with $n + m = n' + m'$, $S_n \cup N_m \leq_m S_{n'} \cup N_{m'}$ iff $n \leq n'$. Thus maximal y satisfying the formula $\mathcal{S} \cup \mathcal{N} subGraph$ contains the largest star occuring as a subgraph of x. We extract the star from this object to obtain the maximum degree of x.

Proof of Lemma 4

Lemma 4: The following predicates are definable in first order arithmetic:

1. $nchoose2(n, x)$ iff $x = \binom{n}{2}$ where $\binom{n}{2} = n \times (n-1)/2$.
   $nchoose2(n, x) := 2 \times x + n = n^2$.

2. $div(n, x, y)$ iff n is the quotient when x is divided by y; denoted $n = x/y$.
   $div(n, x, y) := \exists z \; x = y \times n + z \wedge z < y$

3. $rem(n, x, y)$ iff n is the remainder when x is divided by y; denoted $n = rem(x, y)$. $rem(n, x, y) := \exists z \; x = y \times z + n \wedge n < y$
   We note that the exponentiation relation $x^y = z$ is known to be definable in arithmetic (see [13]).

4. $pow2(i, x)$ iff $x = 2^i$.
   $pow2(i, x) := \exists y \; y = 2 \wedge y^i = x$

5. $bit(i, x)$ iff the ith bit of the binary representation of x is a 1.
   $bit(i, x) := rem(x, 2^i) = rem(x, 2^{i-1})$

6. $length(n, x)$ iff the length of the binary representation of x is n. We will denote this unique n by $|x|$.
   $length(n, x) := bit(n, x) \wedge \forall n' \; n < n' \supset \neg bit(n', x)$

Details of Subsection 6.1

$graphOrder(x, n)$ iff $n \in \mathcal{N}$ and the order of x is $|n|$.

$$graphOrder(x, n) := isGraph(x) \wedge 2 \times |x| = 2 + n \times (n - 1)$$

We will denote by $|g_x|$ the order of the graph represented by x.
$edgeExists(x, i, j)$ iff x denotes a graph and $v_i v_j \in E(g_x)$.

$$edgeExists(x, i, j) := \exists n \; graphOrder(x, n) \wedge$$
$$((1 \leq i < j \leq n \wedge bit((2ni - 2n - i^2 - i + 2j)/2, x))$$
$$\vee (1 \leq j < i \leq n \wedge bit((2nj - 2n - j^2 - j + 2i)/2, x))$$

By doing some counting, we can see that the tuple $\{v_i, v_j\}, i < j$ occurs at bit position $(2ni - 2n - i^2 - i + 2j)/2 = n(n-1)/2 - (n-i)(n-i+1)/2 + j - i$.

Defining Permutations and Isomorphism. Any permutation of vertices of a vertex labelled graph induces a permutation on the edges of a graph. To identify all numbers which represent the same graph under our encoding, we will need to represent permutations on $[n]$ and their actions.

$perm(x, n)$ iff x represents a permutation on $[n]$.

$$perm(x, n) := |x| = 1 + n \times \lfloor log(n) \rfloor \wedge \forall i \; 1 \leq i \leq n \; \exists! j \; 1 \leq j \leq n$$
$$i = (rem(x, 2^{j \, |n|}) - rem(x, 2^{(j-1) \, |n|}))/2^{(j-1)|n|}$$

We represent a permutation by a bit string which is of length $n \times \lfloor log(n) \rfloor + 1$, note that $\lfloor log(n) \rfloor$ is the same as $|n|$ i.e. the length of the string n. The most significant digit is to be ignored, after which every block of $\lfloor log(n) \rfloor$ bits represents a number from 1 to n. In addition, every such block must be unique (in order to guarantee that it is a permutation). The permutation sends $i \in [n]$ to the number represented by the ith block from the left. The formula checks that every $i \in [n]$ is obtained from a unique block $j \in [n]$.

$applyperm(x, i, j, n)$ iff x is a permutation on $[n]$ and sends i to j for $i, j \in [n]$.

$$applyperm(x, i, j, n) := perm(x, n) \wedge$$
$$(rem(x, 2^{(n-i+1)|n|}) - rem(x, 2^{(n-i)|n|}))/2^{(n-i)|n|} = j$$

We can now define the isomorphism of graphs:

$$sameGraph(x, y) := \exists n \; |x| = |y| = 1 + \binom{n}{2} \wedge \exists z \; perm(z, n) \wedge$$

$$\forall i \; \forall j \; 1 \leq i < j \leq n \supset (edgeExists(x, i, j) \iff (\exists i' \exists j' \; applyPerm(z, i, i', n)$$
$$\wedge \; applyPerm(z, j, j', n) \wedge edgeExists(y, i', j')\;))$$

The formula states that for x and y to represent the same graph, there must exist a permutation z such that for any tuple $\{v_i, v_j\}$ of vertices of x, $v_i v_j \in E(g_x)$ iff $v_{z(i)} v_{z(j)} \in E(g_y)$.

References

1. Bienstock, D., Langston, M.A.: Algorithmic implications of the graph minor theorem. Handb. Oper. Res. Manag. Sci. **7**, 481–502 (1995)
2. Bodlaender, H.L.: Treewidth: algorithmic techniques and results. In: Privara, I., Ruzicka, P. (eds.) MFCS '97. LNCS, vol. 1295, pp. 19–36. Springer, Berlin (2003)
3. Diestel, R.: Graph Theory. Springer, Berlin (2005)
4. Enderton, H.: A Mathematical Introduction to Logic. Academic press, Waltham (2001)
5. Ershov, Y.L., Lavrov, I.A., Taimanov, A.D., Taitslin, M.A.: Elementary theories. Russ. Math. Surv. **20**(4), 35–105 (1965)
6. Hatami, H., Norine, S.: Undecidability of linear inequalities in graph homomorphism densities. J. Am. Math. Soc. **24**(2), 547–565 (2011)
7. Immerman, N.: Descriptive Complexity. Springer Science & Business Media, Berlin (2012)
8. Ježek, J., McKenzie, R.: Definability in substructure orderings, iv: finite lattices. Algebra Univers. **61**(3–4), 301–312 (2009)
9. Ježek, J., McKenzie, R.: Definability in substructure orderings, i: finite semilattices. Algebra Univers. **61**(1), 59–75 (2009)
10. Ježek, J., McKenzie, R.: Definability in substructure orderings, iii: finite distributive lattices. Algebra Univers. **61**(3–4), 283–300 (2009)
11. Ježek, J., McKenzie, R.: Definability in substructure orderings, ii: finite ordered sets. Order **27**(2), 115–145 (2010)
12. Kach, A.M., Montalban, A.: Undecidability of the theories of classes of structures. J. Symbolic Logic **79**(04), 1001–1019 (2014)
13. Kaye, R.: Models of Peano Arithmetic. Oxford University Press, USA (1991)
14. Kudinov, O.V., Selivanov, V.L.: Definability in the infix order on words. In: Diekert, V., Nowotka, D. (eds.) DLT 2009. LNCS, vol. 5583, pp. 454–465. Springer, Heidelberg (2009)
15. Kudinov, O.V., Selivanov, V.L., Yartseva, L.V.: Definability in the subword order. In: Ferreira, F., Löwe, B., Mayordomo, E., Mendes Gomes, L. (eds.) CiE 2010. LNCS, vol. 6158, pp. 246–255. Springer, Heidelberg (2010)
16. Kunos, Á.: Definability in the embeddability ordering of finite directed graphs. Order **32**(1), 117–133 (2015)
17. Kuske, D.: Theories of orders on the set of words. RAIRO-Theor. Inform. Appl. **40**(01), 53–74 (2006)
18. Robertson, N., Seymour, P.D.: Graph minors. xx. wagner's conjecture. J. Comb. Theor. B **92**(2), 325–357 (2004)
19. Wires, A.: Definability in the substructure ordering of simple graphs (2012). http://www.math.uwaterloo.ca/awires/SimpleGraphs.pdf

The Complexity of Disjunction
in Intuitionistic Logic

R. Ramanujam[1], Vaishnavi Sundararajan[2], and S.P. Suresh[2(✉)]

[1] The Institute of Mathematical Sciences, Chennai, India
jam@imsc.res.in
[2] Chennai Mathematical Institute, Chennai, India
{vaishnavi,spsuresh}@cmi.ac.in

Abstract. In the formal study of security protocols and access control systems, fragments of intuitionistic logic play a vital role. These are required to be efficient, and are typically disjunction-free. In this paper, we study the complexity of adding disjunction to these subsystems. Our lower bound results show that very little needs to be added to disjunction to get co-NP-hardness, while our upper bound results show that even a system with conjunction, disjunction, and restricted forms of negation and implication is in co-NP. Our upper bound proofs also suggest parameters which we can bound to obtain PTIME algorithms.

Keywords: Intuitionistic logic · Proof theory · Disjunction · Complexity

1 Introduction

Intuitionistic logic is a subject with a rich history, with connections to fundamental aspects of mathematics, philosophy and computer science. What is perhaps surprising is that it also finds application in such concrete areas of computer science as system security and communication security in distributed protocols. Consider the question: given a finite set of formulas X, a formula α in a positive fragment of some propositional logic, and an intuitionistic proof system $\vdash$, does $X \vdash \alpha$? This sounds arcane, but is of practical importance when X is a security policy that specifies permissions and α is the assertion of someone being permitted some action [1,10]. Or it might be the case that X is a set of terms picked by an eavesdropper watching a channel and α is a term to be kept secret [8]. Inference in such situations is typically intuitionistic. Consider a formula A has t for an agent A participating in a cryptographic protocol and a term t. A different agent B might not be able to assert $(A$ has $t) \vee \neg(A$ has $t)$, since it might be that B does not have all the components that go into building the term t and the system does not allow B to assert anything about t in such a case. To consider another example, B cannot assert A has t by assuming $\neg(A$ has $t)$ and then deriving a contradiction. To consider a third example, consider a formula A can read f, where A is a user and f is a file. An access control

© Springer International Publishing Switzerland 2016
S. Artemov and A. Nerode (Eds.): LFCS 2016, LNCS 9537, pp. 349–363, 2016.
DOI: 10.1007/978-3-319-27683-0_24

policy may be silent on whether A can read the file or not. Thus the formula $(A$ can read $f) \vee \neg(A$ can read $f)$ is not a validity in this system. This allows the possibility that even though A cannot read file f according to the current policy, it may be allowed that access in an extension of the policy.

In the applications mentioned above, the complexity of derivability is of prime importance, since a derivability check is often a fundamental component of more detailed security structures [6]. These systems are usually disjunction-free, with a PTIME derivability procedure [2,7,11]. But reasoning about disjunction is also important for security applications, even though it typically increases the complexity of the derivability problem (see [15], for example). In this paper, we explore the effect of disjunction on the complexity of various subsystems of intuitionistic logic.

The PTIME systems referred to above do not include full implication either. This is obvious, since it is well-known that the derivability problem for intuitionistic logic (and even its implication-only fragment) is PSPACE-complete.[1] In this context, [11] considers a restriction of full implication, the so-called *primal implication* which is defined by the following rule.

$$\frac{X \vdash \beta}{X \vdash \alpha \to \beta} \to$$

In this rule, we have the same set of antecedents (set of formulas to the left $\vdash$) both in the premise and conclusion, and this contributes to an efficient solution to the derivability problem.

We show that when we add disjunction to such efficient systems, derivability is in co-NP. The results are similar to those in [4], but while the results there are obtained via a translation to classical logic, we provide an explicit algorithm. Our focus is on the algorithm itself, which is a general procedure to lift a PTIME decision procedure for a logic to a co-NP procedure for the same logic with disjunction. We also provide a modification of the above procedure that runs in PTIME when we restrict the formulas on which disjunction elimination is applied in a proof.

We also show that we cannot do better than co-NP for the above logics. Subsystems involving disjunction are co-NP-hard with such minimal additions as the elimination rule for implication, or the introduction rule for conjunction. We also show that we get co-NP-hardness when we consider a system with rules for disjunction and the elimination rule for negation.

Related Work. As we mentioned earlier, application areas like security typically work with an intuitionistic system, and the complexity of derivability is important in such applications. In the study of cryptographic protocols, the cryptographic primitives are represented as rules in a proof system, following Dolev and Yao [8]. These logics are typically positive and conjunctive. The derivability

---

[1] From now on, whenever we refer to the complexity of a logic, we implicitly mean the complexity of the derivability problem for it.

problem for the basic Dolev-Yao system is in PTIME [16]. Other interesting non-classical conjunctions like blind pairing can make the problem hard when they interact distributively with the standard pairing operator [3].

The results reported in this paper are very close to work done in the realm of authorization logics, specifically primal infon logic and its extensions. It was shown that primal infon logic is in PTIME [2,11] but adding disjunction makes the problem co-NP-complete [4]. Specifically, it was shown that a system with primal implication, conjunction, disjunction and $\perp$ is co-NP-hard, using a translation from classical logic. Our lower bound results can be seen as a refinement of the result in [4], as we show that disjunction with *any one* of these other connectives is already co-NP-hard. The upper bound results are also very similar to those in [4], but we provide an explicit algorithm while the results there are obtained via a translation to classical logic. Our procedures can be seen as a way of lifting PTIME decision procedures for *local theories* [7,14] to co-NP procedures for the same logics with disjunction. More recently, the complexity of primal logic with disjunction was studied in further detail in [13], but the proofs are via semantic methods.

Another important area of study is the **disjunction property** and its effect on complexity. A system is said to have the disjunction property if it satisfies the following condition: whenever $X \vdash \alpha \vee \beta$ and X satisfies some extra conditions (for example, $\vee$ does not occur in any formula of X), then $X \vdash \alpha$ or $X \vdash \beta$. The disjunction property and its effect on decidability and complexity have been the subject of study for many years. For example, it has been proved that as long as any (propositional) logic that extends intuitionistic logic satisfies the disjunction property, derivability is PSPACE-hard, and otherwise it is in co-NP (see Chap. 18 of [5]). Various other papers also investigate extensions of intuitionistic logic with the disjunction property [9,12,17]. In contrast to these results, our paper considers *subsystems* of intuitionistic logic obtained by restricting implication. Further, in our paper, the focus is more on the *left disjunction property*: namely that $X, \alpha \vee \beta \vdash \delta$ iff $X, \alpha \vdash \delta$ and $X, \beta \vdash \delta$.

2   Preliminaries

Assume a countably infinite set of atomic propositions $\mathscr{P}$. The set of formulas Φ is given by

$$\alpha, \beta ::= p \mid \neg \alpha \mid \alpha \wedge \beta \mid \alpha \vee \beta \mid \alpha \to \beta$$

For a set of operators $\mathscr{O}$, we denote by $\Phi^{\mathscr{O}}$ the set of all formulas consisting only of the operators in $\mathscr{O}$. For example, $\Phi^{\{\vee\}}$ is the set of all formulas built only using the $\vee$ operator, $\Phi^{\{\vee, \wedge\}}$ is the set of all formulas built only using the $\vee$ and $\wedge$ operators, etc. For ease of notation, we ignore the braces and instead use $\Phi^{\vee}$, $\Phi^{\vee, \wedge}$, etc.

The set of **subformulas** of a formula α, denoted $\mathsf{sf}(\alpha)$, is defined to be the smallest set S such that: $\alpha \in S$; if $\neg \beta \in S$, $\beta \in S$; and if $\beta \wedge \gamma \in S$ or $\beta \vee \gamma \in S$ or $\beta \to \gamma \in S$, $\{\beta, \gamma\} \subseteq S$. For a set X of formulas, $\mathsf{sf}(X) = \bigcup_{\alpha \in X} \mathsf{sf}(\alpha)$.

$$\frac{}{X, \alpha \vdash \alpha} \; ax$$

$\dfrac{X, \alpha \vdash \beta \quad X, \alpha \vdash \neg \beta}{X \vdash \neg \alpha} \; \neg i$	$\dfrac{X \vdash \beta \quad X \vdash \neg \beta}{X \vdash \alpha} \; \neg e$
$\dfrac{X \vdash \alpha \quad X \vdash \beta}{X \vdash \alpha \wedge \beta} \; \wedge i$	$\dfrac{X \vdash \alpha_0 \wedge \alpha_1}{X \vdash \alpha_j} \; \wedge e$
$\dfrac{X \vdash \alpha_j}{X \vdash \alpha_0 \vee \alpha_1} \; \vee i$	$\dfrac{X \vdash \alpha \vee \beta \quad X, \alpha \vdash \delta \quad X, \beta \vdash \delta}{X \vdash \delta} \; \vee e$
$\dfrac{X, \alpha \vdash \beta}{X \vdash \alpha \rightarrow \beta} \; \rightarrow i$	$\dfrac{X \vdash \alpha \rightarrow \beta \quad X \vdash \alpha}{X \vdash \beta} \; \rightarrow e$

Fig. 1. The system IL

The logic is defined by the derivation system in Fig. 1. By $X \vdash_{IL} \alpha$, we mean that there is a derivation in IL of $X \vdash \alpha$. (For ease of notation, we drop the suffix and use $X \vdash \alpha$ to mean $X \vdash_{IL} \alpha$, when there is no confusion.)

Definition 1 (Derivability Problem). Given X and α, is it the case that $X \vdash_{IL} \alpha$?

Among the rules, ax, $\wedge e$ and $\rightarrow e$ are the *pure elimination rules*, $\neg e$, $\neg i$ and $\vee e$ are the *hybrid rules* and the rest are the *pure introduction rules*. A **normal derivation** is one where the major premise of every pure elimination rule and hybrid rule is the conclusion of a pure elimination rule. The following fundamental properties hold, and the proofs are standard in the proof theory literature.

Proposition 2. *1. (Monotonicity) If $X \vdash \alpha$ and $X \subseteq X'$, then $X' \vdash \alpha$.*
2. (Admissibility of Cut) If $X \vdash \alpha$ and $X, \alpha \vdash \beta$, then $X \vdash \beta$.
3. (Left Disjunction Property) $X, \alpha \vee \beta \vdash \delta$ iff $X, \alpha \vdash \delta$ and $X, \beta \vdash \delta$.
4. (Left Conjunction Property) $X, \alpha \wedge \beta \vdash \delta$ iff $X, \alpha, \beta \vdash \delta$.

Theorem 3 (Weak Normalization). *If there is a derivation π of $X \vdash \alpha$ then there is a normal derivation ϖ of $X \vdash \alpha$. Further, if a formula $\alpha \vee \beta$ occurs as the major premise of an instance of $\vee e$ in ϖ, it also occurs as the major premise of an instance of $\vee e$ in π.*

Theorem 4 (Subformula Property). *Let π be a normal derivation with conclusion $X \vdash \alpha$ and last rule r. Let $X' \vdash \beta$ occur in π. Then $X' \subseteq sf(X \cup \{\alpha\})$ and $\beta \in sf(X \cup \{\alpha\})$. Furthermore, if r is a pure elimination rule, then $X' \subseteq sf(X)$ and $\beta \in sf(X)$.*

3 The Impact of Disjunction: Lower Bounds

To gauge the effect of disjunction, we first consider disjunction in isolation, and show that the derivability problem is in PTIME. This indicates that the lower bound results that appear later in this section are a result of *interaction* between the various logical rules, rather than due to disjunction alone.

3.1 The Disjunction-Only Fragment

Let $IL[\vee]$ denote the fragment of IL consisting of the ax, $\vee i$ and $\vee e$ rules, and involving formulas of $\Phi^{\vee}$.

Theorem 5. *The derivability problem for $IL[\vee]$ is in PTIME.*

Suppose $X = \{\alpha_i^1 \vee \alpha_i^2 \vee \cdots \vee \alpha_i^k \mid 1 \leq i \leq n\}$ is a set of formulas from $\Phi^{\vee}$, with each $\alpha_i^j \in \mathscr{P}$. Let $\beta = \beta^1 \vee \beta^2 \vee \cdots \vee \beta^k \in \Phi^{\vee}$, with each $\beta^j \in \mathscr{P}$. (Note that any input to the derivability problem of $IL^{\vee}$ can be converted to the above form by choosing appropriate k, flattening the disjunctions, and repeating disjuncts). We now have the following claim.

Claim. $X \vdash \beta$ iff there exists an $i \leq n$ such that $\alpha_i^1 \vee \alpha_i^2 \vee \cdots \vee \alpha_i^k \vdash \beta$.

Proof. It is obvious that if $\alpha_i^1 \vee \alpha_i^2 \vee \cdots \vee \alpha_i^k \vdash \beta$ then $X \vdash \beta$ (by Monotonicity).

For proving the other direction, suppose (towards a contradiction) $X \vdash \beta$, but there is no i such that $\alpha_i^1 \vee \alpha_i^2 \vee \cdots \vee \alpha_i^k \vdash \beta$. In particular, from the Left Disjunction Property, for every i, some $\alpha_i^{j_i} \nvdash \beta$. Without loss of generality, assume that $j_i = 1$ for every i. Thus we have $\alpha_1^1 \nvdash \beta$, $\alpha_2^1 \nvdash \beta$, $\ldots$, $\alpha_n^1 \nvdash \beta$.

Now, since $X \vdash \beta$ and $\alpha_i^1 \vdash \alpha_i^1 \vee \cdots \vee \alpha_i^k$ for each $i \leq n$, it follows by Admissibility of Cut that $\alpha_1^1, \ldots, \alpha_n^1 \vdash \beta$ (and there is a normal proof π with that conclusion). Since all the α_i^1s are atomic propositions, the only rules that can appear in π are ax and $\vee i$. Therefore, at some point, one of the α_i^1s must have contributed to a β^j via an ax rule. However, this gives us $\alpha_i^1 \vdash \beta$ (by deriving β^j and then applying $\vee i$), which is a contradiction. Thus we have the required claim. □

Given this claim, we know that it is enough to see if a particular formula on the left (say α_i) derives β. In particular, from the Left Disjunction Property, we get that every disjunct in α_i needs to derive β. Therefore, the derivability problem is equivalent to checking if there is a formula in X all of whose disjuncts occur in β, and thus we obtain the required PTIME procedure.

3.2   Disjunction and Conjunction

We have now confirmed that the ∨-only fragment is in PTIME. It is also known that some other fragments (for example the fragment consisting of primal implication, conjunction, and a restricted negation) give rise to PTIME logics. However, we obtain the following result for the logic with conjunction and disjunction.

Let $\mathsf{IL}[\vee, \wedge]$ denote the fragment of IL consisting of the ax, $\vee i$, $\vee e$, $\wedge i$ and $\wedge e$ rules, and involving formulas of $\Phi^{\vee, \wedge}$.

Theorem 6. *The derivability problem for $\mathsf{IL}[\vee, \wedge]$ is co-NP-hard.*

The hardness result is obtained by reducing the validity problem for boolean formulas to the derivability problem for $\mathsf{IL}[\vee, \wedge]$. In fact, it suffices to consider the validity problem for boolean formulas in disjunctive normal form. We show how to define for each DNF formula φ a set of $\mathsf{IL}[\vee, \wedge]$-formulas S_φ and an $\mathsf{IL}[\vee, \wedge]$-formula $\overline{\varphi}$ such that $S_\varphi \vdash \overline{\varphi}$ iff φ is a tautology.

Let $\{x_1, x_2, \ldots\}$ be the set of all boolean variables. For each boolean variable x_i, fix two distinct atomic propositions $p_i, q_i \in \mathscr{P}$. We define $\overline{\varphi}$ as follows, by induction.

- $\overline{x_i} = p_i$
- $\overline{\neg x_i} = q_i$
- $\overline{\varphi \vee \psi} = \overline{\varphi} \vee \overline{\psi}$
- $\overline{\varphi \wedge \psi} = \overline{\varphi} \wedge \overline{\psi}$

Let $\mathrm{Voc}(\varphi)$, the set of all boolean variables occurring in φ, be $\{x_1, \ldots, x_n\}$. Then $S_\varphi = \{p_1 \vee q_1, \ldots, p_n \vee q_n\}$.

Lemma 7. *$S_\varphi \vdash \overline{\varphi}$ iff φ is a tautology.*

Proof. Recall that a propositional valuation v over a set of variables $\mathscr{V}$ is just a subset of $\mathscr{V}$ – those variables that are set to **true** by v.

For a valuation $v \subseteq \{x_1, \ldots, x_n\}$, define $S_v = \{p_i \mid x_i \in v\} \cup \{q_i \mid x_i \notin v\}$.

By repeated appeal to the Left Disjunction Property, it is easy to see that $S_\varphi \vdash \overline{\varphi}$ iff for all valuations v over $\{x_1, \ldots, x_n\}$, $S_v \vdash \overline{\varphi}$. We now show that $S_v \vdash \overline{\varphi}$ iff $v \models \varphi$. The statement of the lemma follows immediately from this.

- We first show by induction on $\psi \in \mathsf{sf}(\varphi)$ that whenever $v \models \psi$, it is the case that $S_v \vdash \overline{\psi}$.
  - If $\psi = x_i$ or $\psi = \neg x_i$, then $S_v \vdash \overline{\psi}$ follows from the ax rule.
  - If $\psi = \psi_1 \wedge \psi_2$, then it is the case that $v \models \psi_1$ and $v \models \psi_2$. By induction hypothesis, $S_v \vdash \overline{\psi_1}$ and $S_v \vdash \overline{\psi_2}$. Hence, by using $\wedge i$, it follows that $S_v \vdash \overline{\psi_1} \wedge \overline{\psi_2}$.
  - If $\psi = \psi_1 \vee \psi_2$, then it is the case that either $v \models \psi_1$ or $v \models \psi_2$. By induction hypothesis, $S_v \vdash \overline{\psi_1}$ or $S_v \vdash \overline{\psi_2}$. In either case, by using $\vee i$, it follows that $S_v \vdash \overline{\psi_1} \vee \overline{\psi_2}$.

– We now show that if $S_v \vdash \overline{\varphi}$, then $v \models \varphi$. Suppose π is a normal proof of $S_v \vdash \varphi$, and that there is an occurrence of the $\wedge e$ rule or $\vee e$ rule in π with major premise $S' \vdash \gamma$. We denote by ϖ this subproof with conclusion $S' \vdash \gamma$. Note that ϖ ends in a pure elimination rule, since π is normal and every pure elimination rule and hybrid rule has as its major premise the conclusion of a pure elimination rule. By Theorem 4, we see that $S' \subseteq \mathsf{sf}(S_v) = S_v$, and $\gamma \in \mathsf{sf}(S')$. But γ is of the form $\alpha \vee \beta$ or $\alpha \wedge \beta$, and this contradicts the fact that $S_v \subseteq \mathscr{P}$. Thus π consists of only the ax, $\wedge i$ and $\vee i$ rules. We now show by induction that for all subproofs π' of π with conclusion $S_v \vdash \overline{\psi}$, $v \models \psi$.

- Suppose the last rule of π' is ax. Then $\overline{\psi} \in S_v$, and for some $i \leq n$, $\psi = x_i$ or $\psi = \neg x_i$. It can be easily seen that $v \models \psi$ (by the definition of S_v).
- Suppose the last rule of π' is $\wedge i$. Then $\overline{\psi} = \overline{\psi_1} \wedge \overline{\psi_2}$, and $S_v \vdash \overline{\psi_1}$ and $S_v \vdash \overline{\psi_2}$. Thus, by induction hypothesis, $v \models \psi_1$ and $v \models \psi_2$. Therefore $v \models \psi$.
- Suppose the last rule of π' is $\vee i$. Then $\overline{\psi} = \overline{\psi_1} \vee \overline{\psi_2}$, and either $S_v \vdash \overline{\psi_1}$ or $S_v \vdash \overline{\psi_2}$. Thus, by induction hypothesis, either $v \models \psi_1$ or $v \models \psi_2$. Therefore $v \models \psi$. $\qquad\square$

3.3 Disjunction and Implication Elimination

We now consider another minimal system, $\mathsf{IL}[\vee, \to e]$, consisting of the rules ax, $\vee i$, $\vee e$ and $\to e$ and involving formulas from $\Phi^{\vee, \to}$, and prove the following result.

Theorem 8. *The derivability problem for* $\mathsf{IL}[\vee, \to e]$ *is co-NP-hard.*

The proof is by reduction from the validity problem for 3-DNF, as detailed below.

Let φ be a 3-DNF formula with each clause having exactly three literals. Let $\mathrm{Voc}(\varphi) = \{x_1, \ldots, x_n\}$. We define $indx(\varphi)$ to be the set $\{1, \ldots, n\} \cup \{1', \ldots, n'\}$, where $(i')' = i$ for any $i \in indx(\varphi)$. For $i \leq n$, we define $\ell(i) = x_i$ and $\ell(i') = \neg x_i$. We define the following sets.

$$S_\varphi := \{p_a \vee p_{a'} \mid a \in indx(\varphi)\}.$$

$$T_\varphi := \{p_a \to p_b \to p_c \to p_{abc} \mid a, b, c \in indx(\varphi)\}.$$

We define $\overline{\varphi}$ as follows:

$$\overline{\varphi} := \bigvee \{p_{abc} \mid \ell(a) \wedge \ell(b) \wedge \ell(c) \text{ is a disjunct of } \varphi\}.$$

For each valuation $v \subseteq \{x_1, \ldots, x_n\}$, define S_v to be

$$\{p_i \mid x_i \in v\} \cup \{p_{i'} \mid x_i \notin v\}.$$

Lemma 9. $S_\varphi, T_\varphi \vdash \overline{\varphi}$ iff φ is a tautology.

Proof. By repeated appeal to the Left Disjunction Property, it is easy to see that $S_\varphi, T_\varphi \vdash \overline{\varphi}$ iff $S_v, T_\varphi \vdash \overline{\varphi}$ for all valuations v over $\{x_1, \ldots, x_n\}$. We now show that for all such valuations, $v \models \varphi$ iff $S_v, T_\varphi \vdash \overline{\varphi}$.

Let π be a normal proof of $S_v, T_\varphi \vdash \overline{\varphi}$. The last rule of π has to be $\vee i$, since if π ends in an elimination rule, from the Subformula Property it follows that a disjunction is a subformula of $S_v \cup T_\varphi$, which is not the case. Repeating this argument, we see that there is a subproof of π with conclusion $S_v, T_\varphi \vdash p_{abc}$ for some disjunct $\ell(a) \wedge \ell(b) \wedge \ell(c)$ of φ. We now show that for any valuation v, $S_v, T_\varphi \vdash p_{abc}$ iff $v \models \ell(a) \wedge \ell(b) \wedge \ell(c)$.

If $v \models \ell(a) \wedge \ell(b) \wedge \ell(c)$, then we have $p_a, p_b, p_c \in S_v$ (from the definition of S_v), and therefore by applying the $\rightarrow e$ rule to $p_a \rightarrow p_b \rightarrow p_c \rightarrow p_{abc}$ in T_φ, we have $S_v, T_\varphi \vdash p_{abc}$. In the other direction, suppose we have a normal proof π of $S_v, T_\varphi \vdash p_{abc}$. By examining S_v and T_φ, we see that only $p_a \rightarrow p_b \rightarrow p_c \rightarrow p_{abc}$ mentions p_{abc}. So it is clear that p_c must be derivable from S_v, T_φ, and the last rule of π must be $\rightarrow e$, applied to $p_c \rightarrow p_{abc}$. Now in order for this formula to be derivable, p_b must be derivable, and similarly p_a must be derivable. Since p_a, p_b and p_c can only be obtained by ax, it must be that $p_a, p_b, p_c \in S_v$ and therefore $v \models \ell(a) \wedge \ell(b) \wedge \ell(c)$.

Thus we have that $S_v, T_\varphi \vdash p_{abc}$ iff $v \models \ell(a) \wedge \ell(b) \wedge \ell(c)$, and the required claim follows.   □

4   Upper Bounds

We now show that a system with conjunction, disjunction, primal implication, and a restricted version of negation (allowing only negation elimination, but not negation introduction) is in co-NP. We first give a PTIME procedure for the logic without disjunction elimination and then lift it to a co-NP procedure which accounts for disjunction elimination.[2]

Fix a set of formulas X_0 and a formula α_0 for the rest of the section. Let $\mathsf{sf} = \mathsf{sf}(X_0 \cup \{\alpha_0\})$. Let $N = |\mathsf{sf}|$.

Definition 10. For any $X \subseteq \mathsf{sf}$:

– $derive(X) = \{\alpha \in \mathsf{sf} \mid X \vdash \alpha\}$.
– $derive'(X) = \{\alpha \in \mathsf{sf} \mid$ there is a proof of $X \vdash \alpha$ not using the $\vee e$ rule$\}$.

The following properties of *derive* and *derive'* are immediate.

– $X \subseteq derive'(X) \subseteq derive(X)$.
– $derive(X) = derive'(derive(X)) = derive(derive(X))$ (by Admissibility of Cut).

---

[2] It is important to note that we consider only the negation elimination rule. The algorithms in this section do not work in the presence of the $\neg i$ rule. Nor do we know of a straightforward modification to handle the $\neg i$ rule. It is not easy to say without further study whether the complexity stays the same or increases, either.

- $derive'(X) = derive'(derive'(X))$ (by Admissibility of Cut).
- If X is of the form $derive'(Y)$, then $derive'(X) = X$. If X is of the form $derive(Y)$, then $derive(X) = X$.

4.1 A PTIME procedure for *derive'*

In the absence of $\lor e$, there is no branching during proof search. Hence we can compute $derive'(Y)$ bottom-up in PTIME, as detailed below in Algorithm 1.

For $Y \subseteq \mathsf{sf}$, we define $onestep(Y) \subseteq \mathsf{sf}$ to be the set

$$\{\alpha \in \mathsf{sf} \mid \alpha \text{ is the conclusion of a rule } \mathsf{r} \text{ (other than } \lor e \text{) with premises } Z \subseteq Y\}.$$

Two important observations about $onestep(Y)$.

- $Y \subseteq onestep(Y)$, because of the rule ax.
- $onestep(Y)$ is computable in time $O(N^2)$, where $N = |\mathsf{sf}|$. This is because in all the rules other than $\lor e$, the antecedents (formulas occurring to the left of $\vdash$) in the premises are the same as the antecedents in the conclusion. Thus we need to consider only consequents (the formulas to the right of $\vdash$) in a proof. This means that we only need to consider all pairs of formulas in Y to compute $onestep(Y)$.

Algorithm 1. Algorithm to compute $derive'(X)$, for $X \subseteq \mathsf{sf}$

1: $Y \leftarrow \varnothing$;
2: $Y' \leftarrow X$;
3: **while** $(Y \neq Y')$ **do**
4:      $Y \leftarrow Y'$;
5:      $Y' \leftarrow onestep(Y)$;
6: **end while**
7: **return** Y.

Since $|\mathsf{sf}| = N$ and Y increases monotonically, the **while** loop runs only for N iterations. Thus $derive'(X)$ is computable in time $O(N^3)$.

4.2 A co-NP procedure for *derive*

Algorithm 2 checks if $X_0 \nvdash \alpha_0$. It uses the notion of a *down-closed set*. A set X of formulas is *down-closed* if it satisfies the following two conditions:

- $derive'(X) \subseteq X$.
- whenever $\alpha \lor \beta \in X$, then either $\alpha \in X$ or $\beta \in X$.

Y is said to be a *down-closure* of X if Y is down-closed and $X \subseteq Y$.

In Algorithm 2, it is an invariant that $Y = derive'(Z)$ for some Z and hence $derive'(Y) \subseteq Y$. Thus when Y is not down-closed, there exists $\beta_0 \lor \beta_1 \in Y$ such that neither β_0 nor β_1 is in Y.

The algorithm guesses a down-closure Y of X_0 such that $\alpha_0 \notin Y$. The following theorem guarantees that one can successfully guess such a Y iff $X_0 \nvdash \alpha_0$. This ensures the correctness of the algorithm.

Algorithm 2. Algorithm to check if $X_0 \nvdash \alpha_0$

1: $Y \leftarrow derive'(X_0)$;
2: **while** (Y is not down-closed) **do**
3:      guess a formula $\beta_0 \vee \beta_1 \in Y$ such that $\beta_0 \notin Y$ and $\beta_1 \notin Y$;
4:      guess $i \in \{0, 1\}$;
5:      $Y \leftarrow derive'(Y \cup \{\beta_i\})$;
6: **end while**
7: Return "Yes" if $\alpha_0 \notin Y$, and "No" otherwise.

Theorem 11. *For any X and α (with $X \cup \{\alpha\} \subseteq sf$), $X \vdash \alpha$ iff $\alpha \in Y$ for every down-closure Y of X.*

This theorem is a consequence of the following three lemmas. But first we need a general claim related to the Left Disjunction Property.

Claim. Suppose $\varphi_0 \vee \varphi_1 \in Z$ and $i \in \{0, 1\}$. Then $Z \setminus \{\varphi_0 \vee \varphi_1\}, \varphi_i \vdash \theta$ iff $Z, \varphi_i \vdash \theta$.

Lemma 12. *For any X and α (with $X \cup \{\alpha\} \subseteq sf$), $X \vdash \alpha$ iff $Y \vdash \alpha$ for every down-closure Y of X.*

Proof. Suppose $X \vdash \alpha$ and Y is a down-closure of X. Then $X \subseteq Y$ and hence it is immediate that $Y \vdash \alpha$.

Suppose on the other hand that $X \nvdash \alpha$. We show that there is a sequence $Y_0 \subsetneq Y_1 \subsetneq \cdots \subsetneq Y_n \subseteq sf$ of sets such that

- $X \subseteq Y_0$,
- Y_n is down-closed,
- for all $i \leq n$, $derive'(Y_i) \subseteq Y_i$, and
- for all $i \leq n$, $Y_i \nvdash \alpha$.

The sequence is constructed by induction. Y_0 is defined to be $derive'(X)$. Since $X \nvdash \alpha$, it follows that $Y_0 \nvdash \alpha$. Suppose Y_k has been defined for some $k \geq 0$ such that $Y_k \nvdash \alpha$. If Y_k is down-closed, we are done. Otherwise, since $derive'(Y_k) \subseteq Y_k$, there is a $\beta_0 \vee \beta_1 \in Y_k$ such that $\beta_0 \notin Y_k$ and $\beta_1 \notin Y_k$. Since $Y_k \nvdash \alpha$, it follows by the Left Disjunction property that $Y_k \setminus \{\beta_0 \vee, \beta_1\}, \beta_i \nvdash \alpha$ for some $i \in \{0, 1\}$. By Claim 4.2 it follows that $Y_k, \beta_i \nvdash \alpha$ for some $i \in \{0, 1\}$.

$$Y_{k+1} = \begin{cases} derive'(Y_k \cup \{\beta_0\}) & \text{if } Y_k, \beta_0 \nvdash \alpha \\ derive'(Y_k \cup \{\beta_1\}) & \text{otherwise} \end{cases}$$

Clearly $Y_k \subsetneq Y_{k+1}$ and $derive'(Y_{k+1}) = Y_{k+1}$. Assume without loss of generality that $Y_{k+1} = derive'(Y_k \cup \{\beta_0\})$. By construction, $Y_k \cup \{\beta_0\} \nvdash \alpha$. Now suppose $Y_{k+1} \vdash \alpha$. Then, since $Y_k \cup \{\beta_0\} \vdash \varphi$ for every $\varphi \in Y_{k+1}$, it would follow by Admissibilty of Cut that $Y_k \cup \{\beta_0\} \vdash \alpha$, which is a contradiction. Thus $Y_{k+1} \nvdash \alpha$. Thus we can always extend the sequence as desired.

Further, the Y_i's are strictly increasing, and are all subsets of sf. Thus $n \leq |sf|$ and the above construction terminates. Y_n is a down-closure of X that does not derive α. $\qquad \square$

Lemma 13. *Let π be a proof of $X \vdash \alpha$ with at least one occurrence of the $\vee e$ rule. Then there is an occurrence of $\vee e$ in π with major premise $X \vdash \varphi \vee \psi$ such that $\varphi \vee \psi \in derive'(X)$.*

Proof. In any proof of the form

$$
\frac{\begin{array}{ccc} \pi_1 & \pi_2 & \pi_3 \\ \vdots & \vdots & \vdots \\ X_1 \vdash \alpha_1 & X_2 \vdash \alpha_2 & X_3 \vdash \alpha_3 \end{array}}{Y \vdash \delta} \; r
$$

we say that any rule in π_1 is to the left of r, r is to the left of any rule in π_2, and any rule in π_2 is to the left of any rule in π_3.

Now consider the leftmost occurrence of $\vee e$ in π. It is the last rule of a subproof π' of π which looks as follows.

$$
\frac{\begin{array}{ccc} \pi_1' & \pi_2' & \pi_3' \\ \vdots & \vdots & \vdots \\ X' \vdash \varphi \vee \psi & X', \varphi \vdash \theta & X', \psi \vdash \theta \end{array}}{X' \vdash \theta} \; \vee e
$$

Since this is the leftmost occurrence of $\vee e$, there is no occurrence of $\vee e$ in π_1'. Further, if $X' \neq X$, it means that π' is part of the proof of a minor premise of some other $\vee e$ rule in π. But that contradicts the fact that π' ends in the leftmost $\vee e$ in π. Thus $X' = X$, and π_1' witnesses the fact that $\varphi \vee \psi \in derive'(X)$. $\square$

Lemma 14. *For a down-closed Y, $Y \vdash \alpha$ iff $\alpha \in Y$.*

Proof. If $\alpha \in Y$, then it is obvious that $Y \vdash \alpha$.

In the other direction, suppose $Y \vdash \alpha$ via a proof π with k instances of $\vee e$. We prove the required claim by induction on k.

In the base case, $k = 0$, and $\alpha \in derive'(Y)$. Since Y is down-closed, $derive'(Y) \subseteq Y$ and we have $\alpha \in Y$.

In the induction step, suppose there is an instance of $\vee e$ in the proof of $Y \vdash \alpha$. By Lemma 13, we know that there is at least one occurrence of $\vee e$ (say $Y \vdash \delta$) with major premise $Y \vdash \varphi \vee \psi$ such that $\varphi \vee \psi \in derive'(Y) \subseteq Y$, which looks as follows.

$$
\frac{\begin{array}{ccc} \pi_1 & \pi_2 & \pi_3 \\ \vdots & \vdots & \vdots \\ Y \vdash \varphi \vee \psi & Y, \varphi \vdash \delta & Y, \psi \vdash \delta \end{array}}{Y \vdash \delta} \; \vee e
$$

Thus we have $\varphi \vee \psi \in Y$. Since Y is down-closed either $\varphi \in Y$ or $\psi \in Y$. Suppose, without loss of generality, that $\varphi \in Y$. Now consider π_2. Since $\varphi \in Y$, we know that $Y \cup \{\varphi\} = Y$, and we can replace the big proof of $Y \vdash \delta$ by π_2, thereby reducing the number of instances of $\vee e$ in the proof of $Y \vdash \alpha$. By induction hypothesis, $\alpha \in Y$, and the lemma follows. $\square$

Running Time. We now analyze the running time of Algorithm 2. Since Y strictly increases with each iteration of the loop, there are at most $N = |\mathsf{sf}|$ iterations of the loop. In each iteration, we test whether Y is down-closed, which amounts to checking whether there is some $\beta_0 \lor \beta_1 \in Y$ such that neither β_0 nor β_1 is in Y. This check takes $O(N)$ time. We also compute $derive'(Y)$ in each iteration, which takes time $O(N^3)$. Thus the overall running time is $O(N^4)$. This can be improved to $O(N^2)$ by using a linear-time algorithm for $derive'$ like the one given in [11].

4.3 Bounding Resources

As is evident from the lower bound proofs, disjunction elimination contributes heavily to the complexity of the derivation problem. Thus the use of the $\lor e$ rule is an important resource. It makes sense to bound the use of this resource and explore its effect on complexity. In particular, we show that if we bound the set of formulas on which to perform disjunction elimination, we get a procedure whose running time is polynomial in the input size, though exponential in the number of disjunction eliminations allowed. The following definition makes this notion precise.

Definition 15. Let A be a set of disjunctive formulas. We define a proof of α from X using A (denoted $X \vdash_A \alpha$) as a proof where any $\lor e$ rules are applied only to formulas which appear in A.

Recall that we have fixed a set sf of size N, and that we consider the derivability of $X \vdash \alpha$ where $\mathsf{sf}(X \cup \{\alpha\}) \subseteq \mathsf{sf}$. We define $derive_A(X)$ to be $\{\beta \in \mathsf{sf} \mid X \vdash_A \beta\}$. Note that $derive_\varnothing(X)$ is $derive'(X)$. The check for $X \vdash_A \alpha$ is done by using Algorithm 3 to compute $derive_A(X)$ and then testing whether $\alpha \in derive_A(X)$. (For the purposes of the algorithm, we assume that the set A is equipped with a linear order, so we can refer to the least formula in any subset of A.)

Algorithm 3. Algorithm to compute $derive_A(X)$

1: **function** $f(A, X)$
2:    $Y \leftarrow derive'(X)$;
3:    **if** $A \cap Y = \varnothing$ **then**
4:        **return** Y;
5:    **else**
6:        $A' \leftarrow A \setminus \{\alpha \lor \beta\}$, where $\alpha \lor \beta$ is the least formula in $A \cap Y$;
7:        **return** $f(A', Y \cup \{\alpha\}) \cap f(A', Y \cup \{\beta\})$;
8:    **end if**
9: **end function**

In order to prove the correctness of the above algorithm, we require the following claim.

Claim. Suppose A is a set of disjunctions and $\alpha \lor \beta \in A$. Let $A' = A \setminus \{\alpha \lor \beta\}$. Then the following hold:

- If $X \vdash_A \gamma$ then $X, \alpha \vdash_{A'} \gamma$ and $X, \beta \vdash_{A'} \gamma$.
- If $X \vdash_A \alpha \vee \beta$, $X, \alpha \vdash_{A'} \gamma$ and $X, \beta \vdash_{A'} \gamma$, then $X \vdash_A \gamma$.

Proof.

- Suppose $X \vdash_A \gamma$. Then by monotonicity, we obtain a proof π of $X, \alpha \vdash \gamma$, such that the major premise of every instance of the $\vee e$ rule in π is in A. Note that for every sequent $X' \vdash \delta$ in π, $\alpha \in X'$. Consider any subproof π' of π whose conclusion is $X' \vdash \delta$ and last rule is $\vee e$ with major premise $\alpha \vee \beta$ (if there is no such subproof, then π witnesses the fact that $X, \alpha \vdash_{A'} \gamma$). π' has the following form.

$$
\begin{array}{ccc}
\pi'_1 & \pi'_2 & \pi'_3 \\
\vdots & \vdots & \vdots \\
\end{array}
$$

$$
\frac{X' \vdash \alpha \vee \beta \quad X', \alpha \vdash \delta \quad X', \beta \vdash \delta}{X' \vdash \delta} \; \vee e
$$

  But observe that since $\alpha \in X'$, $X' \cup \{\alpha\} = X'$. Thus π'_2 is itself a proof of $X' \vdash \delta$. We can replace π' by π'_2, thereby removing at least one instance of the $\vee e$ rule involving $\alpha \vee \beta$ in π. Repeating this, we obtain that $X, \alpha \vdash_{A'} \gamma$. A similar reasoning gives us the result for $X, \beta \vdash_{A'} \gamma$.
- Performing an or-elimination on $\alpha \vee \beta$ using the given proofs of $X, \alpha \vdash_{A'} \gamma$ and $X, \beta \vdash_{A'} \gamma$ and $X \vdash_A \alpha \vee \beta$ for premises gives us the required result of $X \vdash_A \gamma$. $\qquad \square$

Lemma 16. (Correctness of Algorithm 3). *For all X and A,*

$$
derive_A(X) = f(A, X).
$$

Proof. The proof is by induction on the size of A. The base case is when $A = \varnothing$, when clearly the procedure f returns $derive'(X)$.

For the induction case, suppose $X \vdash_A \delta$, and let $Y = derive'(X)$. Consider a normal proof π witnessing $X \vdash_A \delta$ and assume without loss of generality that there is at least one instance of $\vee e$ in π. From Lemma 13, we see that there is an instance of $\vee e$ in π with major premise $X \vdash \varphi \vee \psi$, where $\varphi \vee \psi \in derive'(X)$. Thus $A \cap Y \neq \varnothing$. Let $\alpha \vee \beta$ be the least formula in $A \cap Y$. Now since $X \subseteq Y$, $Y \vdash_A \delta$. Furthermore, $\alpha \vee \beta \in Y$. Hence, by Claim 4.3, $Y, \alpha \vdash_{A'} \delta$ and $Y, \beta \vdash_{A'} \delta$, where $A' = A \setminus \{\alpha \vee \beta\}$. Since A' is of smaller size than A, by the induction hypothesis, $derive_{A'}(Z) = f(A', Z)$ for any Z. Thus $\delta \in f(A', Y \cup \{\alpha\}) \cap f(A', Y \cup \{\beta\})$. It follows from the definition of f that $\delta \in f(A, X)$. Thus $derive_A(X) \subseteq f(A, X)$.

On the other hand suppose $\delta \in f(A, X)$, and assume without loss of generality that $A \cap Y \neq \varnothing$, where $Y = derive'(X)$. Letting $\alpha \vee \beta$ be the least formula in $A \cap Y$ and $A' = A \setminus \{\alpha \vee \beta\}$, it is clear that $\delta \in f(A', Y \cup \{\alpha\}) \cap f(A', Y \cup \{\beta\})$ from the definition of f. Since A' is of smaller size than A, it follows from the induction hypothesis that $Y, \alpha \vdash_{A'} \delta$ and $Y, \beta \vdash_{A'} \delta$. Since $Y = derive'(X)$, it is the case that $X \vdash' \gamma$ for every $\gamma \in Y$. Thus we can appeal to the admissibility of cut to conclude that $X, \alpha \vdash_{A'} \delta$ and $X, \beta \vdash_{A'} \delta$. It follows from Claim 4.3 that $X \vdash_A \delta$. Thus $f(A, X) \subseteq derive_A(X)$. $\qquad \square$

Theorem 17. *If* $|A| = k$, *then* $derive_A(X)$ *is computable in time* $O(2^k \cdot N)$.

Proof. There are at most 2^k recursive calls to f, and in each invocation we make one call to $derive'$, which takes $O(N)$ time. Thus the overall running time is $O(2^k \cdot N)$. □

5 Discussion

To summarize our results, we have proved that IL[∨] is in PTIME, while even minimal extensions like [∨, ∧], [∨, → e] and [∨, ⊥] are co-NP-hard. On the other hand, even the system with conjunction, disjunction, primal implication and negation elimination is in co-NP.

Of the two rules for negation, ¬e does not modify the assumptions in the sequents, whereas ¬i discharges the assumption α while concluding ¬α. There does not appear to be a straightforward adaptation of either Algorithm 1 or Algorithm 2 to handle ¬i. As we mentioned earlier, it is not clear whether the complexity of the logic changes either. Note that [4] considers a fragment with rules for primal implication, disjunction, and a ⊥ operator. While full implication and ⊥ can express full negation, primal implication and ⊥ can only capture the effect of the ¬e rule, not the ¬i rule. So the complexity of the fragment involving primal implication, conjunction, disjunction and "full" negation is still open. We leave this for future study.

We can also consider adding □-like modalities to the [∧, ∨] fragment of our logic. This system is in co-NP, and the algorithm proceeds along similar lines to the one in [15]. On the other hand, if we add modalities to a logic with implication (even primal implication), the system is PSPACE-complete [4].

There are several interesting ways in which to take this work forward. It is worthwhile to look for logics with restricted forms of disjunction that are efficiently solvable. We also need to identify scenarios in which it suffices to consider a bounded number of disjunction eliminations, wherein our PTIME algorithm in Sect. 4.3 is applicable.

References

1. Abadi, M., Burrows, M., Lampson, B., Plotkin, G.: A calculus for access control in distributed systems. ACM TOPLAS **15**(4), 706–734 (1993)
2. Baskar, A., Naldurg, P., Raghavendra, K.R., Suresh, S.P.: Primal infon logic: derivability in polynomial time. In: FSTTCS 2013, LIPIcs, vol. 24, pp. 163–174 (2013)
3. Baskar, A., Ramanujam, R., Suresh, S.P.: A DEXPTIME-complete Dolev-Yao theory with distributive encryption. In: Hliněný, P., Kučera, A. (eds.) MFCS 2010. LNCS, vol. 6281, pp. 102–113. Springer, Heidelberg (2010)
4. Beklemishev, L.D., Gurevich, Y.: Propositional primal logic with disjunction. J. Log. Comp. **24**(1), 257–282 (2014)
5. Chagrov, A., Zakharyaschev, M.: Modal Logic. Clarendon Press, Oxford (1997)
6. Comon-Lundh, H., Shmatikov, V.: Intruder deductions, constraint solving and insecurity decisions in presence of exclusive or. In: LICS 2003, pp. 271–280 (2003)

7. Comon-Lundh, H., Treinen, R.: Easy intruder deductions. In: Dershowitz, N. (ed.) Verification: Theory and Practice. LNCS, vol. 2772, pp. 225–242. Springer, Heidelberg (2003)
8. Dolev, D., Yao, A.: On the security of public-key protocols. IEEE Trans. Inf. Theory **29**, 198–208 (1983)
9. Gabbay, D.M., de Jongh, D.H.J.: A sequence of decidable finitely axiomatizable intermediate logics with the disjunction property. J. Sym. Log. **39**(1), 67–78 (1974)
10. Gurevich, Y., Neeman, I.: DKAL: Distributed-knowledge authorization language. In: 21st IEEE CSF Symposium, pp. 149–162. IEEE Press, New York (2008)
11. Gurevich, Y., Neeman, I.: Logic of infons: the propositional case. ACM Trans. Comp. Log. **12**(2), 9:1–9:28 (2011)
12. Kurokawa, H.: Hypersequent calculi for intuitionistic logic with classical atoms. APAL **161**(3), 427–446 (2009)
13. Magirius, M., Mundhenk, M., Palenta, R.: The complexity of primal logic with disjunction. IPL **115**(5), 536–542 (2015)
14. McAllester, D.A.: Automatic Recognition of Tractability in Inference Relations. JACM **40**(2), 284–303 (1993)
15. Ramanujam, R., Sundararajan, V., Suresh, S.P.: Extending Dolev-Yao with assertions. In: Prakash, A., Shyamasundar, R. (eds.) ICISS 2014. LNCS, vol. 8880, pp. 50–68. Springer, Heidelberg (2014)
16. Rusinowitch, M., Turuani, M.: Protocol insecurity with finite number of sessions and composed keys is NP-complete. TCS **299**, 451–475 (2003)
17. Sakharov, A.: Median logic. Technical report, St. Petersberg Mathematical Society (2004). http://www.mathsoc.spb.ru/preprint/2004/index.html

Intransitive Temporal Multi-agent's Logic, Knowledge and Uncertainty, Plausibility

Vladimir Rybakov[1,2](✉)

[1] School of Computing, Mathematics and DT,
Manchester Metropolitan University (full time), Chester Street,
Manchester M1 5GD, UK
[2] Mathematical Institute, Siberian Federal University (part time),
Krasnoyarsk, Russia
V.Rybakov@mmu.ac.uk

Abstract. We study intransitive temporal logic implementing multi-agent's approach and formalizing knowledge and uncertainty. An innovative point here is usage of non-transitive linear time and multi-valued models - the ones using separate valuations V_j for agent's knowledge of facts and summarized (agreed) valuation together with rules for computation truth values for compound formulas. The basic mathematical problems we study here are - decidability and decidability w.r.t. admissible rules. First, we study general case - the logic with non-uniform intransitivity and solve its decidability problem. Also we consider a modification of this logic - temporal logic with uniform non-transitivity and solve problem of recognizing admissibility in this logic.

Keywords: Temporal logic · Multi-agent's logic · Non-transitive time · Deciding algorithms · Knowledge · Admissible rules

1   Introduction

In the area of applications logic to Computer Science worthy place is occupied by temporal logic. It works very efficiently in various subdivisions of CS, Information Sciences and KR. An important version of temporal logic for CS, – LTL – linear temporal logic (with UNTIL and NEXT), was introduced by Z. Manna, and A. Pnueli in late 1980'. Since then, many impressive results concerning pure logical properties of LTL (e.g. decidability and axiomatization) were obtained (cf. e.g. Gabbay and Hodkinson [9–11], Vardi [26,27]). An essential component of information sciences is the notion of knowledge - a highly reliable information which is collected up to the moment and has some particular importance.

The approach to concept of knowledge in CS via multi-agent environment, when the knowledge to be obtained via agent's discussions, cooperation, evaluations, computational experiments, etc. formed a solid branch in CS (cf. e.g. [5,14,29–31]). An interpretation of knowledge in a multi-agent logic with distances was offered in Rybakov et al. [22]), an algorithm solving satisfiability

© Springer International Publishing Switzerland 2016
S. Artemov and A. Nerode (Eds.): LFCS 2016, LNCS 9537, pp. 364–375, 2016.
DOI: 10.1007/978-3-319-27683-0_25

problem was found. Concept of Chance Discovery in multi-agent's environment was studied in Rybakov [23,24]), a logic defining uncertainty via agents opinions was studied in McLean et al. [15]). Representation of agent's interaction as a dual of the logical operation formalizing common knowledge (which earlier was suggested in Fagin et al. [7]) was elaborated in Rybakov [20,21].

Approach to model knowledge in terms of symbolic logic appeared (probably first time) in Hintikka [13] in his book *Knowledge and Belief.* Now the field of knowledge representation and reasoning about knowledge in logical framework is very popular area. Various modal and multi-modal logics were used for modeling agents reasoning. In particular, multi-modal logics were used for this purpose in Balbiani et al. [6], Vakarelov [28], Fagin et al. [7], Rybakov [17,20]. Modern study of knowledge and believes in terms of single-modal logic was undertaken in Halpern et al. [12].

Concept of justification in terms of epistemic logic makes an another view-angle on knowledge (cf. e.g.. Artemov et al. [2,3]). The problem of rational agents and its effect to logical omniscience problem is studied recently (cf. Artemov, et al. [1]).

This our paper investigates intransitive temporal logic implementing multi-agent's approach and formalizing knowledge and uncertainty in this framework. An innovative point here is usage of non-transitive linear time and multi-valued models - the ones using separate valuations V_j for agent's knowledge of facts and summarized (agreed) valuation and rules for computation truth valued for compound formulas. We illustrate how the notion of knowledge and uncertainty might be represented in such framework. The basic mathematical problems we study here are the fundamental ones for any logical system - decidability and decidability w.r.t. admissible rules. First we consider very general case - the logic with non-uniform intransitivity and solve its decidability problem. The problem of recognizing admissible rules in this logic remains open. Next, we consider a modification of this logic - temporal logic with uniform non-transitivity and solve problem of recognizing admissibility in this logic.

2   Notation, Logical Language, Brief Motivation

To make our paper easy readable (without looking for external literature) we very briefly recall necessary definitions and notation. The language of Linear Temporal Logic (LTL in sequel) extends the language of Boolean logic by operations **N** (next) and **U** (until).

Formation rules for LTL-formulas built up from a set *Prop* of propositional letters are as follows: any letter of *Prop* is a formula. The set of all formulas is closed w.r.t. applications of Boolean operations, the unary operation **N** (next) and the binary operation **U** (until). Informal interpretation of the formula **N**φ is: φ holds in the next time point (state). A formula φ**U**ψ has meaning: φ will be true until ψ first time will be true. Standard semantic models for LTL are the following infinite linear Kripke structures.

A model is a quadruple $\mathcal{M} := \langle \mathcal{N}, \leq, \text{Next}, V \rangle$, where $\mathcal{N}$ is the set of all natural numbers; $\leq$ is the standard linear order on $\mathcal{N}$, Next is the binary relation, where a Next b is true iff b is the number next to a, that is $b = a + 1$.

The valuation V for a set of letters $P \subseteq Prop$ is a mapping which assigns truth values to elements of S. That is, for any $p \in S$, $V(p) \subseteq \mathcal{N}$. The set $V(p)$ is the set of all n from $\mathcal{N}$ where p is true (w.r.t. V).

The triple $\langle \mathcal{N}, \leq, \text{Next} \rangle$ from the above is said to be a Kripke frame (which we will denote in sequel for short by $\mathcal{N}$). For any Kripke model $\mathcal{M}$, the truth values via V for the propositional letters are extended to arbitrary formulas as follows:

$\forall p \in Prop \ (\mathcal{M}, a) \Vdash_V p \Leftrightarrow a \in \mathcal{N} \wedge a \in V(p);$

$(\mathcal{M}, a) \Vdash_V (\varphi \wedge \psi) \Leftrightarrow (\mathcal{M}, a) \Vdash_V \varphi \wedge (\mathcal{M}, a) \Vdash_V \psi;$

$(\mathcal{M}, a) \Vdash_V \neg\varphi \Leftrightarrow not[(\mathcal{M}, a) \Vdash_V \varphi];$

$(\mathcal{M}, a) \Vdash_V \mathbf{N}\varphi \Leftrightarrow \forall b[(a \text{ Next } b) \Rightarrow (\mathcal{M}, b) \Vdash_V \varphi];$

$(\mathcal{M}, a) \Vdash_V (\varphi \mathbf{U} \psi) \Leftrightarrow \exists b[(a \leq b) \wedge ((\mathcal{M}, b) \Vdash_V \psi) \wedge$

$\forall c[(a \leq c < b) \Rightarrow (\mathcal{M}, c) \Vdash_V \varphi]].$

A formula φ is said to be valid in the model $\mathcal{M}$ (denotation – $\mathcal{M} \Vdash \varphi$) if, for any b from $\mathcal{M}$ ($b \in \mathcal{N}$), $(\mathcal{M}, b) \Vdash_V \varphi$. The linear temporal logic LTL is the set of all formulas which are valid in all models.

The aim of our paper is to investigate linear logic with **intransitive time**. Therefore we briefly motivated our assumption about non-transitivity. Why we may assume that time might be non-transitive, what we mean by that? Here we consider time as a computational resource (e.g. its admitted length), as an individual human perception of time, as a background for collection and elicitation of knowledge.

Let us start from the **individual perception of time** in our human memory. We sense time as a sequence of events which we remember, we perceive it as a linear discrete succession (since we do not memorize very many events within few seconds). Our human memory if limited, finite. This means that what we knew and remembered a year ago might not be in our memory now; what we knew ten years ago may be not remembered by us a year ago. This says about intransitivity of human memory about time events in past.

Consider now **computational aspect** of time as a resource for analysis of results. Time events while computational runs may be recorded in protocols of computation for inspection and references. Any protocol is a finite sequence of records and not all necessary information might be found there. Though protocol may give references to other older protocols recorded in earlier computations. Here the time in applicational (not philosophical) aspect looks as non-transitive.

Assume that we work with extraction of data from **databases**. Data may be recorded in DBs and the storage of any one is finite. Any DB is recorded during a finite interval of time and may be incomplete. The procedure of updating DBs is a sequence of actions in time. This sequence may have terminating points, and what an old DB may contain could be already omitted in the updated one, so DBs knowledge in time is non-transitive.

Is we consider **multi-agents reasoning** than time events may be viewed as an individual ones (with effects as pointed above) and else this multi-agents brings its own effects. E.g. the amount of agents participating in taking decisions may be changed during some intervals of time; it might be not-uniform and to swell the same as to shrink. The priority of experts views may be changed, etc. And if experts view on truth of a statement was affirmative five years ago it may be opposite now. So, agents knowledge in time environment looks as intransitive.

3   Temporal Multi-agent's Modes, Temporal Logic

Our approach is based on non-transitive temporal logic $\mathcal{LTL}_{NT}$ and technique allowing to find its decision algorithm (cf. Rybakov [25]). We start from giving precise definition of our new, modified models and description of rules for computation truth values of formulas. Then we first comment how these new models may represent multi-agent information, knowledge and uncertainty, give some illustrating examples.

Definition 1. *An intransitive linear frame is a tuple*

$$\mathcal{F} := \langle N, \leq, \text{Next}, \bigcup_{j \in N} [R_j] \rangle,$$

components of which are as follows.

- $N = \bigcup_{i \in In \subset N}[i, m_i]$ *($[i, m_i]$ is the interval of all natural numbers situated between i and m_i). The set In is a set of indexes - it is a subset of N;*
- $\forall i_1, i_2 \in In$, $i_1 \neq i_2 \Rightarrow (i_1, m_{i_1}) \cap (i_2, m_{i_2}) = \emptyset$;
- $\forall i \in In$ $(m_i > i)$; *for any $j \in [i, m_i]$ any R_j is the standard linear order on the interval $[j, m_i]$;*
- Next *is the standard NEXT relation on N:* **n** Next **m** *if $m = n + 1$.*

For the sequel we fix notation: $t(i) := m_i$ - boundary of transitivity for i. The multi-agent's models $\mathcal{M}$ on such frames $\mathcal{F}$ are defined by fixing valuations $V_i, i \in A, ||A|| < \infty$ for a set of letters P - agents valuations for truth of letters $p \in P, -$, i.e. $\forall i, \forall p \in P, V_i(p) \subseteq N$.

A is a set of indexes for agents, for each model it may be different (any model may have its own fixed agents, their quantity may be different). For all n, $n \in V_i(p)$ is interpreted as p is true at the state n by opinion of the agent i. Also we consider the agreed (global) valuation V for letters from P: $-$

$$\mathbf{V(p)} = \{\mathbf{n} \mid \mathbf{n} \in \mathbf{N}, ||\{\mathbf{i} \mid \mathbf{i} \in \mathbf{A}, \mathbf{n} \in \mathbf{V_i(p)}\}|| > \mathbf{k}\},$$

where k is a fixed rational number (for this given model), which is bigger than $||A||/2$.

That is k is the **threshold**, which shows that the number of the agents which are sure that p is true in the given state (world) is big enough, bidder than half.

The particular value of k may vary from model to model - each one has its *own threshold*. Now any such model $\mathcal{M}$ is a multi-valued model - with a finite number of different valuations.

The logical language for our logic based at such models is an extension of the one for LTL from previous section. We extend it by agent's knowledge operations $A_i, i \in A$ applied to only letters - for all $p \in P, A_i(p)$ is a formula. We introduce rules for computation truth on models $\mathcal{M}$ for formulas as follows. For letters and boolean operations it is standard: $\forall p \in P, \forall n \in N, (\mathcal{F}, n) \Vdash_V p \Leftrightarrow p \in V(p); (\mathcal{F}, n) \Vdash_V \alpha \wedge \beta \Leftrightarrow [(\mathcal{F}, n) \Vdash_V \alpha$ and $(\mathcal{F}, n) \Vdash_V \beta]$; etc. For operation $\mathbf{N}$ - next, it is standard again:

$$\forall n \in N, (\mathcal{M}, n) \Vdash_V \mathbf{N}\varphi \Leftrightarrow [(n \ Next \ m) \Rightarrow (\mathcal{M}, m) \Vdash_V \varphi].$$

But $\mathbf{U}$ - until operation, and agents operations work in a non-standard way, since the models are intransitive and since agents truth operations work as nominals. We suggest the following rules:

Definition 2. *For any formulas φ and ψ,*

$$\forall n \in N, (\mathcal{M}, n) \Vdash_V (\varphi \ \mathbf{U} \ \psi) \Leftrightarrow$$

$$\exists m[(nR_n m) \wedge ((\mathcal{M}, m) \Vdash_V \psi) \wedge \forall k[(n \leq k < m) \Rightarrow (\mathcal{M}, k) \Vdash_V \varphi]];$$

$$\forall n \in N, \forall i \in A, (\mathcal{M}, n) \Vdash_V A_i(p) \Leftrightarrow n \in V_i(p).$$

The **agent's knowledge operations** A_i, as we see, are applied to only letters but not to temporal compound formulas. The *reason* for it is as follows. The origin of the problem comes from our definition of models as multi-valued models (when each agent has own valuation of propositional letters; but we see that approach is very natural and well corresponding to multi-agents reasoning).

If we would allow the operations A_i to be freely applied to arbitrary compound formulas and sub-formulas, then usage them in temporal (and modal) formulas would immediate cause clash/conflict in computation truth values.

E.g. if we have a formula $A_i(\varphi)$ with temporal formula φ, we either have to redefine truth values for letters p always in now and future (which means to ignore knowledge of other agents), or to resolve what to do with all other possible agent's sub-formulas $A_j(p_m)$ w.r.t. the agent i. Thus, it might be that we need to give some agents a preference, but it is not clear what for to make an advantage to some ones. So, because this uncertainty, we prefer to let these cobwebs for future research and to study first this basic case.

It is easy to accept that this approach correspond well to our intuition about multi-agent information and time. The agents have own knowledge about facts, we code it by $A_i(p)$. But the rules for computation compound formulas are already objective, general and global, the same for all agents. Though, to consider different rules for computation truth values for compound, nested formulas – looks as an attractive and promising idea.

The logic we wish to introduce is the collection of all general statements, formulas which are valid in all models.

Definition 3. *The multi-agent non-transitive logic* TMA$_{Int}$ *is the set of all formulas which are valid in all models* $\mathcal{M}$.

This logic is temporal, and therefore we may define via **U** the modal operations $\Box$ and $\Diamond$ in standard way: $\Diamond\varphi := \top\mathbf{U}\varphi$, $\Box\varphi := \neg\Diamond\neg\varphi$. The logic is intransitive which allows such formulas as e.g.

$$\Box p \wedge \Diamond\mathbf{N}\neg p$$

to be satisfiable. Indeed - it is sufficient to take the model with all $m_i - i = 3$ and p to be true on the interval $[1, 2, 3]$ and to be false elsewhere.

The understanding (formal definition(s)) of knowledge, uncertainty and plausibility may be convincingly interpreted if we will consider the models with time and NEXT directed to past (not to future). We may easy agree that knowledge is coming from past, but not from the future.

The past time - in our human memory (or in storage of information in DBs from previous experience, length of protocols for completed computations, etc.) evidently looks as *non-transitive*. Indeed, any **database** contains records stored in a finite amount of time, though it may contain information where to look for earlier events (so to say - to use NEXT - pointer to a new time interval).

We may understand knowledge as facts, statements which are convincingly true for all period of time which we remember (at least for leading part of experts, agents). This locally, in models, to be expressed by formula $\Box\varphi$: φ was always true in past for dominating parts of experts (agents).

Then, the **uncertainty** (in this approach), may be interpreted as e.g. $\Diamond\varphi \wedge \Diamond\neg\varphi$ - in some time points we remember agent's view for truth of φ was supportive, and in some - the opinion of a majority was against. So, the truth for φ in the interval of time which we remember was uncertain, not stable. Consider some more subtle example:

$$\Box[(\varphi \rightarrow [\Diamond\neg\varphi \vee \mathbf{N}\neg\varphi]) \wedge (\neg\varphi \rightarrow [\Diamond\varphi \vee \mathbf{N}\varphi])].$$

This formula expresses more delicate statement about uncertainty of truth values for φ - it always oscillates.

Plausibility of φ may be interpreted, e.g., as follows: $\neg\varphi \wedge \mathbf{N}\Box\varphi$: today experts hesitate about truth of φ, but always before today (admittedly - long time) they accepted it to be true. More example:

$$\Box\varphi \wedge \Box\Diamond\mathbf{N}^2\neg\varphi.$$

This formula says that φ is very plausible - as long as we remember with confidence, it was true, but in 2 steps above our reliable capacity of memory for past it was a state where it was false with some evidence.

Many similar interpretations reflecting various subtleties of understanding uncertainty and plausibility may be suggested via this approach (e.g. using preference in opinion of most knowledgeable agents, and so forth).

3.1   Technical Part, Decidability Algorithm of TMA$_{Int}$

Now we turn to main technical problems solved in this paper, first we consider decidability of TMA$_{Int}$. We will use here the approach from Rybakov [25] extending it for agent's knowledge operations.

An essential part of this approach is usage of the normal reduced forms for formulas, more exactly for inference rules to which formulas may be converted. It is very useful because it allows to avoid complicated calculations and evaluations for nested formulas. We will use reduction of formulas to inference rules (sequents). An inference rule is a sequent compound from the premises and the conclusion:

$$\mathbf{r} := \frac{\varphi_1(x_1, \ldots, x_n), \ldots, \varphi_l(x_1, \ldots, x_n)}{\psi(x_1, \ldots, x_n)}.$$

Here $\varphi_1(x_1, \ldots, x_n), \ldots, \varphi_l(x_1, \ldots, x_n)$ and $\psi(x_1, \ldots, x_n)$ are formulas constructed out of letters (variables) $x_1, \ldots, x_n$. This rule formalizes simplest reasoning step: $\psi(x_1, \ldots, x_n)$ (which is called conclusion) follows (logically follows) from all formulas $\varphi_1(x_1, \ldots, x_n), \ldots, \varphi_l(x_1, \ldots, x_n)$.

Definition 4. *We say that a rule* $\mathbf{r}$ *is* valid *in a model* $\mathcal{M}$ *if and only if the following holds:*

$$[\forall n \, ((\mathcal{M}, n) \Vdash_V \bigwedge_{1 \leq i \leq l} \varphi_i)] \Rightarrow [\forall m \, ((\mathcal{M}, m) \Vdash_V \psi)].$$

Otherwise we say $\mathbf{r}$ *is refuted in* $\mathcal{M}$*, or refuted in* $\mathcal{M}$ *by* V*, and write* $\mathcal{M} \nVdash_V \mathbf{r}$*. A rule* $\mathbf{r}$ *is* valid *in a frame* $\mathcal{F}$ *(notation* $\mathcal{F} \Vdash \mathbf{r}$*) if it is valid in any model based at* $\mathcal{F}$*.*

Usage of inference rules for decidability problem (verification if a formula is a theorem for our logic) is based at the following simple fact. Given a formula φ, we transform φ into the rule $x \to x/\varphi$. Then it is evident that

Lemma 1. *Formula* φ *is a theorem of* TMA$_{Int}$ *(that is* $\varphi \in$ TMA$_{Int}$*) iff the rule* $(x \to x/\varphi)$ *is valid in any frame* $\mathcal{F}$*.*

Thus, we bring decidability of formulas to decidability of rules; but surprisingly it simplifies the problem. We will use the rules in the reduced form.

Definition 5. *We say that a rule* $\mathbf{r}$ *has* reduced normal form *if*
$\mathbf{r} = \varepsilon/x_1$ *where*

$$\varepsilon := \bigvee_{1 \leq j \leq l} [\bigwedge_{1 \leq i \leq n} x_i^{t(j,i,0)} \wedge \bigwedge_{1 \leq i \leq n} (\mathbf{N}x_i)^{t(j,i,1)} \wedge \bigwedge_{m \in A,\, 1 \leq i \leq n} (A_m x_i)^{t(j,m,i,1)} \wedge$$

$$\bigwedge_{1 \leq i,k \leq n, i \neq k} (x_i \mathbf{U} x_k)^{t(j,i,k,1)}]$$

always $t(j,i,m), t(j,i,k,1), t(j,m,i,1) \in \{0,1\}$ *and, for any formula* α *above,*
$\alpha^0 := \alpha,\ \alpha^1 := \neg\alpha.$

Definition 6. *A rule* $\mathbf{r_{nf}}$ *in reduced normal form is a* normal reduced form *for a given rule* $\mathbf{r}$ *iff, for any frame* $\mathcal{F}$ *for* TMA_{Int}, $\mathcal{F} \Vdash \mathbf{r} \Leftrightarrow \mathcal{F} \Vdash \mathbf{r_{nf}}$.

Theorem 1. *For any given rule* $\mathbf{r}$ *we can construct in (single) exponential time some it's reduced normal form* $\mathbf{r_{nf}}$.

Proof is rather simple and short. It is sufficient to specify the language of our logic to the general algorithm described in e.g. Lemma 5 from [4] and to follow closely its proof. We may consider the rules with only single premise, so let $r = \alpha/\beta$ be an inference rule. For r, $Sub(r)$ be the set of all subformulas of r. We need a set of new variables $Z = \{z_\gamma \mid \gamma \in Sub(r)\}$.

Let us consider the rule in the *intermediate form*:

$$r_{\mathrm{if}} = z_\alpha \wedge \bigwedge_{\gamma \in Sub(r) \backslash Var(r)} (z_\gamma \leftrightarrow \gamma^\#)/z_\beta,$$

where

$$\gamma^\# = \begin{cases} z_\delta * z_\epsilon & \text{when } \gamma = \delta * \epsilon \text{ for } * \in \{\wedge, \vee, \rightarrow, \mathbf{U}\}. \\ *z_\delta & \text{when } \gamma = *\delta \text{ for } * \in \{\neg, \mathbf{N}, A_m, m \in A\}, \end{cases}$$

The rules r and r_{if} are equivalent w.r.t. truth at any model. Indeed, suppose M, be a model with a valuation V over its frame such that $M \not\Vdash_V r$. Then $M \Vdash_V \alpha$ and there exists an element $w \in N$, such that $(M,w) \not\Vdash_V \beta$. Let W be the valuation defined as follows: $W(z_\gamma) = V(\gamma)$. It is straightforward to show that $M \Vdash_W z_\alpha \wedge \bigwedge_{\gamma \in Sub(r) \backslash Var(r)} (z_\gamma \leftrightarrow \gamma^\#)$. In addition, $(M,w) \not\Vdash_W z_\beta$.

For the other direction, suppose $M \Vdash_W z_\alpha \wedge \bigwedge \{z_\gamma \leftrightarrow \gamma^\# \mid \gamma \in Sub(r) \backslash Var(r)\}$ and $(M,w) \not\Vdash_W z_\beta$, for some valuation $W : Z \rightarrow 2^N$ and some $w \in N$. Define $V : Var(r) \rightarrow 2^N$ by $V(x_i) = W(z_{x_i})$. It follows directly that for all $\gamma \in Sub(r)$, $V(\gamma) = W(z_\gamma)$. Thus $M \Vdash_V \alpha$, $(M,w) \not\Vdash_V \beta$, hence $M \not\Vdash_V r$.

Finally, we transform the premise of the obtained rule r_{if} into a perfect disjunctive normal form over primitives of the form x_i, $\mathbf{N}x_i$, $A_m x_i$ and $x_i \mathbf{U} x_j$. This requires no more than exponential time on the number of variables, i.e., on the number of sub-formulas of the original rule (the same as for reduction of any boolean formula to the perfect disjunctive normal form). Q.E.D.

Based at this reduction of formulas to rules in reduced forms and technique borrowed from Rybakov [25] we may prove

Theorem 2. *The satisfiability problem for* TMA$_{Int}$ *is decidable. There is an algorithm which, for any given formula, verifies its satisfiability, and computes a valuation satisfying it in a* **special** *finite model* $\mathcal{F}(N(r))$ *if it is satisfiable (at next stage we can transform this model in a standard infinite model).*

Here we extend the proof from [25] to adopt usage of agents knowledge operations A_m (following closely to the original proof). Thus, the logic TMA$_{Int}$ is decidable; this is first main technical result of our paper.

4   Problem of Admissibility

Far the more complicated decidability problem is decidability w.r.t. admissible inference rules. We would like to study admissibility problem for a logics from suggested background. Recall that a rule

$$\mathbf{r} := \frac{\varphi_1(x_1, \ldots, x_n), \ldots, \varphi_l(x_1, \ldots, x_n)}{\psi(x_1, \ldots, x_n)},$$

is said to be *admissible* in a logic L if, for every tuple of formulas, $\alpha_1, \ldots, \alpha_n$, we have $\psi(\alpha_1, \ldots, \alpha_n) \in L$ whenever $\forall i \; [\varphi_i(\alpha_1, \ldots, \alpha_n) \in L]$.

The solution of the admissibility problem for the logic LTL itself (i.e. finding an algorithm recognizing admissibility of inference rules) was obtained in Rybakov, 2008, [19] (cf. also [18]), basis for rules admissible in LTL was found in Babenyshev and Rybakov, 2011, [4].

We have to specify the notion of admissibility for inference rules in our multi-agent's logics because we use agent's knowledge operations A_i which cannot be used above nested formulas.

Definition 7. *A given rule*

$$\mathbf{r} := \frac{\varphi_1(x_1, \ldots, x_n), \ldots, \varphi_l(x_1, \ldots, x_n)}{\psi(x_1, \ldots, x_n)},$$

is said to be admissible *in the logic* TMA$_{Int}$ *if, for every tuple of formulas, $\alpha_1, \ldots, \alpha_n$, we have $\psi(\alpha_1, \ldots, \alpha_n) \in$ TMA$_{Int}$ whenever $\forall i \; [\varphi_i(\alpha_1, \ldots, \alpha_n) \in$ TMA$_{Int}]$, where for any x_i above if x_i has at least one occurrence in r in form $A_j(x_i)$ then $\alpha_i = x_i$.*

The restriction for substitutions above is necessary since our multi-agent logic cannot admit nested formulas bounded by agent's knowledge operations A_j. A restriction for substitutions in defining admissibility was already considered in literature (cf. for instance, - Odintsov, Rybakov - [16]). We currently cannot answer the question about recognizing admissibility in the logic TMA$_{Int}$ from previous section, but we are able to do it for its restricted version - the one for models with bounded intransitivity.

Definition 8. *A temporal frame $\mathcal{F}$ with uniform non-transitivity m is a particular case of frames for* TMA_{Int}

$$\mathcal{F} := \langle N, \leq, \mathrm{Next}, \bigcup_{i \in N} [R_i] \rangle$$

given in Definition 1 in Sect. 3, when any interval $[i, t(i)]$, has length m, where m is a fixed natural number (measure of intransitivity).

So, the only distinction from our general case in the previous section is that instead of arbitrary measure on intransitivity m_i for any world i, we consider the same and fixed one - m. It looks as we assume that models (objective world), not agents, always must remember the same interval of the time in past - the one with length m. Then we define models on such frames as we did earlier above (bearing in mind the presence multi-agent's valuations for agent's knowledge about truth the facts and agreed truth valuation V).

Definition 9. *The logic $\mathrm{TMA}_{Int,m}$ is the set of all formulas which are valid at any model $\mathcal{M}$ with the measure of intransitivity m.*

The definition of admissibility for inference rules in this logic is exactly the same as we defined above in this section for TMA_{Int}. It seems that to consider and to discuss such logic is reasonable, since we may put limitations on the size of time intervals that agents (experts) may introspect in future (or to remember in past). An easy observation concerning the logic $\mathrm{TMA}_{Int,m}$ itself is that it is decidable: it is **trivial** (since for verification if a formula of temporal degree k is a theorem of $\mathrm{TMA}_{Int,m}$ we will need to check it on only initial part of the frames consisting only $k + 1$ subsequent intervals of length m). One more immediate observation is:

Proposition 1. $\mathrm{TMA}_{Int,m} \not\subseteq \mathrm{TMA}_{Int}$ *for all m.*

Proof is evident since

$$(\bigwedge_{i \leq m} [p \wedge \mathbf{N}^i p] \rightarrow \Box p) \in \mathrm{TMA}_{Int,m}.$$

The main technical result of this section is solution of the admissibility problem for logics $\mathrm{TMA}_{Int,m}$.

Theorem 3. *For any m, the linear temporal logic with* uniform *non-transitivity $\mathrm{TMA}_{Int,m}$ is decidable w.r.t. admissibility of inference rules.*

The proof is essentially other than the one for decidability w.r.t. admissible rules of the linear (transitive) temporal logic LTL itself (given in [19]). Presence of infinite sequence intransitivity intervals in the models makes the case different.

5 Open Problems

We think the following open questions could be of interest:

(i) Decidability of TMA_{Int} itself w.r.t. admissible inference rules.
(ii) Decidability w.r.t. admissible rules of the variant of $TMA_{Int,m}$ with non-uniform intransitivity, when intransitivity intervals are of length at most m, but the length may be different.
(iii) The problems of axiomatization for TMA_{Int} and $TMA_{Int,m}$ are open.

References

1. Artemov, S., Kuznets, R.: Logical omniscience as infeasibility. Ann. Pure Appl. Logic **165**, 6–25 (2014)
2. Artemov, S.: Justified common knowledge. Theor. Comput. Sci. **357**, 4–22 (2006)
3. Artemov, S., Nogina, E.: Introducing justification into epistemic logic. J. Logic Comput. **15**, 1059–1073 (2005)
4. Babenyshev, S., Rybakov, V.: Linear temporal logic LTL, basis for admissible rules. J. Logic Comput. **21**, 1057–177 (2011)
5. Belardinelli, F., Lomuscio, A.: Interactions between knowledge and time in a first-order logic for multi-agent systems: completeness results. J. Artif. Intell. Res. **45**, 1–45 (2012)
6. Balbiani, P., Vakarelov, D.: A modal logic for indiscernibility and complementarity in information systems. Fundam. Inform. **50**, 243–263 (2002)
7. Fagin, R., Halpern, J., Moses, Y., Vardi, M.: Reasoning About Knowledge. MIT Press, Cambridge (1995)
8. Friedman, H.: One hundred and two problems in mathematical logic. J. Symbolic Logic **40**, 113–130 (1975)
9. Gabbay, D.M., Hodkinson, I.M., Reynolds, M.A.: Temporal Logic: Mathematical Foundations and Computational Aspects. Clarendon Press, Oxford (1994)
10. Gabbay, D.M., Hodkinson, I.M.: An axiomatization of the temporal logic with until and since over the real numbers. J. Logic Comput. **1**, 229–260 (1990)
11. Gabbay, D., Hodkinson, I.: Temporal logic in context of databases. In: Copeland, J. (ed.) Logic and Reality, Essays on the legacy of Arthur Prior. Oxford University Press, Oxford (1995)
12. Halpern, J., Samet, D., Segev, E.: Defining knowledge in terms of belief. Modal Logic Perspect. Rev. Symbolic Logic **2**, 469–487 (2009)
13. Hintikka, J.: Knowledge and Belief: An Introduction to the Logic of the Two Notions. Cornell University Press, Ithaca (1962)
14. Lomuscio, A., Michaliszyn, J.: An epistemic halpern-shoham logic. In: Proceedings of the 23rd International Joint Conference on Artificial Intelligence (IJCAI 2013), pp. 1010–1016. AAAI Press, Beijing (2013)
15. McLean, D., Rybakov, V.: Multi-agent temporary logic $TS4_{K_n}^U$ based at non-linear time and imitating uncertainty via agents' interaction. In: Rutkowski, L., Korytkowski, M., Scherer, R., Tadeusiewicz, R., Zadeh, L.A., Zurada, J.M. (eds.) ICAISC 2013, Part II. LNCS, vol. 7895, pp. 375–384. Springer, Heidelberg (2013)
16. Odintsov, S., Rybakov, V.: Inference rules in Nelson's logics, admissibility and weak admissibility. Logica Univers. **9**, 93–120 (2015)

17. Rybakov, V.V.: Refined common knowledge logics or logics of common information. Arch. Math. Logic **42**, 179–200 (2003)
18. Rybakov, V.V.: Logical consecutions in discrete linear temporal logic. J. Symbolic Logic **70**, 1137–1149 (2005)
19. Rybakov, V.V.: Linear temporal logic with until and next, logical consecutions. Ann. Pure Appl. Logic **155**, 32–45 (2008)
20. Rybakov, V.: Logic of knowledge and discovery via interacting agents - decision algorithm for true and satisfiable statements. Inf. Sci. **179**, 1608–1614 (2009)
21. Rybakov, V.: Linear temporal logic LTL_{K_n} extended by multi-agent logic K_n with interacting agents. J. Logic Comput. **19**, 989–1017 (2009)
22. Rybakov, V., Babenyshev, S.: Multi-agent logic with distances based on linear temporal frames. In: Rutkowski, L., Scherer, R., Tadeusiewicz, R., Zadeh, L.A., Zurada, J.M. (eds.) ICAISC 2010, Part II. LNCS, vol. 6114, pp. 337–344. Springer, Heidelberg (2010)
23. Rybakov, V.V.: Chance discovery and unification in linear modal logic. In: König, A., Dengel, A., Hinkelmann, K., Kise, K., Howlett, R.J., Jain, L.C. (eds.) KES 2011, Part II. LNCS, vol. 6882, pp. 478–485. Springer, Heidelberg (2011)
24. Rybakov, V.V.: Logical analysis for chance discovery in multi-agents' environment. In: Graña, M., et al. (eds.) Frontiers in Artificial Intelligence and Applications, vol. 243, pp. 1593–1601. Springer, Heidelberg (2012)
25. Rybakov, V.: Non-transitive linear temporal logic and logical knowledge operations. J. Logic Comput. (2015). doi:10.1093/logcom/exv016
26. Vardi, M.: An automata-theoretic approach to linear temporal logic. In: Moller, F., Birtwistle, G. (eds.) Logics for Concurrency. LNCS, vol. 1043, pp. 238–266. Springer, Heidelberg (1995). http://citeseer.ist.psu.edu/vardi96automatatheoretic.htm
27. Vardi, M.Y.: Reasoning about the past with two-way automata. In: Larsen, K.G., Skyum, S., Winskel, G. (eds.) ICALP 1998. LNCS, vol. 1443, pp. 628–641. Springer, Heidelberg (1998)
28. Vakarelov, D.: A modal characterization of indiscernibility and similarity relations in Pawlak's information systems. In: Ślęzak, D., Wang, G., Szczuka, M.S., Düntsch, I., Yao, Y. (eds.) RSFDGrC 2005. LNCS (LNAI), vol. 3641, pp. 12–22. Springer, Heidelberg (2005)
29. Wooldridge, M.J., Lomuscio, A.: Multi-agent VSK logic. In: Brewka, G., Moniz Pereira, L., Ojeda-Aciego, M., Guzmán, I.P. (eds.) JELIA 2000. LNCS (LNAI), vol. 1919, pp. 300–312. Springer, Heidelberg (2000)
30. Wooldridge, M.: An automata-theoretic approach to multi-agent planning. In: Proceedings of the First European Workshop on Multi-agent Systems (EUMAS 2003). Oxford University (2003)
31. Wooldridge, M., Huget, M., Fisher, M., Parsons, S.: Model checking multi-agent systems: the MABLE language and its applications. Int. J. Artif. Intell. Tools **15**, 195–225 (2006)

Ogden Property for Linear Displacement Context-Free Grammars

Alexey Sorokin[1,2]([✉])

[1] Faculty of Mathematics and Mechanics, Moscow State University, Moscow, Russia
alexey.sorokin@list.ru
[2] Faculty of Innovations and High Technologies,
Moscow Institute of Physics and Technology, Dolgoprudny, Russia

Abstract. It is known that Ogden lemma fails for the class of k-well-nested multiple context-free languages for $k \geq 3$. In this article we prove a relaxed version of this lemma for linear well-nested MCFLs and show that its statement may be applied to generate counterexamples of linear well-nested MCFLs by the method already existing for the stronger variant.

Keywords: Linear discontinuous context-free languages · Pumping lemma · Ogden lemma · Limit ogden lemma · Counterexamples

1 Introduction

It is known since 1980-s that context-free grammars are too weak to capture the syntax processes in natural language ([7]). There were several suggestions how to preserve such properties as polynomial parsing complexity and independence of derivations from context. One of the most interesting candidates was the class of well-nested multiple context-free languages (wMCFLs, [2]), which are argued to be a formalisation of the notion of mildly context-sensitive grammars ([1]).

It is widely accepted that the language MIX $= \{w \in \{a, b, c\}^* || w|_a = |w|_b = |w|_c\}$ is not mildly context-sensitive since it allows too much freedom in word order. Therefore if wMCFLs are indeed the desired formalisation, then MIX should lie outside this family. In any case, the methods to prove that a language is not a wMCFL should be developed. In 2014 Alexey Sorokin ([10]) strengthened the pumping lemma for wMCFLs, extending the earlier result of Kanazawa([2]). He showed that for every well-nested $(k + 1)$-MCFL L there exists a number p such that every word $w \in L$ with $|w| \geq p$ admits a representation $w = x_0 u_0 y_0 v_0 x_1 u_1 x_1 v_1 \ldots x_k u_k y_k v_k x_{k+1}$ satisfying the following conditions:

1. $u_0 v_0 \ldots u_k v_k \neq \epsilon$,
2. $\sum\limits_{i}(|u_i| + |v_i|) \leq p$,
3. For every $n \in \mathbb{N}$ it holds, that $x_0 u_0^n y_0 v_0^n x_1 u_1^n x_1 v_1^n \ldots x_k u_k^n y_k v_k^n x_{k+1} \in L$

The work was partly supported by the grant NSh-1423.2014.1 "Mathematical logic and algorithm theory."

© Springer International Publishing Switzerland 2016
S. Artemov and A. Nerode (Eds.): LFCS 2016, LNCS 9537, pp. 376–391, 2016.
DOI: 10.1007/978-3-319-27683-0_26

Sorokin also tried to prove the following statement, similar to standard Ogden lemma ([5]): for every word w there is a family of decompositions of the form given above, such that all but at most p symbols is located in some u_i, v_i (the "pumpable" segments) in exactly one decomposition and the remaining ones do not participate in u_i, v_i at all. However, his proof contained an error, moreover, Makoto Kanazawa gave a counterexample ([3]): the language including the words $a^{n_1} b^{n_0} c^{n_0} d^{n_1} \$ a^{n_2} b^{n_1} c^{n_1} d^{n_2} \$ \ldots \$ a^{n_r} b^{n_r-1} c^{n_r-1} d^{n_r}$, $r \geq 3$, $n_0, \ldots, n_r \in \mathbb{N}$ is generated by a 3-well-nested-MCFG, but in every possible decomposition the \$-s are outside the pumpable segments. This counterexample is very strong since the given wMCFG is non-branching (linear in our terminology), i.e. right sides of its rules contain at most one nonterminal. In fact, non-branching 3-wMCFLs are the narrowest class of languages which lack Ogden property, because for 2-MCFLs (which are tree-adjoining languages) Ogden lemma was proved in [6].

However, even weaker variants of Ogden lemma may be useful for proving that a particular language is not a k-MCFL. Our goal is the following relaxed formulation: the number of unpumped symbols should be bounded not by a constant but by arbitrarily slow linear function. We say that a language L satisfies limit Ogden property if for any positive $\alpha < 1$ there exists a number p_α such that for any $w \in L$ with $|w| \geq p_\alpha$ at most $\alpha |w|$ symbols do not participate in pumpable segments and the others take part in them exactly once. We prove that linear k-wMCFLs satisfy limit Ogden property for all natural k. Before giving the general proof we explain the construction for linear 3-wMCFLs.

2   Definitions

We use the displacement context-free representation of well-nested MCFLs, basing on [8]. The only difference is that string tuples are replaced by gapped strings. Let Σ be a finite alphabet, then Σ^* denotes the set of all words with letters in Σ, ϵ being the empty string. When Σ is fixed, Θ_k denotes the set of all tuples of the form $(u_0, \ldots, u_k)$, $u_i \in \Sigma^*$ and $\Theta = \bigcup_{k \in \mathbb{N}} \Theta_k$. We call k the rank of the tuple $u = (u_0, \ldots, u_k)$ and denote it by $\mathrm{rk}(u)$. The length $|u|$ of a tuple $|u|$ is the sum of lengths of all its components, we denote by $\Theta^{(l)}$ the set of all tuples of length l. The notations $\Theta^{(\leq l)}$ and $\Theta^{(\geq l)}$ are understood in a natural way.

On the set of tuples we define the concatenation operation $\cdot \colon \Theta_i \times \Theta_j \to \Theta_{i+j}$ and the countable set of intercalation operations $\odot_l \colon \Theta_i \times \Theta_j \to \Theta_{i+j-1}$, $l \geq 1$:

$$(x_0, \ldots, x_i) \cdot (y_0, \ldots, y_j) = (x_0, \ldots, x_i y_0, \ldots, y_j)$$
$$(x_0, \ldots, x_i) \odot_l (y_0, \ldots, y_j) = (x_0, \ldots x_{l-1} y_0, y_1, \ldots, y_j x_l, \ldots, x_i)$$

Let N be a finite ranked set of nonterminals and $\mathrm{rk} \colon N \to \mathbb{N}$ be the rank function. Let $Op_k = \{\cdot, \odot_1, \ldots, \odot_k\}$, the set $Tm_k(N, \Sigma)$ of k-correct terms is defined as follows:

1. $\forall j \leq k \, (\Theta_j \subset \mathrm{Tm}_k(N, \Sigma))$.
2. If $\alpha, \beta \in \mathrm{Tm}_k$ and $\mathrm{rk}(\alpha) + \mathrm{rk}(\beta) \leq k$, then $(\alpha \cdot \beta) \in \mathrm{Tm}_k$, $\mathrm{rk}(\alpha \cdot \beta) = \mathrm{rk}(\alpha) + \mathrm{rk}(\beta)$.

3. If $j \leq k$, $\alpha, \beta \in Tm_k$, $\mathrm{rk}(\alpha) + \mathrm{rk}(\beta) \leq k+1$, $\mathrm{rk}(\alpha) \geq j$, then $(\alpha \odot_j \beta) \in Tm_k$, $\mathrm{rk}(\alpha \cdot \beta) = \mathrm{rk}(\alpha) + \mathrm{rk}(\beta) - 1$.

We assume that all the operation symbols are leftassociative and concatenation has greater priority then intercalation. We may also omit the $\cdot$ symbol, so the notation $A \odot_2 BC \odot_1 D$ means $(A \odot_2 ((B \cdot C)) \odot_1 D)$.

Let $\mathrm{Var} = \{x_1, x_2, \ldots\}$ be a countable ranked set of variables, such that for every k there is an infinite number of variables having rank k. A context $C[x]$ is a term where a variable x occurs in a leaf position, the rank of x must respect the constraints of term construction. Provided $\beta \in Tm_k$ and $\mathrm{rk}(x) = \mathrm{rk}(\beta)$, $C[\beta]$ denotes the result of substituting β for x in C. A valuation function ν maps all the elements of Θ to themselves and variable — to the words of the same rank. Interpreting the connectives from Op_k as corresponding binary operations, we are able to calculate the value of every ground term (i.e. containing no nonterminal occurrences). It is easy to prove that $\mathrm{rk}(\alpha) = \mathrm{rk}(\nu(\alpha))$ for every α. The set of k-correct ground terms is denoted by $\mathrm{GrTm}_k(\Sigma)$. Analogously, a context is ground if it contains no nonterminals.

Definition 1. *A k-displacement context-free grammar (k-DCFG) is a quadruple $G = \langle N, \Sigma, P, S \rangle$, where Σ is a finite alphabet, N is a finite ranked set of nonterminals and $\Sigma \cap N = \varnothing$, $S \in N$ is a start symbol such that $rk(S) = 0$ and P is a set of rules of the form $A \to \alpha$. Here A is a nonterminal, α is a term from $Tm_k(N, \Sigma)$, such that $rk(A) = rk(\alpha)$.*

Definition 2. *The derivability relation $\vdash_G \in N \times Tm_k$ associated with the grammar G is the smallest reflexive transitive relation such that the facts $(B \to \beta) \in P$ and $A \vdash C[B]$ imply that $A \vdash C[\beta]$ for any context C. Let $L_G(A) = \{\nu(\alpha) \mid A \vdash_G \alpha, \alpha \in GrTm_k\}$ denote the set of word, which are derivable from a nonterminal A, then $L(G) = L_G(S)$.*

Example 1. A k-DCFG $G_k = \langle \{S, T\}, \{a_i, b_i \mid i \in [0; k]\}, P, S \rangle$, where the set P is defined below, derives the language $L_k = \{a_0^m b_0^m \ldots a_k^m b_k^m\}$.

$$S \to \underbrace{(\ldots(\ T \odot_1 \epsilon) \ldots)}_{(k-1) \text{ times}} \odot_1 \epsilon$$

$$T \to a_0(T \odot_1 (b_0, a_1)) \ldots \odot_k (b_{k-1}, a_k))b_k$$
$$T \to (\underbrace{\epsilon, \ldots, \epsilon}_{(k+1) \text{ times}})$$

Definition 3. *A term is called linear, if it contains at most one occurrence of nonterminal. A DCFG is linear[1], if all rights sides of the rules are linear terms.*

A linear DCFG is said to be in linear Chomsky normal form, if it has the rules only of the following form:

---

[1] Following the Russian tradition, the author prefers the term "linear" to "nonbranching". However, he acknowledges the extreme ambiguity of this term. May be, some other term should be more proper.

1. $A \to B \odot_j u$, $B \in N - \{S\}$, $u \in \Theta^{(1)}$;
2. $A \to B \cdot u$ or $A \to u \cdot B$, $B \in N - \{S\}$, $u \in \Theta^{(1)}$;
3. $A \to u$, $u \in \Theta^{(1)}$;
4. $S \to \epsilon$.

As proved in [11], every linear grammar may be transformed to linear Chomsky normal form. To generalize our proofs we introduce an auxiliary notion: a grammar is said to be reduced if its rules have only the following form:

- $A \to \alpha * \beta$, $* \in Op_k$, $\alpha, \beta \in (N - \{S\}) \cup (\Theta - \{\epsilon\})$ and α or β belongs to N.
- $A \to u$, $u \in \Theta - \{\epsilon\}$.
- $S \to \epsilon$.

If we extend the valuation to nonterminals, matching every $A \in N$ with some tuple $\mu(A) \in \Theta_{\mathrm{rk}(A)}$, then we are able to calculate the value of arbitrary term as well. Two terms are equivalent if they have the same value under any possible valuation. Note that if we replace the term α in the rule $A \to \alpha$ by some equivalent term α', then the generated language remains the same. Moreover, if it holds that $A \vdash_G \alpha$, then adding a rule $A \to \alpha'$ does not alter the language as well. We will extensively use this argument in the paper. The equivalence of contexts is defined analogously. A term is k-essential, if its rank does not exceed k and all its nonterminal nodes are of rank k or less (internal nodes and terminal leaves may be of greater rank). The lemma below is an analogue of Lemma 7 in [10]. Before proving the lemma we introduce some technical notion.

Definition 4. *A term α is called specialized if it has one of the following forms:*

1. $\alpha = u$, $u \in \Theta^{(1)}$;
2. $\alpha = \beta \cdot u$ or $\alpha = u \cdot \beta$, $u \in \Theta^{(1)}$, β being a specialized term;
3. $\alpha = \beta \odot_j u$, $u \in \Theta^{(1)}$, β being a specialized term.

We want to prove that every term is equivalent to some specialized term using the next statement, which is easily proved by induction.

Statement 1. *For every tuple $w \in \Theta_{\geq 1}$ there exist tuples $w_1, \ldots, w_t$ in Θ_1 with $|w_1| = \ldots = |w_t| = 1$ and indexes $j_1, \ldots, j_t$, such that the context $x \odot_j w$ is equivalent to the context $(\ldots (x \odot_{j_1} w_1) \odot_{j_2} \ldots) \odot_{j_t} w_t$.*

Lemma 1. *For every k, any k-essential specialized linear term α is equivalent to some k-correct specialized linear term α'.*

Proof. The scheme of the proof is taken from Lemma 7 [10]. We use induction on the size of α, maximal rank of its subterms and the number of subterms of maximal rank.

3   Direct Descendance and Pumping Properties

In this section we give introduce the concepts extensively used through the paper. We assume that an arbitrary linear DCFG G in linear Chomsky normal form is fixed and consider only the derivations in this fixed grammar. We additionally assumes that every nonterminal of G derives some word. Notions and statements valid for every reduced grammar, if the contrary is not explicitly mentioned.

Recall the proof of pumping lemma for context-free languages: we consider a grammar in Chomsky normal form and by the pigeon-hole principle find on a sufficiently long branch two nodes labeled by the same nonterminal A. In our terminology, it means that $A \vdash C[A]$ for some ground context C. Repeating this derivation, we obtain that $A \vdash C[C[A]] \vdash C[C[C[A]]]$ and so on. Turning from contexts to strings and using the definition of Chomsky normal form, we obtain the statement of the lemma.

However, a literal application of this construction to DCFGs fails. Consider a grammar with the rules $A \to B(b, c)$ and $B \to A \odot_1 a$, here $\mathrm{rk}(A) = 1, \mathrm{rk}(B) = 0$. Then $A \vdash (A \odot_1 a)(b, c)$, so if $A \vdash (u_0, u_1)$, then also $A \vdash (u_0 a u_1 b, c)$, so both the fragments derived from the lowest A are in the same fragment in the tuple derived from the highest A, which is not the case predicted by the pumping lemma. The key problem is the nonterminal of lower rank between the two occurrences of A in the constructed derivation tree. It is easy to see, that in case of derivation $B \to A \odot_1 a \to (B(b, c)) \odot_1 a \sim Bbac$ we obtain the required statement: $B \vdash u$ implies $B \vdash ubac$. To obtain an analogous pumping for nonterminals of higher rank a grammar should be transformed. The transformation we use improves the one from [10], which bases on the idea from [2].

To describe the required transformation we need some definitions. The derivation trees for DCFGs are defined just like for the context-free grammars: if the derivation has the form $A \vdash C[B] \vdash C[\beta]$ with the rule $B \to \beta$ applied, then its tree is obtained from the tree of $A \vdash C[B]$ by attaching the syntactic tree of β to the leaf corresponding to the distinguished occurrence of B. The tree is said to be terminal if it derives a tuple from Θ. For every tree node we define its rank as the rank of the term derived from this node. Nodes are labeled by nonterminals and binary connectives in a natural way. For example, if the grammar has the rules $A \to (B \odot_1 b)C$, $B \to (a, c)$, $C \to d$, then the derivation $A \vdash (B \odot_1 b)C \vdash ((a, c) \odot_1 b)d \sim abcd$ has the tree shown below.

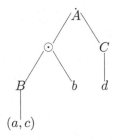

In what follows we usually denote a derivation tree by T and its nodes by u, v, possibly with indexes. If u is a node, then T_u denotes the subtree of T consisting of all descendants of u and $\omega(u)$ the value of the term derived by this tree. Let u be a node of a derivation tree and v be its descendant, then T_{u-v} denotes the tree obtained by removing from T_u all the nodes of T_v except v itself. We also define a function $\iota_u(v)$ which shows, in which fragments of $\omega(u)$ the fragments of $\omega(v)$ are continuous subwords. Note that this function depends only from the derivation between u and v. The formal definition of this function is given below. It works for every reduced grammar; α and β always belong to $N \cup \Theta$.

1. If u is a node of rank l, then $\iota_u(u) = (0, \ldots, l)$.
2. Let v be a node of rank $l + m$, where a rule $A \to \alpha_l \beta_m$ is applied (the subscripts as usual denote the ranks) and $\iota_u(v) = (i_0, \ldots, i_{l+m})$. If v_1 and v_2 are the nodes corresponding to α and β, then

$$\iota_u(v_1) = (i_0, \ldots, i_l), \qquad \iota_u(v_2) = (i_l, \ldots, i_{l+m})$$

3. Let v be a node of rank $l + m$, where a rule $A \to \alpha_{l+1} \odot_j \beta_m$ is applied and $\iota_u(v) = (i_0, \ldots, i_{l+m})$. If nodes v_1 and v_2 correspond to α and β, then

$$\iota_u(v_1) = (i_0, \ldots, i_j, i_{j+m}, \ldots, i_{l+m}), \qquad \iota_u(v_2) = (i_j, \ldots, i_{j+m})$$

4. If v is a node of rank l, a rule $A \to \alpha$ is applied in it and v_1 is the node corresponding to α, then $\iota_u(v_1) = \iota_u(v)$.

Definition 5 [2]. *A descendant v of u is its* direct descendant *if* $\mathrm{rk}(v) = \mathrm{rk}(u)$ *and* $\iota_u(v) = (0, 1, \ldots, \mathrm{rk}(u))$.

Lemma 2. *If v is a direct descendant of u in a terminal tree and* $\omega(v) = (v_0, \ldots, v_l)$, *then* $\omega(u) = (x_0 v_0 y_0, \ldots, x_l v_l y_l)$ *for some strings* $x_0, y_0, \ldots, x_l, y_l$, *which depend only from* T_{u-v}.

Proof. Follows from the definition.

Lemma 3. *Let v be a descendant of u with* $\mathrm{rk}(v) = \mathrm{rk}(u)$, *then*

1. *If* $\mathrm{rk}(v) = \mathrm{rk}(u) = 0$, *then v is a direct descendant of u.*
2. *If there is a node of smaller rank on the branch between u and v, then v is not a direct descendant of u.*
3. *If all the nodes between u and v have the same rank, then v is a direct descendant of u.*

Proof. Easily follows from the definitions. The third statement uses induction on the distance between v and u.

The notion of direct descendance is defined for syntactic trees just as for derivation trees. A context $C[x]$ is called direct, if its variable leaf is a direct descendant of its root.

---

[2] In [10] a more narrow definition of direct descendance was used. However, all the arguments of that paper are still valid for current definition.

Lemma 4. *If $A \vdash C[A]$ for some direct context and* $\mathrm{rk}(A) = r$, *then there exist words* $x_0, y_0, \ldots, x_r, y_r$, *such that for any tuple* $u = (u_0, \ldots, u_r) \in L(A)$ *it also holds that* $(x_0 u_0 y_0, \ldots, x_r u_r y_r) \in L(A)$.

Proof. Because the grammar does not contain useless nonterminals, we may assume that C is a ground context (just continue all the derivations downwards from nonterminals in its yield). Then attach the derivation tree of u to A in the distinguished position of C and obtain another derivation tree T' in the grammar under consideration. Let v be the word derived by T', by Lemma 2 it has the required form.

Let T be a fixed derivation tree, deriving some word w. We call a pumping pair two occurrences u and v of the same nonterminal A, where v is a direct descendant of u. The subtree $T_{\mathsf{u}-\mathsf{v}}$ is called a pump and the symbols of w which are in the yield of this subtree — the scope of this pump. Let a language L be fixed, a word $w \in L$ is said to be r-pumpable, if there is a "pumping" decomposition $w = x_0 u_1 x_1 \ldots u_r x_r$ such that $u_1 \ldots u_k \neq \epsilon$ and for every natural n the word $x_0 u_1^n x_1 \ldots u_r^n x_r$ belongs in L. The statement below follows from the definitions (see [2], Lemma 7).

Lemma 5. *Let G be a reduced grammar and T be a terminal derivation tree in it, which contains a pumping pair of rank t. Then the word w derived by T is $2(t+1)$-pumpable.*

We call a word w r-pumpable with coverage α if at least $\alpha|w|$ symbols of it are inside one of u_j in some pumping decomposition. Our ultimate goal is to prove that in every k-DCFL for any $\alpha < 1$ all the words except a finite number are $2(k+1)$-pumpable with coverage α. We prove the following claim: for every k-DCFL L there is a grammar G, such that $L(G) = L$, and a number p, such that for any $w \in L$ with $|w| \geq p$ exists a derivation tree T_w with at least $\alpha|w|$ symbols of w in scope of some pump of T_w. Note that in this case the pumpable segments of different pumps would be disjoint.

We make terminological remark on derivation trees of linear grammars in Chomsky normal form. They can be either terminal or nonterminal, the latter have one nonterminal node in their yield. In the second case we call this node the foot node of the tree. In both cases internal nodes of the tree form a single branch called the spine. Every leaf of the yield is a son of some node in the spine; moreover, since every leaf contains exactly one alphabet symbol, we may establish a one-to-one correspondence between the letters of the derived word and the spine of its derivation tree (in case of nonterminal tree the foot node is excluded from this bijection). Which is even more important, a symbol is in scope of some pump if and only if its parent lies on the path between two elements of the same pumping pair, which we call the pumping chain. Therefore to measure the fraction of symbols covered by pumps we suffice to measure the number of spine nodes covered by disjoint pumping chains. That is exactly the characteristic we will consider in the rest of the paper.

4   The Main Theorem

Recall, our goal is the following: for every linear DCFL L and every $\alpha < 1$ we construct a grammar $G_{L,\alpha}$, which generates this language and possesses the following property: there is a number p such that for every $w \in L$ with $|w| \geq p$ there exists a derivation tree T such that at least $\alpha|w|$ symbols of w are in scope of some pump of T. Equivalently: there is a number β such that for any word $w \in L$ exists a derivation tree T where at least $\alpha|w| - \beta$ nodes of the spine are covered by a pumping chain. The desired limit Ogden property is an easy consequence of this statement.

The main technique we apply is the replacement of "redundant" subtrees by weakly equivalent trees containing nodes of smaller rank. Two derivation trees are said to be weakly equivalent if they derive equivalent terms. A rule $A \to \alpha$ is said to be derivable in grammar G if $A \vdash_G \alpha$, it is weakly derivable if there exists a term $\alpha' \sim \alpha$, such that $A \vdash_G \alpha'$. Obviously, adding a weakly derivable rule does not change the language generated by the grammar. We call G' a conservative extension of grammar G, if it includes all the rules of G and $L(G') = L(G)$.

Lemma 6. *Let G be a grammar in linear Chomsky normal form, $A \vdash_G \alpha$ be a derivable rule and $\mathrm{rk}(\alpha) = r$. Then there exists a grammar G' in linear Chomsky normal form, which is a conservative extension of G and has all the same nonterminals of rank $r + 1$ and greater, such that some term $\alpha' \sim \alpha$ possesses the following property: the derivation $A \xrightarrow{} \alpha'$ uses only nonterminals of rank r or smaller.*

Proof. Since G is in Chomsky normal form, then α is specialized. Hence, there exists an equivalent r-correct specialized term α'. We create the rule $A \to \alpha'$ and then binarize it, adding a new nonterminal and a new rule for every binary node in the syntactic tree of α'. The rules obtained through binarization are added to G to construct the required conservative extension G'. It is easily observed, that all the new rules satisfy the definition of linear Chomsky normal form.

The previous lemma allows us to avoid redundant usage of nonterminals of high rank (the initial derivation tree of $A \vdash \alpha$ might have internal nodes of rank greater than $\mathrm{rk}(\alpha)$). The next lemma shows that in some conditions we are able to necessitate the existence of lower rank nonterminals.

Lemma 7. *Let G be a grammar in linear Chomsky normal form, $A \vdash_G \alpha$ be a derivable rule and $\mathrm{rk}(\alpha) = r$. Assume that $\alpha = C[B]$ for some indirect ground context C and $\mathrm{rk}(B) = r$. Then there exist a term $\alpha' \sim \alpha$ and a conservative extension G' of G with the following property: the derivation $A \to \alpha'$ in G' does not use nonterminals of rank greater than r and obligatorily uses some nonterminal of smaller rank.*

Proof. We start as in Lemma 6 by constructing an r-correct specialized linear term α', which is equivalent to α. Then α' has the form $C'[B]$ and the context C' has internal nodes only of rank r or smaller. If all its internal nodes are of rank r, then by Lemma 2 C' is a direct context. But in this case it is easy to define a valuation, which maps α and α' to different tuples. This is a contradiction and the lemma is proved.

Consider a derivation tree T, a pump π and let u and v be its top and bottom nodes. We call the collapsing of π the following procedure: the pump is removed from the tree and T_v is attached to the former position of u.

Statement 2. *Let T' be obtained from derivation tree T by collapsing some pump, then:*

1. T' is a correct derivation tree.
2. If v' is a direct descendant of u' in T', then it was such descendant in T.

Proof. We prove only the second statement, the first is trivial. In v is not on the path from u' to v' in T', then there is nothing to prove. Otherwise we prove for every descendant z of u' in T' that $\iota'_{u'}(z) = \iota_{u'}(z)$, where ι' denotes the ι function for the tree T'. This is done by induction on the distance from u' to z. The only nontrivial case is z = v where the direct descendance between u and v is used. The statement is proved.

The next lemma shows which spines are desirable in order to prove a lower bound on the number of covered symbols. We call an (i, j)-tree a derivation tree in a specialized linear grammar, whose root has rank i and whose foot node — rank j. We call a tree direct if its foot node is a direct descendant of its root. Note that whether a tree is direct or not is preserved modulo equivalence. In what follows we denote by $\mathcal{C}(T)$ the number of nodes in the spine of derivation tree T which are covered by pumping chains and by $d(T)$ the depth of this tree. We call "direct" spines the spines of direct derivation trees.

Lemma 8. *Let π be a continuous segment of the spine containing nodes of the same rank r. Then there is a coverage by pumping chains, that does not cover at most d nodes of π, the constant d depends only from the grammar.*

Proof. Through the proof and the rest of the paper we denote by N_t the set of nonterminals of rank t. We claim that the number of uncovered nodes does not exceed $|N_r|$. Indeed, consider all uncovered nodes in the optimal cover. If there are two nodes with the same label outside the cover, then they are direct descendants and we may add their pumping chain to the cover, possibly removing its subchains to preserve disjointness. That increases the number of covered nodes and contradicts optimality, the lemma is proved.

Lemma 9. *Let T be a direct (r, r)-tree, which has internal nodes only of rank r and $r + 1$, then $\mathcal{C}(T) \geq d(T) - D$ for some constant D, depending only from the grammar.*

Proof. Consider a direct (r, r) tree T with the following property: the difference $d(T) - \mathcal{C}(T)$ for it is strictly greater then for all trees of smaller size. If we prove the depth of T to be bounded, the lemma will also be proved.

Statement 3. *The spine of T has no pumping chains.*

Proof. Assume the contrary, let u and v be the top and the bottom node of the corresponding pump π. Collapsing this pump yields a smaller tree T'. Let us prove that the number of uncovered nodes in the spine of T' cannot be greater then in T. Indeed, the spine of T' is obtained from the spine of T by removing the semi-interval $[u; v)$. If v is not covered, then we may add to the cover the pumping chain between u and v and obtain that all the removed nodes were covered. Otherwise the pumping pair (u', v') whose chain covers v in T' remains a pumping pair in T due to Lemma 9. In both cases $d(T') - \mathcal{C}(T') = d(T) - \mathcal{C}(T)$ (all removed nodes were covered) which contradicts the derivation of T.

Observe that all the nodes of rank r in the spine of T are direct descendants. Hence this spine contains at most $|N_r|$ nonterminals of rank r by Dirichlet's principle. By the same argument maximal length of a continuous chain of rank $r + 1$ nodes is $|N_{r+1}|$. Hence, $d(T) \leq |N_r| + (|N_r| - 1)|N_{r+1}|$ which proves the lemma.

Now we turn to the proof of our first main result: the limit Ogden property of linear 2-DCFLs. We fix an arbitrary $\alpha \in (0, 1)$ and a DCFG $G = \langle N, \Sigma, P, S \rangle$ and construct its conservative extension $G' = \langle N', \Sigma, P', S \rangle$ with the following property: for every word w there exists a derivation tree T_w in this grammar with at least $\alpha|w| - \beta$ nodes of the spine covered by pumping chains, where β is a constant, depending only from α and the grammar G'.

Let t be some sufficiently large natural number whose exact value will be determined later. We construct the grammar G' by duplicating all the rules corresponding to indirect $(1, 1)$-derivation trees with depth less than t. For every word $w \in L(G)$ we construct a derivation tree T'_w with desirable properties, starting from an arbitrary derivation tree T_w of this word in the grammar G.

We label the nodes of rank 1 by the following algorithm: firstly a topmost node of rank 1 is labeled. After labeling a node u we either label its furthest direct descendant, if it differs from u, or the next node of rank 1. Then we partition the tree by the labeled nodes, obtaining subtrees $T_0, \ldots, T_r$. We call an indirect tree T_i replacable if its spine does not contain nodes of rank 0 and its depth is $\leq t$. To construct the tree T'_w we replace every replacable T_i by a weakly equivalent T'_i tree whose spine contains a rank 0 nonterminal, which exists by construction and Lemma 7. We call all trees produced in such a way admissible.

Now consider the partition of T'_i's spine by the labeled nodes. We assume that a segment between two labeled nodes includes the upper one, but not the lower. The following segments may occur:

1. The segment before the first labeled node,
2. The segment below the last labeled node.
3. The segments between direct descendants, whose length is greater then t, we call them segments of the first type.
4. The segments of length $> t$, whose last node is not a direct descendant of the first — segments of the second type.

5. The segments between direct descendants, whose length is t or less, we call them segments of the third type.
6. "Old" segments containing nodes of rank 0, — segments of the fourth type.
7. Segments obtained by duplication, they are segments of the fifth type. Each such segment has length $\leq t$ and contains a "new" nonterminal of rank 0.

As earlier, we consider the admissible tree T', for which the difference $\delta(T') = \alpha d(T') - C(T')$ called the defect is greater than for all smaller admissible trees. We prove that $\delta(T')$ is bound by some constant. We process segments of different types separately. Note that to satisfy the definition of T' its spine should contain every nonterminal of rank 0 at most once: otherwise removing a pump between the repeating occurrences leads to a smaller tree with the same defect (the argument we used in Lemma 9). We denote by M_j the number of type j segments occurring in T' and by L_j — the total length of such segments. U_j stands for the number of uncovered spine nodes in the segments of type j.

1. The topmost segment contains only nodes of rank 0, therefore by the remark on repeating occurrences it contains at most $|N_0|$ nodes.
2. Since a node of rank 0 cannot immediately precede a node of rank 2, the lowest segment consists of a continuous chain of rank 2 nodes followed by a sequence of rank 0 nodes. By the pump collapsing argument the whole segment contains at most $|N_2|$ rank 2 nodes, N_0 rank 0 nodes and a single node of rank 1, which bounds its length by $|N_0| + |N_2| + 1$.
3. By Dirichlet's principle there are at most $|N_0'|$ segments of the fifth type and maximal length of such segment is at most t, therefore $U_5 \leq L_5 \leq t|N_0'|$.
4. By the arguments we applied to the lowermost segment, the segment of type 4 consists of a continuous chain of rank 2 nodes, followed by a sequence of rank 0 nodes. As before, the chain of rank 2 nodes cannot contain more than $|N_2|$ elements, and the total number of rank 0 nodes in such segments is at most $|N_0|$. Since every such segment contains a node of rank 0, then $M_4 \leq |N_0|$. Finally, we conclude that $U_4 \leq L_4 \leq 2|N_0| + |N_0||N_2|$.
5. A segment of type 3 cannot be followed by segments of type 1 and 3 by construction (the segment of type 3 will not be maximal contradicting its definition). Any such segment contains at most $|N_2| + 1$ uncovered nodes by Lemma 8, we have $U_3 \leq (|N_2| + 1)(M_2 + M_4 + M_5 + 1) \leq (|N_2| + 1)(M_2 + |N_0'| + |N_0|)$.
6. By the arguments used before, a segment of type 1 contains at most $|N_2| + 1$ uncovered nodes, and a node of type 2 — $|N_0| + |N_2| + 1$ such nodes, so $U_1 \leq M_1(|N_2|+1)$ and $U_2 \leq M_2(|N_2|+|N_0|+1)$. Since every such segment has length t or greater, we obtain that $U_1 \leq L_1\frac{|N_2|+1}{t}$ and $U_2 \leq L_2\frac{|N_2|+|N_0|+1}{t}$.

Calculating the total number U of uncovered nodes, we have $U \leq |N_0| + (|N_0|+|N_2|+1)+U_1+U_2+U_3+U_4+U_5 \leq D+(L_1+L_2)\frac{|N_2|+|N_0|+1}{t}+L_2\frac{|N_2|+1}{t}+ |N_0'|(|N_2|+1+t) \leq L\frac{2(|N_2|+1)+|N_0|}{t} + E$, where D is a constant depending only from the parameters of initial grammar G and E additionally depends from t and G', but not from the derivation itself. Choosing $t \geq \frac{2(|N_2|+1)+|N_0|}{1-\alpha}$ we obtain $U \leq (1-\alpha)L + \beta$ which was required. Using the fact that $\alpha < 1$ was arbitrary we derive the following theorem:

Theorem 1. *For any linear 2-DCFL L and every $\alpha < 1$ there exists a grammar $G_{L,\alpha}$ generating L, which has the following property: there is a number p, such that for every $w \in L$ with $|w| \geq p$ there exists a derivation tree T of G, where at least $\alpha|w|$ symbols of w are in scope of some pump.*

As far as we know, the strongest result of Ogden type known before was Ogden lemma for tree-adjoining grammars (i.e. 1-DCFGs), proved in [6]. Therefore we have already made one step forward in DCFGs hierarchy. The goal of the next section is to generalize the proof to linear DCFGs of arbitrary rank.

5 Ogden Property for Linear DCFGs

In this section we prove the limit Ogden property for the whole family of linear DCFGs. We fix a number k, an arbitrary k-DCFG G in linear Chomsky normal form, generating the language $L = L(G)$ and arbitrary $\alpha < 1$. Our goal is to construct its conservative extension G' with the following property: for every $w \in L$ there exists a derivation tree T'_w, whose spine contains at least $\alpha|w| + \beta$ nodes covered by pumping chains with the constant β depending only from G'. We prove by downward induction on r the following claim: for every (r, s)-derivation tree T' of G' with $s \geq r$, whose spine do not contain nodes of rank less than r, it is possible to construct at equivalent tree T'' with $\mathcal{C}(T') \geq \alpha d(T') + \beta_r$, β_r being a constant independent from T' (but depending from r).

The claim holds for $r = k$ since a spine of a (k, k)-tree is a chain of direct descendants and has at most constant number of uncovered nodes by Lemma 8. Hence the induction base is valid for arbitrary α. Let us prove the induction step for current r, let $G^{(r+1)}$ be the extension satisfying the induction statement for the step $r + 1$ and coverage α' which we define later. To construct the grammar $G^{(r)}$ we duplicate all the indirect (r, r)-derivation trees whose depth is t or less with t being a sufficiently large number also determined later. We denote their sets of nonterminals by $N^{(r+1)}$ and N^r respectively. The structure of the induction step resembles the proof of Theorem 1.

Consider an arbitrary (r, s)-tree T', where duplication was applied to all (r, r)-segments which can be duplicated. Let π be the spine of the tree obtained, we start with the following claim: π contains at most $2(t - 1)$ nonterminals from N^r. Indeed, a spine of a new subtree (i.e. appeared by duplication) contains a node of rank less than r by Lemma 7. Therefore the distance from a "new" nonterminal to the closest nonterminal of smaller rank is at most $t - 1$. Hence new nonterminals may occur only between $(t - 1)$ uppermost or lowermost nonterminals of π. So at the induction step it is enough to consider only the trees containing old nonterminals. Let us prove the following auxiliary statement:

Statement 4. *Let T be a direct (r,r)-tree, then $C(T) \geq \alpha' d(T) + \gamma$ for some constant γ.*

Proof. As earlier, consider a tree with the defect $\alpha' d(T) - C(T)$ greater than for any smaller tree. By standard arguments it does not contain any pumping chains. Consequently, it contains at most $|N_r^{(r+1)}|$ occurrences of rank r nonterminals. Consider the segments between the occurrences of rank r nonterminals, let $L_1, \ldots, L_s$ be their lengths and $C_1, \ldots, C_s$ be the number of covered nodes on each such segment. By induction hypothesis we have $C_j \geq \alpha' L_j + \beta_{r+1}$. So we obtain $C(T) = \sum_j C_j \geq \alpha' \sum L_j + s\beta_{r+1} \geq \alpha'(L - |N_r^{(r+1)}|) + s|N_r^{(r+1)}| = \alpha' L + \gamma$

for a constant γ depending only from $G^{(r+1)}$. The statement is proved.

Now we partition the spine by rank r nodes by the same algorithm as we applied to linear 2-DCFGs: the segments in the partition are either maximal direct (r,r)-spines, or the segments between adjacent (r,r)-nodes which are not direct descendants (adjacency means there are no nodes of rank r between them). As in Theorem 1, we have the segments of the following types:

1. The segment below last rank r node — type 0 segment.
2. Direct (r,r)-segments of length larger than t — type 1 segments.
3. Indirect (r,r)-segments of length larger than t — the segments of type 2.
4. Direct (r,r)-segments of length t or less — type 3 segments. There are no other types since all short indirect (r,r)-segments have been duplicated.

Let as earlier M_j, L_j and U_j be the number of segments of type j, the total length of such segments and the number of uncovered nodes on them. The letters M, L, U stand for the same characteristics of the whole tree. We denote $\kappa = 1-\alpha$ and $\kappa' = 1 - \alpha'$, our goal is to prove that $U \leq \kappa L + \zeta$ for some constant ζ.

1. The lowermost segment is a $(r+1, s')$-segment for some $s' \geq r+1$. Hence by induction hypothesis $U_0 \leq \kappa' L_0 + \zeta_0$ for some constant ζ_0 depending only from $G^{(r+1)}$.
2. As in the proof of Theorem 1, $M_3 \leq M_2 + 1$. Applying Statement 4 to every segment of type 3 and summing the results, we obtain $U_3 \leq \kappa' L_3 + \gamma M_3 \leq \kappa' L_3 + \gamma(M_2 + 1) \leq \kappa' L_3 + \frac{\gamma}{t} L_2 + \gamma$.
3. Analogously applying Statement 4 to every segment of type 1, we have $U_1 \leq \kappa' L_1 + \gamma M_1 \leq (\kappa' + \frac{\gamma}{t}) L_1$.
4. Consider a segment ρ of type 2, all inner nodes of ρ have larger rank. Let U_ρ and L_ρ be its number of uncovered nodes and total length, we have $U_\rho \leq \kappa'(L_\rho - 1) + \zeta_0 + 1 \leq \kappa' L_\rho + \zeta_0 + 1$. It implies $U_2 \leq \kappa' L_2 + (\zeta_0 + 1)M_2 \leq L_2(\kappa' + \frac{\zeta_0+1}{t})$.

So the total number of uncovered nodes can be bounded by $U = U_0 + U_1 + U_2 + U_3 \leq \kappa' L_0 + \zeta_0 + (\kappa' + \frac{\gamma}{t})L_1 + (\kappa' + \frac{\zeta_0+1}{t})L_2 + \kappa' L_3 + \frac{\gamma}{t} L_2 + \gamma \leq (\kappa' + \frac{\gamma+\zeta_0+1}{t})L + \zeta$ for some constant ζ. Since γ and ζ_0 do not depend from t, we first choose $\kappa' = \frac{\kappa}{2}$ and then $t \geq 2\frac{\gamma+\zeta_0+1}{\kappa'}$. Finally we obtain $U \leq \kappa L + \zeta$ which was required. The induction step is proved. Setting $r = 0$ we obtain the theorem.

6   Application of Ogden Lemma

Not to be pure theoreticists, in this section we demonstrate an application of limit Ogden property. Due to the lack of space we give only one simple example just to demonstrate the principal components of the technique. The scheme of the proof closely follows Theorem 5 in [10].

A constituent is the part of the word derived from a node in derivation tree. A constituent of rank r consists of $r + 1$ continuous strings (some of them may be empty). With every constituent we associate a curve as shown below (for $r = 2$). The key argument is that either regions bounded by different curves do not intersect, or one region lies entirely inside another, for the proof see [9].

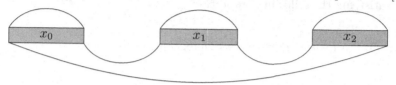

Since a pump is the difference between two constituents, we match every pump with the region corresponding to its outer constituent (we call this region the outer region of the pump). Simple topology shows that there exist only three principal variants for mutual location of two pumps π_1 and π_2: (1) the leftmost element of π_1 lies to the right of the rightmost element of π_2; (2) the first variant does not hold, but the outer regions of pumps do not intersect; (3) the outer region of π_1 includes the one of π_2. In the first case we say pumps form a linear pair, in the second π_1 embraces π_2 and in the third π_1 is an outer pump for π_2. This three variants are schematically illustrated below.

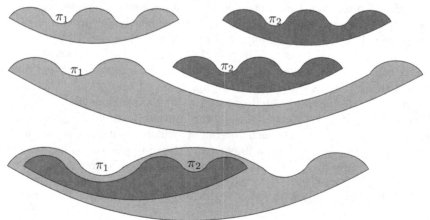

We apply the introduced machinery to show that the language $6\text{-}MIX = \{w \in \{a, b, c, d, e, f\}^* \mid |w|_a = |w|_b = |w|_c = |w|_d = |w|_e = |w|_f\}$ cannot be generated by a linear 2-DCFG. We use the fact that linear well-nested MCFGs are closed under intersection with regular languages which is easily proved by standard means.

Theorem 2. *The language* $6\text{-}MIX = \{w \in \{a, b, c, d, e, f\}^* \mid |w|_a = |w|_b = |w|_c = |w|_d = |w|_e = |w|_f\}$ *cannot be generated by a linear 2-DCFG.*

Proof. It is enough to prove the statement for the language $L = \{(a^+b^+c^+d^+e^+ f^+)^2\} \cap 6\text{-}MIX = \{a^{m_1}b^{m_2}c^{m_3}d^{m_4}e^{m_5}f^{m_6}a^{n_1}b^{n_2}c^{n_3}d^{n_4}e^{n_5}f^{n_6} \mid m_1 + n_1 = \ldots = m_6 + n_6\}$. We enumerate maximal continuous segments, containing the same letter, from left to right. We call a pump intersecting with the segments numbered $i_1, \ldots, i_t$ (and possibly some others) an $(i_1, \ldots, i_t)$-pump. We take some small $\kappa < 1$, which will be determined letter, and consider the grammar G_α generating L, such that every word w with $|w| \geq p_\kappa$ has at most $\kappa|w|$ symbols not covered by a family of disjoint pumps. Consider a word $a^{m_1}b^{m_2}c^{m_3}d^{m_4}e^{m_5}f^{m_6}a^{n_1}b^{n_2}c^{n_3}d^{n_4}e^{n_5}f^{n_6}$ satisfying the following properties:

1. $\min(m_j, n_j) > \kappa|w|$,
2. $m_1 > (5M + 1 + \kappa|w|)$, where $M = \max(m_5, m_6, n_2, n_3, n_4)$,
3. $m_6 \geq (n_1 + \kappa|w| + 1)$.

For example, we may set $\kappa = \frac{1}{100}$, $p = p_\alpha$, $m_5 = n_1 = n_2 = n_3 = n_4 = p$, $m_6 = 2p$, $m_1 = m_2 = m_3 = m_4 = n_5 = 11p$, $n_6 = 10p$. We fix a set of pumps from the statement of limit Ogden lemma and consider only the pumps from this set.

1. *For any $i \in [1, 12]$ an i-pump exists.* Indeed, every segment contains more than $\kappa|w|$ letters.
2. *Every pump is a $[i_1, i_2, i_3, i_4, i_5, i_6]$-pump, where all $i_j \in \{j, j + 6\}$.* Every pump should contain equal positive number of a-s, b-s, c-s and so on, therefore it intersects with an a-segment, b-segment and so on.
3. *There are at most M a-s covered by $[1, 5]$-pumps.* In $[5]$-pumps all the e-s are from the fifth segment (due to the previous statement), therefore there are at most M e-s covered by such pumps, and consequently, at most M a-s.
4. *There is a $[1, 2, 3, 4, 11, 12]$-pump.* π_1 Equivalently, there is a $[1]$-pump, which is not a $[1, 5]$-, $[1, 6]$-, $[1, 8]$-, $[1, 9]$-, $[1, 11]$-pump. By the previous step, there are at most $5M$ a-s, covered by such pumps, so at least $\kappa|w| + 1$ a-s remain. One of them must be covered by a pump.
5. *There is a $[1, 6]$-pump π_2.* Analogously to the two previous steps.
6. *It is in fact a $[1, 6, 12]$-pump.* Observe the mutual position of π_2 and π_1. The outer curve of π_2 intersects with 6-segment, which is outside the region of π_1. Since π_1 and π_2 do not form a linear pair, the only possibility left is that π_2 is an outer pump for π_1. Then it intersects with 12-th segment.

But then π_2 intersects with two f segments, a contradiction. The theorem is proved.

The same method yields that the language $2(k + 1) - \text{MIX} = \{w \in \{a_1, \ldots, a_{2(k+1)}\}^* \mid |w|_{a_1} = \ldots = |w|_{a_{2(k+1)}}\}$ cannot be generated by a linear k-DCFG.

The author thanks the reviewers of LFCS 2016 for their helpful comments.

7   Conclusion and Future Work

We have proved the limit Ogden property for linear DCFGs (i.e., linear well-nested MCFGs). There are several natural ways to extend our results: the first is to prove an analogous statement for the whole class of DCFGs or to show its falsity, the second is to apply the statement of limit Ogden lemma in a more sophisticated manner to give more examples of languages which are not linear DCFLs. For example, the author suggests that the language $3 - \text{MIX} = \{w \in \{a, b, c\}^* \mid |w|_a = |w|_b = |w|_c\}$ lies outside this family (a weak version of Kanazawa-Salvati conjecture, [4]), but do not has even a sketch of the proof of this hypothesis. The problem is that in case of higher rank DCFGs we have to turn from graphic arguments to more systematic use of topology. Nevertheless, the author hopes his technique will help to determine the exact place of well-nested MCFLs in the family of context-sensitive languages.

References

1. Joshi, A.K.: Tree adjoining grammars: how much context-sensitivity is required to provide reasonable structural descriptions? University of Pennsylvania, Moore School of Electrical Engineering, Department of Computer and Information Science (1985)
2. Kanazawa, M.: The pumping lemma for well-nested multiple context-free languages. In: Diekert, V., Nowotka, D. (eds.) DLT 2009. LNCS, vol. 5583, pp. 312–325. Springer, Heidelberg (2009)
3. Kanazawa, M.: The failure of Ogdens lemma for well-nested multiple context-free grammars (2014). http://research.nii.ac.jp/~kanazawa/publications/ogden.pdf
4. Kanazawa, M., Salvati, S.: MIX is not a tree-adjoining language. In: Proceedings of the 50th Annual Meeting of the Association for Computational Linguistics: Long Papers, vol. 1, Association for Computational Linguistics, pp. 666–674 (2012)
5. Ogden, W.: A helpful result for proving inherent ambiguity. Theory Comput. Syst. **2**(3), 191–194 (1968)
6. Palis, M.A., Shende, S.M.: Pumping lemmas for the control language hierarchy. Math. Syst. Theory **28**(3), 199–213 (1995)
7. Shieber, S.M.: Evidence against the context-freeness of natural language. In: Savitch, W.J., Bach, E., Marsh, W., Safran-Naveh, G. (eds.) The Formal Complexity of Natural Language, pp. 320–334. Springer, Heidelberg (1987)
8. Sorokin, A.: Normal forms for multiple context-free languages and displacement lambek grammars. In: Artemov, S., Nerode, A. (eds.) LFCS 2013. LNCS, vol. 7734, pp. 319–334. Springer, Heidelberg (2013)
9. Sorokin, A.: Monoid automata for displacement context-free languages. In: Colinet, M., Katrenko, S., Rendsvig, R.K. (eds.) ESSLLI 2012/2013. LNCS, vol. 8607, pp. 154–173. Springer, Heidelberg (2014)
10. Sorokin, A.: Pumping lemma and ogden lemma for displacement context-free grammars. In: Shur, A.M., Volkov, M.V. (eds.) DLT 2014. LNCS, vol. 8633, pp. 154–165. Springer, Heidelberg (2014)
11. Sorokin, A.: Normal forms for linear displacement context-free grammars (2015). http://arxiv.org/abs/1507.08600

Levy Labels and Recursive Types

Rick Statman[✉]

Department of Mathematical Sciences,
Carnegie Mellon University, Pittsburgh, PA 15213, USA
statman@cs.cmu.edu

Abstract. In this note we introduce a generalization of the Levy label technique which applies easily to lambda calculus with beta-eta conversion and lambda calculus with surjective pairing a' la PSP.

Keywords: Lambda calculus · Recursive types · Levy labels

1 Introduction

In this note we introduce a generalization of the Levy label technique [5] which applies easily to lambda calculus with beta-eta conversion and lambda calculus with surjective pairing a' la PSP [10]. Our technique is based on the algebraic approach to recursive types first described by Dana Scott [8]. This approach is, literally speaking, not a true generalization of Levy labels, but rather Levy labels are an abstraction of the recursive types approach when the latter is applied to the case of lambda calculus with beta conversion.

The Levy label technique is a powerful method, which itself generalizes Hindley's theory of developments [4], for proving the Church-Rosser theorem for lambda calculus with beta, and other fundamental results such as the standardization theorem. However, it is very unclear how to apply this technique to extensions of beta. Already in the case of eta, how does one assign Levy labels to the eta reduction of

$$((\lambda x((O^k x^l)^m))^n y^\circ)^p,$$

where O is Barendregt's big omega term [1], so as to get

$$(O^k y^{\min(l,o,n-1)})^{\min(m,n-1,p)}?$$

Similar problems arise in applying this technique to combinators.

The principal properties of Levy labels are

(a) completeness;
   all finite reductions can be labeled, and
(b) strong normalization;
   all labeled reductions are finite,

in addition to the weak diamond property, which we shall discuss below. The challenge is to make this work for recursive types.

S. Artemov and A. Nerode (Eds.): LFCS 2016, LNCS 9537, pp. 392–406, 2016.
DOI: 10.1007/978-3-319-27683-0_27

2   Preliminaries

We begin with some preliminaries for the untyped calculus with pairing. The atoms of the language are defined by:

(i) the variables $x, y, z, \ldots$ are atoms.
(ii) the constants P, L, R are atoms.

The terms of the language are defined by:

(i) atoms are terms
(ii) if X, Y are terms then so are (XY) and $\lambda x X$.

We shall adopt the customary conventions:

(i) parens are deleted and restored by left association and the use of Church's infixed "dot"notation when necessary.
(ii) parens are added around abstractions for readability.

The axiom and rules of untyped lambda calculus are the following: The first 5 axioms correspond to the classical theory of untyped lambda calculus with surjective pairing $SP+$ eta considered by Colin Mann:

$$
\begin{array}{llll}
\text{(beta)} & (\lambda x X)\ Y & = & [Y/x]\ X \\
\text{(eta)} & X & = & \lambda x.\ Xx \quad x \text{ not free in } X \\
(L/Pa) & L(PXY) & = & X \\
(R/Pa) & R(PXY) & = & Y \\
(P/Dp) & P(LX)(RX) = & & X
\end{array}
$$

The next 6 axioms correspond to the extended theory of Stovring (FP,[12]) and Statman (PSP, [10]),

$$
\begin{array}{lll}
(P/Ap) & PXYZ & = P(XZ)(YZ) \\
(L/Ap) & LXY & = L(XY) \\
(R/Ap) & RXY & = R(XY) \\
(L/Ab) & L(\lambda x X) & = \lambda x(LX) \\
(R/Ab) & R(\lambda x X) & = \lambda x(RX) \\
(P/Ab & P(\lambda x X)(\lambda x Y) & = \lambda x\ PXY
\end{array}
$$

There are certain useful derived rules.

(1) (P/Dp) and (P/Ap) implies (L/Ap) and (R/Ap)
$L(XY) = L(P(LX)(RX)Y) = L(P(LXY)(RXY)) = LXY$
similarly for R
(2) (L/Ap) and (R/Ap) and (P/Dp) implies (P/Ap)
$L(PXYZ) = L(PXY)Z = XZ$ and $R(PXYZ) = R(PXY)Z = YZ$
therefore
$PXYZ = P(L(PXYZ))(R(PXYZ)) = P(XZ)(YZ)$.
(3) (eta) and (P/Ap) implies (P/Ab)
$P(\lambda x X)(\lambda x Y) = \lambda y.\ P(\lambda x X)(\lambda x Y)y = \lambda y.\ P((\lambda x X)y)((\lambda x Y)y) = \lambda\ xPXY$

(4) (eta) and (L/Ap) implies (L/Ab)
$L(\lambda xX) = \lambda y.L(\lambda xX)y = \lambda y.L((\lambda xX)y) = \lambda x(LX)$
similarly for R

(5) (eta) implies $LP = K$ and $RP = K^*$
$L(PX) = \lambda x.\ L(PX)x = \lambda x.\ L(PXx) = \lambda x.X$ thus
$LP = \lambda x.LPx = \lambda x.L(Px) = \lambda xy.\ x$
similarly
$R(PX) = \lambda x.\ R(PX)x = \lambda x.\ R(PXx) = \lambda x.x$ hence
$RP = \lambda yx.x$

The result of Klop [6] is that the Church-Rosser property fails for the following classical reductions for SP:

(beta)	$(\lambda xX)\ Y$	red.	$[Y/x]X$
(L/Pa)	$L(PXY)$	red.	X
(R/Pa)	$R(PXY)$	red.	Y
(P/Dp)	$P(LX)(RX)$	red.	X

Nevertheless, this theory was proved conservative over beta by Roel De Vrijer [3]. Stovring and, later, Statman (for the combinator case) introduced new reductions for the first 5 and an additional three which enjoy Church-Rosser. Stovring Reductions for FP with eta:

(beta)	$(\lambda xX)Y$	fpred.	$[Y/x]X$	
(etae)	X	fpred.	$\lambda x.\ Xx$	x not free in X
(L/Pa)	$L(PXY)$	fpred.	X	
(R/Pa)	$L(PXY)$	fpred.	Y	
(P/De)	X	fpred.	$P(LX)(RX)$	
$P/Ap)$	$PXYZ$	fpred.	$P(XZ)(YZ)$	
(L/Ab)	$L(\lambda xX)$	fpred.	$\lambda x(LX)$	
(R/Ab)	$R(\lambda xX)$	fpred.	$\lambda x(RX)$	

Here we modify these reductions again

(Beta)	$(\lambda xX)Y$	$\hookrightarrow$	$[Y/x]X$	
(Etae)	X	$\hookrightarrow$	$\lambda x.\ Xx$	x not free in X
(Left)	$L(PXY)$	$\hookrightarrow$	X	
(Rght)	$R(PXY)$	$\hookrightarrow$	Y	
(Surj)	$P(LX)(RX)$	$\hookrightarrow$	X	
(Sure)	X	$\hookrightarrow$	$P(LX)(RX)$	
(Appl)	LXY	$\hookrightarrow$	$L(XY)$	
(Appr)	RXY	$\hookrightarrow$	$R(XY)$	
(Appp)	$PXYZ$	$\hookrightarrow$	$P(XZ)(YZ)$	

for a version of $PSP+$ eta in lambda calculus. It is obvious that the congruence generated by $\hookrightarrow$ coincides with FP. There will be several occasions below where we will not want to use Etae. In those cases we shall write "W.W.E." (works without Etae).

3 Properties of $\hookrightarrow$

We shall be interested in a number of subsystems of $\hookrightarrow$:

$$jj \hookrightarrow := \{\text{Surj}\}$$
$$je \hookrightarrow := \{\text{Sure}\}$$
$$oj \hookrightarrow := \hookrightarrow -\{\text{Surj}\}$$
$$oe \hookrightarrow := \hookrightarrow -\{\text{Sure}\}$$
$$oo \hookrightarrow := \hookrightarrow -\{\text{Surj}, \text{Sure}\}$$
$$oa \hookrightarrow := \hookrightarrow -\{\text{Etae}\}$$

Fact: $je \hookrightarrow$ has the strong diamond property W.W.E.: viz, if $Xje \hookrightarrow Y$ and $Xje \hookrightarrow Z$ then there exists W sych that $Yje \hookrightarrow W$ and $Zje \hookrightarrow W$.

Proof. Verify the strong diamond for parallel $je \hookrightarrow$, as in Barendregt 3.2.   ∎

Proposition 1. *(Surj postponement W.W.E.)*
  Suppose that $Xoe \hookrightarrow Y$. Then there exists Z such that $Xoo \hookrightarrow Zjj \hookrightarrow Y$.

Proof. By induction on the number of Surj reductions in an $oe \hookrightarrow$ reduction sequence from X to Y. The basis case is trivial so it suffices to prove that if $Xjj \hookrightarrow Uoo \hookrightarrow Y$ then there exists a V such that $Xoo \hookrightarrow Vjj \hookrightarrow Y$. Toward this end we suppose that certain Surj redexes $P(LW)(RW)$ in X have been beta expanded to $(\lambda x P(Lx)(Rx))W$, and the Surj reduction to U can be simulated

$$(\lambda x\ P(Lx)(Rx))W \quad \text{Surj} \quad (\lambda x\ x)W \text{ Beta } W$$

This can certainly be done in the case of a single Surj redex. In short, we are assuming that the Surj reduction at hand does not send the two different occurrences of W to different terms. Now suppose that $Y\ oo \hookrightarrow Z$, we distinguish 4 cases:

Case 1: Beta. This case follows from Appl, Appr, and Appp.
Case 2: Etae. Immediate.
Case 3: Left, Right. This case follows from Appl, Appr, Appp, and Left resp. Right.
Case 4: Appl, Appr, Appp. This case follows from Appl, Appr, and Appp. End of proof.

4 Recursive Type Paths

First, we "Church"type untyped terms by paths of recursive types. These typings convert to other such typings by some simple operations of clockwise and counterclockwise rotation, and left and Right shift.
  Simple types $A, B, C, \ldots$ are built up from atoms p_i by $\to$. We shall employ the usual ideas associated with such types such as the notions of subtypes, positive and negative occurrences, and strictly positive and negative occurrences.

Let R be a simultaneous recursion, as in 7.3.10 of [2].

$$R = \{p_i = P_i(p_1, \ldots, p_e) | i = 1, \ldots, e\}.$$

There is a corresponding reduction relation:

$$P_i(p_1, \ldots, p_e) \text{ Red. } p_i \quad i = 1, \ldots, e.$$

We make several assumptions about R which can always be arranged:

(i) (non-triviality) each P_i contains $\rightarrow$
(ii) (Knuth-Gross) for distinct i, j.
    P_i is not a subtype of P_j.

There may be type atoms p_j which do not appear in R. These are treated as simple types and will be mostly ignored below. We write $B = A+$ and $A = B-$ if A Red. B. If $B = A+$ then i is uniquely determined. We write $*$ for $+$ or $-$ ambiguously.

A type path is a sequence of types $A_1, \ldots, A_n$ such that $A_i+ = A_{i+1}$ or $A_i- = A_{i+1}$. We denote type paths by $r, s, t, \ldots$. A recursive typing of a term X is an assignment of type paths to subterms of X, where we write

$$Y : A_1, \ldots A_n$$

if Y has been assigned the type path $A_1, \ldots, A_n$, satisfying the following conditions

(1) for subterms (UV):
    $U : r, A \rightarrow B$ and $V : s, A$ and $(UV) : B, t$
(2) for subterms $\lambda u\, U$:
    each occurrence of $u : A, r$ for possibly different r but always the same A
    (we write this as $u = u : A,$)
    $U : s, B$ and $\lambda u U : A \rightarrow B, t$.
(3) for each occurrence of P, $P : A \rightarrow (A \rightarrow A), r$
    for each occurrence of L, $L : A \rightarrow A, r$
    for each occurrence of R, $R : A \rightarrow A, r$.

As said before, these are to be understood as "Church" typings as this notion is discussed in 1.1 of [2]. We shall use the usual superscript notation when useful.

We now define the notion of type conversion $\Leftarrow to \Rightarrow$, a congruence relation on typings, by defining type reduction $to \Rightarrow$ as follows

(1) Clockwise Rotation of an Application:

$$(*) \quad (U^{r, A \rightarrow B, A^* \rightarrow B} V^{s, A^*})^{B, t} \, to \Rightarrow$$
$$(U^{r, A \rightarrow B} V^{s, A^* A})^{B, t}$$

(2) Right Shift of an Application:

$$(*)\ (U^{r,A\to B,A\to B^*}V^{s,A})^{B^*,t}\ to \Rightarrow$$
$$(U^{r,A\to B}V^{s,A})^{B,B^*,t}$$

(3) Counter-clockwise Rotation of an Abstraction:

$$(*)\ (\lambda u^{A,}U^{r,B})^{A\to B,A^*\to B,s}\qquad to \Rightarrow$$
$$(\lambda u^{A^*,A,}[u^{A^*,A,}/u^{A,}]U^{r,B})^{A^*\to B,s}$$

(4) Left Shift of an Abstraction:

$$(*)\ (\lambda u^{A,}U^{r,B})^{A\to B,A\to B^*,s}\ to \Rightarrow$$
$$(\lambda u^{A,}U^{r,B,B^*})^{A\to B^*,s}$$

(5) Counter-clockwise Rotation of L:

$$(*)\ L^{A\to A}(U^{r,A})^{A,A^*,s}\qquad to \Rightarrow$$
$$L^{A^*\to A^*}(U^{r,A,A^*})^{A^*,s}$$

(6) Counter-clockwise Rotation of R

$$(*)\ R^{A\to A}(U^{r,A})^{A,A^*,s}\qquad to \Rightarrow$$
$$R^{A^*\to A^*}(U^{r,A,A^*})^{A^*,s}$$

(7) Counter-clockwise rotation of P

$$(*)\ ((P^{A\to(A\to A)}(U^{r,A}))^{A\to A}(V^{s,A}))^{A,A^*,t}\ to \Rightarrow$$
$$((P^{A^*\to(A^*\to A^*)}(U^{r,A,A^*}))^{A^*\to A^*}(V^{s,A,A^*}))^{A^*,t}$$

Proposition 2. *Every sequence of type reductions of a given term terminates.*

Proof. Suppose that X has been typed and has an infinite reduction sequence S. A subterm PUV, LU or RU of X is said to be marginal if at some stage of S the subterm is a redex of counter-clockwise rotation. Note that the head (P, L, R) of a marginal subterm cannot have a type path of length greater than 1 after its first counter-clockwise rotation. Now we order the subterms of X excluding the heads of its marginal subterms as follows

$$
\begin{array}{lll}
U, V < PUV & PUV & \text{marginal} \\
U\ \ < LU & LU & \text{marginal} \\
U\ \ < RU & RU & \text{marginal} \\
V\ \ < (UV) < U & UV & \text{not marginal} \\
U\ \ < \lambda uU & &
\end{array}
$$

Clearly $<$ is a linear order of height, say, h. Now consider that tail-end of S which begins after the last marginal subterm is counter-clockwise rotated for the first time. A simple ordinal assignment of ordinals $< \omega^{h+1}$ shows this tail-end must be finite. End of proof.

5 Typed Reductions

In addition, we have beta reduction, eta expansion, and reductions corresponding to pointwise surjective pairing on terms with types. This relation on typed terms is denoted $ot \Rightarrow$.

(1) Beta Reduction:

$$((\lambda u^{A,}(U^{r,B}))^{A \to B}(V^{s,A}))^{B,t} \; ot \Rightarrow$$
$$[V^{s,A}/u^{A,}]U^{r,B,t}$$

(2) Etae Reduction (Eta Expansion):

$$U^{r,A \to B, A^* \to B, s} \qquad\qquad ot \Rightarrow$$
$$(\lambda u^{A^*}, (U^{r,A \to B}u^{A^*,A})^B)^{A^* \to B, s}$$

$$U^{r,A \to B, A \to B^*, s} \qquad\qquad ot \Rightarrow$$
$$(\lambda u^{A,}(U^{r.A \to B}u^A)^{B,B^*})^{A \to B^*, s}$$

(3) Left Reduction:

$$(L^{A \to A}((P^{A \to (A \to A)}(U^{r,A})(V^{s,A}))^A))^{A,t} \; ot \Rightarrow$$
$$U^{r,A,t}$$

(4) Rght Reduction:

$$(R^{A \to A}((P^{A \to (A \to A)}(U^{r,A})(V^{s,A}))^A))^{A,t} \; ot \Rightarrow$$
$$V^{s,A,t}$$

(5) Surj Reduction:

$$(P^{A \to (A \to A)}((L^{A \to A}X^{r,A})^A))^{A \to A}$$
$$((R^{A \to A}X^{r,A})^A))^{A,s} \qquad\qquad ot \Rightarrow$$
$$X^{r,A,s}$$

(6) Appl Reduction:

$$((L^{(A \to B) \to (A \to B)}(U^{r,A \to B}))^{A \to B}V^{s,A})^{B,t} \; ot \Rightarrow$$
$$(L^{B \to B}(U^{r,A \to B}(V^{s,A})^B)^B)^{B,t}$$

(7) Appr Reduction:

$$((R^{(A \to B) \to (A \to B)}(U^{r,A \to B}))^{A \to B}V^{s,A})^{B,t} \; ot \Rightarrow$$
$$(R^{B \to B}((U^{r,A \to B})(V^{s,A})^B)^B)^{B,t}$$

(8) Appp Reduction:

$$((P^{(A \to B) \to ((A \to B) \to (A \to B))}$$
$$(U^{r,A \to B})(V^{s,A \to B}))^{A \to B}W^{t,A})^{B,t'} \qquad\qquad ot \Rightarrow$$
$$((P^{B \to (B \to B)}(U^{r,A \to B}(W^{t,A}))^B)))^{B \to B}$$
$$(V^{s,A \to B}(W^{t,A})^B))^{B,t'}$$

Taken together $to \Rightarrow$ and $ot \Rightarrow$ define a reduction relation on typed terms denoted $tt \Rightarrow$. The reduction relation generated by just $to \Rightarrow$, Beta and Etae on typed terms be denoted $tc \Rightarrow$. The reduction relation generated by $to \Rightarrow$, and $tt \Rightarrow$ - {Etae} is denoted $tp \Rightarrow$.

Example: (Barendregt's big omega)

Let $M = \lambda x. \; xx.$ and $R = \{p = p \to q\}$. Set $A_0 = p$ and $A_{n+1} = A_n \to q$. We type M as follows

$$(\lambda x^{A_3}.x^{A_3}x^{A_3,A_2})^{A_4}$$

and extend this to MM by $M^{A_4}(M^{A_4,A_3})$, where for brevity sake we omit the outermost type path.

With this typing

$$
\begin{aligned}
&M^{A_4}(M^{A_4,A_3}) && ot \Rightarrow\\
&(M^{A_4,A_3})(M^{A_4,A_3,A_2}) && to \Rightarrow\\
&(M^{A_4})(M^{A_4,A_3,A_2,A_3}) && ot \Rightarrow\\
&(M^{A_4,A_3,A_2,A_3})(M^{A_4,A_3,A_2,A_3,A_2}) && to \Rightarrow\\
&(M^{A_4,A_3,A_2})(M^{A_4,A_3,A_2,A_3,A_2,A_1}) && to \Rightarrow\\
&(M^{A_4,A_3})(M^{A_4,A_3,A_2,A_3,A_2,A_1,A_2}) && to \Rightarrow\\
&(M^{A_4})(M^{A_4,A_3,A_2,A_3,A_2,A_1,A_2,A_3}) && ot \Rightarrow\\
&(M^{A_4,A_3,A_2,A_3,A_2,A_1,A_2,A_3})(M^{A_4,A_3,A_2,A_3,A_2,A_1,A_2,A_3,A_2}) && to \Rightarrow\\
&M^{A_4}(M^{A_4,A_3,A_2,A_3,A_2,A_1,A_2,A_3,A_2,A_1,A_0,A_1,A_2,A_3}).
\end{aligned}
$$

Note that an atom appears in the type path of the second occurrence of M insuring that the reduction terminates. This illustrates the general case.

6   Completeness

Every untyped term has a typing in our system. Indeed, entire finite reduction trees can be typed; this is similar to case of Levy labels.

Proposition 3. *(Completeness of typing terms)*
Every term has a typing.

Proof. We refer the reader to Sect. 2.3 of [2]. For the most general typing of a simply typable term, equations are associated to the subterms of a term and Robinson's unification algorithm is employed to find a most general solution. Robinson's algorithm actually outputs a simultaneous recursion which solves the equations associated with a simple typing of a given term, and which is trivial if the term actually has a simple typing. The simultaneous recursion just encodes the failures of the "occurs check", and easily can be made to be a "simple" recursion. The process assumes only invertibility $a \to b = c \to d$ implies $a = c$ and $b = d$. For this we refer the reader to 8.3.28 of [2]. Here we require only that for each occurrence of P, $P : A \to (A \to A)$, for each occurrence of L, $L : A \to A$, and for each occurrence of R, $R : A \to A$. The paths assigned to sub-terms are just the sequences of terms resulting from the substitutions of equals for equals which are necessary to solve the corresponding equations. Now suppose that we are given a positive integer n. Then we can arrange a solution such that each atom lies below at least $n \to$'s. For those atoms not in the recursion can be substituted for $[A \to B/p]$ and those that appear in R can be replaced

$$[P_i(p_1, \ldots, p_e)/p_i].$$

This substitution takes the 1 step reduction

$$P_i(p_1, \ldots, p_e) \ \text{Red}.p_i$$

into the many (at least 2) step reduction

$$P_i(P_1, \ldots, p_e) \ \text{Red}.^*$$
$$P_i(p_1, P_2, \ldots, P_e) \ \text{Red}.^* \ p_i.$$

This substitution can be repeated n times. A typing satifying this condition is said to be pumped up. End of proof.

Proposition 4. *(Completeness of typing reductions W.W.E.)*

If we have a finite reduction tree T of an untyped term X then there is typing of all the terms in the tree such that if $Y = Y^r \hookrightarrow Z = Z^s$ then there exists a typing $Y = Y^{r'}, Z = Z^{s'}$ such that

$$Y^r \Leftarrow to \Rightarrow Y^{r'} \ ot \Rightarrow Z^{s'} \Leftarrow to \Rightarrow Z^s$$

except in case the $\hookrightarrow$ is Sure in which case $Y^{r'} \Leftarrow ot \ Z^{s'}$.

Proof. Omitted for space considerations.

Theorem 1. *The Church-Rosser property for $tp \Rightarrow$ implies the Church-Rosser property for $oa \hookrightarrow$.*

Proof. It suffices to establish the strong diamond property for $oa \hookrightarrow$. Suppose that $Y \hookleftarrow oa \ X \ oa \hookrightarrow Z$. By the completeness of typing we can type the reduction tree with X at its root to obtain typings of Y and Z such that $Y \Leftarrow tp \Rightarrow Z$. By the Church-Rosser property for $tp \Rightarrow$ there exists typed W such that

$$Y \ tp \Rightarrow W \ \Leftarrow tp \ Z.$$

Now by Surj postponement there exist U, V such that

$$Y \ oo \hookrightarrow U \ jj \hookrightarrow jj \ W \ \hookleftarrow jj \ V \ \hookleftarrow oo \ Z$$

equivalently

$$Yoo \hookrightarrow U \ \hookleftarrow je \ W \ je \hookrightarrow V \ \hookleftarrow oo \ Z.$$

By the strong diamond property for $je \hookrightarrow$ there exists X' such that $Uje \hookrightarrow X' \hookleftarrow je \ V$ and thus

$$Y oo \hookrightarrow U \ je \hookrightarrow X' \hookleftarrow je \ V \hookleftarrow oo \ Z$$

and

$$Y oa \hookrightarrow X' \hookleftarrow oa \ Z.$$

End of proof.

7 Localization

Definition: A type reduction is said to be local to an untyped $oe \hookrightarrow$ redex if

(1) the redex is Beta and the reduction is a clockwise rotation or right shift where U is the abstraction term and V is the argument of the redex or it is a counter-clockwise rotation or left shift with U and V similarly positioned.
(2) the redex is eta and the reduction is a clockwise rotation or right shift where U is in function position and $u = V$ is the argument or it is a counter-clockwise rotation or left shift with U and u similarly positioned.
(3) the redex is Left or Right and the reduction is any rotation or shift originating from the body of the redex but not from either argument of P.
(4) the redex is Surj and the reduction is any rotation or shift originating from the body of the redex.
(5) the redex is Appl, Appr, or Appp and the reduction is any clockwise rotation or right shift as in the case of Beta.

We note that local type reductions precede the contraction of all redexes except Etae, which they follow.

We have two versions of the localization lemma. The first version is for type reductions and is straightforward. The second version is for type conversions and is very useful for proving strong normalization.

Lemma 1. *(localization)*

(i) *For typed term X^r if $X^r to \Rightarrow Y^s ot \Rightarrow Z^t$ where the $ot \Rightarrow$ is the contraction of the Beta, Left, Right, Surj, Appl, Appr, Appp redex U then there exists $W^{s'}$, $V^{t'}$ such that $X^r to \Rightarrow W^{s'} ot \Rightarrow V^{t'} to \Rightarrow Y^s$ where the first $to \Rightarrow$ is local to the untyped U and the $ot \Rightarrow$ is the contraction of the newly typed U.*
(ii) *For typed term X^r if $X^r \ ot \Rightarrow Y^s to \Rightarrow Z^t$ where the $ot \Rightarrow$ is the contraction of the Etae redex to the Y^s subterm $\lambda u(Uu)$ then there exists $W^{s'}, V^{t'}$ such that*

$$X^r to \Rightarrow W^{s'} ot \Rightarrow V^{t'} to \Rightarrow Y^s$$

where the second $to \Rightarrow$ is local to the untyped $\lambda u(Uu)$ and the $ot \Rightarrow$ is the contraction to the newly typed $\lambda u(Uu)$

Proof. Omitted for space considerations.

8   Strong Normalization

In this section we prove strong normalization for $tt \Rightarrow$: viz, every $tt \Rightarrow$ reduction sequence terminates. The proof is very long because of the number of cases but most are routine. So, we shall take certain liberties; indeed, we shall give an explicit proof only for $tc \Rightarrow$ leaving the rest to the reader. The following lemma is very convenient for the proof.

Lemma 2. *(localization for conversions). If $X^r \Leftarrow to \Rightarrow Y^s$ Beta Z^t, the last by contracting the beta redex $(\lambda u U\ V)$ then there exists a type conversion of X^r local to $(\lambda u U\ V)$, which consists of counter-clockwise rotations or left shifts of $\lambda u U$ and clockwise rotations or right shifts of $(\lambda u U\ V)$, to a term $Y'^{s'}$ such that $Y'^{s'}$ Beta $Z'^{t'} \Leftarrow to \Rightarrow Z^t$.*

Proof. Omitted for space considerations.

Definition: An alternating reduction sequence

$$X_1^{r_1}, \ldots, X_n^{r_n}, \ldots$$

consists of a sequence of typed terms $X_j^{r_j}$ such that $X_j^{r_j}$ type converts to a term which typed beta reduces to $X_{j+1}^{r_{j+1}}$ in one step. X^r is strongly normalizable if there are no infinite alternating reduction sequences beginning with X^r.

We shall give a de Vrijer (Tait) style proof. A van Daalen (Sanchiz) style proof can also be given.

Definition of the sets $HSN(A)$:

We define the set of hereditarily strongly normalizable terms of type A as follows:

$$
\begin{aligned}
X^{r,p_i} &: HSN(p_i) \text{ iff } X^{r,p_i} \text{ is strongly normalizable} \\
X^{r,A \to B} &: HSN(A \to B) \text{ iff for all } Y^{s,A} \\
&: HSN(A) \text{ we have } (X^{r,A \to B} Y^{s,A}) \\
&: HSN(B)
\end{aligned}
$$

Lemma 3. *On $HSN(A)$:*

(i) $x^{r,A}$: $HSN(A)$
(ii) If $X^{r,A}$: $HSN(A)$, then $X^{r,A}$ is strongly normalizable.
(iii) If $X^{r,A}$: $HSN(A)$ and $X^{r,A} \Leftarrow to \Rightarrow Y^{s,A}$ or Beta reduces to $Y^{s,A}$, then $Y^{s,A} : HSN(A)$.
(iv) If $X^{r,A}$: $HSN(A)$, then $X^{r,A,A^*} : HSN(A^*)$.

Proof. Omitted for space considerations.

Definition: A substitution @ is said to be admissible if for any variable $x^{A,r}$ we have @$(x^{A,r})$ has the form $X^{s,A}$ and $X^{s,A} : HSN(A)$.

Lemma 4. *(admissible substitution): If for all* $V^{s,A} : HSN(A)$, $[V^{s,A}/u^{A,}]U^{r,B} :$ $HSN(B)$ *then* $(\lambda u^{A,}U^{r,B})^{A\to B} : HSN(A \to B)$.

Proof. Omitted for space consideration.

Proposition 5. *If* @ *is admissible then* $@(X^{r,A}) : HSN(A)$.

Corollary 1. *Every typed term is strongly normalizable.*

Lemma 5. *(termination of Etae reductions):*
An alternating sequence of Etae reductions and local clockwise rotations or right shifts always terminates.

Proof. We assign a value to each path of types $= 2^{\text{ sum of the lengths of types in the path}}$ and a value to each term $=$ sum of the values of the paths of subterms. Then Etae reductions reduce value since

$$2^{a+b+c+1+b+d+1+e} \geq 8 * 2^{\max\ a+b+c+1,b+d+1+e}$$
$$> 2^{a+b+c+1} + 2^{b+d+1+e} + 2^b + 2^{c+d}$$

and clockwise rotations and right shifts to eta variables clearly reduce value. End of proof.

Remark (standardization of Etae): An alternating sequence of eta expansions and local clockwise rotations or right shifts can always be done from right to left and top to bottom.

Lemma 6 *(short Etae postponement). An Etae reduction, followed by a type reduction, followed by the contraction of a Beta redex can be replaced by a type reduction, followed by the contraction of a Beta redex followed by an alternating sequence of Etae reductions and type reductions.*

Proof. By the localization lemma we can assume that the type reduction is local to the Beta redex. by a second application of localization we can assume that it is local to the Etae reduction. The eta redex has the form

$$(\lambda u^{A,}(U^{r,A^*\to B}u^{A,A^*})^B)^{A\to B,s} \text{ or}$$
$$(\lambda u^{A,}(U^{r,A\to B^*}u^A)^{B^*,B})^{A\to B,s}$$

which results from the Etae reduction of $U^{r,A^*\to B,A\to B,s}$ or $U^{r,A\to B^*,A\to B,s}$ respectively. Now the eta redex may be

(i) in function position as the head of the Beta redex,
(ii) U is the abstraction term of the Beta redex and u is the argument,
(iii) in argument position as the argument of the Beta redex,
(iv) a proper subterm of the head or the argument of the Beta redex, or
(v) disjoint from the Beta redex or the Beta redex is contained in U.

Case (i): In this case all the Etae local clockwise rotations and right shifts can be turned into Beta local clockwise rotations and right shifts preceeded by the single clockwise rotation or right shift affected by the Etae reduction. In this case both the Etae reduction and the Beta reduction disappear.

Case (ii): In this case all the Etae local clockwise rotations and right shifts can be turned into counter-clockwise rotations and left shifts of the abstraction term, U, of the Beta redex. Both the Etae reduction and Beta reduction disappear. The other cases, (iii), (iv), and (v) are straightforward. End of proof.

Lemma 7. *(long Etae postponement): If there is an alternating sequence of type reductions, Beta reductions, or Etae reductions from X^r to Y^s then there is one with no Etae reduction anywhere before a Beta reduction.*

Proof. We first prove this for the case of an alternating sequence of type reductions and Etae reductions followed by an alternating sequence of typed reductions and Beta reductions. Since every term is strongly normalizable we can prove this by induction on the size of the entire tree of alternating type reductions and Beta reductions beginning with the first term in the alternating sequence. This case follows from the lemma on Etae postponement. The result now follows from this case. End of proof.

Theorem 2. *An alternating sequence of type reductions and Beta-Etae reductions terminate.*

Proof. We suppose that we have an infinite such sequence. We repeatedly postpone Etae reductions. By strong normalization of Beta reduction there must be some stage after which any Beta reduction cancels with a previous Etae reduction, and by the lemma on termination of Etae reductions there must be a stage after that after which each additional Etae reduction is cancelled by a following Beta reduction. But these cases of cancellation all involve strict type reductions, which by the proposition on termination of type reduction must eventually terminate. End of proof.

9   Weak Diamond and the Church-Rosser Theorem

Here we verify the weak diamond property for $tc \Rightarrow$, and $tp \Rightarrow$. This property actually fails for $tt \Rightarrow$. For example, for counter-clockwise rotations of L and Etae:

$$L^{(A\to B)\to(A\to B)}(x^{A\to B})^{A\to B, A\to B^*} \qquad\qquad to \Rightarrow$$
$$(\lambda u^{A,}$$
$$((L^{(A\to B)\to(A\to B)}(x^{A\to B})^{A\to B}u^A)^{B,B^*})^{A\to B^*}$$

and

$$L^{(A\to B)\to(A\to B)}(x^{A\to B})^{A\to B, A\to B^*} \qquad\qquad to \Rightarrow$$
$$L^{(A\to B^*)\to(A\to B^*)}(x^{A\to B, A\to B^*})^{A\to B^*}$$

It is interesting to point out here that for both $tc \Rightarrow$ and $tp \Rightarrow$ some cases for $to \Rightarrow$ use $ot \Rightarrow$.

Proposition 6. *(weak diamond)*

(i) If $Y^s \Leftarrow tcX^r\ tc \Rightarrow Z^t$ then there exists $W^{r'}$ such that
$Y^s tc \Rightarrow W^{r'} \Leftarrow tcZ^t$.

(ii) If $Y^s \Leftarrow tpX^r\ tp \Rightarrow Z^t$ then there exists $W^{r'}$ such that
$Y^s\ tp \Rightarrow W^{r'} \Leftarrow tp\ Z^t$

Proof. Omitted for space considerations.

Corollary 2. *(Church-Rosser for typed terms:)*

(i) If $X^r \Leftarrow tp \Rightarrow Y^s$ then there exists a Z^t such that
$X^r tp \Rightarrow Z^t \Leftarrow tpY^s$.

(ii) If $X^r \Leftarrow tc \Rightarrow Y^s$ then there exists a Z^t such that
$X^r tc \Rightarrow Z^t \Leftarrow tcY^s$.

It is worth remarking here what happens if eta reduction is used. First, clockwise rotations and right shifts conflict with eta reduction

$$(\lambda u^{A}.U^{s,A^* \to B, A \to B}{}_u{}^A)^B)^{A \to B,t}\ to \Rightarrow$$
$$((\lambda u^{A}.U^{s,A^* \to B}{}_u{}^{A,A^*})^B)^{A \to B,t}.$$

This suggests using full type conversion instead of type reduction. If this is done then weak diamond for eta reduction fails. For example,

$$\lambda v^{A^*}.(u^{A \to B, A^* \to B, A^* \to B^*}v^{A^*})^{B^*} \quad \Leftarrow to \Rightarrow$$
$$\lambda v^{A,A^*}.(u^{A \to B}(v^{A^*},A))^{B,B^*} \quad \Leftarrow to \Rightarrow$$
$$\lambda v^{A^*}.(u^{A \to B, A \to B^*, A^* \to B^*}v^{A^*})^{B^*}.$$

This suggests a commutative version of paths but we will not pursue this here. Nevertheless eta postponement can be proved. The failure of weak diamond can be overcome by switching to eta expansion at the end of postponed reduction sequences since, with full type conversion, we have the upward Church-Rosser property for eta. This is complicated, round about and appears to have no advantage over simply starting with Etae.

10   Conclusion and Open Problems

A corollary to our work is the theorem of Roel de Vrijer that surjective pairing is conservative over beta conversion. For suppose that X and Y are pure lambda terms and $X \leftarrowtail oa \hookrightarrow Y$. By the Church-Rosser theorem for $\hookrightarrow$ there exists Z such that $X \hookrightarrow Z \leftarrowtail Y$. By the proposition on completeness of typing reductions, and the Church-Rosser theorem for $tp \Rightarrow$ there exist U, V such that $Xtp \Rightarrow Y \Leftarrow tpZtp \Rightarrow V \Leftarrow tpY$ Here U, V are pure terms. Now by the proposition on Surj postponement there exist U', V' such that $U \leftarrowtail jjU' \leftarrow \leftarrowtail oo\ Z\ oo \hookrightarrow V'jj \hookrightarrow V$. Now $\hookrightarrow$ is obviously Church-Rosser so there exists

Z' such that $U'oo \hookrightarrow\!\!\!\!\rightarrow Z' \leftarrow\!\!\!\hookrightarrow ooV'$. Thus by Stovring's proposition 5.4 (actually only a weak version is needed) U' beta converts to V'.

It would be very interesting to extend this method to an extension of Stovring's FP; possibly including the reduction $P(\lambda x X)(\lambda x Y) \hookrightarrow \lambda x (PXY)$.

11   Glossary of Abbreviations

To := types only
ot := only terms
tt := both types and terms
tc := types and classical beta-etae (eta expansion)
tp := types and surjective pairing
jj := only surjectivity
je := only surjectivity expansion
oj := everything except surjectivity
oe := everything except surjectivity expansion
oo := everything except surjectivity and surjectivity expansion
oa := everything except etae(eta expansion)

References

1. Barendregt, H.: The Lambda Calculus, Amsterdam (1981)
2. Barendregt, H., Dekkers, J., Statman, R.: Lambda Calculus with Types. Cambridge University Press, Cambridge (2013)
3. De Vrijer, R.: Extending the lambda calculus with surjective pairing is conservative. In: LICS, pp. 204–215 (1989)
4. Hindley, R.: Reductions of residuals are finite. Trans. A.M.S **240**, 345–361 (1978)
5. Levy, J.J.: An algebraic interpretation of the lambda beta K calculus and a labeled lambda calculus. In: Bohm, C. (ed.) Lambda Calculus and Computer Science Theory. LNCS, vol. 37, pp. 147–165. Springer, Heidelberg (1975)
6. Klop, J.W.: Combinatory Reduction Sytems. Dissertation: University of Utrecht (1980)
7. Robinson, J.A.: Machine oriented logic based on the resolution principle. JACM **12**(1), 23–41 (1965)
8. Scott, D.: Some philosophical issues concerning theories of combinators. In: Bohm, C. (ed.) Lambda Calculus and Computer Science Theory. LNCS, vol. 37, pp. 346–366. Springer, Berlin (1975)
9. Statman, R.: Recursive types and the subject reduction theorem, Technical report 94–164, Deparment of Mathematical Sciences, Carnegie Mellon University (1994)
10. Statman, R.: Surjective pairing revisited. In: Van Oostrom, V., Van Raamsdonk, F., (eds.), Liber Amicorum for Roel De Vrijer, University of Amsterdam (2009)
11. Statman, R.: On polymorphic types of untyped terms. In: Beklemishev, L.D., Queiroz, R. (eds.) WoLLIC 2011. LNCS, vol. 6642, pp. 239–256. Springer, Heidelberg (2011)
12. Stovring, S.: Extending the extensional lambda calculus with surjective pairing. Log. Methods Comput. Sci. **2**(2:1), 1–14 (2006)

Author Index

Printed in the United States
By Bookmasters